Banana: Genomics and Transgenic Approaches for Genetic Improvement

Sukhada Mohandas
Kundapura V. Ravishankar
Editors

Banana: Genomics and Transgenic Approaches for Genetic Improvement

Springer

Editors
Sukhada Mohandas
Division of Biotechnology
ICAR-Indian Institute of Horticultural Research
Bangalore, India

SM Plant Production Technologies Pvt Ltd.
Bangalore, India

Kundapura V. Ravishankar
Division of Biotechnology
ICAR-Indian Institute of Horticultural Research
Bangalore, India

ISBN 978-981-10-1583-0 ISBN 978-981-10-1585-4 (eBook)
DOI 10.1007/978-981-10-1585-4

Library of Congress Control Number: 2016951640

Printed on acid-free paper

This Springer imprint is published by Springer Nature
The registered company is Springer Science+Business Media Singapore Pte Ltd.

Foreword

Bananas and plantains represent a staple food crop for 400 million people. Its world production amounts to more than 145 million tonnes, of which close to 20 % are grown in India. This is because bananas are so popular but also because they are an important part of the culture. For example, it is a tradition to tie banana plants at the entrance of a house where a marriage is taking place to bring the couple good luck. Bananas continue to be vulnerable to pests and diseases but also to abiotic stresses that are becoming increasingly problematic due to climate change. India, with its enormous diversity of bananas, including wild relatives, is well placed to face these challenges.

The field of genomics is picking up speed. After the whole genome sequence was published in *Nature* in 2012, we noticed the doubling of germplasm requests from the Bioversity International *Musa* Germplasm International Transit Centre (ITC). Recent efforts to sequence the banana genome and studies on cloning genes of important traits using next generation techniques, proteomics, transcriptomics and metabolomics have immensely helped to understand the response of banana at the molecular and cellular level.

In this context, the present book, edited by **Dr. Sukhada Mohandas** and **Dr. K.V. Ravishankar** entitled *Banana: Genomics and Transgenic Approaches for Crop Improvement*, is a unique blend of information on banana genomics and transgenic approach for crop improvement. The overview of progress in this area of research has been put together by leading experts who have organized the book into two parts. The first part deals with evolution, taxonomy, classical breeding and understanding of the banana genome through next generation sequencing and molecular markers. Metabolomics and molecular aspects of fruit ripening are also discussed. The second part covers all aspects of transgenic development starting with genes, and gene transfer techniques, regeneration protocol and strategies used for the development of trait-specific transgenic bananas. The latest advances are included, such as the successful field trials conducted on bacterial blight, wilt resistance and insect resistant plants. The review articles included in the book draw from a vast bibliography which is a valuable source for scientists and students. Overall, the book gives the reader comprehensive information and discussion on the banana genome, and its improvement through classical breeding and genetic engineering.

The book poses the two following main challenges in banana research:

First and foremost, *Musa* research relies on the availability of a broad genetic base. Only by continually increasing the diversity conserved in *ex situ* as well as in *in situ* collections can we maximize our progress in selecting and improving the right varieties.

As banana pests and diseases continue to spread across the globe, and with the effects of climatic change becoming evident, new approaches that unlock potential resistance are urgently needed. With the recent great strides in the field of genomics, there is a movement to link novel methods and technologies to ongoing breeding efforts. Promising genomics-based approaches for containing/eradicating threats to production are therefore discussed at length in this book.

The editors and scientists who have contributed to the book have made significant advances in their research. I congratulate the editors for bringing out this compilation, which will have a great impact on *Musa* research, teaching and the transfer of new technologies. This book will be read by a wide range of scientists and donors, not only from India but also from around the world. It is of interest to the public and private sectors from both developed and developing countries and will promote a rapid uptake of the latest technologies, contributing to improving food security and better livelihoods in countries where it is most needed.

Dr. Nicolas Roux, PhD

Senior Scientist, Genetic Resources and
Genomics Group Leader Bioversity International
Parc Scientifique Agropolis II 34397 Montpellier
Cedex 5, France

Preface

Banana and plantain are the important crops of the world with an annual production of 144 million tonnes (FAOSTAT, 2013). Of late, banana production has been severely threatened by both biotic and abiotic stress factors due to rapid change in climate and the evolution of new races of pathogens. Recent advancements in science and technology have helped generate large amount of information on banana genome sequence and transgenic technologies, which have helped evolve strategies to improve banana for various agronomical situations.

Keeping this in view, we have designed this book to collate holistic genomic and biotechnological information on banana crop. We have organized this book into two parts. The first eight chapters deal with evolution, taxonomy, classical breeding, understanding banana genome through next generation sequencing and markers available for this crop. We have also discussed metablomics and molecular aspects of fruit ripening in detail. In the other twelve chapters of the book, we present plant tissue culture, various techniques used for transformation and strategies used for the development of trait specific transgenic. On the whole, we have attempted to provide up-to-date information available on the banana genome and novel transgenic technologies, which are being adapted in banana improvement.

The majority of the cultivated banana are evolved from two wild species *Musa acuminata* and *M. balbisiana* through natural inter-species hybridization. Cultivated banana are sterile and polyploidy in nature, and hence improvement through classical breeding is a herculean task. However, breeders have adopted different strategies to overcome many of the difficulties. The traditional breeding methods involving diploids and the use of molecular markers have helped to develop the high-density linkage map. Further, this has helped assembly of whole genome sequence data in banana. Next-generation sequencing techniques proteomics and metablomics have immensely helped to understand the response of banana at molecular and cellular level. The chapter on abiotic stress studies deals in detail on banana crop response to various abiotic stresses. Similarly, we present molecular and genomic aspects of biotic stress like fungal, bacterial, virus and insect pests in a chapter. The chapter on molecular aspects of ripening deals mainly with gene expression studies and biochemical changes during ripening. Following this chapter, we present a chapter on metablomics, where many biochemical and phytochemical aspects of banana have been discussed. Phytochemical aspects of various parts of banana and their use in traditional medicines are presented.

For genetic improvement of banana and plantains, biotechnological approaches have been widely used without compromising basic characters of the plant. A lot of concerted effort over the years from different groups has helped in obtaining good transformation protocol for banana. In the early years of banana research, regeneration of the crop was mainly through asexual means, and hence gene transfer using apical meristems and its stable integration was found difficult as transformation using meristems was found to result in chimeras. Therefore, methods using embryogenic cell suspensions (ECS) were attempted. Electroporation or *Agrobacterium* co-cultivation of gene carrying vector with ECS was successfully used to transfer different foreign genes into banana. Particle bombardment and co-cultivation of wounded meristems with *Agrobacterium* were also used to get transformants. The introduction of a centrifugation step during co-cultivation was found to improve transformation efficiency significantly. The chapter on somatic embryogenesis and novel tools for banana transformation provides an insight into the latest development in the field of banana transformation.

Identification of organ specific promoters is essential for successful expression of genes in different locations. Keeping this in view, the isolation of promoters using insertional mutagenesis is widely employed method. Advances in genomics led to the genome sequencing and identification of candidate promoters for specific expression patterns. Actually, promoter analysis for activity characterization has been confirmed through experimentation with different techniques, including reporter genes, bioinformatics analysis for candidate *cis*-acting element and promoter prediction and expression analysis of related genes. A chapter devoted to banana promoter analysis reviews characterization of different banana promoters and the analysis of promoter sequences available in GenBank using available bioinformatics tools and a novel method to identify motif sequences. A list of promoters used in the development of genetically modified banana is presented.

Abiotic and biotic factors adversely affect plant growth and development. Plants react to adverse conditions by producing specialized signals and modulate the transcription factors which regulate the genes coding for synthesis proteins and metabolites which are involved in stress tolerance. Several transcription factors involved in abiotic stress resistance and genes like dehydrins and aquaporins upregulted during stress-induced conditions have been identified and their role elucidated. Pathogenicity-related proteins, stress-associated proteins, vegetative storage proteins and several transcriptional factors have been discussed in the genomic chapter. Overexpression of some of these genes and transcription factors in banana and their effectiveness in combating stress have been examined later in the book. Factors involved in regulating gene expression, including microRNAs or miRNAs which are non-coding RNAs and are involved in post-transcriptional regulation of gene expression, have been identified from banana cultivars and their role validated in banana by overexpressing them in the crop. Recent developments in genomics, high-throughput sequencing and phenotyping platforms have given way to molecular breeding. Ecotilling and genome editing are also used to induce new variations and to incorporate new traits. Enhancement of abiotic stress tolerance in banana through transgenic means is discussed in a seperate chapter which gives an account of the recent developments in this area.

Several diseases affecting banana are debilitating and reducing the yield drastically. Genetic modification of banana has been a widely accepted tool due to the limited success of conventional breeding. Panama disease caused by *Fusarium oxysporum* f sp *cubense* (Foc) is the most devastating and causes 100 % yield loss in many cultivars of banana. Foc is known to exist as four important races (race 1, 2, 3 and 4) of which race 1 and 4 are of serious concern as they attack the commercially acceptable banana cultivars across the globe. A review on transgenic banana for Fusarium wilt resistance highlights the application of genetic engineering for imparting resistance against Fusarium wilt and discusses various strategies that have been employed involving PR-related genes (Ace-AMP1 gene and defensin gene), antimicrobial genes, anti-apoptosis gene, RNAi-mediated approach and host-induced gene silencing (HIGS) that confer certain level of tolerance towards pathogen infection. Further, the cisgenic approach utilizing R genes and native cell death genes from Musa sp. have also proven promising. The understanding of host pathogen interaction in terms of defense and signalling related pathways and the study of pathogenecity mechanism which help in identifying critical genes for targeting pathogen are discussed.

Overexpression of antimicrobial peptides, like defensin, antimicrobial peptide MSI-99, magainin, endochitinase gene (TnEn-42) and chitinases stilbene synthase, have been found beneficial in producing fungal resistant crops. The resulted transgenic banana plants were phenotypically normal. RNA interference (RNAi) is another emerging strategy for control of pathogens, through silencing of a vital gene associated with pathogens. The chapter on the development of Sigatoka resistance using transgenic means discusses the merits of such an approach.

Several R genes and AMPs like lysozymes, magainins, cecropins, attacins, thionins and defensins have been identified to control bacterial pathogens. Transgenic bananas expressing either sweet pepper *Pflp* or *Hrap* gene have been developed and are under evaluation for resistance to Xanthomonas wilt disease in field trials in Uganda. The chapter on transgenics for bacterial wilt resistance discusses in detail the recent developments in the area and the management practices adopted and through cultural practices to check the spread of diseases.

Viruses are great limiting factors for banana production. Banana bunchy top caused by *Banana bunchy top virus* (BBTV) is one of the most devastating diseases of bananas in Hawaii and many areas of Asia, Africa and the Pacific. Several groups have investigated this possibility and utilized post-transcriptional gene silencing (PTGS) or RNA interference (RNAi) approaches to generate BBTV resistance in several cultivars. Recent efforts and approaches to develop BBTV-resistant transgenic banana is reviewed in the chapter on viral resistance. Future potentials of using transgenic banana as alternatives for the management of banana virus diseases are also discussed. Another strategy that utilizes a process termed "virus-activated cell death" has been developed by James Dale at the Queensland University of

Technology (QUT) in Australia. Transgenic plants using this strategy have been developed and are under evaluation. Different strategies used for developing virus resistance are discussed in this chapter.

A chapter on molecular farming provides an overview of different plant-derived products currently in the market or are in different stages of development including phases of clinical trials. Special emphasis has been given on banana being used as an expression host, advantages and limitations of using banana in plant molecular farming and the different approaches which can be utilized to overcome those limitations. Iron deficiency anemia (IDA) is a global problem, affecting women and children of the lower strata of society. Banana is considered as a potential fruit crop to become "micronutrient-enriched". Fortification of banana is more advantageous over other plants owing to its ploidy, parthenocarpic fruit development, its reach to the masses at large and availability throughout the year. Studies on the physiology of iron uptake in plants, translocation, storage and redistribution in plants and recent advances made in the understanding of the mechanisms of iron uptake in humans, homeostasis and transgenic approaches for increasing the iron content in bananas are described in a chapter on biofortification for iron deficiency anemia.

Vitamin A is an essential micronutrient required for several physiological functions. Hence, the biofortification of banana to develop a rich source of pro-vitamin A through genetic engineering tool could be an ideal approach. Retinyl esters and pre-vitamin A are the two dietary sources for the body. Retinyl esters are obtained from the meat and dairy products, whereas PVA is obtained from plant sources in the form of carotenoids. Only certain forms of carotenoids such as α-carotene, β-carotene and β-cryptoxanthin known as PVA are converted to vitamin A in the body. Genome engineering tools used to enhance the PVA content and lycopene are discussed in the chapter on pro-vitamin A-enriched banana for tackling malnutrition.

Overall, the present book gives a comprehensive information and discussion on banana genome, its improvement through classical breeding and genetic engineering. We are highly obliged to the experts who contributed to this book. We acknowledge continuous support from Indian Council of Agricultural Research, New Delhi, through project on "Network Project on Transgenic in Crops" for more than a decade to our banana research work. The expertise gained through these projects have made us to embark on editing this book. Finally, we hope the readers will enjoy reading this book and it would be useful to get comprehensive information on banana.

Bangalore, Karnataka, India
May 2, 2016

Sukhada Mohandas
Kundapura V. Ravishankar

Contents

Part I Genomics

1 **History, Origin, Domestication, and Evolution** 3
A. Rekha

2 **Banana Breeding** 13
Rema Menon

3 **Genes and Markers: Application in Banana Crop Improvement** 35
Manosh Kumar Biswas and Ganjun Yi

4 **Current Status of Banana Genome in the Age of Next Generation Sequencing** 51
Megha Hastantram Sampangi-Ramaiah and Kundapura V. Ravishankar

5 **Genomics of Biotic Stress Tolerance in Banana** 61
Pavitra Kotari, V. Swarupa, and Kundapura V. Ravishankar

6 **Abiotic Stress Tolerance Research Using-Omics Approaches** 77
Ewaut Kissel and Sebastien C. Carpentier

7 **Molecular Analysis of Fruit Ripening in Banana** 93
Antara Ghosh, T.R. Ganapathi, and V.A. Bapat

8 **Metabolite Profiling in Banana** 107
K.S. Shivashankara

Part II Genetic Engineering

9 **Novel Gene Transfer Technologies** 127
Harjeet Kaur Khanna and Pradeep Chand Deo

10 **Somatic Embryogenesis as a Tool in Genetic Transformation** 141
H.D. Sowmya, T.R. Usharani, and Sukhada Mohandas

11 **Promoter Analysis in Banana** 157
Efrén Santos, Ricardo Pacheco, Liliana Villao, Luis Galarza, Daniel Ochoa, Carlos Jordán, and José Flores

12 Enhancing Abiotic Stress Tolerance 181
Anjana Rustagi, Shashi Shekhar, Shalu Jain, Deepak Kumar, and Neera Bhalla Sarin

13 Transgenic Technologies for Bacterial Wilt Resistance 197
Leena Tripathi, Jaindra Nath Tripathi, and Jerome Kubiriba

14 Engineering Resistance to *Fusarium* Wilt 211
T.R. Usharani, H.D. Sowmya, C. Sunisha, and Sukhada Mohandas

15 Engineering Resistance to Sigatoka 227
H.D. Sowmya, T.R. Usharani, C. Sunisha, and Sukhada Mohandas

16 Engineering Resistance to Viruses 237
James C. Green, Wayne Borth, and John S. Hu

17 Transgenic Approaches to Improve Resistance to Nematodes and Weevils 247
Hugh Roderick, Leena Tripathi, and S. Poovarasan

18 Molecular Farming: Prospects and Limitation 261
Himanshu Tak, Sanjana Negi, T.R. Ganapathi, and V.A. Bapat

19 Provitamin A Enrichment for Tackling Malnutrition 277
Navneet Kaur, Shivani, Ashutosh Pandey, and Siddharth Tiwari

20 Biofortification for Alleviating Iron Deficiency Anemia 301
Prashanti Patel, Karuna Yadav, and T.R. Ganapathi

Index 339

Part I

Genomics

1 History, Origin, Domestication, and Evolution

A. Rekha

Abstract

Banana, a popular and ancient fruit, has a complex state of evolution in plant systematics. Being one of the highly evolved crops, it has attracted many researchers to study various aspects of its origin, evolution, and domestication. In the present chapter, an account of history of domestication, spread of the species to various continents, and studies on ethnobotanical and linguistic evidence of crop origin and distribution has been presented. Further, their taxonomic status as evidenced by various research works, including numerical taxonomy, cytological studies, and molecular markers to understand the involvement of genomes other than *M. acuminata* and *M. balbisiana* in the evolution of the cultivated bananas, is given.

Keywords

Musa • Taxonomy • Evolution • Domestication • History • Origin

1.1 Introduction

Banana and plantains are the important sources of starch for the millions of people in the developing world; it is a major commercially cultivated fruit crop of India which is as old as the Indian civilization. Bananas are also considered as the poor man's apple and said to be the fruit of heaven (Amalraj et al. 1993). It appears to be one of the earliest fruit crops cultivated by mankind at the beginning of civilization. In India and Africa, bananas are very predominant and popular among people, and they are liked by both poor and rich alike. Unlike other fruits, banana is available throughout the year, and it is the cheapest among all other fruits in the country. In any Indian household, it is an inevitable necessity for all social and cultural occasions as it is considered as a symbol of good omen, fertility, and prosperity. Bananas are put into a variety of uses in India, especially in South India. Almost every part of the plant is used in some way or another; hence it is popularly known as "Kalpatharu." The fruit is easily digestible, a good food for people suffering from gastritis and other stomach ailments

A. Rekha (✉)
Division of Fruit Crops, ICAR-IIHR,
Hessaraghatta, Bengaluru 560089, India
e-mail: arekha@iihr.res.in

S. Mohandas, K.V. Ravishankar (eds.), *Banana: Genomics and Transgenic Approaches for Genetic Improvement*, DOI 10.1007/978-981-10-1585-4_1

(Rao 1984). Apart from its fruit which is eaten either raw as a vegetable or cooked, the leaves are used as plates. The male flower bud and the central axis of the pseudostem are used as vegetable. The juice of the central axis is said to be useful in kidney stone treatment, and the leaf sheaths find their use in fiber and paper industry some Eumusa cultivars are listed in Table 1.1.

1.2 History and Domestication

There is a clear reference to banana in Sanskrit literature like the "Ramayana," Kautilya's "Arthashastra," and the Tamil classic "Shilappadikaram" (Krishnamurthi and V.S.Seshadri 1958). Descriptions of bananas are given in Greek writings on 327 BC in Indus

Table 1.1 Eumusa cultivars

Genomic group	Name	Country
AA	Pisang Mas, Pisang Lilin, Pisang Keladi, Pisang Boyan, Pisang Kapas	Malaysia, Indonesia
	Lady Finger	Hawaii
	Klue Khai, Kulai Lai, Kulai Hom, Kulai Sa	Thailand
	Kadali, Matti, Anaikomban, Namari, Sanna Chen Kadali	India
	Bata-Bata, Lakatan, Pogpogan, Amas	Philippines
	Banana Ouro	Brazil
AAA	Pisang Ambon, Pisang Serendah, Pisnag Thai, Pisang Tualang, Pisang Masak Hijau	Malaysia, Indonesia
	Bluefields, Chinese, Hamakua, Red	Hawaii
	Kulai Nak, Kulai Nang Nuan, Kulai Nam, Kulai Khom, Kulai Hom Thong	Thailand
	Harichal, Jahaji, Pacha Vazhai, Basrai, Lal Kela	India
	Bangan, Inabaca, Manang	Philippines
	Nanicao, Mestica	Brazil
AB	Ney Poovan, Safed Velchi, Kunnan, Vannetu Kunnan	India
	Lady Finger	Hawaii
AAB	Pisang Raja, Pisang Tandok, Pisang Rasthali, Pisang Kelat	Malaysia, Indonesia
	Brazilian, Apple, Dwarf Plantain, Eslesno, Father Leonore	Hawaii
	Klue Nagar Chang	Thailand
	Rasthali, Rajapuri, Virupakshi, Nendran, Thiruvananthapuram	India
	Tundoc, Letondal, Ternate, Canara	Philippines
	Banana Pacova	Brazil
ABB	Pisang Awak, Pisang Kelat Siam, Pisan Abu, Pisang Batu	Malaysia, Indonesia
	Largo, Ice Cream, Chamaluco	Hawaii
	Klue Namawa, Klue Maliong, Klue Hak Muk, Klue Ong	Thailand
	Peyan, Monthan, Boothi Bale, Nalla Bontha Batheesa, Kach Kola	India
	Pitogo, Garango, Bisco, Inabaniko, Pelipita, Sabang Iloco	Philippines

Source: Stover and Simmonds (1987)

Valley, during the expedition of Alexander the Great in India (Reynolds 1927; Kervegant 1935). Most botanists believe that bananas were introduced from India to the Middle East and across North America by Arabs. The Portuguese along with the Spanish were instrumental in the worldwide spread of bananas and plantains especially to America (Price 1995). These evidences suggest the early existence of banana in India. The wild *Musa acuminata* occurs in Assam, Burma, Siam, Indo-China, the Malayan peninsula and archipelago, and the Philippines. The center of diversity of *M. acuminata* lies in the Malayan area where four out of five subspecies were found and is thus considered as the primary center of origin of cultivated bananas. Historical evidences show that the Arabs have introduced the banana from India to Palestine and Egypt, perhaps in the seventh century AD. It soon became popular in those areas and later spread to the east coast of Africa at a very early date and subsequently throughout the African continent. Bananas were reported to have been introduced in Central America in 1516 AD, where it spread rapidly and attained commercial significance. The spread of bananas to the West Indian Islands was through Christian missionaries, and wherever banana reached, it assumed economic importance due to its greater adaptability and commercial value.

Bananas are the source of starch for a sizable population in the tropics and subtropics for millions of years. The studies involving archaeology, genetics, and linguistics have provided an understanding regarding the history of banana domestication (Perrier et al. 2011). Historically, *Musa* species and genotypes have been created by migration of human populations and interaction between the groups which helped in the exchange of genotypes. These interactions created the introduction of species and helped in further generation of natural hybrids, which are parthenocarpic diploids or triploids. Some of these hybrid cultivars were widely adopted and dispersed either by preference or by chance. A group of triploids, maybe because of the environmental adaptability and triploid nature, might have been dispersed by clonal propagation across vast areas. In addition, the dispersal of bananas through humid tropics and subtropical regions, across the Indian Ocean, proves interlinkages, predominantly local, social networks extending from New Guinea to West Africa; these networks may be about 2500 years old.

There are some hints that the banana cultivation was prevalent in Harappan civilization (2500–1900 BC). Such studies would provide exciting proofs about the prehistorical dispersal of banana in Southeast Asia. However, the other evidences, especially that of historical and linguistics, suggested that the main introduction of edible bananas was about 2000 years later. This paved way to hypothesis that the "Kot Diji" *Musa*-like phytoliths might have been in cultivation for fiber or as ornamental plants. However, evolution of local names in relation to different *Musa* genome groups (AAA, ABB, and AAB), involving starchy or sweet varieties, is not well understood. New linguistic and ethnobotanical data from India is lacking, to fill the gap in the understanding of *Musa* names.

It appears that there is close similarity between the proposed history and botanical classification with observed history and linguistic terminologies. As far as bananas in cultures are concerned, introduction of bananas in Africa and the ancient terms in Papuan languages strongly suggest a pre-Austronesian dispersal across the islands of Southeast Asia. The linguistic evidence is a proof to archaeobotanical record, which helped in approximately tracing the dispersal of bananas from New Guinea.

Musa spp. domestication was a highly complex process; it has taken over thousands of years and involved multiple steps, separated by time and place (Carreel et al. 2002; De Langhe and de Maret 1999). Banana also has a unique testimony to the early, long-term, and deep impacts of people in rainforests. The long-term management and manipulation of specific plant resources within rainforests might have influenced the evolution of these plants. The archaeobotanical evidence of *Musa* bananas in areas other than the natural ecological regions of the genus, like Africa, indicates the fact that there would have been introduction, adoption, and dispersal by the people during very early or later years (Neumann and Hildebrand 2009; Vrydaghs and De Langhe 2003; De Langhe et al. 2009; Vrydaghs et al. 2003). Dispersal of bananas from New Guinea to Eastern Indonesia during mid-Holocene was inferred by studies of Denham and Donohue (2009), Donohue and Denham (2009), and Kennedy (2009). *Ensete*, the closely related genus of *Musa*, would have contributed in the evolution of the cultivars in the initial stages of banana domestication which needs further understanding. The domestication of Ethiopian *Ensete ventricosum* (Welw.) Cheesman, as a source of diverse products and as a staple source of starch, is well documented (Brandt et al. 1997; Purseglove 1975). There may also be a possibility that a parallel selection for starch production in the corm and pseudostem of Malaysian *Musa* species would have occurred. The classic example is selection for enhanced starch storage in the rhizome of a New Caledonian *Musa* plant, which is described as having a "glaucous, violet stem and a turnip-like rhizome when cooked, it resembled a yam in taste" (Simmonds 1959). The cultivation of *M. textilis* and *M. balbisiana* for fiber indicates that seedy fruit is not necessarily an indication of "wildness." While parthenocarpy is the key to the edibility of fruit, its reproduction and transmission depended upon vegetative propagation (Simmonds 1962; Daniells et al. 2001).

The evolution of seedless edible bananas from seeded wild species is complex. Recent research on genetic studies revealed that the process involved a long hybridization period including different taxa (Perrier et al. 2009). This could happen with human interventions who took a major role in carrying different taxa to new zones helping in hybridizations. It has been proved by the ancient events where human interventions were observed. The mode of dispersal was verified by phytoliths from archeological sites.

Process of evolution involved thousands of *Musa* species with high genetic diversity which indicates that it may be having multiple origins. Knowledge on functional structural genomics and genes; reproductive physiology; comparative genomics with rice, *Arabidopsis*, and other model species; and cytogenetics has helped in understanding the *Musa* diversity (Heslop-Harrison and Schwarzacher 2007).

The complicate genome constitution and ploidy of banana accessions were determined by the plant and fruit morphological studies since the 1940s; methods of numerical taxonomy were adapted which gave better understanding (Simmonds and Weatherup 1990a, b; Ortiz 1997; Ortiz et al. 1998; Pollefeys et al. 2004). Flow cytometric analysis (Doležel et al. 1999; Doležel and Bartos 2005) provided accurate and rapid surveys at the juvenile stage of the plants to know the ploidy status of the collections and new hybrids. Chromosome preparations with in situ hybridization techniques using DNA probes which can distinctly label the A and B genomes have shown that the full sets of $x=11$ chromosomes of A genome are present (Osuji et al. 1997), and most cultivars have 11 chromosomes with complete genomes. The classic example is the work of d'Hont et al. (2000) who used in situ hybridization technique to show that the variety "Pelipita" ($2n = 3x = 33$) included 8 A genome chromosomes and 25 B genome chromosomes instead of 11 A and 22 B genome chromosomes which is normally expected in the case of ABB type, whereas two other AAB plantain types consisting of 33 chromosomes had more than 11 B genome chromosomes. Their studies also confirmed the presence of complete "S" and "T" genomes, from *Musa schizocarpa* and *M. textilis*, respectively. All these studies indicate probability of back-crossing, or chromosome elimination would have occurred during the process of evolution of some varieties. Molecular analyses have proved the existence of chromosome markers from *Musa* species other than the A and B genome. The diploid variety "Wompa" had AS genome, whereas other genotypes were found to be consisting of AAT and ABBT genome composition.

Cultivar Pelipita (ABB genomic group): $2n = 3x = 33$ chromosomes

Expected – 11A chromosomes + 22 B chromosomes

Observed – 8 A chromosomes + 25 B chromosomes

Other plantain group cultivars (AAB genomic group)
  Expected – 22A chromosomes + 11B chromosomes
  Observed – 22 A chromosomes + > 11 B chromosomes and chromosomes of S and T genomes (i.e., *M. schizocarpa* and *M. textilis*)

1.3 Taxonomy

The bananas are indigenous to warm, humid areas of South Asia, and they are giant herbs of monocotyledons, under the order Zingiberales as classified by Huchinson, but was classified under the order Scitamineae by Bentham and Hooker in *Genera Plantarum. Musa* is placed under the family Musaceae, consisting of two genera *Musa* and *Ensete*. The genus *Musa* is further divided into four sections, two of which contain species with chromosome number of 2n = 20; they are Callimusa and Australimusa. The other two sections Eumusa and Rhodochlymus have a basic chromosome number of 2n = 22. The Rhodochlymus and Callimusa consist of plants with ornamental importance which do not produce edible fruits, whereas Australimusa contains *M. textilis* from which Manila hemp is produced (Figs. 1.1 and 1.2).

Earlier edible bananas were placed under *M. paradisiaca* L. for plantain and *M. sapientum* L. for banana by Linnaeus, which were later considered to be hybrids of two main species *M. acuminata* and *M. balbisiana* based on 15 important morphological characters (Dodds and Simmonds 1948). All banana and plantain cultivars evolved from the section Eumusa and its group of species. This is the biggest section in the genus and geographically widespread. The section contains 11 species which was divided into two subsections Eumusa (1) and Eumusa (2) by Simmonds and Weatherup (1990a, b) based on numerical taxonomy. Shepherd and Ferreira (1982) identified cultivars which may be the result of hybridization involving *M. schizocarpa* among *Musa* germplasm collections of Papua New Guinea. A Philippine clone was found to be the result of an early hybridization between *M. balbisiana* and *M. textilis* and landraces consisting of three

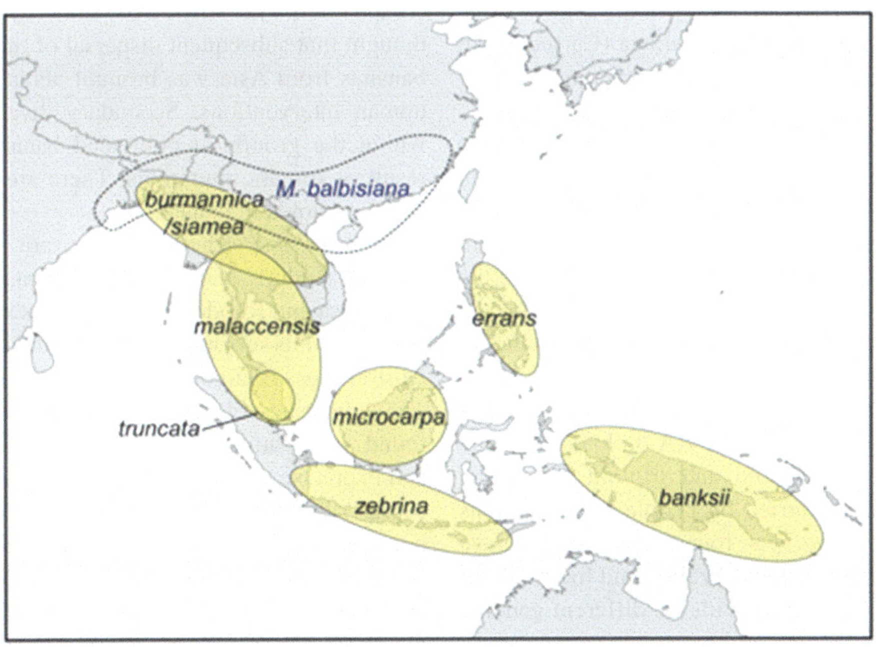

Fig. 1.1 Distribution of subspecies of *Musa acuminata* in Southeast Asia (Source Perrier et al. 2011. From PROMUSA website)

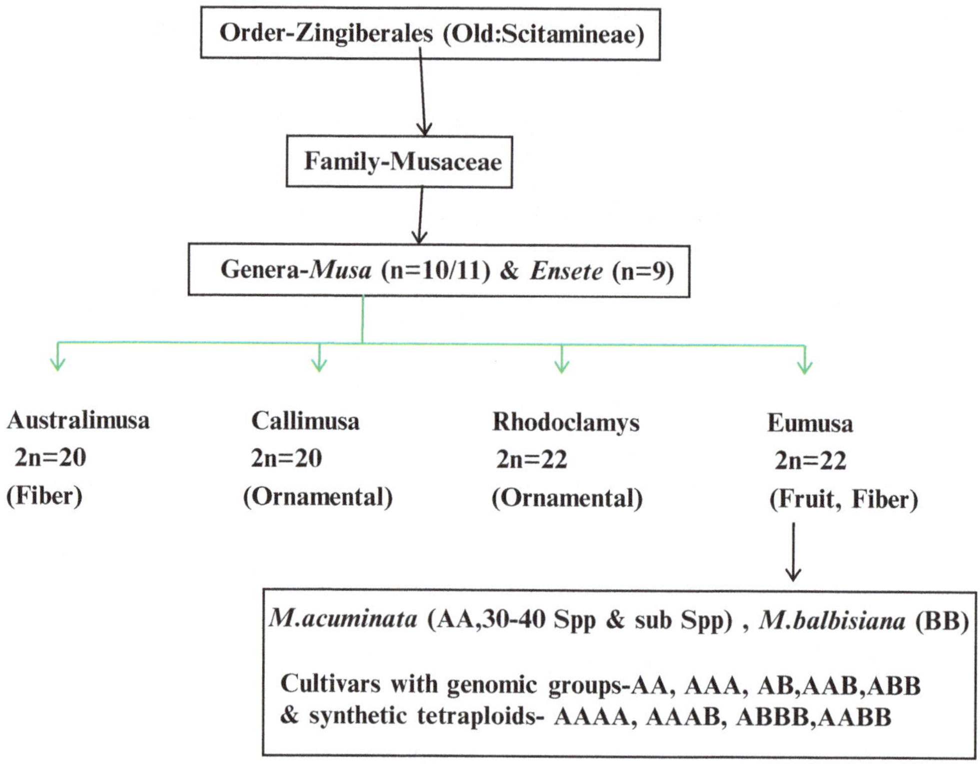

Fig. 1.2 Classification of Musaceae (Simmonds 1962)

genomes, *acuminata*, *balbisiana*, and *textilis*, which was in Papua New Guinea (Carreel et al. 1994). These evidences prove the complexity of origin of banana cultivars and their taxonomy.

The edibility of fruits of diploid *M. acuminata* (AA) originated as a result of two mutation events, involving induction of female sterility and parthenocarpy. Triploid AAA cultivars were derived from these diploids which might be the result of crosses between edible diploids and wild *M. acuminata* subspecies, which resulted in a wide variation among AAA cultivars. These triploids are highly vigorous and have larger fruits, which replaced the AA diploids of Asian countries. The diploid and triploid *acuminata* cultivars were carried by people to the regions where *balbisiana* was found which might have resulted in natural hybridization and formation of new hybrid progenies with the different genome composition like AB, AAB, and ABB. It was thought that subsequent dispersal of these edible bananas from Asia was brought about again by human interventions. Secondary diversification within the groups of cultivated bananas is the result of somatic mutations. There are allo- and autopolyploids in banana.

Eumusa and Rhodochlymus are found in Assam (India) and Thailand area, whereas Callimusa and Rhodochlymus groups are seen in Borneo and its surrounding islands and Indonesia. Australimusa is largely found in Malayan islands. The subspecies of *Musa acuminata* are largely found in Assam, Indo-China, Malayan islands, and Papua New Guinea which is also the primary center of cultivated AA types. *M. balbisiana* occurred in Ceylon, India, Burma, Siam, and Malayan islands where the A × B hybrids have evolved.

1.4 Molecular Evidences of Evolution

The *Musa* genome sequence provided an invaluable source for studies on plant gene and genome evolution studies (D'Hont et al. 2012). It also gave new insight in studies related to evolution of monocotyledons and their relations with each other. Characters specific to the family Poaceae could be highlighted, which helped in analyzing the emergence of this family. Identification of several deeply conserved regions within monocotyledons and between monocotyledons and eudicotyledons was possible with the available *Musa* genome sequence. This paved a way for detecting novel motifs with a functional gene regulation which provides valuable information on conserved genes. It was observed that there could have been three steps of polyploidization in the *Musa* lineage, followed by gene loss (deletion) and chromosome rearrangements (translocations or inversions); such changes might have resulted in little synteny conservation between lineages retaining some gene or group of genes; thus, the groups would have had an opportunity for independent diversification.

Several molecular methods were adapted to analyze diversity and evolution; in the early 1980s, analysis of isozyme and anthocyanins confirmed that *Musa* germplasm was genetically diverse (Jarret and Litz 1986; Horry and Jay 1988). As soon as DNA markers along with PCR-based techniques were available, several markers like RFLP, AFLP, RAPD, IARP, and SSR microsatellite markers are being used to analyze diversity across different countries and research groups with their germplasm collections.

De Jesus et al., in 2013, used flow cytometry and PCR-RFLP to characterize the Brazilian accessions. They studied 221 accessions by flow cytometry and confirmed the correct ploidy for 212 (95.9 %); however, genomic constitution could not be identified with flow cytometry, whereas the genomic constitution of 209 (94.6 %) accessions could be confirmed by digestion of the ITS region. Neighbor-joining cluster analysis from SSR binary data helped in detection of two major groups, which was distinguished by the presence or absence of the B genome, and subgroups were found to be as per the genomic composition and commercial classification of the cultivars.

Controlled reciprocal crosses were made among *Musa* species by Faure et al. (1994) to demonstrate maternal/cytoplasmic transmission of chloroplast DNA but showed an unusual phenomenon of paternal transmission of mitochondrial DNA in *Musa acuminata*. The study was further confirmed by Carreel et al. (2002) who analyzed the origins of more than 300 *Musa* genotypes and concluded that most cultivars show mitochondrial genome linkage to two subspecies of *M. acuminata*, *M. acuminata* ssp. *banksii* and *M. acuminata* ssp. *errans*. It was found that some cultivars of AB, AAB, and ABB need not be simple allopolyploids, but most cultivars have different proportions of A and B genome chromosomes and/or possess different doses of recombinant chromosomes. All the research results published so far involving studies on cytoplasmic and nuclear DNA and at chromosomal, as well as protein, levels proved this concept. The hypothesis that hybrid banana cultivars would have evolved through backcrossing of interspecific hybrids with parental species, which would have led to formation of a complex spectra of genotypes/cultivars, seems to be true. De Langhe et al. (2010) debated that AAB accessions could have arisen as a result of either AB × AA or AA × AB cross, as contribution of A or B genome in this group does not agree with simple allopolyploid genome formulae of Simmonds and Shepherd (1955). Deviations were observed in the required total score of 35–37 based on morphological characterization, in "Pome" and "Silk" group cultivars. This was further strengthened by the study of Carreel et al. (2002) where inheritance of organelle (chloroplast and mitochondria) DNA was used. They also described the probable evolution of interspecific hybrids through (BB × AA) × AA or (AA × BB) × AA cross combinations to get BAA- or AAB-type cultivars where the cultivars have "B-type" or "A-type" cytoplasm and hence dosage of A or B genome varies.

1.5 Conclusions

Banana is one of the fruits available throughout the year in tropical and subtropical humid regions of the world. It is the most difficult crop to the researchers as the evolution of the present-day cultivars is still a mystery. Various researchers have studied the evidences through ethonobotanical and linguistic observations to derive its place of origin and spread. Further evidences are required to know the origin and speciation with reference to an ample number of cultivars available at present with special reference to Indian varieties. The cytological and molecular studies involving mitochondrial and chloroplast-specific markers have indicated the complexity of the evolution of banana cultivars. It appears that further research is necessary to understand the correct genomic status of the many cultivars.

References

Amalraj VA, Velayudhan KC, Agarwal RC (1993) Banana genetic resources. NBPGR, Technical Bulletin

Brandt SA, Spring A, Hiebsch C, McCabe JT, Tabogie E, Diro M, Wolde-Michael G, Yntiso G, Shigeta M, Tesfaye S (1997) The tree against hunger: ensete-based agricultural systems in Ethiopia. American Association for the Advancement of Science, Washington, DC with Awassa Agricultural Research Center, Ethiopia, Kyoto University Center for African Area Studies, Japan, and University of Florida, Gainesville. www.aaas.org/international/africa/enset/index.shtml

Carreel F, Faure S, Gonzalez De Leon D, Lagoda PJL, Perrier X, Bakry F, Tezenas Du Montcel H, Lanaud C, Horry JP (1994) Evaluation de la diversité genetique chez les bananiers diploides (Musa spp.). Genet Sel Evol 26(1):125–136

Carreel F, Gonzalez de Leon D, Lagoda P, Lanaud C, Jenny C, Horry J-P, Tezenas du Montcel H (2002) Ascertaining maternal and paternal lineage within Musa chloroplast and mitochondrial DNA RFLP analyses. Genome 45:679–692

d'Hont A, Paget-Goy A, Escoute J, Carreel F (2000) The interspecific genome structure of cultivated banana, Musa spp, revealed by genomic DNA in situ hybridization. Theor Appl Gene 100:177–183

D'Hont A et al (2012) The banana (Musa acuminata) genome and the evolution of monocotyledonous plants. Nature 488:213–217

Daniells J, Jenny C, Karamura D, Tomekpe K (2001) Musalogue: a catalogue of Musa germplasm. Diversity in the Genus Musa. Compiled and edited by E. Arnaud and S. Sharock. International Network for the Improvement of Banana and Plantain, Montpellier

de Jesus et al (2013) Genetic diversity and population structure of Musa accessions in ex situ conservation. BMC Plant Biol 13:41. doi:10.1186/1471-2229-13-41

De Langhe E, de Maret P (1999) racking the banana: its significance in early agriculture. In: Gosden C, Hather J (eds) The prehistory of food: appetites for change. Routledge, London, pp 377–396

De Langhe E, Vrydaghs L, de Maret P, Perrier X, Denham T (2009) Why bananas matter: an introduction to the history of banana domestication. Ethnobotany Res Appl 7:165–177

De Langhe E, Hřibová E, Carpentier S, Doležel J, Swennen R (2010) Did backcrossing contribute to the origin of hybrid edible bananas? Ann Bot 106:849–857

Denham TP, Donohue M (2009) Pre-Austronesian dispersal of banana cultivars west from New Guinea: linguistic relics from eastern Indonesia. Archaeology in Oceania 44:18–28

Dodds KS, Simmonds NW (1948) Genetical and cytological studies of Musa. IX. The origin of an edible diploid and the significance of interspecific hybridization in the banana complex An addendum on the nomenclature of edible banana. by E.E.Cheesman. 48:285–296

Doležel J, Bartos J (2005) Plant DNA flow cytometry and estimation of nuclear genome size. Ann Bot 95:99–110

Doležel J, Lysak MA, Doleželova M, Valarik M (1999) Analysis of Musa genome using flow cytometry and molecular cytogenetics. In: Proceedings of international symposium on the molecular and cellular biology of banana, New York (USA), 1999/03/22-25, pp 8

Donohue M, Denham TP (2009) Banana (Musa spp.) domestication in the Asia-Pacific region: linguistic and archaeobotanical perspectives. Ethnobot Res Appl 7:293–332

Faure S, Noyer J-L, Carreel F, Horry J-P, Bakry F, Lanaud C (1994) Maternal inheritance of chloroplast genome and paternal inheritance of mitochondrial genome in bananas (Musa acuminata). Curr Genet 25:265–269

Heslop-Harrison JS, Schwarzacher T (2007) Domestication, genomics and the future for banana. Ann Bot 100(5):1073–1084

Horry JP, Jay M (1988) Distribution of anthocyanins in wild and cultivated banana varieties. Phytochemistry 27:2667–2672

Jarret RL, Litz RE (1986) Isozymes as genetic markers in bananas and plantains. Euphytica 35:539–549

Kennedy J (2009) Bananas and people in the homeland of genus Musa: not just pretty fruit. Ethnobotany Res Appl 7:179–197

Kervegant D (1935) Le bannanier et sin exploitation. Societe d' edition geographques. Maritimes et colonials, Paris

Krishnamurthi S, Seshadri VS (1958) Origin and evolution of cultivated bananas. Indian J. Hort 15:135–45

Neumann K, Hildebrand E (2009) Early bananas in Africa: the state of the art. Ethnobot Res Appl 7:353–362

Ortiz R (1997) Morphological variation in Musa germplasm. Gene Resour Crop Evol 44:393–404

Ortiz R, Madsen S, Vuylsteke D (1998) Classification of African plantain landraces and banana cultivars using a phenotypic distance index of quantitative descriptors. Theor Appl Genet 96:904–911

Osuji JO, Harrison G, Crouch J, Heslop-Harrison JS (1997) Identification of the genomic constitution of Musa L. lines (bananas, plantains and hybrids) using molecular cytogenetics. Ann Bot 80:787–793

Perrier X, Bakry F, Carreel F, Jenny C, Horry JP, Lebot V, Hippolyte I (2009) Combining biological approaches to shed light on evolution of edible bananas. Ethnobot Res Appl 7:199–216

Perrier X, Langhe E, Donohue M, Lentfer C, Vrydaghs L, Bakry F, Carreel F, Hippolyte I, Horry J-P, Jenny C, Lebot V, Risterucci A-M, Tomekpe K, Doutrelepont H, Ball T, Manwaring J, Maret P, Denham T (2011) Multidisciplinary perspectives on banana (Musa spp.) domestication. Proc Natl Acad Sci U S A 5:1–8

Pollefeys P, Sharrock S, Arnaud E (2004) Preliminary analysis of the literature on the distribution of wild Musa species using MGIS and DIVA-GIS. INIBAP, Montpellier, http://bananas.bioversityinternational

Price NS (1995) In: Gowen S (ed) Bananas and plantains. Chapman and Hall, London

Purseglove JW (1975) Tropical crops: monocotyledons. 2nd impression. Halsted Press, New York

Rao VNM (1984) Banana. A monograph. ICAR (Indian Council of Agricultural Research) Publication, New Delhi

Reynolds PK (1927) The banana, its history, cultivation and place among staple food. Riverside Press, Cambridge

Shepherd K, Ferreira FR (1982) Biological foundation's banana collection at Laloki Port Moresby Papua New Guinea IBPGR/SEAN. News Lett 8(4):28–34

Simmonds NW (1959) Bananas. Longmans, Green & Co., London

Simmonds NW (1962) Evolution of the bananas. Longmans, Green & Co., London

Simmonds NW, Shepherd K (1955) Taxonomy and origins of cultivated bananas. Bot J Linn Soc 55:302–312

Simmonds NW, Weatherup STC (1990a) Numerical taxonomy of the wild bananas. New Phytol 115:567–571

Simmonds NW, Weatherup STC (1990) Numerical taxonomy of the cultivated bananas. Tropical Agriculture (Trinidad) 67(1):90–92.

Stover RH, Simmonds NW (1987) Bananas. Longman, London

Vrydaghs L, De Langhe E (2003) Phytoliths: an opportunity to rewrite history. In INIBAP Annual Report 2002. International Network for the Improvement of Banana and Plantain, Montpellier, pp 14–17

Vrydaghs L, Swennen RR, Mbida C, Doutrelepont H, De Langhe E, de Maret P (2003) The banana phytolith as a direct marker of early agriculture: a review of the evidence. In: Hart DM, Wallis LA (eds) Phytolith and starch research in the Australia; Pacific – Asian Regions: the state of the art. Terra Australis 19. Pandanus Books, Research School of Pacific and Asian Studies, Australian National University, Canberra, pp 177–185

Banana Breeding

2

Rema Menon

Abstract

The first attempt at breeding bananas through hybridization, in the 1920s was triggered by the devastation of the export banana 'Gros Michel' by *Fusarium* wilt. Conventional breeding despite being hindered by several plant based constraints has been responsible for the production of a whole range of hybrids resistant to diseases and pests. The approach commonly adopted aims at developing new tetraploid varieties by crossing triploids with wild or improved diploid clones with resistance, or secondary triploids derived from crosses between the developed tetraploids and the diploid clones. The limitations of the 3x/2x strategy are low gametic fertility of the triploid variety to be improved. This has led to the development of an alternate pathway, which targeted the development of triploid hybrids directly from crosses involving diploid and doubled diploid varieties. The method exploited the male and female fertility status of doubled diploids which otherwise are sterile at the diploid level. Following the 3x/2x method, hybrids with yield advantage and resistance to biotic stresses developed by major breeding programmes have been adopted in many countries, which was facilitated by the International *Musa* Testing Programme. The new insight gained in *Musa* genetic diversity through molecular tools provided key inputs, useful for the selection of parental combinations. Development of molecular markers linked to major traits is expected to speed up the improvement process. Mutation breeding is offering a unique, alternative approach for the improvement of banana has also been employed to develop new cultivars.

Keywords

Musa • Breeding strategies • Sterility • Parthenocarpy • Fertility

R. Menon (✉)
Banana Research Centre (KAU),
Kannara 680652, Kerala, India
e-mail: rmenon.brs@gmail.com

S. Mohandas, K.V. Ravishankar (eds.), *Banana: Genomics and Transgenic Approaches for Genetic Improvement*, DOI 10.1007/978-981-10-1585-4_2

2.1 Introduction

Bananas originated in South East Asia and are one of the most widely cultivated crops in over 130 countries throughout the tropical and subtropical regions of the world. The annual world production of banana is around 103 million metric tons from an area of 5.14 million ha (FAO 2014). Bananas and plantains are popular and cheap multipurpose crops by virtue of their ability to grow under a wide range of environments, producing fruits year-round and nutritional and have therapeutic values. A valuable source of nutrition for more than 400 million people in the tropics, bananas are rated as the developing world's fourth most important crop after rice, wheat and maize. In the tropics, where most of the population depends on carbohydrate foods as dietary staples, plantains and bananas along with root crops play a significant role as additional source of dietary energy and protein. The fact that global production of bananas and plantains increased by 70 % in the last 40 years representing the fastest growth among starchy staples in developing countries testifies their potential in strengthening food security and the alleviation of poverty in rural areas where 75 % of the population is concentrated.

Bananas and plantains are highly remunerative and play a pivotal role in the livelihood security of farmers. Around 85 % of global banana production is managed by small-scale farmers and most of it is being consumed locally. The livelihood of these farmers depends on the availability of a wide range of cultivars that are adapted to local environmental conditions. The existing diversity is mainly composed of farmer-selected cultivars. The vast gene pool provides immense possibility to use the material in response to environmental changes and continuous emergence of biotic threats. Increasing population pressure and consequent increase in the demand for food on the one hand and depletion of arable land on the other have placed new emphasis on conventional plant breeding. Conventional or classical plant breeding has been responsible for improvement in numerous cultivated plant species in the twentieth century (Johnson 2000). An important component of classical breeding is the extensive testing of the material in various environments in comparison with existing cultivars. Banana production is hampered by a wide range of pests and diseases (Jones 2000), and introducing host plant resistance is the most economical and sustainable way to manage them (Lorensen et al. 2010). A considerable progress has been made in the introgression of resistance to many of these biotic stresses and hybrids evolved (Tomekpe et al. 2004; Lorensen et al. 2010).

2.2 Taxonomy

The family Musaceae comprises of two genera, *Ensete* Haran and *Musa* L. The genus *Musa* has four sections, namely, *Callimusa* and *Australimusa* which have a basic chromosome number of 10 and *Eumusa* and *Rhodochlamys* having the chromosome number 11. *Callimusa* has many non-domesticated members with a lot of ornamental value. Multiplication is fast with their rhizomatous stems and has erect inflorescence with bright-coloured bracts. *Australimusa* has five to six species, but the most important are fibre-yielding *Musa textilis* and fruit-yielding *Musa fe'i*. The distribution of *M. textilis* is mostly in S.E. Asia with reports of occurrence even in Indo-Burma border, while Fe'i bananas have a commercial distribution in the Pacific Islands. *Rhodochlamys* consists of members which are ornamental in nature. They are distinguished by their slender stature and erect inflorescence with brightly coloured bracts. They are female fertile and cross freely with *Eumusa* members with F1 generation exhibiting a dominance of *Rhodochlamys* traits.

Eumusa is the only source of present-day edible bananas except for Fe'i bananas of *Australimusa*. They are characterized by robust pseudostem; horizontal, angular or pendulous bunches; and in most cases parthenocarpic fruits. All edible bananas are believed to have originated from two species, *Musa acuminata* designated by A genome and *Musa balbisiana* designated as B genome. The relative contribution of each genome has resulted in various combinations, viz. AA, AAA, AB, AAB, ABB, BB, BBB, ABBB, etc. (Stover and Simmonds 1987).

2.3 Origin, Distribution and Diversity in Bananas and Plantains

Virtually all banana varieties of today have evolved from intra- and interspecific hybridization of the two diploid ($2n = 2x = 22$) wild species *Musa acuminata* Colla and *Musa balbisiana* Colla. These diploid *Musa* species have seeded fruit with a little starch and a small amount of pulp and of no value as a crop. Evolution of edibility in the *Eumusa* series of cultivars started with wild *Musa acuminata* subspecies which are predominant in South East Asia. The appearance of vegetative parthenocarpy and female sterility in wild *acuminata* facilitated the production of seedless diploid *Musa acuminata*. Another key event was hybridization with *Musa balbisiana*, a hardier species. Both *Musa acuminata* and *Musa balbisiana* are diploids and represented, respectively, as AAw and BBw. Further, triploids evolved through fertilization of viable diploid cells formed due to breakdown of meiosis at the second division with haploid pollen (Simmonds 1962; Jones 2000).

India harbours a rich diversity of wild *balbisiana* types and few subspecies of *acuminata*. Even though the present-day bananas are the result of natural introgression between the two diploid wild species, the presence of *M. schizocarpa* and *M. textilis* represented by S and T genomes in cultivated and wild types has opened new avenues to the understanding of the evolution theory (Jones 2000). Molecular analyses have also confirmed the presence of S and T genomes (D'Hont et al. 2000).

Of 11 wild species of *Musa* reported across the globe, six are present in India. *M. acuminata and M. balbisiana*, the progenitors of cultivated banana, are widely distributed in the banana-growing regions, viz. South Indian states, north-eastern states and Andaman and Nicobar Islands. *M. itinerens*, *M. nagensium*, *M. sikkimensis* and *M. cheesmani* are the other wild species, mainly concentrated in the north-eastern region (Simmonds 1962).

Of the eight subspecies under *Musa acuminata* reported from Asian countries, only three have been reported in India – *banksii*, *burmanica* and *burmannicoides*.

Musa balbisiana originated in the drier parts of India and is widely distributed from this region to the Philippines and New Guinea, but is absent in Central Malaysia. It is hardier, drought and disease tolerant than *M. acuminata*. No subspecies are recognized.

Banana has been subjected to a rigorous course of evolution to transform itself from seedy non-pulpy wild progenitors to present-day parthenocarpic edible and high-yielding banana (Simmonds 1962). The four major factors, viz. parthenocarpy, sterility, polyploidy and vegetative propagation for perpetuation of useful traits, have contributed alone or in combination to the evolution of present-day bananas. Apart from the nuclear genome, the significance of plastid and mitochondrial genomes contributing to important agronomic traits has been suggested. Studies in this direction by Faure et al. (1994) not only demonstrated maternal transmission of chloroplast DNA but also showed the occurrence of paternal transmission of mitochondrial DNA in *Musa acuminata*. Carreel et al. (2002) concluded, based on genetic analyses of 300 *Musa* genotypes, that most cultivars are linked through their mitochondrial genomes to *banksii* and *errans*, two subspecies of *Musa acuminata*.

Most cultivated bananas and plantains are highly female sterile and cannot reproduce sexually. Clonal propagation is only possible and survival in nature and geographical dispersal cannot happen without human intervention. Consequently, secondary diversification in areas devoid of wild *Musa* plants has been attributed to somatic mutations of introduced materials (Purseglove 1975).

Diversity in the *Eumusa* series of edible banana is made up of varieties with AA, AB, AAA, AAB, ABB, AAAA and ABBB genomes. There are over a thousand types of bananas in existence, subdivided into 50 groups of varieties. The most important bananas are categorized, commonly as the AAA dessert bananas, the AAA highland cooking bananas and beer bananas of East Africa, the AAB plantains and the ABB cooking bananas (Simmonds 1962; Jones 2000).

The taxonomic method followed to determine the genome of banana cultivars based on morphological characters (pseudostem colour, shape of petiolar canal and bract features) gives a good estimate of the genetic composition and ploidy of cultivars (Simmonds and Shepherd 1955). Numerical taxonomy has refined the approach (Ortiz et al. 1998). Flow cytometry provides accurate estimates without growing mature plants (Dolezel and Bartos 2005). A good correlation with results of molecular method is also obtained. Clones/cultivars within each genomic group are identified based on additional morphological characters. Many clones have given rise to morphotypes that differ in fruit and bunch morphology, pigmentation and height. The same clone/ cultivar can have different names in different locations and synonyms, adding to taxonomic complexity (Menon 2000a; Uma et al. 2005a). For a better identification and classification of a cultivar or a landrace, a basic knowledge about their evolution, taxonomic status and description of traits is essential. For this purpose, the banana descriptor published (IPGRI/INIBAP/CIRAD 1996) is utilized for distinguishing the important traits.

Cultivars and landraces within a genome are referred to as 'group 'and 'subgroups'. Wild accessions are denoted as 'types'. Diploids AA or AB are characterized by their more slender pseudostems and more upright leaves. Triploid cultivars are classified under three genomic groups, viz. AAA, AAB and ABB. Triploids are bigger, sturdier plants than diploids with increased fruit size. Tetraploid cultivars are few in number and belong to the AAAA, AABB, AAAB and ABBB genomic groups. They possess robust pseudostem and leaves that tend to droop. Tetraploids have formed from fertilization of triploid egg cells by haploid pollen (Jones 2000).

Fruit development in banana is characterized by vegetative parthenocarpy, the development of pulp without pollination. In the case of wild seeded bananas, pollination precedes normal fruit development without which fruits remain unfilled and shrivelled. A normally mature fruit contains a mass of hard black seeds surrounded by a scanty sweetish pulp which develops from the ovary walls and septa. If ovaries of seeded bananas are protected against pollination, they do not develop. In contrast, edible bananas are vegetatively parthenocarpic, i.e. development of pulp without pollination. The ovules which shrivel early can be seen embedded in the edible pulp. The physiology of parthenocarpic development in banana is apparently mediated by an autonomous production of auxin in the mature ovary (Purseglove 1975; Stover and Simmonds 1987). Seeds vary with respect to size, shape and colour. The size varies from 4 to 20 mm and shape can be spherical, triangular or ovoid. Brown to pitch black-coloured seed coat may be warty or smooth.

2.3.1 Major Cultivar Groups

2.3.1.1 Cavendish Subgroup

Cavendish cultivars may have originated in the South China-Vietnam area and Malaysian region. They are high yielding than all other natural clones (Robinson 1996). The varieties in the Cavendish subgroup are separated mainly by differences in height and bunch and finger characteristics. There is a gradation in height from the shortest (Dwarf Cavendish) to the tallest. The Cavendish subgroup is responsible for 30 % of the world's production of banana fruit. Stover and Simmonds (1987) recognized four major clone sets distinguished on height: 'Dwarf Cavendish' types are the shortest in stature; 'Grand Nain' types are medium dwarfs; and intermediate in height are between 'Giant Cavendish' and 'Dwarf Cavendish' groups. 'Giant Cavendish' types are taller and 'Pisang Masak Hijau' types, the tallest. Apart from plant height, clones in the Cavendish subgroup also differ in other morphological characters such as petiole length, bract persistence, bunch grade and pseudostem colour.

2.3.1.2 Plantain Subgroup

The plantain subgroup is very important as plantain cultivars provide food for millions of people in the West and Central Africa and Latin America-Caribbean regions. Cultivars are also found in East Africa and South and South East Asia. A total of 23 % of the world's production of banana fruit comes from plantain. The term plantain has been used in the past to describe all cooking banana types, but now it refers only to clones belonging specifically to the plantain subgroup within the AAB genomic group of banana where fruit remains starchy at ripeness. They are characterized by the orange yellow colour of the compound tepal in the flower and of pulp at ripeness. Fruits are long and slender, angular, pointed and unpalatable when raw (Stover and Simmonds 1987).

Cultivars in the plantain subgroup are placed in four main clone sets, which are distinguished on bunch and inflorescence characteristics (Tezenas du Montcel and Davos 1978). 'French' plantain types have many hands with comparatively small fingers and an inflorescence axis covered with persistent hermaphrodite and male flowers. The large male bud is also persistent. 'Horn' plantain types have few hands of very large fingers, no hermaphrodite flowers and no male axis. 'French Horn' and 'False Horn' plantain types are intermediate classification categories between 'French' and 'Horn' plantains. The male bud is absent at maturity in both of these types, but there are many hermaphrodite flowers on 'French Horn' cultivars and a few on 'False Horn'.

Though South India is believed to be the centre of origin of French plantain group (Simmonds 1966), wide diversity has been reported from Central Africa (De Langhe 1964). Nendran, the French plantain cultivar, is the predominant variety in India and is represented by several clones recognizable through variation in plant stature, pseudostem colour, bunch morphology and development of the male phase (Jacob 1952). Menon et al. (2002b) collected, characterized and catalogued the variability of Nendran in South India and based on morphological characterization recognized ten morphotypes. A key was developed for their identification.

2.3.1.3 Mysore Subgroup

Mysore is grown on a large scale in South Asia. However, outsidc India and Sri Lanka, it is only occasionally encountered. Exceptions are Trinidad, where it is used to shade cocoa, and Western Samoa. 'Poovan' is a popular cultivar of this group grown all over the India in a perennial cropping system and is the leading commercial cultivar of southern and north-eastern states and easily available all through the year. It is known by different names viz. Mysore Poovan, Palayankodan, Champa, Alpan, Karpurachakkarakeli in India (Singh et al. 2001). It is a large and extremely vigorous cultivar, easily recognized by its pinkish-purple midribs and large cylindrical bunches of tightly packed short, plump bottlenecked fruits which ripen to an attractive bright yellow, and the flesh is agreeably sweet acid (Stover and Simmonds 1987).

2.3.1.4 Silk Subgroup

Silk is a very popular dessert cultivar in South and South East Asia, East Africa and Latin American-Caribbean regions characterized by fruit with a sweet-acid taste (Stover and Simmonds 1987).

The ripe fruits are thin skinned and fall of easily from the bunches (highly deciduous) weighing 15–18 kg and it takes 13–15 months to come to harvest. Hard lumps in fruits occur, especially when grown in acidic soils. *Fusarium* wilt is a major threat to this cultivar resulting in restricted cultivation. It is tolerant to leaf spot.

2.3.1.5 Pome Subgroup

Cultivars in the Pome subgroup are important dessert banana types in India and Brazil, where their subacid flavour is much appreciated. The taste is also popular in Australia and Hawaii (Stover and Simmonds (1987)). However, cultivars are generally not very productive. The Pome subgroup has several cultivars and mutants. In general the bunch is semi-horizontal and fruits are slightly angular and the peel is thick (Stover and Simmonds 1987).

2.3.1.6 Pisang Awak Subgroup

Pisang Awak is a widely disseminated, high-yielding, cooking/dessert cultivar group, which is also used as a beer banana in East Africa. It is very common in Thailand, Vietnam and elsewhere in the Indo-China region. This triploid group has high inherent fertility which yields seedless edible fruits if un-pollinated, but bears seeds when pollinated by pollen from fertile diploids growing in the neighbourhood or by wild bananas. This group is represented by many cultivars in India, viz. 'Karpooravalli', 'Pey Kunnan', 'Vella Palayankodan', 'Dakshinsagar' and 'Paloor' (Uma and Sathiamoorthy 2007).

2.3.2 Conservation of Genetic Resources

Banana improvement programmes depend on the full range of diversity and therefore warrant effective conservation and documentation of existing cultivars and wild relatives. Banana biodiversity is conserved as full-size plants in field gene banks (Amalraj et al. 1993; Menon et al. 2000a; Uma et al. 2005a) or as in vitro plantlets (Van den Houwe et al. 1995) or cryopreserved (Panis et al. 2007). The diversity of *Musa* germplasm is conserved in about 60 national collections. An in vitro collection comprising 1,200 *Musa* accessions are held at the global banana collection managed by Bioversity International and hosted by the Belgian university KU Leuven at the International Transit Centre (ITC). Characterization and evaluation data from 22 collections are available in the Musa Germplasm Information System (MGIS) (Channeliere et al. 2011). Widespread diseases such as banana bunchy top virus, banana bract mosaic virus, weevil borers, Sigatoka leaf spot and *Fusarium* wilt are serious threats to germplasm, small-scale production systems and field repositories. Several of these uncommon or rare homestead/backyard clones are on the verge of extinction. The various species of *Musa* that occur wild in the forests (Western Ghats and north-eastern region) also face a threat from deforestation, wild animals and shifting cultivation. Hence there is an urgent need to conserve the threatened wild species and rare genetic resources for their utilization in the future. In vitro conservation methods are routinely applied for banana germplasm conservation and exchange at the NBPGR, New Delhi (Agrawal et al. 2007). Field collections are maintained at NRCB, Trichy and some State Agricultural Universities. Germplasm conservation activity is financially supported by Indian Council of Agricultural Research through the All India Coordinated Research Project on Tropical Fruits.

2.3.3 Breeding Targets

Major threats to banana and plantain production include fungal, bacterial and viral pathogen, several insect pests and a complex of plant parasitic nematodes (Table 2.1). Among fungal diseases, black leaf streak disease (BLSD) or black Sigatoka caused by *Mycosphaerella fijiensis* is the most important in the world. Sigatoka leaf spot or yellow Sigatoka caused by *M. musicola* is a related disease causing similar damage. *Fusarium* wilt caused by the fungus *Fusarium oxysporum* f.sp. *cubense* is widely regarded as the most destructive plant disease in the recorded history (Moore et al. 1995). Being soilborne, it is also described as a disease that refuses to go away (Ploetz and Churchill 2011). The disease was responsible for the destruction of Gros Michel-based export trade and its replacement by the resistant Cavendish (AAA) group cultivars. Tropical race 4 attacking Cavendish is an extremely virulent form of the pathogen reported more recently (Hwang and Ko 2004). Since fungicidal sprays or cultural practices cannot contain the disease, long-term option of developing resistant varieties to replace susceptible one is suggested (Hwang and Ko 2004). Viruses also constitute major production constraints besides being an impediment to germplasm enhancement and movement (Jones 2000). Four major viruses encountered in banana include banana bunchy top virus (BBTV), banana bract mosaic virus (BBMV), cucumber mosaic virus (CMV) and banana streak virus (BSV). BSV-related sequences are integrated within the nuclear genome and can cause a viral infection (Harper et al. 1999).

Among insect pests, banana corm weevil (*Cosmopolites sordidus*) has attracted worldwide attention resistance to this pest that has been a breeding target. Banana stem weevil (*Odoiporus longicollis*) is a devastating pest of plantains in India (Padmanabhan and Sathiamoorthy 2007).

Table 2.1 Major diseases and pests of banana/plantain

I. *Diseases*	
i. *Fungal diseases*	
Sigatoka leaf spot	*Mycosphaerella musicola*
Black leaf streak	*Mycosphaerella fijiensis*
Eumusae leaf spot	*Mycosphaerella eumusae*
Panama wilt	*Fusarium oxysporum* f.sp. *cubense*
ii. *Bacterial diseases*	
Rhizome rot	*Erwinia* sp.
iii. *Viral diseases*	
Banana bunchy top virus	BBTV
Banana bract mosaic virus	BBMV
Cucumber mosaic virus	CMV
Banana streak virus	BSV
II. *Insect pests and nematodes*	
Rhizome weevil	*Cosmopolites sordidus*
Pseudostem weevil	*Odoiporus longicollis*
Banana aphid	*Pentalonia nigronervosa*
Nematodes	*Radopholus similis*, *Meloidogyne incognita*, *Pratylenchus* sp.

Plant parasitic nematodes are also a major constraint to *Musa* production (Stover and Simmonds 1987) and include *Radopholus similis*, *Meloidogyne incognita*, *Pratylenchus coffeae* and *Helicotylenchus multicinctus*.

The development of cultivars with high yield and resistance/tolerance to these production constraints is the primary focus of the banana improvement programme. The overall strategy has been to incorporate resistance to *Fusarium* wilt, Sigatoka and other pests in existing cultivars rather than aiming for genetic materials that are drastically different. Even though initial focus was on breeding for resistance to Sigatoka leaf spot, objectives have gradually widened to include resistance to nematodes and several other pests as well as modifying plant architecture, growth habit and fruit quality (Tomekpe et al. 2004).

2.3.4 Breeding Challenges in *Musa*

Musa is a polyploid crop with ploidy ranging from diploid ($2n = 2x = 22$) to tetraploids ($2n = 4x = 44$). Most cultivated bananas are triploids ($2n = 3x = 33$) and sterile harbouring various combinations of either one, two or three A, B, S or T genomes. New banana cultivars are exceptionally cumbersome to develop. Selection for desirable characters is time consuming and it may take up to 12 years to develop a new cultivar. *Musa* breeding is based mainly on the phenotypic mass recurrent selection. The high levels of heterozygosity make identification of ideal parental material difficult, and very large populations are required for the selection of individual clones with good agronomic traits. This is virtually impossible to attain due to the low seed set in crosses. Generally, few seeds are obtained (an average of 1–1.5) and acquiring large numbers of seeds is a labour-intensive and tedious process (Ortiz and Vuylsteke 1995a; Ssebuliba et al. 2006, 2009). The genes for resistance to diseases and pests are introgressed from wild diploid species which also carry many undesirable traits, e.g. low yield and non-parthenocarpy. The process of eliminating the unwanted traits requires several backcrosses that lengthen the breeding process (Rowe and Rosales 1992). The multigenic nature and low heritability of some traits also slow down the breeding process. *Musa* breeding is also problematic due to the narrow genetic diversity of the germplasm (Pillay et al. 2004) and the lack of information about wild species that carry useful agronomic traits. Only a few wild diploids have been used as male parents in majority of the breeding programmes. As cultivated banana is propagated asexually, its genetic base is narrow, as diversity dependent on somatic mutation. Limited genetic variation has resulted in a crop lacking resistance to fungal, bacterial and viral pathogens and numerous pests (Miller et al. 2011).

2.3.5 History of *Musa* Breeding

The destruction of the export banana 'Gros Michel' by *Fusarium* wilt led to the development of the first banana breeding programme at the Imperial College of Tropical Agriculture (ICTA) in Trinidad in 1922 and later in Jamaica in 1924 (Jones 2000). These programmes aimed at producing 'Gros Michel' resistant to *Fusarium* wilt. Subsequently with the appearance of Sigatoka leaf spot in accessions, resistance to leaf spot was also included as a breeding objective. Two *Fusarium* wilt-resistant tetraploids developed initially were 'Bodles Altafort' from a cross between 'Gros Michel' × 'Pisang Lilin' and 2390-2 derived from 'Highgate', a dwarf mutant of 'Gros Michel' and 'Pisang Lilin'. The establishment of a breeding programme by the United Fruits Company in 1959 was driven by the need to breed resistant dessert bananas. The programme was brought under FHIA (Fundación Hondureña de Investigación Agricola), Honduras, in 1984. FHIA has developed a wide range of hybrids (Table 2.2) presently being grown in several countries. CIRAD (Centre de Coopération International en Recherche Agronomique pour le Développement), France, concentrated on the creation of new triploid dessert banana utilizing existing diversity in diploid dessert varieties (Bakry et al. 1997). Plantain breeding has been the main focus of breeding efforts of IITA (International Institute of Tropical Agriculture), Ibadan, in Nigeria, Cameroon and Uganda and more recently including improvement of East African highland bananas (Lorensen et al. 2010). Plantain breeding formed the major objective of CARBAP (Centre de Recherches Régionales sur Bananiers et Plantains), Cameroon, and has the largest collection of plantains in the world (Tomekpe et al. 2004). EMBRAPA (Empresa Brasileira de Pesquisa Agropecuária), Brazil, worked on Pome and Silk subgroups important in Brazil. Sterility conferred by triploidy limits the use of hybridization in banana improvement and constitutes a challenge to conventional breeding methods. In spite of the inherent difficulties, significant progress has been achieved over the last two decades (Tomekpe et al. 2004).

2.3.6 Breeding Methods

2.3.6.1 Diploid Breeding

Diploids comprising wild, semiwild or parthenocarpic accessions are characterized by high fertility and considerable genetic diversity and are rich sources of resistance/tolerance to diseases and pests. They serve as good parents and are ideal material for genetic and cytogenetic studies (Tomekpe et al. 2004). The development of agronomically superior disease-resistant diploids was identified as a priority area in banana and plantain breeding (Rowe and Rosales 1994). They outlined the diploid breeding efforts at FHIA. A breakthrough in this direction was the development of SH-2095, the first diploid hybrid selected during the first 10 years of the programme and which formed the basis of subsequent diploids selected from segregating populations. SH-3142 is another diploid hybrid resistant to burrowing nematode through extensive pollination of 'Pisang Jaribuaya' which is a natural AA diploid resistant to burrowing nematodes. In contrast to the sterile 'Pisang Jaribuaya', the hybrid was both male and female fertile and integrated in breeding schemes. Its resistance to *Fusarium* wilt was exploited in the development of FHIA-01 tetraploid hybrid. The subsequent series of crosses focused on the development of an improved diploid with multiple resistance for which SH-3723 formed the parental line. A wide range of improved diploids evolved have been integrated in breeding schemes.

Table 2.2 Characteristics of hybrids developed by FHIA

Hybrid	Genealogy	Characteristics
FHIA- 01 (AAAB)	'Prata Anã' (AAB) × SH-3142 (AA)	Resistance to black sigatoka, races 1 and 4 of *Fusarium* wilt and nematode (*Radopholus similis*), tolerance to low temperatures and unfavourable conditions of raining and soil fertility, strong plants with good architecture, large bunches and fruits with good texture and taste
FHIA-02 (AAAA)	'Williams' (AAA) × SH-3393 (AA)	High resistance to black Sigatoka, present the same plant height of 'Valery', however the bunch characteristics are inferior to this cultivar
FHIA-03 (AABB)	SH-3386 (ABB) × diploid SH-3320 (AA)	Generated from the clone Gaddatu ABB, it is more rustic and productive than the Bluggoes. Present low stature and bunches that reach to 50 kg and resistance to moko disease and drought
FHIA-15 (AAAB) FHIA-17 (AAAA) FHIA-18 (AAAB)	Descendent of Maqueño Highgate hybrid (AAA) (mutant of 'Gros Michel'), Prata Anã hybrid (AAB)	Resistance to race 1 of the *Fusarium* presents tolerance to black Sigatoka. Good characteristics of bunch, resistance to race 1 of *Fusarium* wilt and more tolerant to black Sigatoka than the Grande Naine, however its plants are higher than that cultivar, mainly in the second cycle. High production, resistant to yellow and black Sigatokas and fruits present good taste
FHIA-21 (AAAB)	French plantain hybrid (AAB)	Resistance to black Sigatoka and produce larger bunches than to the other False Horn cultivars
FHIA-23 (AAAA)	Highgate hybrid (AAA)(mutant of 'Gros Michel')	Good characteristics of bunch, resistance to race 1 of *Fusarium* wilt and more tolerant to black Sigatoka than the Grande Naine, however its plants are higher than that cultivar, mainly in the second cycle
AVP-67 (AAAB)	Crosses involving plantain French type (AAB) and Maqueño (ABB)	Good characteristics to use unripe or ripe, resistant to race 1 Panama disease and tolerance to black Sigatoka
SH-3640	Prata Anã (AAB) × SH-3393	High production and resistance to black Sigatoka, partial resistance to yellow Sigatoka and ripe fruits present very good taste

Rowe and Rosales (1994)

At IITA, evaluation of several subspecies of wild *acuminata* (*banksii*, *burmannica*, *malaccensis*, *microcarpa*) led to the observation that disease slowing model of partial resistance was durable than a gene for gene hypersensitive reaction (van den Planck 1982; Foure et al. 2000). Two resistant improved diploid parents, TMB2X51051 and TMB2X91283 (AA), were developed from crosses with resistant sources which were also resistant to *R. similis* (Tenkouano et al. 2003). Seven high-yielding diploid hybrid selections developed and resistant to black Sigatoka and nematodes were utilized in recurrent diploid breeding and in 4x 2x crosses (Tenkouano and Swennen 2004).

Vuylsteke et al. (1993) observed that plantain-derived F1 population comprise 80 % diploids and the proportion was influenced by the plantain cultivars. The plantain-derived diploids designated as TMP2x have been considered significant in germplasm enhancement at the 2× level and as a source of plantain alleles (Vuylsteke and Ortiz 1993). Since the plantain-derived diploids possessed 50 % plantain genes in their genome, the probability of recovering progeny with plantain-like characteristics was more than those derived from Calcutta 4 as pollen donor. Many plantain-derived diploids resistant to leaf spot and with a fruit quality similar to that of plantains have been evolved.

M53 is an improved diploid developed with resistance to leaf spot caused by *Mycosphaerella* and to *Fusarium* wilt. Numerous monospecific AA and interspecific AB diploid hybrids have been evolved by the National Research Centre for Banana and Tamil Nadu Agricultural University, India (Tomekpe et al. 2004).

2.3.6.2 The 3x/2x Strategy

The 3x/2x strategy has been widely applied to create tetraploid hybrids resistant to diseases and with good agronomic value by pollinating susceptible triploids with male fertile diploids that are resistant. Low fertility of triploid cultivars results in only small populations and necessitates a considerably higher number of crosses. The hybrid tetraploids have male and female fertility inherited from the fertile diploid and set seed on pollination which has a bearing on the quality. Therefore the primary tetraploids produced by this method were considered as intermediate products and subsequently crossed with diploids to obtain triploids close to the original triploid cultivars.

Tomekpe et al. (2004) pointed out that the genetic gain obtained by nuclear restitution in the first stage (production of tetraploids) is reduced by the recombination which occurs during meiosis of the tetraploid in the second stage. Thus the choice of the diploid used in the F1 and F2 stage has a direct bearing on the reconstitution of secondary triploid hybrids. The AAAB hybrids resulting from crosses between AAB and AA express symptoms of BSV due to activation of viral sequences integrated in the B genome of these cultivars (Tenkouano and Swennen 2004).

Based on this strategy FHIA, IITA and EMBRAPA could select good quality secondary triploid hybrids. CARBAP has applied this method to generate sizeable population of secondary triploids from primary tetraploids derived from dwarf plantains (Tomekpe et al. 2004).

2.3.6.2.1 Breeding Work at FHIA

The FHIA programme is the oldest and aimed at the development of dwarf dessert banana, plantain and cooking banana hybrids with resistance to black Sigatoka, burrowing nematodes and race 4 of *Fusarium* wilt. Though the breeding work was initiated with the use of 'Highgate' (dwarf mutant of 'Gros Michel') as a seed fertile triploid to produce tetraploid hybrids, 'Dwarf Prata' (AAB Pome) turned out to be a useful triploid line for breeding dessert banana (Rowe 1990). The discovery that French plantain (AAB plantain) can produce tetraploid progenies with improved bunch features provided opportunities to breed disease-resistant plantains (Rowe 1987). 'Gaddatu' (ABB Gubao) proved to be a suitable parental line for developing cooking bananas (Rowe and Rosales 1994). The breeding efforts resulted in the development of three outstanding hybrids.

The development of the tetraploid (AAAB) hybrid FHIA-01 was possible by crossing the burrowing nematode-resistant SH-3142 diploid (AA) with 'Dwarf Prata'. This is considered a significant achievement of the FHIA programme (Rowe and Rosales 1994). The hybrid is resistant to race 4 of *Fusarium* wilt and could be grown without taking control for black Sigatoka. Its suitability to marginal growing conditions, good finger size and better green life in comparison to Cavendish are the reported advantages. FHIA-0 1 was officially released as Goldfinger in Australia during 1995 (Daniells et al. 1995). Among several plantain hybrids selected by the FHIA programme, FHIA-21 derived from AVP-67 (French plantain) and SH-3142 has been the most promising. It has plantain-type shape and pulp colour and flavour (Rowe and Rosales 1994). When compared with local plantain, FHIA-21 produced bigger bunches and is resistant to black Sigatoka. FHIA-03 is a semidwarf hybrid (ABBB) based on the cultivar 'Gaddatu'. It is more hardy and productive than the widely grown cooking banana 'Bluggoe'. Resistance to black Sigatoka and tolerance to drought and poor soils are its other advantages.

Following reports by Simmonds (1962), Stover and Buddenhagen (1986) and Stover and Simmonds (1987) that Cavendish cultivars are female sterile, improvement of this group of dessert bananas through conventional breeding received very little attention till 2002. But subsequent seed set was observed in the wild *Musa acuminata* ssp. *malaccensis* when pollinated with 'Valery' pollen. Several dwarf diploids obtained were integrated in later crosses. Another milestone was the development of a tetraploid plant derived from a cross of 'Williams' as female parent and the improved diploid SH-3142 as male parent that was named FHIA-02. These inputs formed the basis for the development of a Cavendish replacement resistant to BLS and/or Foc TR4 through conventional breeding, and several resistant triploid hybrids could be selected (Moran 2013).

2.3.6.2.2 Breeding Work at IITA

Low reproductive fertility, triploidy and slow propagation were the factors limiting breeding progress in plantains. For imparting resistance to black Sigatoka, screening of germplasm to identify available sources of resistance and female fertile plantain cultivars formed the breeding priorities at IITA. Several French and False Horn plantain cultivars were identified as partially female fertile based on production of viable seed after pollinating with wild diploid banana 'Calcutta 4'. 'Bobby Tannap' and 'Obino l'Ewai' were the two plantains used as female parents. Seed set was influenced by the cultivar used as a female parent and seasonal variations. Higher temperature combined with high solar radiation and low relative humidity improved the seed set (Ortiz and Vuylsteke 1995b). Tetraploid progenies derived from plantain × Calcutta 4 and showing partial resistance to black Sigatoka were further improved by crossing with other resistant diploids. Also plantain-derived diploids were crossed with resistant diploids to broaden the genetic base of resistant population (Ortiz and Vuylsteke 1994a).

The wild *Musa acuminata* ssp. *burmannicoides* clone Calcutta 4 used as male parent was highly resistant to black Sigatoka (Swennen and Vuylsteke 1990), and its resistance was highly heritable, while its poor bunch characteristics were not (Table 2.3). Secondary triploids were successfully generated through crosses between tetraploid hybrids and diploid *Musa* accessions to restore female/male sterility and thereby avoid seed set in the hybrids. 'PITA-14' (AAAB) is a superior secondary triploid hybrid identified in the plantain improvement programme.

The progenies of 3x/2x crosses comprised diploids, triploids and tetraploids. Ploidy was ascertained based on morphological characteristics, stomatal size/density and chromosome counts. Flow cytometry had the advantage of being faster than cytological techniques. Ortiz and Vuylsteke (1994) observed that the formation of 2n eggs by second division restitution (SDR) enabled the occurrence of segregation and recombination of 3x plantain genome allowing the recovery of variability from crosses with plantains. This was contrary to the earlier assumption that 3x female genome is fixed and recombination occurs only in 2x male parent because of which plantain improvement would be limited to diploid breeding (Rowe 1984; Simmonds 1986).

East African highland bananas (AAA) endemic to East African highlands with a narrow genetic base comprise over 80 triploid clones primarily used for cooking and beer production. Screening for female fertility using the wild Calcutta 4 showed that 50 % of the land races are female fertile with potential for genetic improvement through conventional breeding. It was con-

Table 2.3 Phenotypic traits and types of gene action in *Musa*

Apical dominance	One major recessive gene in plantains	Ortiz and Vuylsteke (1994a)
Male and female fertility	Recessive genes interacting with cytoplasm sensitive	Ortiz (1995)
Dwarfism in type French plantains	One major recessive gene for short false internodes with modifiers affecting plant height	Ortiz and Vuylsteke(1995b)
Bunch orientation	Three loci with threshold effect of dominant genes	Ortiz (1995)
Fruit parthenocarpy	Three independent complementary dominant genes. One segregating locus in plantain hybrids and Calcutta 4	Ortiz (1995)
Persistence of male bracts	Two loci with complementary dominant genes, which are independent of the genes for persistence of hermaphrodite flowers in plantains	Ortiz (1995)
Persistence of hermaphrodite flowers and male bracts	Two independent loci with complementary and dominant genes in bananas and plantain-banana hybrids	Ortiz (1995)
Red pigmentation in leaves	Modifier gene interaction due to recessive suppressor	Ortiz (1995)
Pollen presence with 2n chromosomes	Dominant genes	Ortiz (1997)
Bacterial wilt (moko disease) resistance	Several recessive genes	Vakilii (1965b) and Rowe and Richardson (1975)
Fusarium wilt resistance	One major dominant gene for race 1. Polygenic system for race 4	Vakilii (1965a) and Rowe (1991)
Burrowing nematode resistance	One or more dominant genes	Rowe (1991)
Yellow Sigatoka resistance	Recessive genes in *M. acuminata* ssp. *burmanica*, dominant gene in *M. acuminata* ssp. *malaccensis*	Shepherd (1990)
Black Sigatoka resistance		Rowe (1984)
	Multiple gene with dosage effects in *M. acuminata* ssp., *Microcarpa* ssp. *errans*	Vakilii (1968)
	One major recessive gene and two additive minor genes with dosage effect in plantain-banana hybrids	Ortiz and Vuylsteke (1994b)

cluded that by creating population through breeding, genetic diversity of the group could be increased by enabling new allelic recombination (Pillay et al. 2004).

2.3.6.2.3 Breeding Work at CARBAP and EMBRAPA

The tetraploid plantain hybrid CRPB-39 is the outcome of the breeding efforts at CARBAP, Cameroon. It was developed by crossing local plantain variety 'French Clair' and the synthetic diploid M53. The hybrid is agronomically superior to 'French Clair' and displays resistance to BSLD (Cohan et al. 2003).

The Brazilian breeding programme at EMBRAPA has bred many clones with resistance to *Fusarium* wilt (Shepherd et al. 1994). These include PA 03-22 (AAAB), PV 03-44 (AAAB) and PC12 05 (AAAB) that are derived from 'Prata Anã', 'Pacovan' and 'Prata', respectively. These clones are resistant to race 1 and are being tested in many countries

2.3.6.3 4x/2x Strategy

The 4x/2x strategy consists of first producing a tetraploid parent by chromosome doubling of an ancestral diploid or an improved diploid using colchicine followed by hybridization of a diploid parent with the tetraploid for producing triploid hybrids. In the course of evolution of banana cultivars, natural triploid cultivars arose from ancestral diploids through the accidental production of unreduced gametes in one of the diploid parents by hybridization (Simmonds 1962). Therefore this strategy imitates the natural process of evolution. The meiotic error that resulted in unreduced gamete is replaced by chromosome doubling of one of the parents (Bakry et al. 1997; Stover and Buddenhagen 1986; Vakili 1967). The sterility of the triploid hybrid obtained by this method hampered its improvement by conventional methods (Tomekpe et al. 2004). The 4x/2x method as such does not improve existing varieties, but creates new improved varieties and overcomes the limitations of the triploid x diploid crosses limited by the low gametic fertility of the triploid parents and its failure to produce desired end products.

This method explores to identify good specific combining abilities between diploids and doubled diploids as donors of diplogametes and aims to maximize heterosis in the triploid progenies. The criteria for selecting natural or improved diploid varieties include the type of banana to be developed (cooking or dessert), their agronomic characteristics, their behaviour with respect to diseases and pests and their male and female fertility at diploid and tetraploid level. Detailed information about the available genetic resources and relationships between ancestral and cultivated varieties are taken into account to facilitate need-based selection of parental lines (Jenny et al. 2013).

2.3.6.3.1 Breeding Work at CIRAD

Applying this strategy, at CIRAD, progress has been made in the development of AAA dessert bananas, and promising hybrids evolved have been released for large-scale evaluation to banana growers in the French West Indies and in the Caribbean (Horry 2011). This new triploid strategy was validated by the subsequent development of AAB/ABB banana hybrids (Jenny et al. 2013). The major inputs from the study include the following: (i) Differences are noticed in the gametic fertility of the doubled diploids. (ii) Some of the AA diploids used in the development of AAA hybrids were fertile at both diploid and tetraploid levels, while others were completely sterile at the tetraploid level. (iii) All interspecific AB clones used, though sterile at the diploid level, displayed male and female fertility at the tetraploid level. The high fertility of the doubled AB also resulted in large population size. Parthenocarpic (edible) fruit development in the hybrids despite their wild parentage and the positive heterosis effect exhibited within the progenies evaluated are some of the valuable information. Triploidy of the hybrids from crosses involving Kunnan 4X was established using flow cytometry (Dolezel and Bartos 2005). The promising hybrid selections include the AAB '2006-22/III9' (*M. acuminata* ssp. *malaccensis*, × Kunnan 4X) with a Mysore morphology, good bunch weight and a taste similar to Silk subgroup, the sweet-acid AAB banana hybrid '2005/25-L9' (Kunnan 4X × 'IDN110/AAcvRose') with very high yield potential. Among ABB hybrids, two hybrid selections include the Pisang Awak-like '2008/12-I6' (Kunnan 4X × *balbisiana* 'CMR') and another one derived from a cross involving Kunnan 4X and 'Pisang Klutuk'. The resistance to Sigatoka leaf spot shown by the hybrids under field conditions suggested that the resistance is transmitted in a dominant way by the diploid-resistant parent. Further, in view of their pedigree, the authors concluded that the ABB hybrids could also serve as alternatives to Pisang Awak clones or other ABB natural clones in traditional cropping systems where banana production is constrained by *Fusarium* wilt (Jenny et al. 2013).

2.3.6.4 Breeding Work in India

2.3.6.4.1 Tamil Nadu Agricultural University (TNAU)

The breeding programme at Central Banana Research Station, Aduthurai, during 1949 was the earliest among the banana improvement efforts in India and later continued at Tamil Nadu Agricultural University, Coimbatore. In an attempt to improve Kallar Ladan, a Pome cultivar, an AB hybrid from the cross Kallar Ladan × *Musa balbisiana* clone Sawai (BB) was crossed with the diploid Kadali (AA) to develop an AAB hybrid. This was later released for commercial cultivation as Co1 (Azhakiamanavalan et al. 1985). The hybrid closely resembles Virupakshi (AAB), popular Pome banana in the hills of Tamil Nadu. The plants of Co1 are medium tall (2.7 m). The bunch weighs 10.5 kg on an average with 7 hands and 80–85 fruits. The crop duration is 14–15 months.

Krishnamurthy et al. (2004) reported wide variation in the extent of seed germination and parthenocarpy in hybrid progenies from crosses involving diploids and triploids. Many potential synthetic diploids have been generated with resistance/tolerance to Sigatoka leaf spot, *Fusarium* wilt and nematodes. Breeding strategies have been reoriented at the Department of Fruit Crops, by integrating the bred synthetic diploids in hybridization of commercial cultivars like Rasthali, Poovan and Karpooravalli. Few hybrid progenies using Karpooravalli as female parent are promising in terms of yield and resistant reaction against *Radopholus similis* and Sigatoka leaf spot (Kumar and Soorianathasundaram 2007). The potential diploids and hybrids developed were crossed with commercial triploids to develop primary tetraploids and improved diploids. The susceptible check cultivar used was 'Rasthali' (AAB), while the resistant reference cultivar used was 'Pisang Lilin' (AA). The reactions of 19 new synthetic banana phase II hybrids to *M. incognita* were studied under field conditions as well as in controlled inoculation tests in pots. Hybrid H 531 ('Poovan' × 'Pisang Lilin') was found to be resistant, and six hybrids, H-02-34, H-03-05, H-03-13, H-04-12, H-04-24 and NPH-02-01, were found to be tolerant to the root-knot nematode, *M. incognita*.

2.3.6.4.2 Kerala Agricultural University (KAU)

Kerala endowed with a warm humid tropical climate is home to a wide spectrum of edible bananas which include 'Nendran'(AAB), 'Poovan'(AAB), 'Rasthali'(AAB), 'Chenkadali'(AAA), 'Neypoovan'(AB), 'Karpooravalli'(ABB) and 'Monthan' (ABB). Their cultivation is constrained by susceptibility to various diseases and pests, such as Sigatoka leaf spot, *Fusarium* wilt, rhizome and stem weevils, nematode and viral diseases. Conventional breeding directed at the development of resistant hybrids is in progress at the Banana Research Station (BRS), Kannara. Based on earlier breeding initiatives, two dessert banana hybrids, BRS-1 from Agniswar (AAB) × Pisang Lilin (AA) and BRS-2 from Vannan (AAB) × Pisang Lilin (AA), have been developed and released for cultivation in Kerala. These hybrids are extremely resistant to Sigatoka, a trait inherited from the male parent (Menon et al. 2011). *BRS-1* has a short cropping cycle with faster ratoonability, completing four crop cycles in 2.5 years. Plants are medium statured producing 14–16 kg bunch. The fruits are straight and slightly angular turning to yellow on ripening. It also displayed resistance to Sigatoka leaf spot, *Fusarium* wilt, nematodes and other insect pests. *BRS-2* is medium statured completing crop cycle in 10–11 months. Average weight of bunch ranges from 15 to 20 kg with short, stout, dark green fruits compactly arranged and ripening to golden yellow colour. The hybrid has a striking resemblance to Poovan in plant and fruit characters. Fruits are slightly acidic and sweet. Like BRS-1, this hybrid has multiple resistance.

Current breeding strategies are focused on the improvement of 'Nendran', the commercial French plantain cultivar of the state, and envisage imparting higher productivity, dwarf stature and resistance to Sigatoka, weevil borers and nematodes utilizing wild/natural/bred diploids. The development of plantain hybrids resistant to black leaf streak through triploid/diploid crosses (3x/2x strategy) reported (Tomekpe et al. 2004) formed the basis. Female fertile clones of Nendran identified in the gene bank were pollinated with the wild diploid, *Musa acuminata* ssp. *burmannicoides* 'Calcutta 4', highly resistant to Sigatoka. Low seed set and germination hampered the recovery of hybrids for evaluation. One hybrid progeny resulting from the cross

'Nendran' clone 'Chengalikodan' × 'Calcutta 4' with close resemblance to the female parent recorded a plant height of 360 cm, pseudostem girth of 62 cm and 12 leaves at shooting. It registered a bunch weight of 13 kg with 8 hands and 90 fruits. Unlike its susceptible female parent, the hybrid displayed very high resistance to Sigatoka, imparted by the male parent. Being male and female fertile, the hybrid is being backcrossed with 'Nendran' and crossed with selected diploids. Embryo culture is being standardized to improve the recovery of hybrid progeny (Menon et al. 2011).

Development of fertile, agronomically superior diploids also formed a breeding objective. Based on the evaluation of AA diploids for growth and yield characteristics (Menon et al. 2002a), 'Matti', 'Pisang Jaribuaya' and 'Tongat' were selected as female parents. Pisang Lilin and *Musa acuminata* ssp. *burmanicoides* clone 'Calcutta 4' were identified as pollen fertile, potential sources of resistance to Sigatoka leaf spot. These varieties registered an infection index of zero for Sigatoka leaf spot, besides recording pollen fertility of 73–90 %. Of the ten different cross-combinations tried, seed set was observed in 'Matti' × 'Pisang Lilin', 'Matti' × 'Calcutta 4', 'Tongat' × 'Pisang Lilin' and 'Pisang Jaribuaya' × 'Calcutta 4'. The number of seeds varied from 10 to 95. Seeds germinated faster in a mist chamber and seed germination ranged from 42 to 60 %. Out of the 150 hybrid seedling progenies initially evaluated, 15 hybrids from the four crosses were selected. Progenies of 'Matti' × 'Calcutta 4' were all seeded except one which was selected. All the progeny showed resistance to Sigatoka leaf spot. All the 'Matti' × 'Pisang Lilin' hybrid progenies were parthenocarpic and recorded higher bunch weight (8.3–11.0 kg) than the parents. Plant no. 6 which registered a bunch weight of 11.0 kg had a good bunch and fruit characteristics and pollen fertility of 70 % besides leaf spot resistance inherited from the pollen parent. The female parent 'Matti' is partially susceptible to leaf spot (Menon et al. 2011).

2.3.6.4.3 National Research Centre for Banana (NRCB)

Breeding work at NRCB focuses on the improvement of Pisang Awak (ABB) group which in spite of its utility as a dual purpose cultivar is susceptible to *Fusarium* wilt and bacterial rot. Thirty-four Pisang Awak accessions were evaluated for their ability to produce hybrid seeds and tested for embryo viability and embryo regeneration followed by evaluation of progenies. Only 13 accessions (38 %) were found to be excellent female parents with seed production ranging from 4 to 459 seeds/bunch. Progenies are under evaluation (Uma et al. 2011b). Evaluation of Pisang Awak accessions resulted in the selection of promising ones. From the evaluation of Pisang Awak accession, a promising selection Udayam was released by NRCB and Trichy during 2005. The average bunch weight is 37 kg having a potential up to 45–50 kg which is 40 % higher than local Karpooravalli. Crop duration is 13 months and gives high yield in ratoons also. Its cylindrical bunches with well-spaced hands are suitable for long-distance transport. The fruits are high in sugar with 31^{0}B with good blend of acidity and sweetness making it suitable for processing into value-added products like figs, banana juice, wine, etc. It performs well in Tamil Nadu, Kerala, Andhra Pradesh, Karnataka, Bihar and Tripura. The variety has field tolerance to Sigatoka leaf spot and nematode infestation.

Uma et al. (2011a) evaluated 15 diploids for their fertility status, breeding behaviour and compatibility and came up with information on per cent seed set, germination time and success rate in 79 combinations. Anaikomban with good seed set and highest germination capacity proved to be a good female parent for gene pyramiding programme.

2.3.6.4.4 Banana Breeding at Indian Institute of Horticultural Research (IIHR)

At ICAR-Indian Institute of Horticultural Research (IIHR), Bengaluru basic work related to banana breeding was carried out. An analysis of cross-compatibility in banana varieties revealed that selection of female parents with a high tendency to seed set, pollen fertile male parents and triploidy may result in successful cross-combinations (Rekha and Prasad 2001). Barriers were observed in seed set and germination which affected the breeding programmes and development of hybrids (Rekha et al. 2007).

For the development of molecular markers linked to *Fusarium* wilt resistance, a population of F_1 hybrids involving highly contrasting accessions of the AA genomic group that is tolerant (wild *Musa acuminata* ssp. *burmanicoides* Calcutta 4) and highly susceptible (cv. Kadali) were generated. The population is useful in functional genomic studies as segregation is observed for various morphological characteristics (Rekha et al. 2011a, b). A population of AA × BB (wild *M. acuminata* × *M. balbisiana*) showed segregation for pseudostem colour, bunch orientation, male flower colour (Rekha et.al. 2009) and water stress-related traits like stomata and wax content. Mapping populations developed with the highly contrasting types of *M. balbisiana*, i.e. BB × BB cross-combination is available for the development of the BB linkage map.

In an attempt to produce intergeneric hybrid between *Ensete* × *M. balbisiana*, it was observed that apomictic seed set in *Ensete* was also confirmed by molecular studies (Ravishankar et al. 2011).

2.4 Molecular Markers in *Musa* Breeding

Diverse molecular marker methods have been employed to characterize *Musa* germplasm. Earlier studies involving isozymes and anthocyanin analyses threw light on genetically diverse nature of *Musa* germplasm (Jarret and Litz 1986; Horry and Jay 1988). The use of non-DNA-based biochemical markers is reported to give faster and accurate estimates without the requirement for advanced laboratory facilities (Heslop-Harrison and Schwarzacher 2007). *Musa* germplasm including wild species and cultivars has been extensively characterized utilizing various classes of DNA markers (Tenkouano et al. 1999) which include RFLP, AFLP (Ude et al. 2002), PCR-RFLP (Nwakanma et al. 2003), microsatellites (Teo et al. 2005; Ravishankar et al. 2013; Saraswathi et al. 2011) and random amplified polymorphic DNA (RAPD) (Jones et al. 1997). Of the different methods applied to study interrelationships between *Musa* accessions and to group diverse germplasm, PCR-based techniques can efficiently detect levels of polymorphism. Simple sequence repeats (SSRs) which provide codominant chromosome-specific markers have been used extensively in studies to assess genetic relations between accessions (Heslop-Harrison and Schwardacher 2007; Buhariwalla et al. 2005; Ravishankar et al. 2012). Inter-retroelement amplified polymorphisms (IRAP) (Nair et al. 2005) and Diversity Arrays Technology (DArT) markers (Channeliere et al. 2011) are also routinely used.

One of the most important applications of molecular biology in plant breeding is the use of molecular markers linked to important traits. Identification of such markers would facilitate gene introgression and marker-assisted selection. The use of molecular markers facilitates indirect selection of improved cultivars, speeding up the selection process than direct screening under greenhouses or field conditions (Ortiz 1995). A limited number of markers associated with diseases have been identified. Methylation-sensitive amplification polymorphism (MSAP) markers were found associated with resistance to *Mycosphaerella fijiensis* toxin (black Sigatoka) (Gimenez et al. 2006) and could be used as molecular indicators of tolerance to *M. fijiensis* toxins and resistance to black Sigatoka. Lin et al. (2010) developed a highly sensitive molecular method to detect *Fusarium oxysporum* f.sp. *cubense* (foc) race 4 isolates in Taiwan.

A putative RAPD marker for Sigatoka resistance has been identified at the National Research Centre for Banana (NRCB), India. The marker has been cloned, sequenced and converted into a sequence-characterized amplified region (SCAR) marker and is being validated using contrasting parents for the expression of Sigatoka (*M. musicola*) disease resistance and their progenies. Parallel studies have led to the identification of a putative random amplified polymorphic DNA (RAPD) marker for nematode resistance (S. Uma, pers. comm.). Tenkouano and Swennen (2004) reported a putative AFLP marker for fruit parthenocarpy (distinguishing between low and high pulp content) and seedlessness. An RAPD marker has been identified for salt tolerance among clones of cv. 'Dwarf Cavendish' that were obtained through induced mutagenesis (Miri et al. 2009).

2.5 Evaluation of Improved Hybrids

Seven secondary triploid hybrids (PITA PITA 21–26 and BITA 7) from IITA selected for superior agronomic and fruit processing attributes were distributed to farmers in Nigeria. Similarly a wide range of hybrids have been deployed in farmers' field in other West African countries. The improved hybrids are two to five times more productive than local plantains without Sigatoka control measures (Tenkouano and Swennen 2004).

The International *Musa* Testing Programme (IMTP) started in 1989 represented a worldwide collaborative effort coordinated by International Network for the Improvement of Banana and Plantain (INIBAP) now known as Bioversity International to evaluate elite *Musa* hybrids as well as promising germplasm accessions. A total of 25 hybrids were evaluated in diverse locations all over the world in phases I, II and III in comparison to reference cultivars and local check (Molina et al. 2007).

Since 1994 exotic germplasm comprising natural cultivars, wild species and improved hybrids from world *Musa* collection held at INIBAP Transit Centre (ITC), Leuven, Belgium, are available for evaluation in India. The accessions in the form of proliferating in vitro cultures initially quarantined at National Bureau of Plant Genetic Resources (NBPGR, New Delhi) and conserved in its in vitro gene bank (Agrawal et al. 2007) are supplied to NRCB and various centres of All India Coordinated Research Project on Tropical Fruits where need-based evaluations are carried out. The exotic accessions are now conserved in field gene banks in India. Based on field evaluation, a number of natural cultivars have been selected for adoption in Kerala (Menon 2000a; Menon et al. 2005).

The performance of exotic hybrids was evaluated in the International *Musa* Testing Programme phase II and phase III along with reference clones in AICRP centres with NRC for banana as the coordinating centre. During phase II among four hybrids evaluated, SH-3436-9(AAAA) and FHIA-23(AAAA) recorded higher bunch yield and displayed tolerance to Sigatoka leaf spot (Menon et al. 2002c). In phase III evaluations, nine hybrids were evaluated. All the hybrids fell under medium to tall category in respect of plant stature and displayed higher bunch weight than reference cultivars ranging from 16 kg in FHIA-18 to 36.5 kg in FHIA-25. The hybrids also exhibited partial to very high resistance to Sigatoka leaf spot. SH-3640, TMBX 5295-1, FHAI-03, FHIA-17, FHIA-25 and FHIA-21 and CRPB-39 were selected for further evaluations. At BRS, Kannara, three of these hybrids, viz. SH-3640, TMBX 5295-1 and FHIA-03 and FHIA-21, have been advanced to on farm evaluations in selected farmers' plot (Menon et al. 2005). Based on studies at NRCB, Trichy's three hybrids, FHIA-01, FHIA-03 and FHIA-23, have been multiplied and supplied to farmers for testing. FHIA-01 was more suitable for processing with high sugar-acid ratio and low polyphenol oxidation of the pulp which results in pulp browning. FHIA-03 had acceptability as cooking banana (Uma et al. 2005b).

2.6 Mutation Breeding

Most of the accepted banana and plantain cultivars being vegetatively propagated and generally sterile triploids bearing parthenocarpic fruits and genetic improvement through conventional breeding are met with difficulties. Mutation breeding offers a unique alternative approach for the improvement of such vegetatively propagated crops and can change one or more characters of a popular cultivar without altering the genetic make-up. Mutation breeding methods have been reviewed by Krikorian and Cronauer (1984).

Mutagenesis combined with in vitro selection is an appropriate tool for enhancing induction, selection and recovery of desirable mutants (Suprasanna and Rao 1997). Mutation breeding by selection among somaclonal variants is a major approach for banana improvement in Taiwan, and somaclonal variants with resistance to *Fusarium* wilt, shorter plant stature, early flowering or higher bunch weight were selected (Tang and Hwang 1994). Mutation breeding gives better results in diploids than triploids because of the reduced genomic size and also induced mutations being generally recessives. In heterozygous triploids on the other hand, the probability of obtaining homozygous recessives is limited and thereby the manifestation of improved traits (Roux et al. 1994). Roux et al. (2004) has described two commercially released lines derived from gamma-ray-induced mutations. Novaria derived from Grand Naine of Cavendish group was released in 1995. It flowered 10 weeks earlier than the original parental clone. The Thai variety Klue Hom Thong had large bunch size, cylindrical shape and larger fruits. Further, the components of a successful in vitro mutagenesis programme involve the establishment of efficient in vitro regeneration procedure, optimization of mutagenic treatments of cultures and finally screening of treated populations for desired variations. Experiments on in vitro induction of mutations using gamma irradiation in six elite banana cultivars followed by field evaluation resulted in the isolation of mutants (Bapat et al. 2007). Isolation of induced mutations is also limited by the fact that chimera formation usually occurs after mutagenic treatment of a multicellular apex, while a specific mutation is a single cellular event. To overcome these shortcomings, methods of treatment of plant structures of single-cell origin or procedures that result in induction of large mutated sectors in multicellular structures are suggested.

In vitro mutation breeding utilizing gamma rays and EMS to create variability is being attempted at TNAU, Coimbatore, with Robusta, Rasthali, Nendran, Poovan, Karpooravalli, red banana and Monthan (Kumar and Soorianathasundaram 2007). Four mutants in Robusta have been advanced to V4M1 generations with bunch yield of more than 28.0 kg under field conditions, while in Rasthali no useful mutants could be isolated. Induction of tetraploidy in potential diploids like Sannachenkadali (AA), Anaikomban (AA), Kunnan (AB) and Thattillakunnan (AB) has been taken up in vitro utilizing antimitotic agents, viz. colchicine ($C_{22}H_{25}NO_6$) and Oryzalin (3,5-dinitro-N4,N4-dipropylsulfanilamide). A total of 24 tetraploids obtained are presently under evaluation and being integrated in breeding schedule. At NRCB and Trichy, improvement of Rasthali (AAB Silk group) through induced mutagenesis is being attempted. A total of 54 mutants were tested for their genetic fidelity using molecular markers and polymorphism ranged from 30.2 to 50%. The mutants were also 19–21 % different from non-mutated Rasthali plants.

2.7 Conclusion

Genetic improvement of banana offers an ecologically sustainable solution for pest and disease management. Focused efforts by major banana breeding programmes have been responsible for the development of a number of hybrids. The requirement of improved hybrids to confirm to standards in terms of taste, colour and quality to be accepted by consumers has been a major challenge all over the world. The improved yields and the possibility of growing the hybrids without plant protection measures have changed the perception leading to increased adoption by farmers in many countries. With increasing pest and disease pressure, evidence of climate change and the current emphasis placed on the role of banana on human nutrition and health, breeding strategies warrant reorientation to meet new demands. Broadening the genetic base of varieties with more resistant sources, understanding the scientific basis of abiotic stress resistance and application of molecular tools would generate new avenues for genetic improvement of banana through conventional approaches.

References

Agrawal A, Sharma H, Dhillon BS (2007) *In vitro* conservation of *Musa* germplasm in India. In: Singh HP, Uma S (eds) Banana-technological advancements. Association for the Improvement in production and Utilization of Banana (AIPUB), National Research Centre for Banana, Trichy, pp 87–108

Amalraj VA, Velayudhan KC, Agarwal RC, Rana RS (1993) Banana genetic resources. National Bureau of Plant Genetic Resources, New Delhi

Azhakiamanavalan RS, Bhakthavatsalu CM, Sathiamoorthy S, Kulasekaran M (1985) Co.1 banana. South Indian Hortic 32:51–52

Bakry F, Carreel F, Caruana ML, Cote FX, Jenny C, Tezenas du Montcel H (1997) Les bananiers. In: Charrier A, Jacquot M, Hamon S, Nicolas D (eds) L'amelioration des plantes tropicales. CIRAD and ORSTOM, pp 109–140

Bapat VA, Ganapathi TR, Kulkarni VM, Sunil Kumar GB, Srinivas L, Suprasanna P (2007) Cellular and molecular approaches for improvement of banana. In: Singh HP, Uma S (eds) Banana-technological advancements. Association for the Improvement in production and Utilization of Banana (AIPUB),National Research Centre for Banana, Trichy, pp 160–173

Buhariwalla HK, Jarret RL, Jayashree B, Crouch JH, Ortiz R (2005) Isolation and characterization of microsatellite markers from *Musa* balbisiana. Mol Ecol Notes 5:327–330

Carreel F, Gonzalez de Leon D, Lagoda P, Lanaud C, Jenny C, Horry J, Tezenas du Montcel H (2002) Ascertaining maternal and paternal lineage within *Musa* by chloroplast and mitochondrial DNA RFLP analyses. Genome 45:679–692

Channelie're S, Van den houwe I, Arnaud E, Horry JP, Ruas M, Roux N (2011) Standardized procedure for Musa Germplasm characterization. Acta Horticult 897:113–121

Cohan JP, Abadie C, Tomekpe K, T chango T chango J (2003) Agronomic performance and resistance to black leaf streak of the hybrid 'CRBP-39'. Infomusa 12(1):29–32

d'Hont A, Paget-Goy A, Escoute J, Carreel F (2000) The interspecific genome structure of cultivated banana. *Musa* spp. revealed by genomic DNA *in situ* hybridization. Theor Appl Genet 100:177–183

Daniells J, Pegg K, Searle C, Smith M, Whiley T, Langdon P, Bryde N, O'Hare T (1995) Goldfinger in Australia: a banana variety with potential. Infomusa 4(1):5–6

De Langhe E (1964) The origin of variation in the plantain banana. Ghent Belgium State Agric Univ Ghent 39:45–80

Dolezel J, Bartos J (2005) Plant DNA flow cytometry and estimation of nuclear genome size. Ann Bot 95:99–110

FAO (2014) Banana market review and banana statistics 2012–2013, p 39 http://www.Fao.org

Faure' S, Noyer JL, Carreel F, Horry JP, Bakry F, Lanaud C (1994) Maternal inheritance of chloroplast genome and paternal inheritance of mitochondrial genome in bananas (*Musa*). Curr Genet 25:265–269

Four'e E, Mourichon X, Jones DR (2000) Black leaf streak: host reaction, evaluating germplasm for reaction to black leaf streak. In: Jones DR (ed) Diseases of banana, abaca and enset. CABI Publishing, Wallingford, pp 62–67

Giménez C, Palacios G, Colmenares M (2006) *Musa* methylated DNA sequences associated with tolerance to *Mycosphaerella fijiensis* toxins. Plant Mol Biol Rep 24:33–43

Harper G, Osuji JO, Heslop-Harrison JS, Hull R (1999) Integration of banana streak badnavirus into the *Musa* genome: molecular and cytogenetic evidence. Virology 255:207–213

Heslop- Harrison JS, Schwarzacher T (2007) Domestication, genomics and the future of banana. Ann Bot 100(5):1073–1084

Horry JP (2011) The use of molecular markers in the CIRAD banana breeding programme. Acta Horticult 897:235–241

Horry JP, Jay M (1988) Distribution of anthocyanins in wild and cultivated banana varieties. Phytochemistry 27:2667–2672

Hwang SC, Ko WH (2004) Cavendish banana cultivars resistant to Fusarium wilt acquired through somaclonal variation in Taiwan. Plant Dis 88:580–588

IPGRI INIBAP/CIRAD (1996) Descriptors for Banana (Musa Spp) International Plant Genetic Resources Institute, Rome, Italy/International network for the Improvement of Banana and Plantain, Montpellier, France. Centre de Coop'eration internationale en recherché agronomoque pourle d'eveloppement, Montpellier

Jacob KC (1952) Madras bananas a monograph. Government Press, Madras

Jarret RL, Litz RE (1986) Isozymes as genetic markers in bananas and plantains. Euphytica 35:539–549

Jenny C, Holtz Y, Horry JP, Bakry F (2013) Synthesis of new interspecific hybrids from AB germplasm in banana (Musa sp). Acta Hortic 986:209–218

Johnson R (2000) Classical plant breeding for durable resistance to diseases. J Plant Pathol 82(1):3–7

Jones DR (2000) Diseases of banana, abaca and enset. CAB International, Wallingford, p 544

Jones CJ, Edwards KJ, Castiglione S, Winfield MO, Sala F (1997) Reproducibility testing of RAPD, RFLP and SSR markers in plants by a network of European laboratories. Mol Breed 3:381–390

Krikorian AD, Cronauer SS (1984) Handbook of plant cell culture. In: Sharp WR, Evans DA, Ammirato P, Yamada Y (eds) Crop species, vol 2. Macmillan, New York, pp 327–348

Krishnamurthy V, Kumar N, Soorianathasundaram K (2004) Influence of male and female parent on parthenocarpy. Infomusa 13(1):7–9

Kumar N, Soorianathasundaram K (2007) Progress and current strategies in banana improvement in Tamil Nadu. In: Singh HP, Uma S (eds) Banana-technological advancements. Association for the Improvement in production and Utilization of Banana (AIPUB), National Research Centre for Banana, Trichy, pp 128–137

Lin YH, Chang JY, Liu ET, Chao CP, Huang JW, Chang PFL (2010) Development of a molecular marker for specific detection of *Fusarium oxysporum f.sp cubense* race 4. Eur J Plant Pathol 123:353–365

Lorenzen J, Tenkouano A, Bandyopadhyay R, Vroh B, Coyne D, Tripathi L (2010) Overview of banana and plantain (Musa spp.) improvement in Africa: past and future, proc. IC on banana & plantain in Africa. Acta Horticult 879:595–603

Menon R (2000a) *Ex situ* conservation of *Musa* genetic resources in Kerala. In: Abstracts of international conference on science and technology for managing plant genetic diversity in the 21st century, 12–16 June 2000, Kuala Lumpur

Menon R (2000b) Preliminary evaluation of some banana introductions in Kerala (S. India). Infomusa 9(2):27–28

Menon R, Suma A, Cherian AK, Nair S, Sudheesh MV (2002) Conservation and utilization of *Musa* AA diploids in breeding. In: Global conference on banana and plantain, Bangalore, 28–31 Oct 2002. Abstract. pp 52

Menon R, Aravindakshan K, Chandralekha CT (2002a) Evaluation of plantain (*Musa* AAB) variability in South India. In: Global conference on banana and plantain, Bangalore, 28–31 Oct 2002. Abstract. pp 22

Menon R, Cherian AK, Uma S (2002b) Evaluation of four tetraploid banana hybrids in South India. In: Global conference on banana and plantain, Bangalore, 28–31 Oct 2002. Abstract. pp 31

Menon R, Agrawal A, Reddy BMC (2005) Status of introduced *Musa* germplasm at Banana Research Station, Kannara. Indian J Plant Genet Resour 18(1):81–83

Menon R, Cherian AK, Suma A, Mathew MP, Nair S, Aipe KC (2011) Developing resistant banana and plantain cultivars through conventional breeding techniques. Acta Horticult 897:207–214

Miller RNG, Passos MAN, Emediato FL, de Damargo Teixeira C, Pappas Junior GJ (2011) Candidate resistance gene discovery: resistant gene analog characterisation and differential gene expression analysis in *Musa-Mycosphaerella* host-pathogen interactions. Acta Horticult 897:179–185

Miri SM, Mousavi A, Naghavi MR, Mirzaii M, Talaei AR, Khiabani BN (2009) Analysis of induced mutants of salinity resistant banana (*Musa acuminata* cv. Dwarf Cavendish) using morphological and molecular markers. Iran J Biotechnol 7:86–92

Molina AB, Escalant VJ, Van Den Bergh I (2007) International Musa Testing Programme (IMTP: a worldwide programme to evaluate *Musa* varieties. In: Singh HP, Uma S (eds) Banana. Technological advancements. Association for Improvement in Production and Utilization of Banana (AIPUB), NRCB, Trichy, pp 54–61

Moore NY, Bentley S, Pegg KG, Jones DR (1995) Musa disease fact sheet no.5.Fusarium wilt of banana. INIBAP, Montpellier

Morán JFA (2013) Improvement of cavendish banana cultivars through conventional breeding, Fundación Hondureña de Investigación Agrícola (FHIA) La Lima, Honduras. In: Van den Bergh I et al (eds) Proceedings of the International ISHS-ProMusa Symposium on bananas and plantains: towards sustainable global production and improved uses. Acta Hort 986:205–208

Nair AS, Teo CH, Schwarzacher T, Heslop-Harrison JS (2005) Genome classification of banana cultivars from South India using IRAP markers. Euphytica 144:285–290

Nwakanma DC, Pillay M, Okoli BE, Tenkouano A (2003) PCR-RFLP of the ribosomal DNA transcribed spacers (ITS) provides markers for the A and B genomes in *Musa* L. Theor Appl Genet 108:154–159

Ortiz R (1995) Musa genetics. In: Gowen S (ed) Bananas and plantains. Chapman and Hall, London, pp 84–109

Ortiz R (1997) Occurrence and inheritance of 2n pollen in *Musa*. Ann Bot 79(4):449–453

Ortiz R, Vuylsteke D (1994a) Plantain breeding at IITA. In: Jones DR (ed) The improvement and testing of Musa: a global partnership. Proc. Conf. FHIA, Honduras, 27–30, April,1994. INIBAP, Montpellier

Ortiz R, Vuylsteke D (1994b) Genetics of apical dominance in plantain(Musa spp., AAB group) and improvement of suckering behaviour. J Am Soc Horticult Sci 119:1050–1053

Ortiz R, Vuylsteke D (1994c) Inheritance of black Sigatoka resistance in plantain banana (*Musa* spp.) hybrids. Theor Appl Genet 89:146–152

Ortiz R, Vuylsteke D (1995a) Inheritance of dwarfism in plantain (Musa spp.AAB group). Plant Breed 114:466–468

Ortiz R, Vuylsteke D (1995b) Factors influencing seed set in Triploid Musa spp. L and production of diploid hybrids. Ann Bot 75:151–155

Ortiz R, Madsen S, Vuylsteke D (1998) Classification of African landraces and banana cultivars using a phenotypic distance index of quantitative descriptors. Theor Appl Genet 96:904–911

Padmanabhan B, Sathiamoorthy S (2007) Present status of banana stem weevil in India. In: Singh HP, Uma S (eds) Banana. Technological advancements. Association for Improvement in Production and Utilization of Banana (AIPUB), NRCB, Trichy, pp 351–355

Panis B, Van den houwe I, Piette B, Swennen R (2007) Cryopreservation of the banana germplasm collection at the International Transit Centre Biodiversity International. Adv Hortic Sci 21(4):235–238

Pillay M, Ssebuliba R, Hartman J, Vuylstake D, Talemgera D, Tushemreirwe W (2004) Conventional breeding strategies to enhance the sustainability of biodiversity conservation for endemic cultivar. Afr Crop Sci J 12(1):59–65

Ploetz RC, Churchill ACL (2011) Fusarium wilt: the banana disease that refuses to go away. Acta Horticult 897:519–526

Purseglove JW (1975) Tropical crops monocotyledons, vol 1. Halsted Press Division, Wiley, p 519

Ravishankar KV, Rekha A, Malarvizhi M, Ambika DS (2011) Apomictic seed development in *Ensete superbum* induced by pollen of wild banana sp. *Musa balbisiana*. Curr Sci 101(4):493–495

Ravishankar KV, Vidhya L, Cyriac A, Rekha A, Goel R, Singh NK, Sharma TR (2012) Development of SSR markers based on a survey of genomic sequences and their molecular analysis in banana (*Musa* spp.). J Hortic Sci Biotechnol 87:84–88

Ravishankar KV, Raghavendra KP, Athani V, Rekha A, Sudeepa K, Bhavya D, Srinivas V, Ananad L (2013) Development and characterisation of microsatellite markers for wild banana (*Musa balbisiana*). J Hortic Sci Biotechnol 88(5):605–609

Rekha A, Prasad MBNV (2001) Crossability in banana. Diamond jubilee symposium of Indian society of genetics and plant breeding on 'Hundred years of post-mendelian genetics and plant breeding- Retrospect and prospects. 6–9 Nov 2001, New Delhi. Abstract. p 115

Rekha A, Ambika DS, Dinesh MR (2007) Barriers for seed set and germination in banana. Presented at national conference on banana 'Production and Utilization of Banana for Economic livelihood and Nutritional Security'. 25–28 Oct 2007, Tiruchirapalli

Rekha A, Ravishankar KV, Ambika DS (2011a) Generation of segregating F_1 populations for mapping the *Musa* genome. Acta Hort 897:277–278

Rekha A, Ravishankar KV, Ambika DS (2011b) Evaluation of segregating populations from *Musa acuminata* ssp. *burmannicoides* (Calcutta-4). ISHS/ProMusa symposium-Bananas and plantains: toward sustainable global production and improved uses. Bahia Othon Palace Hotel, Salvador, Bahia, Brazil 10–14 Oct 2011. Abstract. p 150

Rekha A, Ravishankar KV, Ambika DS (2009) Generation of mapping populations for segregation of 'B' genome. Presented in ISHS/ProMusa banana symposium-Global Perspectives on Asian Challenges, Phoenix City Hotel, Guangzhou, China, 14–18 Sept 2009. Abstract. p 103

Robinson JC (1996) Bananas and plantains. Crop production Science in horticulture 5. CAB International, Wallingford, 238 pp

Roux NS, Jain SM, Swennen R (2004) Mutation induction in Musa review. In: Banana improvement: cellular, molecular biology, and induced mutations. Proceedings of a meeting held in Leuven, Belgium, 24–28 September 2001. Science Publishers, Inc., pp 23–32

Roux N, Afza R, Brunner H, Morpurgo R, van Duren M (1994) Complementary approaches to cross breeding and mutation breeding for *Musa* improvement. In: Jones DR (ed) The improvement and testing of *Musa*: a global partnership-proceedings of the first global conference of the International *Musa* Testing Program at FHIA, Honduras, 27–30 April 1994. INIBAP, Montpellier, pp 213–218

Rowe PR (1984) Breeding bananas and plantain. Plant Breed Rev 2:135–155

Rowe PR (1987) Breeding plantains and cooking bananas. In: International cooperation for effective plantain and banana research: Proceedings of the 3rd meeting of the International Association for Research on Plantain and Bananas. Abidjan, 23–31 May 1985, pp 21–23

Rowe PR (1991) Advances gen'eticos em banana e pl'atano. Augura Bogota 17(1):19–83

Rowe PR, Richardson DL (1975) Breeding bananas for disease resistance, fruit quality and yield. Tropical Agricultural Research Service, La Lima. (SIATSA BULL.2)

Rowe PR, Rosales F (1992) Genetic improvement of bananas, plantains and cooking bananas in FHIA, Honduras. In: Ganry J (ed) Breeding bananas and plantains: proceedings of an international symposium on genetic improvement of bananas for their resistance to diseases and pests. CIRAD -FLHOR, Montpellier, pp 243–266

Rowe PR (1990) New genetic combination in breeding bananas and plantains resistant to diseases. In: Identification of genetic diversity in the genus *Musa*: proceedings of an international workshop held at Los Banos, 5–10 Sept 1988, pp 114–123

Rowe P, Rosales F (1994) Musa breeding at FHIA. In: Jones DR (ed) The improvement and testing of *Musa*, a global partnership, Proceedings of the global conference of the international testing programme, FHIA, Honduras, 27–30 Apr 1994, pp 117–129

Saraswathi MS, Uma S, Vadivel E, Durai P, Siva SH, Rajgopal G, Sathiamoorthy S (2011) Diversity analysis in Indian cooking bananas (*Musa* ABB) through morphotaxonomic and molecular characterization. Acta Horticult 897:123–129

Shepherd K (1990) Observations on Musa taxonomy. In: Proceedings identification of genetic diversity in the genus Musa, 1988, Los banos, INIBAP, Montpellier, pp 158–165

Simmonds NW (1962) The evolution of the bananas. Longmans, Green, 170pp

Simmonds NW (1966) Bananas, 2nd edn. Longman, London

Simmonds NW, Shepherd K (1955) The taxonomy and origins of the cultivated bananas. J Linn Soc Lond 55:302–312

Simmonds NW (1986) Classification and breeding of bananas. In: Persley GJ, De Langhe EA (eds) Banana and plantain breeding strategies. Proc. of International Workshop, Cairns, pp 69–73

Singh HP, Uma S, Sathiamoorthy S (2001) A tentative key for identification and classification of bananas and plantains in India. National Research Center for Banana, Trichy, 61 pp

Ssebuliba R, Talengera D, Makumbi D, Namanya P, Tenkouano A, Tushemereirwe W, Pillay M (2006)

Reproductive efficiency and breeding potential of East African highland (*Musa* AAA-EA) bananas. Field Crops Res 95:250–255

Ssebuliba R, Makumbi D, Pillay M (2009) Patterns of seed set in East African highland banana (*Musa* sp.) hybrids. J New Seeds 10:160–170

Stover RH, Buddenhagen IW (1986) Banana breeding: polyploidy, disease resistance and productivity. Fruits 41:175–191

Stover RH, Simmonds NW (1987) Bananas, 3rd edn. Longman Scientific and Technical, Harlow, 468pp

Suprasanna P, Rao PS (1997) Selection of mutants using plant cell and tissue culture techniques. In: Hemantarajan A (ed) Advances in plant physiology, vol 1. Scientific Publishers, Jodhpur, pp 103–122

Swennen R, Vuylsteke D (1990) Aspects of plantain breeding at IITA. In: Fullerton RA, Stover RH (eds) Sigatoka leaf spot diseases of bananas. INIBAP, Montpellier, pp 252–266

Tang CY, Hwang SC (1994) *Musa* mutation breeding in Taiwan. In: Jones DR (ed) The improvement and testing of *Musa*: a global partnership-proceedings of the first global conference of the International *Musa* Testing Program at FHIA, Honduras, 27–30 Apr 1994, pp 219–227

Tenkouano A, Swennen RL (2004) Progress in breeding and delivering improved plantain and banana to African farmers, horticultural science focus. Chronica Hortic 44(1):9–15

Tenkouano AJH, Crouch HK, Crouch J, Vuylsteke D, Ortiz R (1999) Comparison of DNA marker and pedigree-based methods of genetic analysis of plantain and banana (*Musa* spp.) clones. I. Estimation of genetic relationships. Theor Appl Genet 98: 62–68

Tenkouano A, Vuylsteke D, Okoro J, Makumbi D, Swennen R, Ortiz (2003) Diploid banana hybrids TMB2x5105-1 and TMB2x9128-3 with good combining ability, resistance to black Sigatoka and nematodes. HortSci 38:468–472

Teo CH, Tan SH, Ho CL, Faridah QZ, Othman YR, Heslop-Harrison JS, Kalendar R, Schulman AH (2005) Genome constitution and classification using retrotransposon-based markers in the orphan crop banana. J Plant Biol 48:96–105

Tezenas du Montcel H, Devos P (1978) Proposal for establishing a plantain determination card. Paradisiaca (Ibadan Nigeria) 3:14–17

Tomekpe K, Jenny C, Escalant JV (2004) A review of conventional improvement strategies for *Musa*. Infomusa 13(2):2–6

Ude G, Pillay M, Nwakanma D, Tenkouano A (2002) Analysis of genetic diversity and sectional relationships in *Musa* using AFLP markers. Theor Appl Genet 104:1239–1245

Uma S, Sathiamoorthy S (2007) Status of genetic diversity in banana in India. In: Singh HP, Uma S (eds) Banana-technological advancements. Association for the Improvement in production and Utilization of Banana (AIPUB), National Research Centre for Banana, Trichy, pp 71–84

Uma S, Sathiamoorthy S, Dorai P (2005a) Banana-Indian genetic resources and catalogue. National Research Centre for Banana, Trichy, p 268

Uma S, Sathiamoorthy S, Saraswathi MS, Dorai P, Sharma HO, Agrawal A (2005b) Evaluation and utilization of introduced *Musa* Germplasm in India. Indian J Plant Genet Resour 18(1):79–81

Uma S, Mustaffa MM, Saraswathi MS, Dorai P (2011a) Exploitation of diploids in Indian banana breeding programmes. Acta Horticult 897:215–223

Uma MM, Mustaffa K, Arun MS, Saraswathi S, Backiyarani P, Durai (2011b) Breeding Pisang Awak – screening of best female parents, donor parents and compatibility studies. ISHS/ProMusa symposium: bananas and plantains: toward sustainable global production and improved uses, Brazil, 10–14 Oct 2011, Abstract. p 212

Vakili NG (1965a) Fusarium wilt resistance in seedlings and mature plants of Musa species. Phytopathology 55:135–140

Vakili NG (1965b) Inheritance of resistance in *Musa acuminata* to bacterial wilt caused by the tomato race of *Pseudomonas solanacearum*. Phytopathology 55:1206–1209

Vakili NG (1967) The experimental formation of polyploidy and its effect in the genus *Musa*. Am J Bot 54:24–36

Vakili NG (1968) Responses of *Musa acuminata* species and edible cultivars to infection by *Mycosphaerella musicola*. Trop Agric 45:13–22

Van den Houwe I, De Smet K, T'ezenas de Montcel H, Swennen R (1995) Variability in storage potential of banana shoot cultures under medium term storage conditions. Plant Cell Tissue Org Cult 42(269):274

Van der Plank JE (1982) Disease resistance in plants, 2nd edn. Academic, New York

Vuylsteke D, Ortiz R (1993) Diploid plantains with black Sigatoka resistance. Mus Africa 2:1–2

Vuylsteke D, Ortiz R, Swennen R (1993) Genetic improvement of plantain and banana at IITA. Infomusa 2(1):10–12

3 Genes and Markers: Application in Banana Crop Improvement

Manosh Kumar Biswas and Ganjun Yi

Abstract

Banana is one of the most economically important horticultural crops. Its improvement is a big challenge to breeders due to complex genomic nature and lack of advanced knowledge of its genetics. In the recent years, banana scientists around the globe have made considerable progress in its improvement through gene identification and subsequently its transformation into cultivars and have developed trait-linked markers to speed up the selection process in the cross-breeding program. Two genomes of the diploid banana *Musa acuminata* (A genome) and *Musa balbisiana* (B genome) have been sequenced and made publicly available. This facilitated the identification of agronomically important trait-linked genes and studies on their function, mode of inheritance, evolution and development of gene-linked markers, the landmark of the gene on the chromosome. Identification and characterization of important genes is the key to any breeding program. In bananas, several genes associated with disease resistance and fruit quality have been isolated, and their molecular characterization reported. Several molecular markers have been developed, including RAPD, ISSR, AFLP, DArT, and SSR, for the genomic group identification, estimation of genetic stability of somaclonal variation, disease-resistant cultivar identification, etc,. This chapter summarizes the progress of the application of agronomically important gene and molecular markers in banana breeding programs around the world.

Keywords

Genetic markers • Diversity • *Musa*

M.K. Biswas (✉) • G. Yi
Institution of Fruit Tree Research, Guangdong Academy of Agricultural Sciences, Guangzhou, Guangdong, China

The College of Life Science, South China Agricultural University, Guangzhou, China
e-mail: mkbcit@ymail.com

3.1 Introduction

Banana is one of the commercially important crops that belongs to the family Musaceae and essential for food security in many tropical and

S. Mohandas, K.V. Ravishankar (eds.), *Banana: Genomics and Transgenic Approaches for Genetic Improvement*, DOI 10.1007/978-981-10-1585-4_3

subtropical countries (D'Hont et al. 2012). Due to its nutritional value, it has become one of the vital foodstuffs in our daily life (Biswas et al. 2015). As a consequence, the banana-producing area has increased, the processing industry has strengthened, and globally banana trade has expanded. A special emphasis has now been given to its improvements in many countries according to the consumer demand. In recent years, several research groups around the globe have initiated banana improvement projects using conventional and advance breeding technology such as cross-breeding, marker-assisted selection, transgenic breeding, biotechnological approach (micropropagation, somaclonal variation, protoplast culture), radiation breeding, etc. There are lots of obstacles in banana breeding due to its complex taxonomy, ploidy nature, low level of seed fertility and seed viability, and genome complexity. Recently, two genomes of banana have been sequenced which has facilitated the understanding of its genomics. A deep insight into banana genetics provides an enormous opportunity to elucidate the function of genes of interest and also to detect variable (or polymorphic) regions in the genome that are associated with agronomic traits (Liu 1997). Therefore, the genome sequence information of banana is the valuable resource to identify genes of interest and develop markers for the tagging of important agronomical traits. Banana scientists around the world are concentrating their view on genomics of banana to pull out pieces of vital information for the development of a marker and to identify genes responsible for disease, stress, yield, quality of fruits, etc.

In this chapter, we illustrate the present scenario in banana breeding programs as well as banana research around the world and also try to address how to overcome the difficulties in banana breeding.

3.2 Gene

Gene is the basic molecular unit of heredity that contains a series of nucleotides in which instructions are contained for the synthesis of RNA that translates into protein. Understanding its gene structure, function, evolution, and mode of inheritance is the key for improvement of any crop species. Crop improvement generally involves the insertion of important genes from cultivated or wild species to a cultivar. Most of the wild banana cultivars rich with disease-resistant, stress-tolerant, and pest- and nematode-resistant genes carry many other undesirable traits such as low yield, seedy fruits, nonedible fruits, etc., while edible banana cultivars carry only good fruit qualities. Therefore, important traits such as disease resistance need to be transferred to the cultivars for its improvement. There are two different strategies commonly used in banana crop improvements viz., conventional cross-breeding and transgenic breeding approach.

3.2.1 Identification and Characterization of Agronomical Important Genes in Banana

Genes are the functional parts of the genome and are directly linked to the agronomical traits of the crops. Identification and functional characterization of agronomically important genes is really a big challenge. It is essential to know the information on its molecular structure, position on the genome, evolution, segregation patterns, etc., for understanding how genes control traits. Recently, two banana genomes (A and B genomes) have been sequenced that speeds up the identification of important genes for subsequent application in the improvement of traits of the banana. Annotation of these genomes identified 36542 and 36638 protein-coding gene models in A and B genomes, respectively (D'Hont et al. 2012; Davey et al. 2013). This protein-coding gene model data set is the valuable resource for further identification and characterization of agronomically important genes in banana. In banana, disease-resistant genes are considered to be more important agronomical traits because several diseases such as black sigatoka, *Fusarium* wilt and bunchy top greatly hamper banana production around the world. Furthermore, most of the cultivated bananas are highly susceptible to the disease and they are vegetatively propagated. So, there is a high possibility of destruction of banana plantation in any particular region once the disease breaks out. Banana breeders have hence given more importance to disease resistance breeding. Several disease-resistant genes have been identified and characterized in banana; among them, NBS-RGCs, *Pto*-RGCs, and NPR1 are notable. Zhao et al. (2009) isolated and characterized *NPR1*-like gene from a local banana cultivar Dongguan Dajiao (*Musa* spp. ABB); this cultivar is well known to be resistant to *Fusarium oxysporum* f.sp. *cubense* (FOC) race 4. In this study, the author noted that this gene is induced by exogenous application of salicylic acid in cv. Dongguan Dajiao but not in cv. Fenjiao which is an FOC race 4 susceptible variety (Zhao et al. 2009). Therefore, this gene has a significant role in the combat FOC race 4 in banana, and it might be a promising candidate for a resistant variety breeding in banana through a transgenic approach. Carotenoids' biosynthesis gene (*PSY*) was isolated from the banana cv. Asupina and cv. Cavendish for understanding the mechanism of pro-vitamin A carotenoid (pVAC) accumulation in banana cultivars. This study showed that banana *PSY* gene is encoded by two paralogs (*PSY1* and *PSY2*) where in *PSY2* is highly expressed in fruit pulp than in the leaf. The enzymatic activity of Asupina PSYs is double to that of Cavendish PSYs; that is why Asupina and Cavendish accumulate different levels of fruit pVAC content (Mlalazi et al. 2012). Consequently, *A-PSY2a* is a good candidate for the genetic improvement of fruit pVAC content in bananas. In addition, gene discovery in banana is in progress using cDNA library sequencing and functional annotation.

3.2.2 Introduce or Transfer Important Genes for Improvement of Banana

In order to introduce/transfer new genes (or traits) into cultivars from their wild relatives of banana, two different strategies have been widely used, conventional breeding and transgenic breeding. Improvement of commercial variety of banana with important agronomic traits through cross-breeding approach is inefficient due to its triploidy and sterility. So, transgenic breeding approach overcomes the limitation of commercial banana variety improvement. A well-characterized gene could be cloned from the wild species of banana or other plant species and transferred to commercial cultivars. Furthermore, this approach is significantly effective only for the traits controlled by a single gene. Genetic transformation and transgenic regeneration method are well established for the banana plant. But to date, there is no significant progress of transgenic breeding in banana compared to other crop plants. In order to improve black sigatoka disease resistance in banana, several genes from different organisms (gene *ThEn*-42 from *Trichoderma harzianum*, *StSy* from grape, *Cu, Zn-SOD* from tomato) have been cloned and transferred to banana. After a 4-year field trial, some of the transgenic lines showed improved tolerance to black sigatoka disease (Vishnevetsky et al. 2011). The replication initiation protein-encoded gene (*Rep*) was transferred to the banana to develop a banana bunchy top virus-resistant cultivar (Tsao 2008). A plant gene that encoded the protein called cystatin is effective against nematodes. This gene which has been transferred to the Cavendish banana improved its tolerance to the nematode (Atkinson et al. 2004). Genes *pflp* and *hrap* were cloned from sweet potato and transferred to banana cultivars. *In vitro* assays showed that the transgenic lines were completely resistant to banana *Xanthomonas* wilt disease (Tripathi et al. 2009). A gene *AMP1* encoding antimicrobial peptide was cloned from onion seeds and transformed in banana cultivar Rasthali (AAB, Silk gp). In this study, embryogenic cells of cv. "Rasthali" were transformed with *Agrobacterium* strain LB4404 harboring binary vector pCAMBIA2301 containing gene *Ace*-AMP1. Transgenic plants were then tested with their performance against *Fusarium oxysporum* f.sp. *cubense* race1 (Foc), and results suggest that *Ace*-AMP1 gene is able to improve resistance of banana against Fusarium wilt disease (Mohandas et al. 2013) under glass house conditions. To facilitate the improvement in nutritional content of banana with vitamin A, vitamin E, or iron, an Austrian scientist transferred vitamin A, vitamin E, or iron accumulation genes in banana cv. "Nakinyika," "Mpologoma," "Nakasabira," and "Sukalindizi" (Pillay et al. 2012b).

3.3 Genetic Markers

A genetic marker is a visible phenotype or biochemical compound or gene or a segment of DNA that can be used to identify individuals or species. Genetic markers are inherited and segregated in Mendelian manner. Based on the principle or methodology, genetic markers are of three types: morphological markers, biochemical markers, and molecular markers. These genetic markers further classified into several subgroups according to the technology and principle.

3.3.1 Overview of Different Marker Technology Applied in Banana Research

3.3.1.1 Morphological Markers

A morphological marker is the visible, prominent phenotype that can be used to distinguish one or more individuals from a population. Agronomical traits such as plant height, leaf architecture, seed coat color, flower color, fruit flavor, leaf aroma, etc., are used as morphological markers to select individual in plant breeding. In banana leaf, architecture and texture are commonly used as morphological markers. For example, the leaf arrangement of diploid banana genotypes is upright (around 45° angle), triploid genotype is horizontal, and tetraploid genotype is downward. Parthenocarpy is one of the most important agronomic traits of banana. Most of the edible banana cultivars are parthenocarpic, they don't produce seeds, and they are triploids. Consequently, this trait was also used to identify breeding materials in the banana breeding program. In 1955, Simmonds and Shepherd discovered banana hybrid identification system using 15 prominent traits, that are highly polymorphic between *M. acuminata* and *M. balbisiana* (Simmonds and Shepherd 1955). Among these, 13 traits were related to the reproductive organ, such as peduncle of inflorescence, pedicels of flower, male flower color, free tepal, stigma color, and arrangement of ovules, and the other seven were related to the bract of the male bud (shape, apex, shoulder, curling, color, color fading, and scars), while the other two were pseudostem color and petiolar canal of the leaves. Based on these phenotypic scoring systems, the *M. acuminata* × *M. balbisiana* hybrids could be characterized into different ploidy levels (diploid, triploid, and tetraploid) and five genetic groups (AA, AB, AAA, AAB, and ABB). Most of the morphological markers are polygenic and highly influenced by the environment. That is why the utility of morphological markers in plant breeding is limited. However, this type of marker is still used in banana breeding, only where other markers' assay facilities are not available.

3.3.1.2 Biochemical Markers

Biochemical markers are the molecules or the products of genes such as proteins/enzymes which are polymorphic in their structure as well as their amino acid composition. For example, isozymes are encoded by the homologous gene located in the different loci in the genome, while allozymes are allelic variants of the same gene. Allozymes and isozymes move at different speeds through a gel because they differ from each other in size, structure, and charge. So, allelic variation can be detected among the individual in a population by using this type of markers. Bonner et al. (1974) first used peroxidase isoenzyme to distinguish banana clones and species. Unfortunately, they failed to distinguish banana clones and species. Thereafter, Rivera (1983) successfully distinguished ABB/BBB (Saba) and ABB (Bluggoe) genomic group using peroxidase and polyphenoloxiydase. After that several researchers around the globe used biochemical markers to characterize banana clones, hybrids, and species, and it was popular until the molecular markers were employed in banana (Rivera 1983; Jarret and Litz 1986a, b; Bhat et al. 1992a, b; Mandal et al. 2001; Megia et al. 2001; Dhanya et al. 2006). The biochemical markers are sensitive to the environment and have a reproducibility problem.

3.3.1.3 Molecular Markers

A molecular marker is a fragment of DNA that is located in a particular region of the genome and variable between two individuals. In the past decade, advances have been made in the molecular marker technology, and several marker technologies have been invented by a molecular biologist. The use of these marker technologies for crop improvement is now a routine work. Based on the principle of molecular marker technology, it can be categorized into several subgroups. The most useful and popular marker technologies that are commonly used in the banana crop improvement program are discussed herewith.

3.3.1.3.1 Restriction Fragment Length Polymorphism (RFLP)

RFLP marker is non-PCR-based molecular marker technique. The variation among the genotypes obtained from this technique is based on restriction site variation of the studied individuals. Restriction site is frequent in the genome, and it is variable between two individuals because the presence of restriction enzyme cleavage site in the genome of two different individuals may not be the same. The presence of restriction site variation between two sister individuals might be due to mutation, crossing over, genome duplication, insertion, or deletion of the DNA fragments. Restriction fragment length polymorphism (RFLP) is a robust, codominant in nature, highly reproducible, and able to characterize heterozygote and homozygote condition of the individual. RFLP markers successfully applied in banana genetics and breeding are useful in estimating variation in the chloroplast genome of the banana (Gawel and Jarret 1991; Baurens et al. 1997), phylogenetic relationship of the banana and its relatives (Gawel et al. 1992; Jarret et al. 1992; Wong et al. 2001), mapping (Miller et al. 2008), etc,. RFLP markers are often used in combination with other types of markers, such as RAPD, SSR, etc., in banana breeding (Bhat et al. 1995), although RFLP markers are not suitable for large-scale utilization in the breeding program because of their complicated assay technique, their expensive nature, and their requirement of sophisticated equipments.

3.3.1.3.2 Variable Number of Tandem Repeats (VNTR)

The VNTR marker also known as non-PCR-based marker technology is used in plant genetics and breeding research. This technique involves restriction digestion followed by the hybridization with the probe containing minisatellite sequences. In principle, the VNTR marker is the fragment of the genome composed of tandem repeat units of a 10–50-base motif (known as minisatellite) and flanked by the conserved DNA restriction site. Many reports showed that tandem repeats are highly abundant and randomly distributed throughout the plant genome (Biswas et al. 2015; Manzo-Sánchez 2008). Therefore tandem repeat containing fragments of the genome could be the vital source for development of a large number of VNTR markers. This marker is highly polymorphic and reproducible, but its assay technique is complex and costly. So this marker is not popular with banana researchers. Several VNTR markers have been developed in *Musa* as reported by Crouch et al. (1998, 1999a) and Kaemmer et al. (1997), and used in genotyping and characterization of *Musa* breeding populations. Recent advances in the genome sequence technology and availability of the whole genome sequence of *Musa* spp. provided the opportunity to develop *Musa* genome-specific VNTR marker. This marker can be useful in banana research for DNA fingerprinting, identification of varieties, cultivars, and individuals, and also at population-level study.

3.3.1.3.3 Random Amplified Polymorphic DNA (RAPD)

This marker is well known for its simple assay system. But a major drawback of this marker is reproducibility. Evidences showed that different laboratories achieved different results when tested under identical PCR parameters and conditions (Pillay et al. 2012a). The common uses of RAPD marker in banana research are to:

Detect diverse genotypes (Kaemmer et al. 1992; Howell et al. 1994; Bhat and Jarret 1995; Uma et al. 2006; Jain et al. 2007).

Identify duplicate accessions from the *in vitro* and *ex vitro* germplasm collection (Ray et al. 2006).

Identify somaclonal variation (Dhanapal et al. 2014).

Differentiate *Musa* genomic group (Howell et al. 1994; Rekha et al. 2001), genetic diversity study, etc. (Bhat and Jarret 1995).

The first RAPD marker was applied for fingerprinting of wild and cultivated species of banana in 1992 (Kaemmer et al. 1992). A and B genome-specific RAPD marker has been identified for *Musa*, and it is very useful to estimate genomic composition of banana cultivars (Pillay et al. 2000; Oselebe et al. 2006; Pillay et al. 2006). Uma et al. (2006) applied 80 RAPD primers to estimate intraspecific variation and relation of Indian wild *Musa balbisiana* Colla collection. In this study, only 4 primes produced 31 polymorphic bands, and all the 16 accessions were clustered into four as against seven clusters obtained through morphotaxonomic characterization. Some of the accessions failed to group in the same cluster as revealed by morphotaxonomy data. This result can be further improved using more RAPD markers and also using other types of markers such as ISSR, ITS, AFLP, etc. In recent years, RAPD has been used with other markers for the *Musa* genetics and breeding study (Howell et al. 1994; Lamare and Rao 2015). Bhat et al. (1995) reported that RAPD primer OPC-15 (5′-GACGGATCAG-3′) had the potential for distinguishing 55 of the cultivars by producing 24 bands but failed to characterize the clones of Gros Michel and Venkel.

3.3.1.3.4 Intersimple Sequence Repeats (ISSR)

ISSR is a PCR-based marker technique, and its assay system is similar to RAPD. This technique does not require any prior knowledge of genome sequences of the organism. Like RAPD, a versatile set of primers is used in the ISSR assay. Furthermore this marker technique is simple, fast, and cost effective. The main drawbacks of this technology are the dominant nature of the marker, that the homology of the bands is uncertain, and that they do not allow the identification of heterozygous from homozygous dominance. This marker involves amplification of the DNA fragments nearby at an amplifiable distance between two identical microsatellite repeat regions oriented in reverse orders (Spooner et al. 2005). ISSR has been widely used for varietal identification, genetic diversity analysis, genetic stability of the tissue culture-derived clone, etc. The use of ISSR in banana research has not received wide attention so far. Several studies using ISSR marker in banana research were mainly limited in small-scale germplasm characterization (Venkatachalam et al. 2007; HaiFei et al. 2010; Poerba and Ahmad 2010; QianJie et al. 2010; Khatri et al. 2011; Yao et al. 2012; Dhanapal et al. 2014; Lamare and Rao 2015) and in genetic fidelity testing of cultivated banana (Rout et al. 2009; Dhanapal et al. 2014). Choudhary et al. (2014) used 40 ISSRs with 60 RAPD markers to estimate molecular variability of 12 plantain ecotypes and found that ISSR is a better tool than RAPD for assessment of genetic diversity in plantain ecotypes. Another study reported the use of ISSR with RAPD to test genetic stability of three micropropagated banana (*Musa* spp.) cultivars and found that ISSR detected more polymorphism than RAPD (Ray et al. 2006). Similarly, ISSR were used for finding genetic uniformity of micropropagated banana plantlets (Rout et al. 2009), *in vitro* mutagenesis, and variance (Khatri et al. 2011). ISSRs were also employed to evaluate the genetic diversity and classification of 27 wild banana accessions collected in Guangxi, China (Qin et al. 2011). Padmesh et al. (2012) explored the pattern of genetic variation of 32 wild *M. acuminata* Colla collected from the Southwestern Ghats in the peninsula of India using ISSR markers.

3.3.1.3.5 Internal Transcribed Spacers (ITS)

The nuclear ribosomal genes (rDNA) encoded three rRNA subunits, viz., 18S, 5.8S, and 26S. These subunits are present in the genome as clusters of tandemly repeated units of 250–20,000 copies (Rogers and Bendich 1987). rDNA is transcribed as a single unit along with the two spacers known as the internal transcribed spacers (ITS) that separate the 5.8S subunit from the 18S and the 26S subunits. The sequences of these ITS regions are variable among closely related clones or species. Therefore, this variation could be used as a marker, and it is a powerful tool for assessing phylogenetic relationships at the lower taxonomic levels. ITS markers are used for genetic diversity, population structure and clone identification. In *Musa*, ITS marker was used to distinguish A and B genomes. Nwakanma et al. (2003) amplified ITS region of seven *M. acuminata* (A genome) and five *M. balbisiana* (B genome) accessions. The result revealed that all the accessions produced 700-bp fragments; subsequently this fragment was digested with ten different restriction enzymes (*Alu*I, *Cfo*I, *Dde*I, *Hae*III, *Hin*fI, *Hpa*II, *Msp*I, *Rsa*I, *Sau*3AI, and *Taq*I). Only *Rsa*I produced a consistent polymorphic banding pattern between *M. acuminata* and *M. balbisiana*. The *Rsa*I produces four fragments (350 bp, 180 bp, 120 bp, and 50 bp) in the *M. balbisiana* accessions and three fragments (530 bp, 120 bp, and 50 bp) in *M. acuminata* accessions. The fragment of 530 bp was unique to the A genome, while two fragments of 350 bp and 180 bp were specific to the B genome. Further 56 accessions from different genomic combinations of *Musa* spp. (viz., AA, AAA, AAB, AB, and ABB) were amplified with ITS primer and subsequently digested with *Rsa*I. Results showed that only a 530-bp fragment was present in A genome-containing individuals but absent in B genome-containing individuals. On the other hand, a B genome-containing individual produced 350-bp and 180-bp fragments that were absent in A genome-containing individuals. This finding was a breakthrough for characterization of *Musa* spp. according to their genomic combination using the ITS marker.

3.3.1.3.6 Inter-retrotransposon Amplified Polymorphism (IRAP)

IRAP is a PCR-based dominant marker. This marker detects retrotransposon (RT) insertion-based polymorphism among the individuals. RTs are abundant in the plant genome, and they compose over 50 % of the total nuclear genome. RT elements act as a mutagenic agent by insertion or deletion in the genome thereby making them a good source of genomic variation (Heslop-Harrison 2000); thus RT can be used as a genetic marker for various purposes in plant genetics research. It has been used in fingerprinting, genetic diversity, and population structure studies in plants (Biswas et al. 2010a, b; Khadivi-Khub et al. 2015; Singh et al. 2015). IRAP markers are also used in *Musa* for classifying the genome constitution (Nair et al. 2005; Teo et al. 2005; Häkkinen et al. 2007; Pachuau et al. 2014), genetic diversity (Saraswathi et al. 2011), and identification of wild *Musa* spp. (Häkkinen et al. 2007; Häkkinen and Teo 2008).

Ty3-gypsy-like LTR sequence of banana (*Musa acuminata* monkey retrotransposon, AF 143332) was used to design IRAP primer, and it was used to identify the B genome in the banana cultivars. Nair et al. (2005) used this marker to classify 36 banana cultivars and observed multiple polymorphic bands. Among these bands, a specific band of 350 bp was observed in all the cultivars with the B genome. In ABB genomes, the band intensity was highly observed in AAB genomes. But there was an exceptional result found in the cultivar "Manoranjitham" (AAA), in which B-specific band was present with similar intensity observed in ABB ("Kosta bontha"). The author suggested that "Manoranjitham" was misidentified. Teo et al. (2005) applied IRAP methods to generate molecular markers for estimation of diversity, genome constitution, and relationships of Malaysian banana cultivars. IRAP is used with an RAPD marker to study intragroup diversity among Cavendish (AAA) clones of banana (Saraswathi et al. 2011). Singh et al. (2015) used IRAP and morphological markers to characterize *Musa* germplasm from Northeast India.

3.3.1.3.7 Amplified Fragment Length Polymorphism (AFLP)

AFLP is a PCR-based multilocus marker technology employed in many plant studies relating to genetic identity, phylogenetic relation, parentage identification of clones and cultivars, etc. It is a robust, highly reproducible, and dominant marker, but in a segregating population, AFLP can be detected as codominant. The AFLP marker is capable of generating 80–500-bp fragments from different genomic sites at the same time; therefore this marker detects high polymorphisms in different genomic regions simultaneously. There are many advantages of AFLP compared with other markers such as RAPD, RFLP, SSR, etc. AFLP has become particularly useful in the study of plant taxa, where genomic information is not available.

Thirty-nine accessions from four main sections of the genus *Musa* were evaluated using eight AFLP markers (Ude et al. 2002a). A wide range of variability was observed among the species within the sections of the genus *Musa*. This study proved that AFLP is a very useful tool in determining taxonomic relationships in the genus *Musa* and potentially useful to resolve some of the complicated taxonomic questions in the genus *Musa*. According to morphological data, Simmonds and Weatherup (1990) place *M. peekeli ssp. peekeli* as a close relative of *M. peekeli ssp. angustigemma*, while Argent (1976) suggests that *M. angustigemma* should rank under *M. peekeli* as a subspecies. AFLP data suggest that *M. peekeli ssp. angustigemma* is clearly distinct from *M. peekeli ssp. peekeli*, and therefore the specific rank of *M. angustigemma* should be retained.

Several studies suggested that the AFLP marker is effective for genetic diversity analysis in *Musa* and the level of polymorphism comparatively higher than other markers used in the *Musa* diversity analysis (Crouch et al. 1999; Loh et al. 2000; Wong et al. 2001, 2002). The relationship of cultivated banana with *M. acuminata* and *M. balbisiana* is precisely classified by AFLP markers (Wongniam et al. 2010). Youssef et al. (2011a) reported that AFLP is able to discriminate A, B, S, and T genomes within *Musa* species. In order to detect genetic stability of somaclonal variation in somatic embryo-derived plants of two banana cultivars, namely, "Grand Naine" and "Williams" (*Musa acuminata* Colla, AAA), AFLP marker technology was applied (Youssef et al. 2011b). Thirty primer combinations were used to detect polymorphism among somaclonal variants, and the result showed that 1.4 % and 1.6 % bands were polymorphic in "Grand Naine" and "Williams," respectively. In addition, 8 and 16 bands were specific to the observed regenerated plants of "Grand Naine" and "Williams," respectively, which were absent in their parents, while ten and five bands are completely absent in the regenerated plant of "Grand Naine" and "Williams," respectively, which were exclusively present in their parents. AFLP markers were also used for the genetic analysis of *Musa* species in combination with other molecular markers (Opara et al. 2010; Youssef et al. 2011a) and also to identify somaclonal variation (James et al. 2004; Vroh-Bi et al. 2011; Youssef et al. 2011b).

3.3.1.3.8 Simple Sequence Repeat (SSR) or Microsatellite Marker

An SSR marker is well known for its genome-wide distribution, codominant inheritance, reproducibility, multiallelic nature, and easy assay technique. Recent progress in DNA sequencing technology has provided an opportunity to routinely develop large sets of SSR markers. Transferability of a genic SSR marker is higher than genomic SSRs. This feature helps to design anchor markers for comparative mapping studies. Since they are often more conserved, genic SSRs may provide an insufficient degree of polymorphism to discriminate between closely related germplasm. Therefore genomic SSRs may be valuable complements. In the recent past decade, SSR markers are extensively used in a wide range of breeding applications. Several hundred EST-SSR markers were developed by in silico EST sequence mining of several *Musa* spp. (Li et al. 2012; Passos et al. 2012, 2013; Backiyarani et al. 2013; Ravishankar et al. 2015). SSR markers were developed from B genome of *Musa* (Buhariwalla et al. 2005; Ravishankar et al. 2013); 226 were developed from *Musa* GSS (survey of the genomic sequence) data (Ravishankar et al. 2012), and 41 were from Calcutta 4 using BAC sequences (Miller et al. 2010). Most of these markers are not freely accessible; some are redundant with alternate IDs or names, while their physical positions and functional natures are unknown. Consequently, the use of these markers in *Musa* spp. improvement is limited. In the whole genome sequences of two banana varieties, more than 0.1 million EST and several thousand GSS sequences were mined for SSR marker discovery, and 119,540 non-redundant SSR markers were developed (Biswas et al. 2015). Subsequently these markers were characterized, classified, and stored in a searchable database known as Musa marker database and it can be access through the following link http://www.agrogene.ac.cn:8088/mumdb/mumdb_home.html. A part of these markers was validated by wet lab assay and their potentiality was estimated for a genetic diversity study in *Musa* population.

3.3.1.3.9 DArT

Diversity arrays technology (DArT) is DNA/DNA hybridization-based molecular marker technology, extensively used for a quick assessment of the structure of germplasm collections. This is the cost-effective genotyping technology that can detect all types of DNA variation including SNP, indel, CNV, and methylation. It also can detect simultaneously variation at numerous genomic loci without any sequence information of the organism. In recent years, this technology become popular and is regularly used in the crop improvement including *Musa* (Risterucci et al. 2009). A total of 836 DArT markers were developed and used for *Musa* spp. genotyping. Ten percent of these markers were A genome specific and able to target this genome portion in a related analysis, among diverse ploidy constitutions (Risterucci et al. 2009). Risterucci et al. (2009) in their study clearly demonstrated the usefulness of DArT markers for *Musa* spp. genotyping and genetic diversity analysis. They used four complexity reduction methods of DArT for *Musa* spp. and tested their performance on 48 *Musa* genotypes and finally invented two methods that produced more polymorphic information content than others. Subsequently, selected methods were used for large-scale *Musa* spp. genotyping.*Musa* accessions, around 168 in number that were collected from CIRAD (Neufchateau, Guadeloupe) and IITA (Ibadan, Nigeria) were used. All these accessions were derived from *M. acuminata* (A genome) and *M. balbisiana* (B genome). And the result reveals that DArT markers classify them according to their origin and genomic combinations. Kilian (2007) developed 1,500 DArT markers using a wide array of *Musa* accessions, which were used for *Musa* framework map. Further these markers with additional 380 markers have been used to construct a map at CIRAD.

3.3.1.3.10 Ecotilling

Ecotilling is a new class of molecular marker technology that was developed as a high-throughput and low-cost platform for the SNPs discovery and small indels. This technology was first used in *Arabidopsis* ecotypes and then adopted for many species including humans, switchgrass, poplar, melon, banana, etc. (Till et al. 2010). Since its discovery, ecotilling has been used for more than 20 plant species for genetic diversity, population structure, mapping, and QTL analysis. In principle, ecotilling is an enzymatic mismatch cleavage-based DNA hybridization technology. Around 700–1,600-bp genic regions are amplified by PCR using fluorescent labeled gene-specific primers. Subsequently, samples are denatured and annealed; then heteroduplexed molecules are produced through the hybridization of polymorphic amplicons. Mismatched regions or double-stranded duplexes are then cleaved using crude extract of celery juice containing the single-strand specific nuclease CEL I. Cleaved products are then resolved in denaturing polyacrylamide gel electrophoresis (PAGE) for observing banding pattern.

In *Musa*, ecotilling method was used for the discovery and characterization of nucleotide polymorphisms of diploid and polyploid accessions. Over 800 novel alleles in 80 accessions were identified as polymorphic using 14 gene-specific primers by Till et al. (2010). In this study, more than 6,000 polymorphisms were detected in over 800 alleles in 80 *Musa* accessions. Further, sequencing-based validation was performed for the detection of SNPs variation among the accessions. Consequently, sequencing and banding patterns reveal that ecotilling is the perfect platform for discovery of polymorphisms in homologous gene targets in *Musa* accessions.

3.3.2 Utility of Marker Technology in Banana Breeding

3.3.2.1 Genotype Identification

Genotype identification is known as genotyping. It is one of the routine works for plant breeders prior to selection of breeding material. Environment has great impact on phenotypic expression on gene or traits. Selection of breeding material based on the phenotypic traits from the population or germplasm collection maintained in a different environment may be misleading. Therefore, genotyping is important in a breeding programme. Genotyping is a DNA-based technology commonly used to characterize or identify any living organism. It is also known as fingerprinting or DNA fingerprinting. This methodology is extensively used in plant breeding for identification of individuals in a population, for discrimination between individuals in an inbred line, or for determination of genetic distance between genotypes in general. Different types of markers such as biochemical, RAPD, ISSR, RFLP, AFLP, SSR, etc., are commonly used for *Musa* spp. genotyping. In the early 1980s, biochemical markers are used for *Musa* spp. genotyping, for example, peroxidase and polyphenoloxidase used to identify "Saba" (ABB/BBB) and "Bluggoe" (ABB) types of banana (Rivera 1983). Dhanya et al. (2006) and Dhanya et al. (2006) used isozymes to identify banana cultivars resistant to *banana bract mosiac virus* (BBrMV). An RAPD marker has been developed for genotyping dwarf off-type Cavendish banana (*Musa* spp. AAA) cultivar. The primer OPJ-04 (5′-CCGAACACGG-3′) was found to amplify an approximately 1.5-kb band which is consistently present in all normal Cavendish cultivars but absent in dwarf Cavendish cultivars (Damasco et al. 1996). Bhat et al. (1995) and Bhat and Jarret (1995) used 60 RAPD markers for identification of 57 *Musa* cultivars; among these RAPD markers, 49 produced consistent results and were able to identify 55 cultivars but failed to characterize Gros Michel and Venkel clones. A total of 33 SSR markers were used to characterize 35 cultivated banana (*Musa* spp.) genotypes, including triploid cultivars and tetraploid hybrids (Creste et al. 2003). Christelova et al. (2011) used 19 sets of fluorescently labeled SSR primers for genotyping 70 diploid and 38 triploid banana accessions.

3.3.2.2 Germplasm Managements

Germplasm collection and its systematic management are essential for quick access of a right individual for the breeding program. In general, genebank collection consists of multiple copies of same genotypes; these duplicates may occur for various regions, for example, documentation error, the sampling of multiple individuals from genetically homogeneous collections, exchange of identical accessions between genebanks, etc. (Spooner et al. 2005). The redundant accessions of the genebank have no significant impact, but maintaining them in the genebank requires time, space, and resource. So the elimination of redundant accessions from a large collection of genebank is a big challenge. Correct classification and identification of unique accessions allows solving germplasm management problems. There is a high possibility of inclusion of redundant accessions in the vegetative propagated crop plant genebank such as banana, because the germplasms of this crop are collected based on its morphological traits and a similar clone is maintained in a different geographical region with a different name. Further, cultivars of the banana are evolved through human selection. Therefore efficient germplasm management is an important factor in banana breeding. In the last few decades, a significant number of studies have been conducted for the germplasm characterization of *Musa* spp. around the world using different kinds of molecular markers including RAPD, AFLP, SSR, etc. (Bhat and Jarret 1995; Creste et al. 2003; Christelová et al. 2011).

3.3.2.3 Genome Composition Determination

Most of the banana cultivars are triploid and derived from interspecies hybridization between *M. acuminata* (A genome) and *M. balbisiana* (B genome), while other banana and plantain genomes are derived from the combination of A, B, T, and S species genomes. The knowledge of genomic combination of the cultivars or a clone of banana and plantain is essential for its improvement. Several molecular markers have already been developed to identify genomic combination of banana and plantain cultivars. Three RAPD primers (A17, A18, D10) from OPERON Technologies (Alameda, CA, USA) were able to identify A and B genome of *Musa* spp. Primer A17 generated two fragments (600 bp, 100 bp) and primer D10 one fragment (320 bp) that was unique to *M. acuminata* (A genome), while primer A18 produced three fragments (200 bp, 250 bp, 300 bp) in *M. balbisiana* (B genome). The fragments B18250 and B18300 were always present in genotypes with at least one B genome; on the other hand, fragment A18200 was present in clones with two B genomes. These three RAPD primers are extensively used for the determination of banana genomic composition (Jones 2000; Pillay et al. 2000). A PCR-RFLP-based ITS marker is also applicable for banana A and B genome identification (Nwakanma et al. 2003). The ITS marker produced one fragment of 530 bp that is A genome specific and two B genome-specific fragments of 350 bp and 180 bp. An interspecific hybrid of A and B genome possessed all three fragments.

3.3.2.4 Genetic Diversity and Population Structure Estimation

Molecular marker technology is widely used for study genetic diversity, population structure, and phylogenetic relationships of the crop plants. Many studies have been attempted to estimate genetic diversity using different types of markers in banana and plantain wild and cultivar collections. For example, the genetic diversity of 100 Indonesian *Musa* cultivars from the different genomic groups of Musa (AA, AAA, AAB, ABB, and BB) was estimated using isozymes of *malate dehydrogenase* (*MDH*), *peroxidase* (*PRX*), and *glutamate oxaloacetate transaminase* (*GOT*) (Megia et al. 2001). The result reveals that *MDH* and *PRX* were more useful than *GOT* for genetic diversity study of Indonesian *Musa* cultivars.

AFLP marker technology was applied to estimate the genetic diversity of banana and plantain population collected from different geographical regions and also different cultivars and wild accessions (Ude et al. 2002a, b; Wong et al. 2002; Wang et al. 2007; Opara et al. 2010; Wongniam et al. 2010; Youssef et al. 2011a). Wong et al. (2001) used an AFPL marker to assess the genetic diversity of 32 Malaysian wild bananas *Musa acuminata* Colla, and results showed that AFLP efficiently classified Malaysian wild bananas based on genetic distance. Ude et al. (2002b) assessed the genetic diversity and phylogenetic relationships of *M. acuminata* and *M. balbisiana* and their natural hybrids and noted that AFLP markers were able to produce enough information about the genetic diversity of *M. balbisiana* accessions to further classify them into two subgroups.

Seven SSR markers were used to estimate genetic diversity and population structure of six *M. ornate* populations (Burgos-Hernández et al. 2013). A low level of genetic diversity was observed in the *M. ornate* populations. This low diversity may occur due to recent fragmentation of events, which meant that there was not enough time gap between populations to detect differences. It may also be the cause of self-pollination, clonal reproduction, bottleneck selection, or the decline in pollinator in population. This study was proposed to conserve and maintain all the remaining *M. ornate* population to maintain gene flow and increase the genetic diversity. The ex situ collection of 224 *Musa* spp. at Embrapa, Brazil, was analyzed by 16 SSR markers for the estimation of genetic diversity and population structure. The findings of this study showed that structure analysis might be useful in identification of ancestry of recently developed tetraploid hybrids and triploid cultivars by breeding programs.

3.3.2.5 Mapping and Marker-Assisted Breeding

Genome mapping sets up the roadmap of a genome that helps to locate important genes, manipulate them, identify the molecular environment of both coding and noncoding DNA sequences, etc. The advancements in the area of sequencing technology and molecular marker developments open the door for tagging agronomically important traits via mapping and marker-assisted breeding. Although several linkage maps have been developed in *Musa* spp., their quality and marker saturation remain poor; that is why their utility in banana breeding is still limited (Faure et al. 1993; Baurens et al. 1997; Vilarhinos 2004; Hippolyte et al. 2010). Faure et al. (1993) developed first partial molecular linkage map based on 77 markers, in which a significant number (36 %) of markers were deviated from Mendelian segregation. The second map was developed for *M. acuminata* cv "M53," in which it exhibited 11 linkage groups with a significant number of markers distorted from the Mendelian segregation ratio. The linkage map developed by Vilarhinos (2004) is composed of 14 linkage groups; 59 % of markers are skewed from the Mendelian segregation ratio; the mapping population is derived from F1 and pseudo-testcross strategy used to construct the linkage group. The oversized linkage groups of this map may be the cause of distorted markers that are involved in the structural rearrangements of chromosomes. The high-density linkage map was developed in *Musa* sp. using DArT and SSR markers (Hippolyte et al. 2010). In this study, two parental maps and one reference map were constructed. The female parent map consists of 11 linkage groups with 261 markers (125 SSRs and 136 DArTs). The map spanned 920 cM, with one marker per 3.8 cM.; 59 and 9 markers were comprised in the largest and smallest linkage group, respectively. The male parent map is obtained at LOD 5; a total of 359 markers consisted the 9 linkage groups with a total map length of 1,081 cM., in which one marker is distributed per 2.9 cM., but the marker distribution in each linkage group was not uniform. Further a synthetic linkage reference map was developed using 489 markers (167 SSRs, 322 DArTs), among which 132 were anchored markers. This map contains 11 linkage groups covering 1,197 cM., at an average of 38 markers per linkage group and one marker per 2.8 cM.

Marker-assisted breeding is useful for identifying markers linked to important traits; these markers can be used for the direct screening under greenhouse condition at early stage of growth. This approach reduces time, cost, and gaps in the breeding program that dramatically speeds up the selection process. There are a few markers identified in *Musa* linked with some important agronomical trait that is related to disease resistance. Four methylation-sensitive amplification polymorphism (MSAP) markers were identified as linked markers for black sigatoka disease-resistant gene of banana (Gimenez et al. 2006). There are some RAPD markers used for selecting black sigatoka disease-resistant, nematode-resistant, and salt-tolerant banana clones (Miri et al. 2009). A banana *SERK*-related marker is associated with somatic embryogenic competence and disease-resistant response in *Musa sp.* (Huang et al. 2010; Cunha et al. 2015).

3.4 Conclusions

In the last few decades, gene discovery and marker technology have revolutionized plant breeding. Genes and their related functions were identified; they were cloned and utilized in bred crop improvement. Scientists around the globe are identifying agronomically important genes for banana varietal improvement. Molecular marker technology has made a great impact on banana breeding, including cultivar identification, parent selection for cross-breeding, identifying genomic group of wild species, germplasm management, characterizing somaclonal variation, etc.

Acknowledgments The author wishes to thank the National Natural Science Foundation of China for the research grant Category: International (Regional) Cooperation and Exchange Program. Program: The Research Fund for International Young Scientist, FY2015. The research grants No: 31550110217 and Dr. Yi Ganjun for his valuable suggestion and modification of MS.

References

Argent G (1976) The wild bananas of Papua New Guinea. Note Roy Bot Gard Edinb 35:77–114

Atkinson HJ, Grimwood S, Johnston K et al (2004) Prototype demonstration of transgenic resistance to the nematode Radopholus similis conferred on banana by a cystatin. Transgenic Res 13(2):135–142

Backiyarani S, Uma S, Varatharj P et al (2013) Mining of EST-SSR markers of Musa and their transferability studies among the members of order the Zingiberales. Appl Biochem Biotechnol 169(1):228–238. doi:10.1007/s12010-012-9975-2

Baurens F, Noyer J, Lanaud C et al (1997) Sequence tagged site markers to draft the genomic structure of the banana chloroplast. Fruits 52(4):247–259

Bhat K, Jarret R (1995) Random amplified polymorphic DNA and genetic diversity in Indian Musa germplasm. Genet Resour Crop Evol 42(2):107–118

Bhat K, Bhat S, Chandel K (1992a) Survey of isozyme polymorphism for clonal identification in Musa. I. Esterase, acid phosphatase and catalase. J Hortic Sci 67(4):501–507

Bhat K, Bhat S, Chandel K (1992b) Survey of isozyme polymorphism for clonal identification in Musa. II. Peroxidase, superoxide dismutase, shikimate dehydrogenase and malate dehydrogenase. J Hortic Sci (U K) 67:737–743

Bhat KV, Jarret RL, Rana RS (1995) DNA profiling of banana and plantain cultivars using random amplified polymorphic DNA (RAPD) and restriction fragment length polymorphism (RFLP) markers. Electrophoresis 16(1):1736–1745

Biswas MK, Baig M, Cheng Y-J et al (2010a) Retrotransposon based genetic similarity within the genus Citrus and its relatives. Genet Resour Crop Evol 57(7):963–972

Biswas MK, Xu Q, Deng X-X (2010b) Utility of RAPD, ISSR, IRAP and REMAP markers for the genetic analysis of Citrus spp. Sci Hortic 124(2):254–261

Biswas MK, Liu Y, Li C et al (2015) Genome-wide computational analysis of Musa microsatellites: classification, cross-taxon transferability, functional annotation, association with transposons & miRNAs, and genetic marker potential. PLoS ONE 10(6), e0131312. doi:10.1371/journal.pone.0131312, PONE-D-14-57877 [pii]

Bonner JW, Warner RM, Brewbaker JL (1974) A chemosystematic study of Musa cultivars. Hortic Sci 9:325–328

Buhariwalla HK, Jarret RL, Jayashree B et al (2005) Isolation and characterization of microsatellite markers from Musa balbisiana. Mol Ecol Notes 5(2):327–330. doi:10.1111/j.1471-8286.2005.00916.x

Burgos-Hernández M, Hernández D, Castillo-Campos G (2013) Genetic diversity and population genetic structure of wild banana Musa ornata (Musaceae) in Mexico. Plant Syst Evol 299(10):1899–1910. doi:10.1007/s00606-013-0846-

Choudhary R, Keshavachandran R, Menon R et al (2014) Molecular variability of plantain ecotypes from the genus Musa (Musaceae). Turk J Bot 38(5):827–834

Christelová P, Valárik M, Hřibová E et al (2011) A platform for efficient genotyping in Musa using microsatellite markers. AoB Plants. doi:10.1093/aobpla/plr024

Creste S, Tulmann Neto A, de Oliveira Silva S et al (2003) Genetic characterization of banana cultivars (Musa spp.) from Brazil using microsatellite markers. Euphytica 132(3):259–268. doi:10.1023/a:1025047421843

Crouch HK, Crouch JH, Jarret RL et al (1998) Segregation at microsatellite loci in haploid and diploid gametes of Musa. Crop Sci 38(1):211–217

Crouch J, Crouch H, Constandt H et al (1999) Comparison of PCR-based molecular marker analyses of Musa breeding populations. Mol Breed 5(3):233–244

Cunha CM, Hinz RH, Pereira A et al (2015) A SCAR marker for identifying susceptibility to Fusarium oxysporum f. sp. cubense in banana. Sci Hortic 191:108–112

Damasco O, Graham G, Henry R et al (1996) Random amplified polymorphic DNA (RAPD) detection of dwarf off-types in micropropagated Cavendish (Musa spp. AAA) bananas. Plant Cell Rep 16(1–2):118–123. doi:10.1007/bf01275464

Davey MW, Gudimella R, Harikrishna JA et al (2013) A draft Musa balbisiana genome sequence for molecular genetics in polyploid, inter-and intra-specific Musa hybrids. BMC Genomics 14(1):683

Dhanapal S, Sekar DS, Satheesh PM (2014) Efficiency of RAPD, SSR and ISSR markers in evaluating the genetic fidelity for micropropagated Musa accuminata plant exposed to coal extracted humic acid and commercially available products. Int J Agric Sci Res (IJASR) 4(4):77–86

Dhanya MK, Rajagopalan B, Umamaheswaran K et al (2006) Isozyme variation in banana (Musa sp.) in response to bract mosaic virus infection. Indian J Crop Sci 1(1–2):1

D'Hont A, Denoeud F, Aury J-M et al (2012) The banana (Musa acuminata) genome and the evolution of monocotyledonous plants. Nature 488(7410):213–217, doi http://www.nature.com/nature/journal/v488/n7410/abs/nature11241.html#supplementary-information

Faure S, Noyer J, Horry J et al (1993) A molecular marker-based linkage map of diploid bananas (Musa acuminata). Theor Appl Genet 87(4):517–526

Gawel N, Jarret R (1991) Chloroplast DNA restriction fragment length polymorphisms (RFLPs) in Musa species. Theor Appl Genet 81(6):783–786

Gawel NJ, Jarret RL, Whittemore AP (1992) Restriction fragment length polymorphism (RFLP)-based phylogenetic analysis of Musa. Theor Appl Genet 84(3–4):286–290. doi:10.1007/bf00229484

Gimenez C, Palacios G, Colmenares M (2006) Musa methylated DNA sequences associated with tolerance to Mycosphaerella fijiensis toxins. Plant Mol Biol Report 24(1):33–43

HaiFei M, GuiMei L, Yu Z et al (2010) Genetic diversity analysis of banana (Musa spp.) based on ISSR molecular marker. Southwest China J Agric Sci 23(4):1206–1210

Häkkinen M, Teo C (2008) Musa rubinea, a new Musa species (Musaceae) from Yunnan, China. Folia Malaysiana 9.1:23–33

Häkkinen M, Teo CH, Othman YR (2007) Genome constitution for Musa beccarii (Musaceae) varieties. Acta Phytotaxonomica Sin 45(1):69–74

Heslop-Harrison J (2000) Comparative genome organization in plants: from sequence and markers to chromatin and chromosomes. Plant Cell 12(5):617–635

Hippolyte I, Bakry F, Seguin M et al (2010) A saturated SSR/DArT linkage map of Musa acuminata addressing genome rearrangements among bananas. BMC Plant Biol 10(1):65

Howell EC, Newbury HJ, Swennen RL et al (1994) The use of RAPD for identifying and classifying Musa germplasm. Genome 37(2):328–332

Huang X, Lu X-Y, Zhao J-T et al (2010) MaSERK1 gene expression associated with somatic embryogenic competence and disease resistance response in banana (Musa spp.). Plant Mol Biol Report 28(2):309–316

Jain PK, Saini ML, Pathak H et al (2007) Analysis of genetic variation in different banana (Musa species) variety using random amplified polymorphic DNAs (RAPDs). Afr J Biotechnol 6(17):1987–1989

James A, Peraza-Echeverria S, Herrera-Valencia V et al (2004) Application of the amplified fragment length polymorphism (AFLP) and the methylation-sensitive amplification polymorphism (MSAP) techniques for the detection of DNA polymorphisms and changes in DNA methylation in micropropagated bananasBanana improvement: cellular, molecular biology, and induced mutations Proceedings of a meeting held in Leuven, Belgium, 24–28 September 2001. Science Publishers, Inc., pp 287–305

Jarret RL, Litz RE (1986a) Enzyme polymorphism in Musa acuminata Colla. J Hered 77(3):183–188

Jarret RL, Litz RE (1986b) Isozymes as genetic markers in bananas and plantains. Euphytica 35(2):539–549. doi:10.1007/bf00021863

Jarret RL, Gawel N, Whittemore A et al (1992) RFLP-based phylogeny of Musa species in Papua New Guinea. Theor Appl Genet 84(5–6):579–584

Jones DR (2000) Introduction to banana, abaca and enset. Diseases of banana, abaca and enset. CAB International, Wallingford, pp 1–36

Kaemmer D, Afza R, Weising K et al (1992) Oligonucleotide and amplification fingerprinting of wild species and cultivars of banana (Musa spp.). Nat Biotechnol 10(9):1030–1035

Kaemmer D, Fischer D, Jarret R et al (1997) Molecular breeding in the genus Musa: a strong case for STMS marker technology. Euphytica 96(1):49–63

Khadivi-Khub A, Shabanian N, Alikhani L et al (2015) Genotypic analysis and population structure of Lebanon oak (Quercus libani G. Olivier) with molecular markers. Tree Genet Genome 11(5):1–10

Khatri A, Bibi S, Dahot MU et al (2011) In vitro mutagenesis in banana and variant screening through ISSR. Pak J Bot 43(2011):2427–2431

Kilian A (2007) Towards effective deployment of diversity arrays technology (dart) in banana genomics and sequencing

Lamare A, Rao S (2015) Efficacy of RAPD, ISSR and DAMD markers in assessment of genetic variability and population structure of wild Musa acuminata colla. Physiol Mol Biol Plants 21(3):349–358. doi:10.1007/s12298-015-0295-1

Li WJ, Ma H, Li ZH et al (2012) Thirty-four Musa (Musaceae) expressed sequence tag-derived microsatellite markers transferred to Musella lasiocarpa. Genet Mol Res 11(3):2094–2098. doi:gmr1898 [pii]10.4238/2012.August.6.13

Liu BH (1997) Statistical genomics: linkage, mapping, and QTL analysis. CRC Press, Boca Raton

Loh JP, Kiew R, Set O et al (2000) Amplified fragment length polymorphism fingerprinting of 16 banana cultivars (Musa cvs.). Mol Phylogenet Evol 17(3):360–366

Mandal AB, Maiti A, Chowdhury B et al (2001) Isoenzyme markers in varietal identification of banana. Vitro Cell Dev Biol-Plant 37(5):599–604. doi:10.1007/s11627-001-0105-z

Manzo-Sánchez G (2008) Construction of a genetic linkage map of the fungal pathogen of banana Mycosphaerella fijiensis, causal agent of black leaf streak disease. Curr Genet 53(5):299–311

Megia R, Caecelia Y, Sulistyaningsih N et al (2001) Isozyme polymorphisms for cultivar identification in Indonesia bananas Hayati 8:81–85

Miller RN, Passos MA, Menezes NN et al (2010) Characterization of novel microsatellite markers in Musa acuminata subsp. burmannicoides, var. Calcutta 4. BMC Res Notes 3:148. doi:1756-0500-3-148 [pii]10.1186/1756-0500-3-148

Miri SM, Mousavi A, Naghavi MR et al (2009) Analysis of induced mutants of salinity resistant banana (Musa acuminata cv. Dwarf Cavendish) using morphological and molecular markers. Iran J Biotechnol 7(2):86–92

Mlalazi B, Welsch R, Namanya P et al (2012) Isolation and functional characterisation of banana phytoene synthase genes as potential cisgenes. Planta 236(5):1585–1598. doi:10.1007/s00425-012-1717-8

Mohandas S, Sowmya H, Saxena A et al (2013) Transgenic banana cv. Rasthali (AAB, Silk gp) harboring Ace-AMP1 gene imparts enhanced resistance to Fusarium oxysporum f. sp. cubense race 1. Sci Hortic 164:392–399

Nair AS, Teo CH, Schwarzacher T et al (2005) Genome classification of banana cultivars from South India using IRAP markers. Euphytica 144(3):285–290

Miller RN, Bertioli DJ, Baurens FC, Santos CM, Alves PC et al (2008) n-TIR NBS-LRR resistance gene analogs in Musa acuminata Colla: isolation, RFLP marker development, and physical mapping. BMC Plant Biol 8(1):15

Nwakanma D, Pillay M, Okoli B et al (2003) PCR-RFLP of the ribosomal DNA internal transcribed spacers (ITS) provides markers for the A and B genomes in Musa L. Theor Appl Genet 108(1):154–159

Opara UL, Jacobson D, Al-Saady NA (2010) Analysis of genetic diversity in banana cultivars (Musa cvs.) from the South of Oman using AFLP markers and classification by phylogenetic, hierarchical clustering and principal component analyses. J Zhejiang Univ Sci B 11(5):332–341. doi:10.1631/jzus.B0900310

Oselebe H, Tenkouano A, Pillay M, Obi I, Uguru M (2006) Ploidy and genome segregation in Musa breeding populations assessed by flow cytometry and randomly amplified polymorphic DNA markers. J Am Soc Hortic Sci 131:780–786

Pachuau L, Atom A, Thangjam R (2014) Genome classification of Musa cultivars from Northeast India as revealed by ITS and IRAP markers. Appl Biochem Biotechnol 172(8):3939–3948. doi:10.1007/s12010-014-0827-0

Padmesh P, Mukunthakumar S, Vineesh P et al (2012) Exploring wild genetic resources of Musa acuminata Colla distributed in the humid forests of southern Western Ghats of peninsular India using ISSR markers. Plant Cell Rep 31(9):1591–1601

Passos MA, de Oliveira Cruz V, Emediato FL et al (2012) Development of expressed sequence tag and expressed sequence tag-simple sequence repeat marker resources for Musa acuminata. AoB Plants 2012:pls030. doi:10.1093/aobpla/pls030pls030 [pii]

Passos MA, de Cruz VO, Emediato FL et al (2013) Analysis of the leaf transcriptome of Musa acuminata during interaction with Mycosphaerella musicola: gene assembly, annotation and marker development. BMC Genom 14:78. doi:1471-2164-14-78 [pii]10.1186/1471-2164-14-78

Pillay M, Nwakanma D, Tenkouano A (2000) Identification of RAPD markers linked to A and B genome sequences in Musa L. Genome 43(5):763–767

Pillay M, Ashokkumar K, James A et al (2012a) Molecular marker techniques in Musa genomic research. In: Genetics, genomics, and breeding of bananas. CRC Press, Boca Raton, pp 70–90

Pillay M, Ashokkumar K, Shunmugam ASK et al (2012b) A case for molecular breeding in Musa. In: Genetics, Genomics, and Breeding of Bananas 281. CRC Press

Pillay M, Ogundiwin E, Tenkouano A, Dolezel J (2006) Ploidy and genome composition of Musa germplasm at the International Institute of Tropical Agriculture (IITA). Afr J Biotechnol 5:1224–1232

Poerba YS, Ahmad F (2010) Genetic variability among 18 cultivars of cooking bananas and plantains by RAPD and ISSR markers. Biodiversitas 11(3):118–123

QianJie W, YongHua Q, HouBin C et al (2010) Establishment of an inter-simple sequence repeats reaction system (ISSR) for banana (Musa AAA). J South China Agric Univ 31(1):13–16

Qin X, Peng H, Long X et al (2011) Preliminary study on ISSR analysis and classification of wild Musa germplasm in Guangxi, China. Acta Hortic 897:259

Ravishankar KV, Vidhya L, Cyriac A, Rekha A, Goel R, Singh NK, Sharma TR (2012) Development of SSR markers based on a survey of genomic sequences and their molecular analysis in banana (*Musa* spp.). J Hortic Sci Biotechnol 87:84–88

Ravishankar KV, Raghavendra KP, Athani V, Rekha A, Sudeepa K, Bhavya D, Srinivas V, Ananad L (2013) Development and characterisation of microsatellite markers for wild banana (*Musa balbisiana*). J Hortic Sci Biotechnol 88(5):605–609

Ravishankar KV, Megha HS, Rekha A, Khadke GN, Veerraju CH (2015) Insights into Musa balbisiana and Musa acuminata species divergence and development of genic microsatellites by transcriptomics approach. Plant Genet 4:78–82

Ray T, Dutta I, Saha P et al (2006) Genetic stability of three economically important micropropagated banana (Musa spp.) cultivars of lower Indo-Gangetic plains, as assessed by RAPD and ISSR markers. Plant Cell Tissue Org Cult 85(1):11–21

Rekha A, Ravishankar K, Anand L et al (2001) Genetic and genomic diversity in banana (Musa spp. and cultivars) based on D2 analysis and RAPD markers. Infomusa 10:29–34

Risterucci AM, Hippolyte I, Perrier X et al (2009) Development and assessment of diversity arrays technology for high-throughput DNA analyses in Musa. Theor Appl Genet 119(6):1093–1103. doi:10.1007/s00122-009-1111-5

Rivera F (1983) Protein and isoenzyme banding patterns among Philippine cooking bananas and their wild parents (Musa species). Paradisiaca 6:7–12

Rogers SO, Bendich AJ (1987) Ribosomal RNA genes in plants: variability in copy number and in the intergenic spacer. Plant Mol Biol 9(5):509–520

Rout G, Senapati S, Aparajita S et al (2009) Studies on genetic identification and genetic fidelity of cultivated banana using ISSR markers. Plant Omics 2.6:250

Saraswathi M, Uma S, Prasanya Selvam K et al (2011) Assessing the robustness of IRAP and RAPD marker systems to study intra-group diversity among Cavendish (AAA) clones of banana. J Hortic Sci Biotechnol 86(1):7

Simmonds NW, Shepherd K (1955) The taxonomy and origins of the cultivated bananas. J Linn Soc Lond Bot 55(359):302–312. doi:10.1111/j.1095-8339.1955.tb00015.x

Simmonds N, Weatherup S (1990) Numerical taxonomy of the wild bananas (Musa). New Phytol 115:567–571

Singh W, Singh N, Handique P et al (2015) Morpho-taxonomical and molecular assessment of Musa genotypes from north-east India by morphological and inter-retrotransposon amplified polymorphism markers. Plant Syst Evol 301(2):563–575. doi:10.1007/s00606-014-1094-9

Spooner DM, Rob van Treuren M, van Treuren R et al (2005) Molecular markers for genebank management. Bioversity Int. No. 10. Bioversity International

Teo C, Tan S, Ho C et al (2005) Genome constitution and classification using retrotransposon-based markers in the orphan crop banana. J Plant Biol 48(1):96–105. doi:10.1007/bf03030568

Till BJ, Jankowicz-Cieslak J, Sági L et al (2010) Discovery of nucleotide polymorphisms in the Musa gene pool by Ecotilling. Theor Appl Genet 121(7):1381–1389

Tripathi L, Mwangi M, Abele S et al (2009) Xanthomonas wilt: a threat to banana production in East and Central Africa. Plant Dis 93(5):440–451

Tsao TT-H (2008) Towards the development of transgenic banana bunchy top virus (BBTV)-resistant banana plants: interference with replication. Phd Thesis. Queens land University of technology, Australia

Ude G, Pillay M, Nwakanma D et al (2002a) Analysis of genetic diversity and sectional relationships in Musa using AFLP markers. Theor Appl Genet 104(8):1239–1245. doi:10.1007/s00122-001-0802-3

Ude G, Pillay M, Nwakanma D et al (2002b) Genetic diversity in Musa acuminata Colla and Musa balbisiana Colla and some of their natural hybrids using AFLP markers. Theor Appl Genet 104(8):1246–1252. doi:10.1007/s00122-002-0914-4

Uma S, Siva SA, Saraswathi MS et al (2006) Variation and intraspecific relationships in Indian Wild Musa balbisiana (BB) population as evidenced by random amplified polymorphic DNA. Genet Resour Crop Evol 53(2):349–355. doi:10.1007/s10722-004-0576-y

Venkatachalam L, Sreedhar R, Bhagyalakshmi N (2007) Genetic analyses of micropropagated and regenerated plantlets of banana as assessed by RAPD and ISSR markers. Vitro Cell Dev Biol-Plant 43(3):267–274

Vilarhinos A (2004) Cartographie genetique et cytogenetique chez le bananier: caracterisation des translocations. Montpellier: Ecole Nationale Supérieure Agronomique

Vishnevetsky J, White TL Jr, Palmateer AJ et al (2011) Improved tolerance toward fungal diseases in transgenic Cavendish banana (Musa spp. AAA group) cv. Grand Nain. Transgenic Res 20(1):61–72

Vroh-Bi I, Anagbogu C, Nnadi S et al (2011) Genomic characterization of natural and somaclonal variations in bananas (Musa spp.). Plant Mol Biol Rep 29(2):440–448

Wang XL, Chiang TY, Roux N et al (2007) Genetic diversity of wild banana (Musa balbisiana Colla) in China as revealed by AFLP markers. Genet Resour Crop Ev 54(5):1125–1132. doi:10.1007/s10722-006-9004-9

Wong C, Kiew R, Loh JP et al (2001) Genetic diversity of the wild banana Musa acuminata Colla in Malaysia as evidenced by AFLP. Ann Bot Lond 88(6):1017–1025

Wong C, Kiew R, Argent G et al (2002) Assessment of the validity of the sections in Musa (musaceae) using AFLP. Ann Bot 90(2):231–238

Wongniam S, Somana J, Swangpol S et al (2010) Genetic diversity and species-specific PCR-based markers from AFLP analyses of Thai bananas. Biochem Syst Ecol 38(3):416–427

Yao J-A, Cai H-J, Shi N-N et al (2012) Genetic relationship among banana (Musa spp.) germplasms revealed by ISSR analysis. Fujian J Agric Sci 1:008

Youssef M, James AC, Rivera-Madrid R et al (2011a) Musa genetic diversity revealed by SRAP and AFLP. Mol Biotechnol 47(3):189–199. doi:10.1007/s12033-010-9328-8

Youssef M, Ku-Cauich R, James A et al (2011b) Genetic analysis of somatic embryogenesis derived plants in banana. Assiut Jout Agric Sci 42:287–300

Zhao J-T, Huang X, Chen Y-P et al (2009) Molecular cloning and characterization of an ortholog of NPR1 gene from Dongguan Dajiao (Musa spp. ABB). Plant Mol Biol Report 27(3):243–249. doi:10.1007/s11105-008-0074-z

Current Status of Banana Genome in the Age of Next Generation Sequencing

4

Megha Hastantram Sampangi-Ramaiah
and Kundapura V. Ravishankar

Abstract

Banana (*Musa* spp.) is the "queen of tropical fruits." It is one of the major staple fruits in many countries. The banana improvement program is challenging due to its complex evolutionary events, human selection, and parthenocarpy. *Musa acuminata* and *Musa balbisiana* are the progenitor species for majority of the modern cultivated bananas. The only way to accelerate the banana breeding program is to understand its genome and employing marker-assisted selection. Recently sequencing of the 523 Mb genome of a *Musa acuminata* – DH-Pahang provided a great fillip to the banana improvement program. Banana genome sequencing revealed the presence of around 36,000 protein coding regions, and transposable elements accounted for more than half of the genome. Earlier attempts of Bacterial artificial chromosome (BAC) sequencing of both these species showed a high degree of collinearity. In this chapter, we have summarized the current status of our understanding of the banana genome with respect to classical linkage mapping approach as well as modern next-generation sequencing approach.

Keywords

Banana genome • Next generation sequencing • Classical mapping • Marker assisted selection (MAS) • Whole genome sequencing • Global Musa Genomics Consortium

4.1 Introduction

Bananas are major staple food and export products in many countries with ~145 million tons of production per year worldwide (FAO 2015). Bananas provide starch, vitamins, and minerals across some of the poorest parts of tropical and

M.H. Sampangi-Ramaiah • K.V. Ravishankar (✉)
Division of Biotechnology, ICAR-Indian Institute of Horticultural Research, Hessaraghatta Lake Post, Bengaluru 560089, India
e-mail: kundapura.ravishankar@gmail.com

S. Mohandas, K.V. Ravishankar (eds.), *Banana: Genomics and Transgenic Approaches for Genetic Improvement*, DOI 10.1007/978-981-10-1585-4_4

subtropical regions, while dessert bananas are a major cash crop in many countries (Heslop-Harrison and Schwarzacher 2007). Back in the first half of the twentieth century, an attempt was made to characterize the nuclear genome of *Musa* at the chromosomal level (Cheesman 1932). This led the researchers to determine the *Musa* chromosome numbers which enabled the researchers to divide the genus into four sections – *Eumusa* ($x=11$), *Rhodochlamys* ($x=11$), *Callimusa* ($x=9$ or 10), and *Australimusa* ($x=10$) (Cheesman 1947). Modern edible bananas are parthenocarpic sterile diploid seed ($2n=2x=22$) and also allopolyploid clones ($3x$, $4x$) derived from crosses between *M. acuminata* and *M. balbisiana*, as well as autopolyploid ($3x$, $4x$) clones of *M. acuminata* (Dolezel et al. 2004). Till date more than a thousand banana landraces have been recognized, and majority of which are derived from the progenitor species *Musa balbisiana* and *Musa acuminata* (Pollefeys et al. 2004; Ravishankar et al. 2015). In the 1940s–1960s, there was a worldwide outbreak of Panama disease which destroyed a principal sweet banana variety "Gros Michel." The "Cavendish" subgroup was found to be resistant to the *Fusarium oxysporum* f. sp. *cubense* (Foc) Race 1; thus, it replaced the "Gros Michel" variety. The cultivated banana varieties are parthenocarpic and vegetatively propagated, and hence, there is a lack of genetic variation in cultivars. But in the intervening years, even the *Fusarium oxysporum* f. sp. *cubense* has evolved, and a new strain, "Tropical Race 4," has emerged to challenge the natural Cavendish resistance (Price 1995; Hwang and Ko 2004). Another major threat to banana would be by *Mycosphaerella fijiensis* which causes black sigatoka disease. Other than the biotic stress, banana is threatened by climate change and will be affected by abiotic stress like temperature and drought. *Musa balbisiana* harbors many traits for abiotic and biotic stress tolerance. Bananas and Plantains are important staple for food security; hence, its improvement and conservation program must be on the high priority. Hence, there is an immediate need to develop alternative resources for the "Cavendish" subgroup and develop molecular resources for *Musa balbisiana* to exploit its resistance characteristics to various stresses.

The technique of flow cytometry was employed to study banana chromosome in the year 1991 (Dolezel 1991), and our understanding of DNA content and ploidy levels in *Musa* species has increased with this technique (Dolezel et al. 1994, 1997). The size of the *Musa* nuclear genome and its karyological stability were confirmed over the time (Lysak et al. 1999). The physical location of middle repetitive sequences in the *Musa* genome (ribosomal RNA genes, retrotransposons, and integrated banana streak badnavirus sequences) was revealed by in situ hybridization experiments (D'Hont et al. 2000; Dolezelova et al. 1998; Balint-Kurti et al. 2000; Harper et al. 1999). Cytogenetics is the preliminary method of choice for studying gross chromosome structure and its changes (Shepherd 1999). However, localization of sequences and its organization is important, and techniques such as Fluorescenece in situ hybridization (FISH) cannot give us sequence-level information, though its resolution is satisfactory (Kahl et al. 2004).

Bananas are mostly sterile and vegetatively propagated; hence, banana breeding is complicated and time consuming. But the breeding process can be accelerated with the help of developing technologies for marker-assisted selection (MAS) (Davey et al. 2011). But the dilemma for a banana researcher would be whether to depend on classical linkage analysis or to adapt to the emerging next-generation sequencing technologies. In this chapter, we examine the current status of banana genome research and discuss its utility for banana improvement.

4.2 Classical Mapping Approach

A molecular map is nothing but the orderly placing of markers with respect to its relative distances between them and assigning them to their linkage groups (LGs) (Jones et al. 1997). Linkage maps are constructed based on distance between points and its recombination frequencies. Linkage map is important as it provides information on the genomic organization of a species (Pillay and Tenkouano 2011). Modern cultivated bananas are typically seedless, and it is a product of two important evolutionary events followed by its domestication, i.e., parthenocarpy and sterility (Simmonds 1962). Hence, the edible bananas are propagated vegetatively, thus increasing the complexity of its breeding to develop new varieties. The first step in any linkage map analysis is to obtain mapping population by crossing contrasting parents.

The first linkage map of diploid *Musa acuminata* was reported early in the year 1993 by Faure and his coworkers, and it was based on a cross between SF265 (AA) × *Banksii* (AA) segregating for parthenocarpy (Faure et al. 1993). It was based on 28 Random Amplification of Polymorphic DNA(RAPD) markers, 58 Restriction Fragment Length Polymorphism (RFLPs), and 4 isozymes. A total of 90 loci were detected, 77 of which were placed in 15 linkage groups while 13 segregated independently (Faure et al. 1993). Segregation distortion of 36 % in all loci was observed, and the researchers explained that it is due to chromosomal structural rearrangement.

The second map was reported in the year 1997 by Noyer et al. (1997) where the map was drawn from a selfed *M. acuminata* diploid "M53" population. The map consisted of more than 300 markers (one third was codominant, RFLPs and simple sequence repeats (SSRs); two thirds were dominant markers, Amplified fragment length polymorphism (AFLPs)) linked in 11 linkage groups. A large number of markers showed high distortion levels, and it could be due to *acuminata* subspecies-specific translocations. The third map for *Musa acuminata* was reported in the year 2004 which was based on 20 RFLPs, 81 AFLPs, and 19 SSRs distributed over 14 LGs and covering 597 cM (Vilarinhos 2004). Out of the 120 markers, 71 of them were skewed ($p<0.05$). The oversized linkage groups comprising distorted markers were supposed to be involved in the structural rearrangements and could have been due to pseudo-linkages (Udall et al. 2005).

In the year 2009, an F1-based linkage map was reported for *Musa acuminata* ssp. *malaccensis*, segregating for *Fusarium oxysporum* f. sp. *cubense* resistance. Here the two parental maps were reported based on AFLP, RAPD, and Sequence-tagged microsatellite site markers (STMS) markers. The two maps consisted of 172 loci in 32 LGs covering 1037.7 cM and 195 loci in 37 LGs covering 1183.3 cM (Kayat et al. 2009).

The fifth *Musa acuminata* map consisted of 489 markers (167 SSRs, 322 DArTs), anchoring 11 LGs covering 1197 cM (Hippolyte et al. 2010). This map was based on F1 progeny (180 individuals) obtained from the cross between "Borneo" and "Pisang Lilin" which are genetically distinct accessions of *M. acuminata*. This is a first saturated map, proposed as a "reference *Musa* map." The researchers also proposed two complete parental maps with interpretations of structural rearrangements localized on the linkage groups. The structural heterozygosity in Pisang Lilin is hypothesized to result from a duplication, which is likely accompanied by an inversion on another chromosome.

The most recent and sixth *Musa acuminata* map consisted of two maternal maps and a combined paternal map with 231, 152, and 361 markers, respectively (DArTs, SSRs, and AS-PCRs) (Mbanjo et al. 2012). The first mapping population (P1; 81 individuals) was created by crossing TmB2 × 6142–1 (susceptible to *Radopholus similis*) (♀) and TmB2 × 8075–7 (resistant) (♂); TmB2 × 6142–1 originated from a cross between East African highland banana Nyamwihogora (AAA) and the wild banana Long Tavoy (AA) (susceptible to *Radopholus similis*), whereas TmB2 × 8075–7 was selected from a cross between the hybrid SH-3362 and the *M. acuminata* ssp. *burmannica* Calcutta 4 (resistant to the nematode) (Dochez 2004; Dochez et al. 2009) (Table 4.1).

Till date there has been no report on *Musa balbisiana*linkage map, which is an urgent need, as it harbors many abiotic and biotic stress resistance characteristics for future banana improvement programs (Simmonds 1962; Thomas and Turner 2001; Robinson and Sauco 2010; D'Ocan et al. 2008).

Table 4.1 Molecular marker-based linkage map developed using segregating mapping population in banana

Genotypes used for obtaining mapping population	Population size	F_1/F_2	Total number of markers used	Number of linkage groups	Type of markers	Genome coverage (cM)	Year of publication
SF265(AA) × *Banksii* (AA)	92	F_2	90	15	RFLP, RAPD, isozyme	606	Faure et al. (1993)
Selfed M53	89	F_1	>300	18	RFLPs, SSRs, and AFLPs	1200	Noyer et al. (1997)
Musa acuminata "Calcutta 4" × *Musa acuminata* "Madang"	n/a	F2	120	14	RFLPs, AFLPs, and SSRs	597	Vilarinhos (2004)
Musa acuminata "Borneo" × *Musa acuminata* "Pisang Lilin"	180	F1	489	11	SSRs and DArTs	1197	Hippolyte et al. (2010)
TmB2x 6142-1 × TmB2x 8075-7	81	F1	744	15 -♀map 16 -♂map	DArTs, SSRs, and AS-PCRs	1004	Mbanjo et al. (2012)

4.3 Next-Generation Sequencing Approach

Before the emergence of next-generation sequencers (NGS), *Musa* genomic nuclear DNA (1.8 Mb) from 13 Bacterial artificial chromosome (BAC) BAC clones showing the microsynteny between *Musa*, *Arabidopsis*, and rice was reported. The 443 predicted genes revealed that Zingiberales genes share GC content and distribution characteristics with eudicot and Poaceae genomes. The comparative study of agronomically important genomic regions of rice and banana revealed a very high level of collinearity between the two genomes. Many evolutionary events such as insertions and deletions have occurred ~4.6 Mya which have led to the divergence of *Musa balbisiana* and *Musa acuminata* (Lescot et al. 2008). This work also estimates that the *Musa* sp. has undergone polyploidization ~60 Mya. Amalgam of these two genomic regions can give rise to new *Musa* sp. with novel phenotypes exhibiting biotic and abiotic stress resistance. Comparison with rice revealed microsynteny regions that have persisted since the divergence of the commelinid orders Poales and Zingiberales at least 117 Mya (Janssen and Bremer 2004). The Sanger sequencing was introduced by Fred Sanger in the year 1977 and is considered as the first-generation sequencing in the present "Genomic Era." Till date Sanger sequencing produces longer and less error-prone reads, but it cannot be applied to genome-scale sequencing, since it is costly and time consuming.

This led to the emergence of second/next-generation sequencers (SGS/NGS). Next-generation sequencing is nothing but non-Sanger, based on high-throughput sequencing, yielding billions of DNA/RNA strands in parallel. The millions of short reads generated by NGS sequencers can be de novo assembled, or if the reference genome is available, it can be a reference assembled into longer contigs. Recently it has been widely used for whole genome sequencing (WGS), whole genome re-sequencing (WGRS), and transcriptome analysis in banana.

Till the year 2012, banana researchers had very little genomic information on banana until D'Hont et al. reported the draft sequence of 523 Mb genome of a doubled-haploid *Musa acuminata* – DH-Pahang (D'Hont et al. 2012). This provided a crucial stepping stone in the banana improvement program. The 523 Mb *Musa*WGS data consisted of 27.5 million Roche/454 single reads and 2.1 million Sanger reads representing 20.5X coverage, and finally 50X Illumina data

Table 4.2 Raw sequencing data overview of *Musa acuminata* – DH Pahang

	Number of reads	Number of bases	Coverage	Insert size (bp)
Sanger	2,049,457	1,537,092,750	2.94	10,000
Sanger (BAC ends)	90,542	67,839,000	0.13	110,000
Single Roche/454	27,495,411	8,952,303,336	17.12	NA
Illumina	553,276,222	26,557,258,656	50.78	500

D'Hont et al. (2012), http://banana-genome.cirad.fr/musa_acuminata

was used to correct the final assembly of length 472.2 Mb (Table 4.2). The currently available *Musa acuminata* reference genome represents 90 % of the original DH Pahang genome. They identified 37 Micro RNA (*MIR*) families and 36,542 protein-coding gene models in the *Musa* genome. The transposable elements accounted for almost half of the *Musa* sequence.

In the year 2013, first draft genome was reported for *Musa balbisiana* diploid variety "Pisang Klutuk Wulung" (PKW) (B-genome) (Davey et al. 2013). The researchers employed Illumina HiSeq 2000 II technology to generate 281 million, 100 bp paired-end Illumina reads, and they assembled the generated reads using the available reference genome (A-genome). The expected PKW genome was ~440 Mb (Cizkova et al. 2013), and currently 78 % of the PKW genome has been sequenced (341.4 Mb). 39,914 unique gene models have been identified on the consensus genome. Further functional annotation of which identified 3,276 Transposable elements (TEs), thus leaving a total of 36,638 protein-coding genes which were identical to the "DH Pahang." Variant analysis revealed a total of 20,657,389 Single nucleotide polymorphism (SNPs), of which 8,738,760 were homozygous and the remaining 10,130,236 heterozygous SNPs, thus representing the degree of heterozygosity in PKW. The researchers also detected 4,880,516 SNPs in the coding regions of the genome. Further by microsatellite mining, 30,559 SSRs were identified which corresponded to a frequency of 5.7 SSRs/kb genome. Di-repeats dominated all the other types of SSRs (http://banana-genome.cirad.fr/content/musa-balbisiana-pisang-klutuk--). Most recent update on the banana genome is the release of the *Musa acuminata* genome v 2 wherein the paired-end sequencing and genotyping by sequencing (GBS) of a segregating population were utilized to detect and correct scaffold misassemblies. The *Musa acuminata* genome v 2 was improvised compared to version 1 as there was reduction in the total *Musa* scaffolds from 7,513 to 1,532, with an N50 3.0 Mb (26 scaffolds) which was 1.3 Mb previously. Unknown sites (N) were reduced from 17.3 % to 10.0 %. Previously the assembly anchoring was 70 % which was increased to 89.5 % in the latest version (Martin et al. 2016). A comparative study of transcriptome from *M. acuminata* and *M. balbisiana*, the comparison of synonymous base substitutions (Ks) and non-synonymous base substitutions (Ka), was calculated using 77 randomly selected transcripts without a stop codon. Around 29 genes which contributed mainly to the divergence of these two *Musa* species were identified. These genes are mostly involved in stress tolerance and fruit quality. The effective number of codons (ENC)and GC3 content in *M. balbisiana* was 57.34 and 0.034 and that of *M. acuminata* was 48.72 and 0.039, respectively. The patterns of codon usage in these genes, studied as the effective number of codons (Nc), showed differences, whereas the GC content in synonymous sites (GC3) was similar in both *M. balbisiana* and *M. acuminata*. Thus the codons were more randomly used in case of *M. balbisiana* compared to *M. acuminata*. For GC content, both *M. balbisiana* and *M. acuminata* showed an intermediate distribution, between unimodal and bimodal types of distribution (Ravishankar et al. 2015).

Other than the WGS of banana, there have been numerous RNA-Seq studies reported to understand the different genes and its role in various stress adaptation like low-temperature stress, *Fusarium oxysporum* resistance, and salt stress

(Molina et al. 2012; Yang et al. 2015; Li et al. 2012, 2013; Bai et al. 2013; Wang et al. 2015; Lee et al. 2015). These studies shed light on understanding the resistance mechanism adapted by banana. This information can be utilized by banana breeders, for mapping the resistant genes and further application in MAS. Though next-generation sequencing does give us a huge amount of data, however, it is not an independent domain.

4.4 Efficient Approach: Old Roots with New Leaves

The success of WGS of DH-Pahang is because the researchers utilized both the advantages of classical linkage mapping and high-throughput technologies. The researchers anchored the scaffolds generated from NGS technology to Pahang linkage groups using 652 markers (SSR and DArT), hence representing an efficient method to accurately assemble the banana genome (D'Hont et al. 2012) (Fig. 4.1). The latest release of *Musa acuminata* genome v 2 also improvised the previous release since it utilized the NGS approach of genotype by sequencing technology of the segregating population, and further the data was used to correct the misassembled scaffolds, thus reducing its number, further increasing the N50 values and decreasing the unknown "N" data in the assembly (Martin et al. 2016).

Whereas the draft genome available for *Musa balbisiana* is obtained by reference assembly to the *Musa acuminata* genome since there is no linkage map available till date to arrange the scaffolds in accurate order (Davey et al. 2013), thus the assembly lacks the information on structural rearrangement differences seen among the A-genome and the B-genome. Hence, there is an urgent need to integrate the old technology of linkage map analysis with the modern NGS technology to develop accurate *Musa balbisiana* genomic resources. One such approach would be developing genic markers using NGS technology and in turn utilizing it to construct a linkage map, further utilizing the developed linkage map for final assembly of the whole genome sequence. The recent report on development of genic SSR markers using transcriptomic approach was reported in both *Musa balbisiana* and *Musa acuminata*. The large number of gene-based SSR markers developed can facilitate molecular marker breeding strategies, development of a genetic linkage map, and QTL analysis (Ravishankar et al. 2015).

To bring all the expertise under one roof, a consortium was created. "The Global Musa Genomics Consortium" (GMGC) is an international network of scientists who are working toward improving banana breeding and management by applying genomics tools. The consortium currently brings together expertise from various institutions in different countries (http://musagenomics.org/). It keeps the banana researchers updated on the activities such as genetic diversity, genome sequencing, genetic mapping, proteomics, bioinformatics, and gene expression and also on all the publications.

It also offers genome tools having all the banana databases and bioinformatics tools. The consortium also consists of genomic resources, comprising of genomic DNA, BAC clones, and cDNA/gDNA. It is a product of the Generation Challenge Programme (GCP) consisting of 52 *Musa* accessions taken from the CIRAD collection located in the French West Indies. Upon request, these DNA samples are distributed to the banana research community via the Musa Genome Resource Centre (MGRC) based at IEB, Olomouc, the Czech Republic, upon request.

The GMGC also provides a compilation of genetic and genomic data (e.g., BAC, EST, and markers) generated till date and have made it easily accessible for banana researchers. Other than the database, GMGC also offers many web-based softwares like TropGeneDB, Cmap, GBrowse, GnpAnnot, and ESTtik for easy access of the available data. It also offers pipeline for comparative genomics via phylogenomic analysis through GreenPhyl tool.

4.5 Conclusion

For a successful banana genome resource development and crop improvement, we need to integrate the old strategies into new technologies.

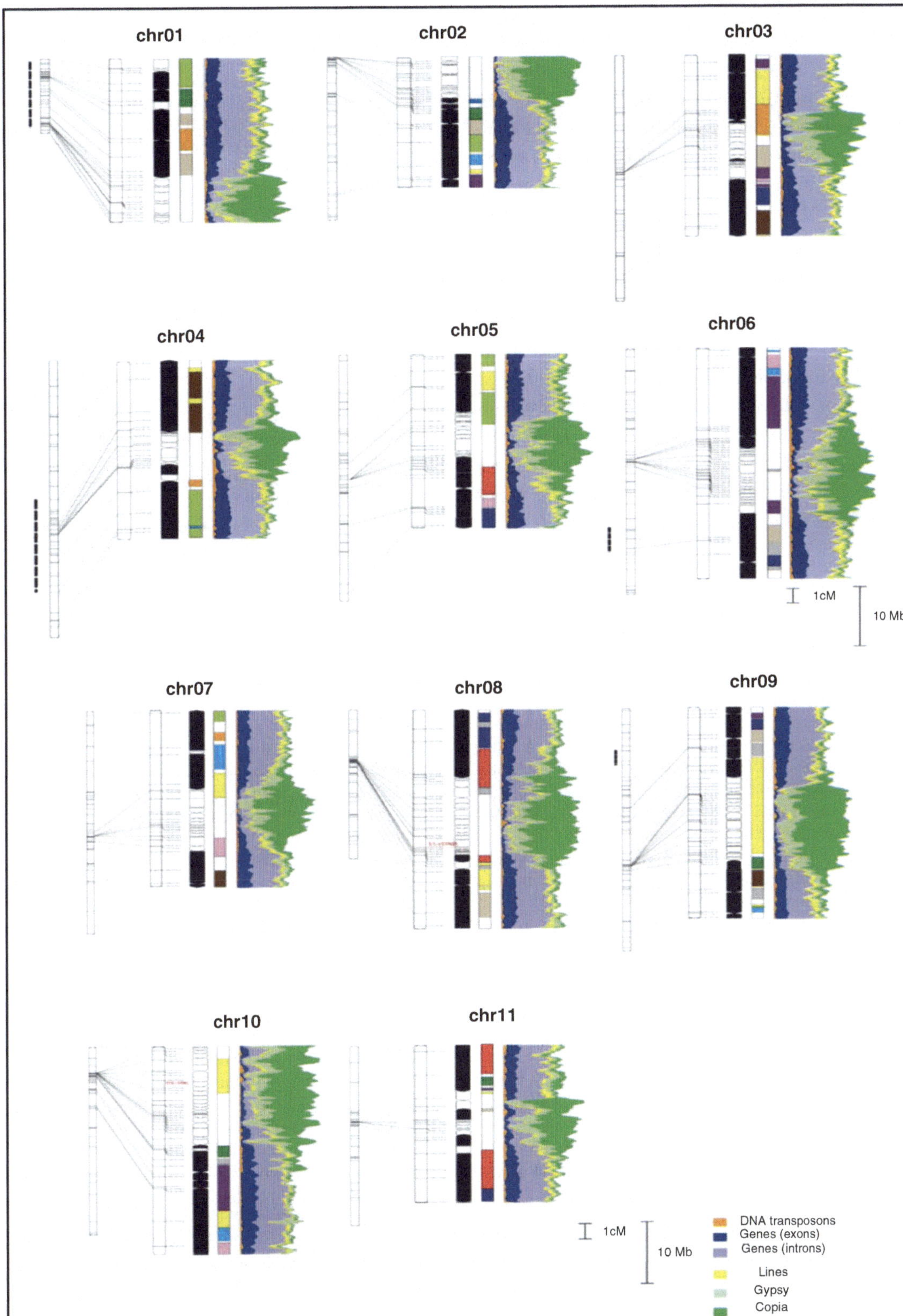

Fig. 4.1 Graphical representation of DH Pahang scaffolds with reference to its genetic map (D'Hont et al. 2012)

There is still scope of improvement in the *Musa acuminata* DH Pahang genome since the assembly covers 90 % of the genome. Further utilization of genomic information for identifying QTLs has to be carried out which will accelerate the banana breeding program. There is an urgent need for a high-density linkage map to be developed for *Musa balbisiana* as it harbors many biotic and abiotic stress-tolerant characteristics. Finally the generated genomic information has to be successfully utilized by the banana breeders to develop alternative resources for the "Cavendish" group and integration of B-genomic regions into cultivars for various stress resistance, nutrition, and vigor characteristics in the new banana varieties.

References

Bai TT, Xie WB et al (2013) Transcriptome and expression profile analysis of highly resistant and susceptible banana roots challenged with Fusarium oxysporum f. sp. cubense tropical race 4. PLoS ONE 8(9):e73945

Balint-Kurti PJ et al (2000) Identification and chromosomal localization of the monkey retrotransposon in Musa sp. Mol Gen Genet 263:908–915

Cheesman EE (1932) Genetical and cytological studies in Musa. I. Certain hybrids of the Gros Michel banana; II. Hybrids of the Mysore banana. J Genet 26:219–312

Cheesman EE (1947) The classification of the bananas. Kew Bull 2:97–117

Cizkova J, Hribova E et al (2013) Molecular analysis and genomic organization of major DNA satellites in banana (Musa spp.). PLoS ONE 8:e54808

D'Hont A et al (2000) The interspecific genome structure of cultivated banana, Musa spp., revealed by genomic DNA in situ hybridisation. Theor Appl Genet 100:177–183

D'Hont A, Denoeud F et al (2012) The banana (*Musa acuminata*) genome and the evolution of monocotyledonous plants. Nature 488(7410):213–217

D'Ocan, Mukasa H et al (2008) Effects of banana weevil damage on plant growth and yield of East African Musa genotypes. J Appl Biol Sci 9:407–415

Davey MW, Van Den BI, Roux N (2011) Vitamin A biofortification in Musa: status, bottlenecks and prospects. Acta Horticult 897:169–178

Davey MW, Gudimella R et al (2013) draft *Musa balbisiana* genome sequence for molecular genetics in polyploid, inter-and intra-specific Musa hybrids. BMC Genomics 14:683

Dochez C (2004) Breeding for resistance to Radopholus similis in East African highland bananas (*Musa spp.*). Dissertation Katholieke Universiteit, Leuven

Dochez C, Tenkouano A et al (2009) Host plant resistance to Radopholus similis in a diploid banana hybrid population. Nematology 11:329–335

Dolezel J (1991) Flow cytometric analysis of nuclear DNA contents in higher plants. Phytochem Anal 2:143–154

Dolezel J et al (1994) Flow cytometric estimation of nuclear DNA amount in diploid bananas (*Musa acuminata* and M. balbisiana). Biol Plant 36:351–357

Dolezel J et al (1997) Use of flow cytometry for rapid ploidy determination in Musa species. Infomusa 6:6–9

Dolezel J, Jain S et al (2004) Cytogenetic and cytometric analysis of nuclear genome in Musa. Banana improvement: cellular, molecular biology, and induced mutations. In: Proceedings of a meeting held in Leuven, Belgium, 24–28 September 2001, 245–249

Dolezelova M et al (1998) Physical mapping of the 18S–25S and 5S ribosomal RNA genes in diploid bananas. Biol Plant 41:497–505

Faure S, Noyer J et al (1993) A molecular marker-based linkage map of diploid bananas (*Musa acuminata*). Theor Appl Genet 87:517–526

Food and Agriculture Organization of the United Nations (FAO) (2015) http://faostat.fao.org

Harper G et al (1999) Integration of banana streak badnavirus into the Musa genome: molecular and cytogenetic evidence. Virology 255:207–213

Heslop-Harrison JS, Schwarzacher T (2007) Domestication, genomics and the future for banana. Ann Bot 100:1073–1084

Hippolyte I, Bakry F, Seguin M et al (2010) A saturated SSR/DArT linkage map of *Musa acuminata* addressing genome rearrangements among bananas. BMC Plant Biol 10:65

Hwang SC, Ko WH (2004) Cavendish banana cultivars resistant to fusarium wilt acquired through somaclonal variation in Taiwan. Plant Dis 88(6):580–588

Janssen T, Bremer K (2004) The age of major monocot groups inferred from 800 + rbcL sequences. Bot J Linn Soc 146:385–398

Jones CT, Morrice DR et al (1997) Gene homologs on human chromosome 15q21–q26 and a chicken microchromosome identify a new conserved segment. Mamm Genome 8:436–440

Kahl G, Jain Set al (2004) The banana genome in focus: a technical perspective. Banana improvement: cellular, molecular biology, and induced mutations. In: Proceedings of a meeting held in Leuven, Belgium, 24–28 September 2001, pp 263–270

Kayat F, Bonar N et al (2009) Development of a genetic linkage map for genes associated with resistance and susceptibility to Fusarium oxysporum f. sp. cubense from an F1 hybrid population of *Musa acuminata* ssp. Malaccensis. Acta Hortic 828:333–340

Lee WS, Gudimella R et al (2015) Transcripts and MicroRNAs responding to salt stress in *Musa acuminata* Colla (AAA Group) cv. Berangan roots. PLoS ONE 10(5):e0127526. doi:10.1371/journal.pone.0127526

Lescot M, Piffanelli P et al (2008) Insights into the Musa genome: syntenic relationships to rice and between Musa species. BMC Genomics 9:58

Li C, Deng G et al (2012) Transcriptome profiling of resistant and susceptible Cavendish banana roots following inoculation with Fusarium oxysporum f. sp. cubense tropical race 4. BMC Genomics 13:374

Li C, Shao J, Wang Y et al (2013) Analysis of banana transcriptome and global gene expression profiles in banana roots in response to infection by race 1 and tropical race 4 of Fusarium oxysporum f. sp. Cubense. BMC Genomics 14:851

Lysak MA et al (1999) Flow cytometric analysis of nuclear DNA content in Musa. Theor Appl Genet 98:1344–1350

Martin G, Baurens FC et al (2016) Improvement of the banana "Musa acuminata" reference sequence using NGS data and semi-automated bioinformatics methods. BMC Genomics 17:243

Mbanjo EGN, Tchoumbougnang F et al (2012) Molecular marker-based genetic linkage map of a diploid banana population (*Musa acuminata* Colla). Euphytica 188:369–386

Molina AB, Williams RC et al (2012) Mitigating the threat of banana Fusarium wilt: understanding the agroecological distribution of pathogenic forms and developing disease management strategies, Volume report number FR 2010–12. Australia: ACIAR, GPO Box 1571, Canberra ACT 2601. ISBN 978 1 921738 17 3

Noyer JL, Dambier D et al (1997) The saturated map of diploid Banana (*Musa acuminata*). Plant and Animal Genome: January 12–16 1997, San Diego 1997. Abstract P335

Pillay M, Tenkouano A (2011) Banana breeding: progress and challenges. CRC Press, Boca Raton

Pollefeys P, Sharrock S, Arnaud E (2004) Preliminary analysis of the literature on the distribution of wild Musa species using MGIS and DIVA-GIS. INIBAP, Montpellier. http://bananas.bioversityinternational.org/files/files/pdf/publications/wildspecies_pollefeys.pdf. Accessed July 2007

Price NS (1995) The origin and development of banana and plantain cultivation. In: Bananas and plantains. Chapman and Hall, London

Ravishankar KV, Megha HS, Rekha A, Khadke GN, Veerraju CH (2015) Insights into Musa balbisiana and Musa acuminata species divergence and development of genic microsatellites by transcriptomics approach. Plant Genet 4:78–82

Robinson JC, Sauco VG (2010) Bananas and plantains. CABI Publishing, Wallingford

Shepherd K (1999) Cytogenetics of the genus Musa. INIBAP, Montpellier

Simmonds NW (1962) The evolution of the bananas. Longmans, London

The banana genome hub. http://banana-genome.cirad.fr/content/musa-balbisiana-pisang-klutuk--

The Global Musa consortium. http://musagenomics.org/

Thomas DS, Turner DW (2001) Banana (Musa sp.) leaf gas exchange and chlorophyll fluorescence in response to soil drought, shading and lamina folding. Sci Hortic 90:93–108

Udall JA, Quijada PA et al (2005) Detection of chromosomal rearrangements derived from homeologous recombination in four mapping populations of Brassica napus L. Genetics 169(2):967–979. doi:10.1534/genetics.104.033209

Vilarinhos AD (2004) Cartographie ge'ne'tique et cytoge'ne'tique chez le bananier: caracte'risation des translocations. Dissertation, Ecole Nationale Supe'rieure Agronomique Montpelier, Montpellier

Wang Z, Jia C et al (2015) Activation of salicylic acid metabolism and signal transduction can enhance resistance to Fusarium wilt in banana (*Musa acuminata* L. AAA group, cv. Cavendish). Funct Integr Genom 15(1):47–62

Yang QS, Gao J et al (2015) Comparative transcriptomics analysis reveals difference of key gene expression between banana and plantain in response to cold stress. BMC Genomics 16(1):446

Genomics of Biotic Stress Tolerance in Banana

5

Pavitra Kotari, V. Swarupa, and Kundapura V. Ravishankar

Abstract

Biotic stresses are a severe threat to crop yield. A few pathogens and insects have succeeded in affecting the plants severely and ultimately leading to the vast economic loss. Various control measures such as the application of pesticides, insecticides and chemical treatments and introgression of R genes and quantitative trait loci (QTLs) into cultivars have contributed to some extent to combat the loss occurring due to biotic stress. However, over the years, there is an emergence of new pathotypes/strains/races of pathogens and insects. Hence, there is a continuous search and need to identify genomic regions and new genes that can contribute towards resistance against these pathogens. Recently, the availability of genomic resources for a wide range of crop species has helped us to improve our knowledge towards understanding the concepts on insect-plant or pathogen-plant interactions. In this chapter, we discuss the usage of genomic tools for major biotic stresses employed for banana and highlight various aspects that are contributing towards an understanding of biotic stress tolerance.

Keywords

Banana disease • Biotic stresses • Genomics • Plant-pathogen interactions • Pest

5.1 Introduction to Biotic Stress

In the natural environment, plants get exposed to various types of stresses simultaneously rather than a particular stress at a time. Interaction effects of multiple stresses are more severe to plants. Biotic stress is a challenge on plants through the damage by pathogen or herbivore

P. Kotari • V. Swarupa
ICAR-Indian Institute of Horticultural Research, Bengaluru 560089, India

K.V. Ravishankar (✉)
Division of Biotechnology,
ICAR-Indian Institute of Horticultural Research, Hessaraghatta Lake Post, Bengaluru 560089, India
e-mail: kundapura.ravishankar@gmail.com

S. Mohandas, K.V. Ravishankar (eds.), *Banana: Genomics and Transgenic Approaches for Genetic Improvement*, DOI 10.1007/978-981-10-1585-4_5

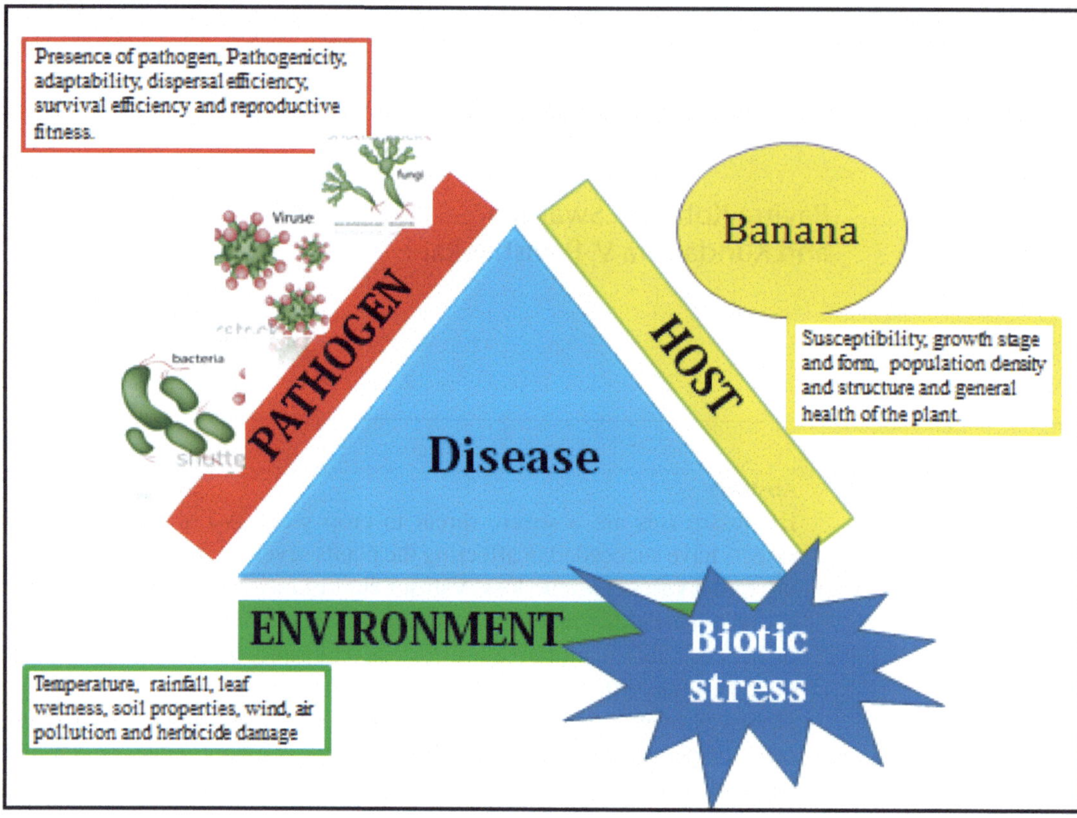

Fig. 5.1 Plant disease triangle. This figure depicts interactions of the three main components essential for any disease to develop. It describes about the conditions required by the three factors for a disease incidence to occur

attack (Rejeb et al. 2014). These extreme environmental conditions (stress factors) bring about functional changes in plants resulting in inhibited growth, reduced bioproduction, physiological acclimatisation and adaptation of species (Ots et al. 2011) (Fig. 5.1). The interaction between the host and the pathogen will also lead to changes in the metabolism of the host plant, such as morphological modifications and eventually death of the infected tissues. Therefore, unravelling the mechanisms involved in disease resistance is important to understand the interactions between plants and pathogens and later to develop strategies to control (Sergeant and Renaut 2010).

The major constraints on present and future production of bananas and plantains are diseases and pests. A range of diseases (black Sigatoka (BS), *Fusarium* wilt, bunchy top virus) and pests (burrowing nematode, weevil borer, etc.) attack banana. Bananas are considered to be one of the top most important crops that are found to be least researched among all the other major food crops. This is mainly because of its long life cycle, sterility and parthenocarpy. These factors hindered banana breeding and made it difficult to tackle virulent diseases and pests. Few pests and diseases are highly destructive and easily spread making it impossible to eradicate. The occurrence and degree of severity of pest outbreaks and plant damage depend upon several factors: environmental conditions, banana varieties(genotype) and disease or pest. Outbreaks of banana disease are more commonly observed where banana plants are grown in a large areas rather than planted in small numbers and spatially separated.

Moist environments favour fungal and bacterial diseases, while dry weather favours many

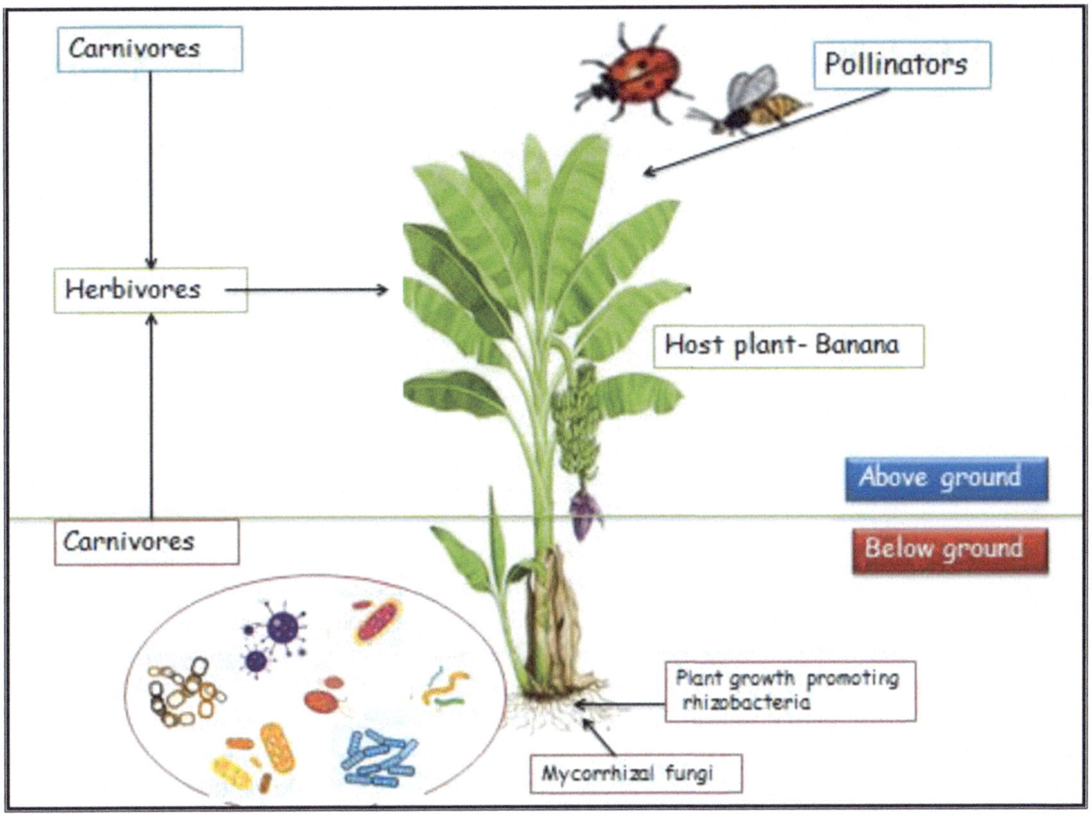

Fig. 5.2 Plant-pathogen interaction: Plants exist in a world filled with bacteria, fungi, nematodes and possibly parasitic plants. They may be infected with viruses during feeding by insects or by other vectors (insects/pest, water, air, humans, etc.). Plant pathogens have made many adaptive mechanisms to enable themselves to invade plants and overcome plant defence mechanisms and colonise plant tissues for growth, survival and reproduction. Pathogens accomplish these activities mostly through secretions of chemical substances that affect certain components or metabolic mechanisms of their hosts. Penetration and invasion are the two activities aided by the mechanical force exerted by certain pathogens on the cell walls of the plant

types of insect outbreaks (e.g. mites). Sometimes, normal weather conditions also affect banana production such as banana virus diseases and scab moths. Based on banana varieties, disease infection also varies: *Fusarium* wilt destroying many commercial plantations and banana bunchy top virus (BBTV) affecting the cavendish subgroups. Based on specific disease and pests, several parasites of *Musa* species have been classified that affect wide host ranges and cause significant damage to *Musa* spp. which target ants and some sap-feeding insects (e.g. aphids). Therefore, they may be a concern for certain vegetables intercropping with *Musa* spp. (Fig. 5.2).

5.1.1 Introduction to Genomics to Understand the Mechanism and Tolerance

The advancement of efficient and upcoming techniques and methodologies for cloning and sequencing has lead to the development of new platforms to reveal about the overgrowing genomic data (Moore et al. 2002).

In short 'genomics' can be described as an approach to study the genome of an organism – the DNA and the organelles and the genes. The study of genomics will give us an understanding of the genome organisation and interaction

Table 5.1 List of plant pathogens and respective disease-causing organisms with the genome size

Sl. no.	Pathogen type	Banana disease	Organism	Genome size	Reference
1	Bacterial	Moko	*Ralstonia solanacearum* (phylotype II)	5.43 Mb	Xu et al. (2011)
2	Fungal	Sigatoka	*Mycosphaerella fijiensis*	74.1 Mb	Kema (2013)
		Panama disease (*Fusarium* wilt)	*Fusarium oxysporum f. sp. cubense* (race 1)	47.84 Mb	Guo et al. (2014)
		Panama disease (*Fusarium* wilt)	*Fusarium oxysporum f. sp. cubense*(race 4)	53.12 Mb	Guo et al. (2014)
3	Viral	Bunchy top	Banana bunchy top virus	Genome components (DNA-R, U3, S, M, C and N) each of approximately 1.1 kbp	Kumar et al. (2011)
		Streak	Banana streak virus	~7.5 kb	Stainton et al. (2015)
4	Nematodes	Root knot	*Meloidogyne incognita*	86 Mb	Bird et al. (2009)

between the different elements within it. The results obtained from genomic study can be used to measure the diversity in different germplasms, to determine the relationships among these accessions and linkage mapping and to determine large-scale genomic or chromosomal rearrangements (Osuji et al. 1997, 1998).

Genomic analysis of plant-associated microorganisms also has become a major part of plant-pathogen interaction because it gives us an idea on the development and suppression of plant diseases. Analyses of microbial genomes will add on to the analyses done on plant genomes by providing new insights into the nature of plant-microbe interactions (Microbial Genomic Sequencing Perspectives of the American Phytopathological Society 2008) (Table 5.1). By genomic studies, we would get a clear picture on the disease susceptibility and resistance between plant hosts and pathogen and interaction with beneficial microorganisms. Sequence data obtained from such studies is essential to improve our knowledge on infection and the interaction of pathogens and host gene products. This will further improve our understanding on how these organisms infect, reproduce and spread.

5.2 Banana Diseases

The major biotic stresses affecting banana are fungi, insects, nematodes, viruses, bacteria and parasitic weeds that drastically decrease banana production (Dita et al. 2006).

Among the various diseases that affect banana, rhizome rot has become a serious problem in the banana cultivation which is caused by *Erwinia* spp., in association with *Fusarium* spp. Under infection, the infected young plants show yellowing of leaves and rotting of rhizome, whereas in older plants it appears as collapse of the infected plants from the middle portion to pseudostem. Several workers have reported the symptoms of the disease as rhizome rot of banana (Tomlinson et al. 1987; Robinson and Mannicom, 1991; Akiew 1998; Thwaites et al. 2000), soft rot of the pseudostem (Chattopadhyay and Mukharjee, 1986; Pereira and Nunes 1988; Guzman and Sandoval 1996), watery rot of the pseudostem (Rivera and Garcia 1981), soft rot of banana (Chattopadhayay 1987), root rot of banana and plantain (Figueroa 1987) and vascular rot and wilt (Gomez-Caicedo et al. 2001). Bacterial wilts (Moko disease), Bugtok disease and blood disease are caused by members of the *Rhizoctonia solanacearum* species complex (Fegan and Prior 2006).

However, the present chapter focuses on the major diseases that threaten banana production, which are:

1. *Bacterial diseases*: bacterial wilt
2. *Fungal diseases*: *Fusarium* wilt, Sigatoka disease, anthracnose and crown rot
3. *Viral diseases*: banana bunchy top virus and cucumber mosaic virus
4. *Insect pests*: leaf and fruit beetle and weevils
5. *Nematodes*: Migratory endoparasites and root lesion nematodes

5.2.1 Bacterial Diseases

Banana bacterial wilt more often called as banana *Xanthomonas* wilt (BXW) is one of the most significant diseases of banana caused by *Xanthomonas campestris* pv. *musacearum* (*Xcm*) (Vurro et al. 2010). Currently it is viewed as the major threat to banana production especially in East and Central Africa. It infects all most all the banana varieties including cooking and dessert types (Ssekiwoko et al. 2006). The disease incidence was first reported on *Ensete*, which is nearly related to banana in Ethiopia (Yirgou and Bradbury 1968) and subsequently on banana (Yirgou and Bradbury 1974). External to Ethiopia, it was first discovered in Uganda in 2001 (Tushemereirwe et al. 2004). The disease symptoms are characterised by a successive yellowing and wilting of leaves, often resembling *Fusarium* wilt disease. However, the secretion of bacterial ooze from the excised tissue is the typical feature of banana bacterial wilt. Fruits tend to show premature and uneven ripening. Blackening and withering of male buds and the occurrence of yellow or brown streaks in the vascular bundles result in the complete wilting and rotting of infected plants.

The first study on the systemicity of *Xcm* was done on the flower-infected plants to determine the efficiency of cutting the diseased stem at soil level was sufficient for the analysis instead of uprooting the entire mat (Ssekiwoko et al. 2006, 2008). In contrast to this study, bacteria were detected in the rhizome of the East African highland banana (EAHB) plants that were at the stage of shrivelled bracts, suggesting that in the case of an infected EAHB cultivar, removing only the infected plant would not be efficient in precluding the disease from entering the rhizome (Nakato et al. 2014). Further, a study conducted in Central Uganda, confirmed that none of the rhizomes of 'Kayinja' and EAHB cultivar had bacteria 21 days after inoculation, but that 7 days later 55 % rhizomes of the EAHB plants and 17 % of the 'Kayinja' ones had bacteria.

Application of genomic tool has been extensively used for screening and studying the genetic nature of the virulent and avirulent *Xanthomonas* strains. Efforts for the development of molecular markers for the detection of *Xcm*, which helps to screen infected banana plants, are being attempted (Aritua et al. 2007; Lewis Ivey et al. 2010; Studholme et al. 2010; Adikini et al. 2011). Wasukira et al. (2012) have used whole-genome sequencing to examine genetic diversity between isolates of *Xcm*. Molecular markers were used to rebuild the phylogenetic relationships between *Xcm* isolates from different geographical locations within the known isolates of the pathogen in Africa. It was observed that the isolates diverged into two major sub-lineages which suggest at least two separate introductions of *Xcm* into the banana-producing regions around Lake Victoria.

However, only limited application of genomics in the case of host banana studies with respect to *Xcm* has been reported. Studies have shown the effective control of BXW by transgenic approach. By utilising banana embryogenic cell suspensions, transgenic bananas carrying ferredoxin-like protein (Pflp) gene were developed and characterised, of which 67 % of the lines were shown to be completely resistant against *X. oryzae* pv. *oryzae* (Namukwaya et al. 2012). Tripathi et al. (2014) have examined 25 banana transgenic lines of the cultivar 'Gonjamanjaya' (AAB) containing rice Xa21 gene and found 12 lines showing resistance against *Xanthomonas* wilt.

5.2.2 Fungal Diseases

Fungi, the most prevailing and destructive pathogens of banana, attack almost all parts of banana plant organs causing both pre- and postharvest production losses.

Presently about 40 fungal pathogens have been identified that cause disease in banana (Jones 1999). The major foliar pathogens causing Sigatoka disease are *Mycosphaerella fijiensis* causing black leaf streak disease, *Mycosphaerella musicola* for Sigatoka leaf spot disease and *Mycosphaerella eumusae* for *eumusae* leaf spot.

The hemibiotrophic fungus *Mycosphaerella fijiensis* is the causal agent of black Sigatoka (BS), the most important devastating foliar disease of banana (*Musa* spp.). By nature it is a haploid, hemibiotrophic ascomycete belonging to class *Dothideomycetes*, order *Capnodiales* and family *Mycosphaerellaceae*. This causes reddish-brown streaks that form a water-soaked border and eventually merge to cause extensive leaf necrosis. The production loss due to this disease can be up to 35–100 % (Gasparotto et al. 2006) and is controlled generally using costly fungicides (Arias et al. 2003). Presently, the genome was of this fungus was sequenced and made available on the Joint Genome Institute website (http://www.jgi.doe.gov/webcite). The genome size is 74.1 Mb long and half of which is estimated to be packed by repetitive element sequences. A study reported that the presence of transposable elements in the genome of *M. fijiensis* has a role in the development of black Sigatoka (Santana et al. 2012). In silico identification of glycosylphosphatidylinositol (GPI) protein family in *M. fijiensis* and analysis of two β-1,3-glucanosyltransferases that play a role in bringing about the pathogenesis were analysed for its expression at different stages of black Sigatoka disease by comparing expression data with black Sigatoka symptoms and fungal biomass inside leaves (Kantun-Moreno et al. 2013).

There is another study which reported on the analysis of compatible interaction of *M. fijiensis* and *Musa* spp. by employing suppression subtractive hybridisation (SSH), for which cDNA library was constructed to identify the transcripts that induced at late stages of infection in the host and the pathogen and also to study the genes that were involved in the synthesis of phenylpropanoids and pathogenesis-related proteins. It was observed and confirmed that antifungal genes encoding PR proteins and GDSL-like lipase were induced 30 days after postinoculation (dpi), indicating that the fungus is actively repressing plant defence. They also reported the presence of single fungal gene that got induced at 37 dpi encoding UDP-glucose pyrophosphorylase (enzyme involved in the biosynthesis of trehalose) (Portal et al. 2011).

Next-generation sequencing (NGS) of the transcriptome of *M. acuminata* to *M. musicola* interaction was carried out to provide a useful information on genes that got expressed during plant immune responses in this pathosystem here. This study could gather information about 36,384 and 35,269 unigene identified for contrasting *M. acuminata* genotypes. Through this study, they could identify a total of 4068 genic simple sequence repeat (SSR) loci in Calcutta 4 and 4095 in cavendish Grande Naine. They also reported 95 defence-related simple sequence repeats that were validated across *M. acuminata* accessions. This data would help to identify candidate genes expressed during infection and to understand the host defence mechanisms (Passos et al. 2013).

Anthracnose caused by *Colletotrichum musae* is considered as the most important and widely distributed diseases of ripe banana fruit (Su et al. 2011). The severity of this disease increases when the banana fruits are wounded during handling and transportation. Banana anthracnose starts as an inactive infection on green fruits in the field. At the later stages, the successful penetration of the fungus is restricted by accumulation of phytoalexins as the fruits ripen (Jegger et al. 1995; Turner 1995; Peeran et al. 2014). Therefore, symptoms generally can be seen only in overripe fruits. Anthracnose becomes a serious problem when bananas are shipped as bunches for a long time and ripened under high temperature (Meredith 1960a). Study reported on isolation and identification of *Colletotrichum* species associated with anthracnose on different banana cultivars using RAPD marker analysis revealed that the RAPD banding patterns of *C. musae* isolates were highly similar but showed intraspecific variations (Zakaria et al. 2009). The management and control of this disease involves the application of fungicides and use of resistant banana cultivars (Abang et al. 2003). Therefore, gaining insights on the genetic diversity of the pathogen population would be useful to have effective programmes for screening banana cultivars for resistance to anthracnose and to avoid development of fungicide resistance.

Fusarium wilt (Panama disease) is the most important disease of banana. This disease is caused by fungus *Fusarium oxysporum* spp. *cubense* (E. F. Smith) (Snyd and Hansen Smith 1940). This fungus has been the cause of one of the most destructive epidemics in history as of 1960; its race 1 had destroyed approx. 40,000 ha of commercial plantation. Until recently, *Foc* tropical race 4 (*Foc* TR4) has been restricted to cavendish-producing Asian countries (Deng et al. 2015). In India and few other countries like Pakistan, Sri Lanka, Bangladesh, etc., only *Foc* race 1 was widely reported. Now there are few reports stating that there is an immediate risk of *Foc* TR4 emergence in these countries (www.promedmail.org; 2015). The fungus makes its entry through the lateral roots and brings about the blockage of the host vascular system resulting in typical wilt symptoms. Initially, there were only few methods for controlling this disease which included physical and chemical measures, but now there are few effective methods being used, which include development of resistant cultivars through molecular breeding and genomics knowledge, to combat this disease (Li et al. 2013a, b). The genome size of *F. oxysporum* was estimated around 18–51 Mb in size (Roncero et al. 2003). In 2007, the genome sequence of Fol was completed at the Broad Institute, where the genome size was estimated to be 59.9 Mb (Michielse et al. 2009). The genome consisted of 15 chromosomes with an estimated 17,735 genes. Pathogenicity studies in *F. oxysporum* initially involved methods such as in planta screening and microscopy of host-pathogen interactions (Gold et al. 2001a; Aboul-Soud et al. 2004). Efforts are being made to have an idea about the mechanisms operating during *Foc* infections (Wang et al. 2012) and further to gain knowledge on the genetics of resistance to *Fusarium oxysporum* f. sp. *cubense* tropical race 4 through transcriptome profiling of banana roots (Li et al. 2012).

The characterisation of tolerance to *Fusarium oxysporum* f. sp. *cubense* infection in banana identified 68 nonredundant gene sequences using suppression subtractive hybridisation and gene expression analysis. These sequences were checked for homology to identify defence-related genes. They observed that the genes like PGIP2, peroxidase and catalase that would help in cell wall strengthening are expressed prominently (Swarupa et al. 2013). A study reported by Guo et al. (2014) gives a clear distinctions about the gene contents and transcriptional regulation between *Foc*1 and *Foc*4 infecting the banana 'Brazil'. As a part of the study, transcriptome analysis done revealed a major difference in transcriptional responses of *Foc*1 and *Foc*4 at 48 h postinoculation in comparison with the vegetative growth stage. It was also observed that virulence-associated genes and signalling pathway genes got upregulated in *Foc*4 than in *Foc*1.

Rossmann et al. (2012) made an attempt to analyse banana-associated microorganisms and select the fungal pathogens responsible for yield losses. For this, they studied the structure and function of microbial communities collected from three different traditional farms by cultivation-independent and cultivation-dependent methods and analysed and observed that the microbial communities were significant across the studied sites and treatments given. It was reported that a high number of indigenous antagonists towards identified fungal pathogens were observed in the microhabitats, and also the bacterial antagonists were enriched in banana plants, while the fungal antagonists were less and mainly found in soil.

Transcriptome studies have shown the involvement of general pathways or genes like ROS-scavenging genes and several genes that are responsive to infection by other pathogenic microorganisms like PR genes, the genes involved in synthesis of phytoalexins and phenylpropanoids (PAL) and cell wall strengthening (Li et al. 2013a, b). Six *I* loci in tomato which are R genes and six dominant resistant loci called RFO in *Arabidopsis* conferring resistance to *Fusarium oxysporum* f. sp. *lycopersici* and *Fusarium oxysporum* f. sp. *matthioli* (*Fom*) infection, respectively ,were identified (Diener and Ausubel 2005). However, several efforts are being done to identify and characterise specific gene/genes that might significantly contribute to banana fungal disease resistance.

5.2.3 Viral Diseases

The causal agent, banana bunchy top virus (BBTV), is an ssDNA virus which may constitute a new group of plant virus (Harding et al. 1993). BBTV is persistently transmitted by the aphid *Pentalonia nigronervosa* Coquerel or by vegetative propagation (Drew et al. 1989; Thomas et al. 1995). This virus is widely dispersed in the Asia-Pacific region.

Banana bunchy top virus (BBTV) disease is one of the most deleterious diseases of banana. It is a major constraint to banana production in India, Australia, South Asia, the South Pacific, Africa and Hawaii (Blomme et al. 2013; Dale et al. 1992; Kumar et al. 2011). Symptoms of bunchy top include chlorosis of leaves, narrowing and dark green streaks in petioles referred as 'Morse code streaking'. Successive infected leaves look abnormal and lead to bunching of leaves at the upper part of the plant establishing typical bunchy top appearance. The virus travels consistently to infect an entire corm portion and proceeds to newly developed shoots derived from the infected corm. Severely infected plants fail to fruit or produce twisted or distorted banana hands. The symptoms are expressed around 25–85 days after primary infection with the virus (Watanabe et al. 2013).

BBTV belongs to the family *Nanoviridae*. The viral genome is 18–20 nm in diameter and is composed of 12 circular single-stranded DNA components (Vetten et al. 2005; Stevens 2010; Qazi 2016). The virus is transmitted by aphids (*Pentalonia nigronervosa*). The only known controlling strategy being applied is identification and destroying them in the field and by the use of pesticides. Currently, biotechnological tools like tissue culture and transgenics are being tried to develop the BBTV-resistant banana plants. In order to identify the presence of virus before planting, especially for indexing virus-free mother plants raised through micropropagation, a PCR-based diagnosis has been standardised. Recently, Mahadev et al. (2013) has applied PCR by designing primers for viral gene-specific primers for detecting BBTV in tissue culture plantlets of *Musa* spp. cultivars 'Virupakshi' and 'Sirumalai' (AAB). Also, multiplex reverse transcription-PCR (mRT-PCR) technique was employed for detecting two viral diseases caused by episomal banana streak Mysore virus and banana bunchy top virus (Selvarajan et al. 2011).

In vitro mutagenesis by gamma irradiation of abaca lines (*Musa textilis* Nee) has helped to produce resistant lines to both BBTV and banana bract mosaic virus (Dizon et al. 2012). Descalsota et al. (2015) have characterised potential gamma-irradiated abaca resistant lines derived from two cultivars, Tinawagan Pula (TP) and Tangongon (TG), using resistance gene analogue (RGA) and simple sequence repeat (SSR) markers. It has been reported that the increase in polymorphism information content (PIC) value in gamma-irradiated plants than control plants, which is indicative of the higher genetic variability, was created by exposure to gamma irradiation.

In order to minimise the soil pollution and health constraints that are led by pesticides, recently bioformulation method is being used efficiently to enhance or induce the host resistance to pathogens. In this effort, Kavino et al. (2008) have found the systemic resistance induction in banana against BBTV by the combined effect of chitin and *Pseudomonas fluorescens* strain CHA0. During the tri-tropic interaction between growth-promoting endophytic bacteria, BBTV and *Pentalonia nigronervosa*, the presence of differential accumulation of pathogenesis-related proteins and other defence-associated enzymes was identified by Harish et al. (2009). Development of disease resistance by transgenic approaches in banana is being employed for various diseases. Using RNAi phenomenon, Shekhawat et al. (2012) and Elayabalan et al. (2013) have generated transgenic resistant banana against BBTV.

5.2.3.1 Banana Streak Disease

Banana streak disease caused by pararetrovirus belongs to *Caulimoviridae* family and genus *Badnavirus* (Ploetz 2015). Currently, four banana-infecting *Badnavirus* (BIB) species have been identified (Stainton et al. 2015). They spread through infected propagules and are transmitted by several mealybug species, including

Planococcus ficus, *Dysmicoccus brevipes* and *Planococcus citri* (Borah et al. 2013). The infection leads to chlorotic streaks in leaves extending the death of the plant. They survive in two forms: episomal form with a double-stranded circular DNA genome (~7.5 kb) or the integrated genome into the host as a pararetrovirus endogenous sequence (EPRVs) form. These EPRVs restructure into new infectious form through homologous recombination leading to introduction of new infective viral genome which has been assigned for the recent outbreaks (Chabannes and Iskra-Caruana 2013). These EPRVs are found in various cultivated genotypes of banana which have been originated from three *Musa* species, *M. acuminata* (A genome), *M. balbisiana* (B genome) and *M. schizocarpa* (S genome) (Geering et al. 2001, 2005). The presence of three open-reading frames on one strand was found in the viral genome.

In banana, micropropagation and genetic hybridisation are known to trigger the production of episomal banana streak virus (BSV) from EPRVs (Dallot et al. 2001). Studies have revealed the two responsible factors that are postulated in BSV expression. The presence of the B genome in the *Musa* genotypes is the first, where the second factor is the presence of BSV-expressed locus (BEL) (Lheureux et al. 2003). The complete structural and functional analysis of two EPRVs of BSV Goldfinger (BSGfV) present in the genome of the diploid *M. balbisiana* cv. Pisang Klutuk Wulung (PKW) was studied by Gayral et al. (2008). They showed the integrants as allelic insertion, and despite the extensive rearrangements, the two EPRVs contain full-length viral genome. Triploid F1 population of *Musa* (AAB) produced by interspecific cross between *Musa balbisiana* (BB) and tetraploid *Musa acuminata* (AAAA) parents was used to study the incidence of banana streak disease by Lheureux et al. (2003). It was identified that half of the progeny carried BSV particles, and the PCR analysis showed that the endogenous sequence was specific *to M. balbisiana* genome confirming B genome as carrier. Phylogenetic study of banana streak virus carried by Gayral and Iskra-Caruana (2009), by comparing episomal and endogenous viral sequences, revealed that there are around 27 independent integration events, suggesting viral integration is a frequent process.

Molecular tools have helped to identify, to classify and to study the structure and function of EPRV sequences present in the banana genome. Possible risks related with EPRVs have been well established. Hence, extensive attempts to eliminate EPRV sequences from B genome of banana and screening and selection of new interspecific hybrids from integrated sequences within A × B progenies with the help of molecular breeding are necessary. This would help to protect banana from BSV and further spread. Transcriptome studies need to be carried out to identify and characterise the specific genes that are modulated by either BBTV or BSV. This would help in designing breeding strategy to improve the banana.

5.3 Insect Pests

Beetles, weevils and nematodes are considered to be the major pests of banana that affect the development of the corm and root system.

5.3.1 Banana Leaf and Fruit Beetle

Leaf and fruit beetles are one of the major pests of banana found in almost all the varieties of banana (Zheng et al. 2013). It is basically observed that the adult feed on the young unfurled leaves of the banana plant entering into the middle portion of leaf feeding on the very young yellow leaf by scrapping. After feeding, black spots are found on the matured leaf. While, in severe infestation the leaf loses its vigour and the growth of the banana plant is retarded. After fruit set, the beetles leave the leaf and start feeding on the smooth surface of young fruits. The fruit becomes larger day by day, and spot scrapping on the surface of fruit becomes prominent and turned blackish brown; ultimately, the matured fruit is spotted and loses its market value (Alam and Mean 2000).

5.3.2 Weevils

In case of weevils, it was observed that the larval stage was the most destructive stage for infection, where the adults lay eggs at the base of plant and bring about the destruction of the corm or rhizome tissue (Abera et al. 2000; Gold et al. 2001b). This infection affects root development and nutrient uptake resulting in weakening of the entire plant (Acland 1971). In addition, weevil damage reduces plant vigour and enhances sucker mortality (Gold et al. 2001b), and also yield loss up to 100 % has also been reported (Sengooba 1986).

Many studies were done to assess and analyse a wide collection of *Musa* germplasm for weevil resistance in Africa and Asia and concluded that plantains and East African highland bananas (EAHB) were the most susceptible banana types (Fogain and Price 1994; Pavis and Lemaire 1997; Kiggundu et al. 1999; Kiggundu et al. 2003b).

Another report on the effect of weevil infestation on root, corm, shoot and bunch characteristics of eight East African *Musa* spp. genotypes suggested that several ratoon cycles are required for a high weevil population density to cause reduction in yield and plant growth. It was also shown that genotypes with 'B' genome are more tolerant to banana weevils than genotypes with only 'A' genomes (Ocan et al. 2008).

5.3.3 Nematodes

Plant-parasitic nematodes, the 'unseen enemies' of plants, are considered to be important and serious pests to banana production. The most serious and damaging nematodes that attack bananas are the migratory endoparasites like *Radopholus similis* (Cobb) Thorne and the root lesion nematodes *Pratylenchus coffeae* (Zimmerman) Filipjev. The degree of damage and yield loss in banana depends on the nematode species involved, the interaction between nematodes and other soil pathogens, the susceptibility of the cultivar and the environment under study (Sarah 1989; Gowen and Queneherve 1990; Davide et al. 1996). Basically the effect of nematodes are seen on anchorage of banana plants, lengthening of vegetative cycle, yield reduction and irreversible reduction of plantation longevity (Blake 1961; Queneherve 1989; Sarah 1989; Gowen and Queneherve 1990).

Radopholus similis or burrowing nematodes specially attacking the cavendish types are more commonly found attacking the plantain and cooking bananas, while the *Pratylenchus* species have a wide host range (Moens and Perry 2009). These nematodes make their entry through any portion of the roots occupying the cortical parenchyma feeding on the cytoplasmic cells, leading to their destruction and resulting in the formation of cavities. There are few other species of nematodes that keep affecting the roots in different ways either taking resources from the plant or burrowing through the roots causing necrosis and root rot.

A study using seven banana genotypes ranging from susceptible to resistant was conducted to evaluate the screening method for resistant bananas to the burrowing nematode (*Radopholus similis*). For this, the total root necrosis of plant roots was compared between greenhouse and field trials. It was observed that nematode-infested roots were more severely affected in the greenhouse than in the field trials. It was also observed that root necrosis was found to be high in the two resistant cultivars tested in the greenhouse trial (Marin et al. 2000).

The study carried out by Srinivasan et al. (2011) revealed that the presence of root lesion nematodes (*Pratylenchus coffeae*) was maximum followed by spiral nematodes (*Helicotylenchus multicinctus*) and root-knot nematode (*Meloidogyne incognita*) in the root samples taken up for the study, and also the study of soil could reveal that the presence of nematodes was higher in clay soil followed by sandy soil and alluvial soil.

A study, as reported at *African Crop Science Journal*, explains the effect of nematode infection and damage on the absorption of major mineral nutrients in bananas commonly grown in East African highlands. It was observed that the leaf nutrient concentrations were low when compared to established critical nutrient levels (CNLs) for both inoculated and uninoculated plants. Nematode inoculation decreased nutrient concentrations in leaves except for magnesium and calcium. The correlation coefficient values showed that percentage of dead roots and root necrosis significantly decreased with increase in leaf potassium and increased with increase in leaf calcium and magnesium (Talwana et al. 2005).

Another study as reported by Backiyarani et al. (2015) states that chitinase isoforms that were involved in resistant mechanisms were identified based on differential gene expression analysis carried out in resistant and susceptible *Musa* genotypes. Class II chitinases are overexpressed sixfolds in resistant genotype than susceptible Class I chitinases. This confirms that the role played by chitinase genes responds differently at different stress conditions.

These studies show that weevils and nematodes also cause major loss to the banana production. Till now only traditional methods are being employed to screen for resistant cultivars. The genes involved and mechanism underlying in resistance using genomics to pests and nematodes are yet to be examined. This would help in developing pest and resistant cultivars through breeding or other approaches.

5.4 Conclusion

It is well known that plant defence mechanism is complex, and the evolution of new strains of pests makes it a very difficult task to study. This arms race between host and the pathogen demands understanding their interaction at molecular level. Recently, by employing the transcriptomic approach, studies have shown the involvement of general pathways or genes like ROS-scavenging genes, cell wall strengthening genes and PR proteins against fungal pathogens *Foc* and *M. fijiensis* in banana.

Recently, the substantial number of genomics and other studies on banana pathogens revealed their vast diversity, evolution of competing virulent strains/races, complex pathogenicity mechanism and genes and distribution of strains to new locations. The advancement and use of next-generation sequencing to reveal the role of genes and genome of various crop species improve our knowledge for breeding and also facilitate identification of genes and their functional role in contributing towards plant immunity. Transgenic approaches and gene silencing such as RNA interference (RNAi) and virus-induced gene silencing (VIGS) have also contributed towards this aspect. In fact, the possible reasons for the slow progress in *Musa* spp. are their complex genetic system, i.e. its heterozygosity, different genomic organisations, polyploidy and long generation time. Hence, banana breeding is considered as a difficult task. However, presently, utilisation of advanced genomic approaches and information from other monocotyledonous crops like rice and wheat give hope for banana.

Acknowledgements The authors acknowledge the funding support through ICAR NPTC project on *Fusarium* wilt and drought tolerance in banana.

References

Abang MM, Winter S, Mignouna HD, Green KR, Asiedu R (2003) Molecular taxonomic, epidemiological and population genetic approaches to understanding yam anthracnose disease. Afr J Biotechnol 2(12):486–496

Abera KAM, Gold CS, Kyamanya S, Karamura EB (2000) Banana weevil Cosmopolites sordidus Germar ovipositional preferences, timing of attack and larval survival survivorship in a mixed cultivar trial in Uganda. Acta Hortic 540:487–496

Aboul-Soud MA, Yun BW, Harrier LA, Loake GJ (2004) Transformation of Fusarium oxysporum by particle bombardment and characterisation of the resulting transformants expressing a GFP transgene. Mycopathologia 158(4):475–482

Acland JD (1971) East African crops. Longman, London, p 252

Adikini S, Tripathi L, Beed F, Tusiime G, Magembe EM, Kim DJ (2011) Development of a specific molecular tool for detecting Xanthomonas campestris pv. musacearum. Plant Pathol 60(3):443–452

Akiew E (1998) An evaluation of the relative importance of banana corn rot in North Queensland. HRDC Project No.FR 97030

Alam MM, Mean MB (2000) Postharvest fruit rots of banana in Bangladesh and their management. Philippine agriculture. Naya Prokash 206:130–134

Arias P, Dankers C, Liu P, Pilkauskas P (2003) The world banana economy 1985–2002. Food and Agriculture Organisation of the United Nations, Rome

Aritua V, Nanyonjo A, Kumakech F, Tushemereirwe W (2007) Rep-PCR reveals a high genetic homogeneity among Ugandan isolates of Xanthomonas campestris pv musacearum. Afr J Biotechnol 6(3):179–183

Backiyarani S, Uma S, Nithya S, Chandrasekar A, Saraswathi MS, Thangavelu R, Singh NK (2015) Genome-wide analysis and differential expression of chitinases in banana against root lesion nematode (Pratylenchus coffeae) and eumusae leaf spot (Mycosphaerella eumusae) pathogens. Appl Biochem Biotechnol 175(8):3585–3598

Bird DM, Williamson VM, Abad P, McCarter J, Danchin EG, Castagnone-Sereno P, Opperman CH (2009) The genomes of root-knot nematodes. Annu Rev Phytopathol 47:333–351

Blake CD (1961) Root rot of bananas caused by Radopholus similis (Cobb) and its control in New South Wales. Nematologica 6(4):295–310

Blomme G, Ploetz R, Jones D, De Langhe E, Price N, Gold C, Tinzaara W (2013) A historical overview of the appearance and spread of Musa pests and pathogens on the African continent: highlighting the importance of clean Musa planting materials and quarantine measures. Ann Appl Biol 162(1):4–26

Borah BK, Sharma S, Kant R, Johnson AM, Saigopal DVR, Dasgupta I (2013) Bacilliform DNA-containing plant viruses in the tropics: commonalities within a genetically diverse group. Mol Plant Pathol 14(8):759–771

Chabannes M, Iskra-Caruana ML (2013) Endogenous pararetroviruses—a reservoir of virus infection in plants. Curr Opin Virol 3(6):615–620

Chattopadhayay PK, Mukherjee N (1986) A pseudostem rot of banana due to *Erwinia chrysanthemi* pv. *paradisica*. Curr Sci 55(6):789–790

Chattopadhyay PK (1987) Soft rot: a threat to banana cultivation in West Bengal. Indian Hortic 31(4):9

Dale JL, Burns T, Oehlschlager S, Karan M, Harding R (1992) Banana bunchy top virus: prospects for control through biotechnology. In: Biotechnology applications for banana and plantain improvement. International Network for the Improvement of Banana and Plantain, San Jose, pp 85–92

Dallot S, Acuna P, Rivera C, Ramirez P, Cōte F, Lockhart BEL, Caruana ML (2001) Evidence that the proliferation stage of micropropagation procedure is determinant in the expression of banana streak virus integrated into the genome of the FHIA 21 hybrid (Musa AAAB). Arch Virol 146(11):2179–2190

Davide RG, Frison EA, Horry JP, De Waele D (1996) Overview of nematodes as a limiting factor in Musa production

Deng GM, Yang QS, He WD, Li CY, Yang J, Zuo CW, … Yi GJ (2015) Proteomic analysis of conidia germination in Fusarium oxysporum f. sp. cubense tropical race 4 reveals new targets in ergosterol biosynthesis pathway for controlling Fusarium wilt of banana. Appl Microbiol Biotechnol 99(17):7189–7207

Descalsota GIL, Lalusin AG, Viña CB, Mendioro MS, Dizon TO (2015) Screening for BBTV resistance and molecular characterization of gamma irradiated putative mutants of abaca using gene-specific markers. Philipp J Crop Sci 40(1):8–16

Diener AC, Ausubel FM (2005) Resistance to Fusarium oxysporum 1, a dominant Arabidopsis disease-resistance gene, is not race specific. Genetics 171(1):305–321

Dita MA, Rispail N, Prats E, Rubiales D, Singh KB (2006) Biotechnology approaches to overcome biotic and abiotic stress constraints in legumes. Euphytica 147(1–2):1–24

Dizon TO, Damasco OP, Lobina IT, Pinili MS, Lalusin AG, Natsuaki KT (2012) Induction of putative resistant lines of abaca (Musa textilis nee) to banana bunchy top virus and banana bract mosaic virus through in vitro mutagenesis. J ISSAAS 18(1):87–99

Drew RA, Moisander JA, Smith MK (1989) The transmission of banana bunchy top virus in micropropagated bananas. Plant Cell Tissue Org Cult 16:187–193

Elayabalan S, Kalaiponmani K, Subramaniam S, Selvarajan R, Panchanathan R, Muthuvelayoutham R, Balasubramanian P (2013) Development of Agrobacterium-mediated transformation of highly valued hill banana cultivar Virupakshi (AAB) for resistance to BBTV disease. World J Microbiol Biotechnol 29(4):589–596

Fegan M, Prior P (2006) Diverse members of the Ralstonia solanacearum species complex cause bacterial wilts of banana. Australas Plant Pathol 35(2):93–101

Figueroa L (1987) Control of diseases and nematodes of banana and plantain. Rev Mex-de-Fitopathologia 5(2):181–184

Fogain R, Price NS (1994) Varietal screening of some Musa cultivars for susceptibility to the banana borer weevil. Fruits (Paris) 49(4):247–251

Gasparotto L, Pereira JCR, Hanada RE, Montarroyos AVV (2006) Sigatoka negra da bananeira. Embrapa Amazo nia Ocidental, Manaus

Gayral P, Iskra-Caruana ML (2009) Phylogeny of banana streak virus reveals recent and repetitive endogenization in the genome of its banana host (Musa sp.). J Mol Evol 69:65–80

Gayral P, Noa-Carrazana JC, Lescot M, Lheureux F, Lockhart BE, Matsumoto T, Iskra-Caruana ML (2008) A single banana streak virus integration event in the banana genome as the origin of infectious endogenous pararetrovirus. J Virol 82(13):6697–6710

Geering AD, Pooggin MM, Olszewski NE, Lockhart BE, Thomas JE (2005) Characterisation of banana streak Mysore virus and evidence that its DNA is integrated in the B genome of cultivated Musa. Arch Virol 150:787–796, http://dx.doi.org/10.1007/s00705-004-0471-z

Geering AD, Olszewski NE, Dahal G, Thomas JE, Lockhart BE (2001) Analysis of the distribution and structure of integrated banana streak virus DNA in a range of Musa cultivars. Mol Plant Pathol 2(4):207–213

Gold CS, Pena JE, Karamura EB (2001a) Biology and integrated pest management for the banana weevil, Cosmopolites sordidus (Germar) (Coleoptera: Curculionidae). Int Pest Manag Rev 6:79–155

Gold SE, García-Pedrajas MD, Martínez-Espinoza AD (2001b) New (and used) approaches to the study of fungal pathogenicity. Annu Rev Phytopathol 39(1):337–365

Gomez-Caicedo LE, Echeverry NE, Gonzalez SR (2001) Evaluation of cultural, chemical and biological control of vascular rot and wilt in plantain (Musa AAB Simmonds). Montpelier, France: International Network for the Improvement of Banana and Plantain (INIBAP). Infomusa 10(1):17–21

Gowen S, Quénéhervé P (1990) Nematode parasites of bananas, plantains and abaca. Plant parasitic nematodes in subtropical and tropical agriculture:431–460

Guo L, Han L, Yang L, Zeng H, Fan D, Zhu Y, Zhou D (2014) Genome and transcriptome analysis of the fungal pathogen Fusarium oxysporum f. sp. cubense causing banana vascular wilt disease. PLoS One 9(4), e95543

Guzman M, Sandoval J (1996) Symptoms of pseudostem soft rot in the hybrids FHIA-01 and FHIA-02. Corbana 21(46):145–150

Harding RM, Burns TM, Hafner G, Dietzgen RG, Dale JL (1993) Nucleotide sequence of one component of the banana bunchy top virus genome contains a putative replicase gene. J Gen Virol 74:323–323

Harish S et al (2009) Differential expression of pathogenesis-related proteins and defense enzymes in banana: interaction between endophytic bacteria, banana bunchy top virus and Pentalonia nigronervosa. Biocontrol Sci Tech 19(8):843–857

Jeger MJ, Eden-Green S, Thresh JM, Johannson A, Waller JM, Brown AE (1995) Banana diseases. In: Gowen SR (ed) Banana and plantains. Chapman and Hall, London, pp 3117–3381

Jones DR (1999) Diseases of banana, abacá and enset. CABI Publishing, Wallingford

Kantún-Moreno N, Vázquez-Euán R, Tzec-Simá M, Peraza-Echeverría L, Grijalva-Arango R, Rodríguez-García C, Canto-Canché B (2013) Genome-wide in silico identification of GPI proteins in Mycosphaerella fijiensis and transcriptional analysis of two GPI-anchored β-1, 3-Glucanosyltransferases. Mycologia 105(2):285–296

Kavino M, Harish S, Kumar N, Saravanakumar D, Samiyappan R (2008) Induction of systemic resistance in banana (Musa spp.) against banana bunchy top virus (BBTV) by combining chitin with root-colonizing Pseudomonas fluorescens strain CHA0. Eur J Plant Pathol 120(4):353–362

Kema GH (2013) The genome sequence of the banana black sigatoka pathogen Mycosphaerella fijiensis reveals an extraordinary species specific expansion and is applied to analyze avirulence and fungicide resistance. In: Plant and animal genome XXI conference. Plant and Animal Genome

Kiggundu A, Vuylsteke D, Gold CS (1999) Recent advances in host plant resistance to banana weevil, Cosmopolites sordidus (Germar). In: Frison EA, Gold CS, Karamura EB, Sikora RA (eds) Mobilizing IPM for sustainable banana production in Africa proceedings of a workshop on banana IPM held in Nelspruit, South Africa -23–28th November 1998. INIBAP, Montpellier, pp 87–96

Kiggundu A, Pillay M, Viljoen A, Gold C, Tushemereirwe W, Kunert K (2003) Enhancing banana weevil (Cosmopolites sordidus) resistance by plant genetic modification. A perspective. Afr J Biotechnol 2:563–569

Kumar PL, Hanna R, Alabi OJ, Soko MM, Oben TT, Vangu GHP, Naidu RA (2011) Banana bunchy top virus in sub-Saharan Africa: investigations on virus distribution and diversity. Virus Res 159(2):171–182

Lewis Ivey ML, Tusiime G, Miller SA (2010) A polymerase chain reaction assay for the detection of Xanthomonas campestris pv. musacearum in banana. Plant Dis 94(1):109–114

Lheureux F, Carreel F, Jenny C, Lockhart B, Iskra-Caruana M (2003) Identification of genetic markers linked to banana streak disease expression in inter-specific Musa hybrids. Theor Appl Genet 106(4):594–598

Li CY, Deng GM, Yang J, Viljoen A, Jin Y, Kuang RB, Wei YR (2012) Transcriptome profiling of resistant and susceptible Cavendish banana roots following inoculation with Fusarium oxysporum f. sp. cubense tropical race 4. BMC Genomics 13(1):1

Li C, Shao J, Wang Y, Li W, Guo D, Yan B, Peng M (2013a) Analysis of banana transcriptome and global gene expression profiles in banana roots in response to infection by race 1 and tropical race 4 of Fusarium oxysporum f. sp. cubense. BMC Genomics 14(1):1

Li X, Bai T, Li Y, Ruan X, Li H (2013b) Proteomic analysis of Fusarium oxysporum f. sp. cubense tropical race 4-inoculated response to Fusarium wilts in the banana root cells. Proteome Sci 11(1):1

Mahadev SR, Thamilarasan SK, Kathithachalam A (2013) PCR detection of Banana Bunchy Top Virus (BBTV) at tissue culture level for the production of virus free planting materials. Int Res J Biol Sci 2:22–26

Marin DH, Barker KR, Kaplan DT, Sutton TB, Opperman CH (2000) Development and evaluation of a standard method for screening for resistance to Radopholus similis in bananas. Plant Dis 84(6):689–693

Meredith DS (1960) Studies on *Gloeosporium musarum* (Gke and massee) causing storage rots of Jamaican bananas. Anthracnose and its chemical control. Ann Appl Biol 48(2):279–290

Michielse CB, Rep M (2009) Pathogen profile update: Fusarium oxysporum. Mol Plant Pathol 10(3):311–324

Microbial Genomic Sequencing Perspectives of the American Phytopathological Society (2008)

Moens M, Perry RN (2009) Migratory plant endoparasitic nematodes: a group rich in contrasts and divergence. Annu Rev Phytopathol 47:313–332

Moore S, Vrebalov J, Payton P, Giovannoni J (2002) Use of genomics tools to isolate key ripening genes and analyse fruit maturation in tomato. J Exp Bot 53(377):2023–2030

Nakato GV, Ocimati W, Blomme G, Fiaboe KKM, Beed F (2014) Comparative importance of infection routes for banana Xanthomonas wilt and implications on disease epidemiology and management. Can J Plant Pathol 36(4):418–427

Namukwaya B, Tripathi L, Tripathi JN, Arinaitwe G, Mukasa SB, Tushemereirwe WK (2012) Transgenic banana expressing Pflp gene confers enhanced resistance to Xanthomonas wilt disease. Transgenic Res 21(4):855–865

Ocan D, Mukasa HH, Rubaihayo PR, Tinzaara W, Blomme G (2008) Effects of banana weevil damage on plant growth and yield of East African Musa genotypes. J Appl Biosci 9(2):407–415

Osuji JO, Harrison G, Crouch J, Heslop-Harrison JS (1997) Identification of the genomic constitution of Musa L. Lines (bananas, plantains and hybrids) using molecular cytogenetics. Ann Bot 80(6):787–793

Osuji JO, Crouch J, Harrison G, Heslop-Harrison JS (1998) Molecular cytogenetics of Musa species, cultivars and hybrids: location of 18S-5.8 S-25S and 5S rDNA and telomere-like sequences. Ann Bot 82(2):243–248

Ots, K., Indriksons, A., Varnagiryte-Kabasinskiene, I.,Mandre, M., Kuznetsova, T., Kloseiko, J., … Kikamagi, K. (2011). Changes in the canopies of Pinus sylvestris and Picea abies under alkaline dust impact in the industrial region of Northeast Estonia. For Ecol Manag 262(2):82–87

Passos MA, de Cruz VO, Emediato FL, de Teixeira CC, Azevedo VCR, Brasileiro AC, Pappas GJ (2013) Analysis of the leaf transcriptome of Musa acuminata during interaction with Mycosphaerella musicola: gene assembly, annotation and marker development. BMC Genomics 14(1):1

Pavis C, Lemaire L (1997) Resistance of Musa germplasm to the banana borer weevil, Cosmopolites sordidus Germar (Coleoptera: Curculionidae). InfoMusa 5:3–9

Peeran MF et al (2014) Water in oil based PGPR formulation of Pseudomonas fluorescens (FP7) showed enhanced resistance against Colletotrichum musae. Crop Prot 65:186–193

Pereira LV, Nunes RAS (1988) Soft rot of the rhizome and pseudostem of banana (*Musa acuminata*). Fitopathologia Bras 13(1):70–71

Ploetz RC (2015) Fusarium wilt of banana. Phytopathology 105(12):1512–1521

Portal O, Izquierdo Y, De Vleesschauwer D, Sanchez-Rodriguez A, Mendoza-Rodriguez M, Acosta-Suarez M, Ocana B, Jimenez E, Hofte M (2011) Analysis of expressed sequence tags derived from a compatible Mycosphaerella fijiensis-banana interaction. Plant Cell Rep 30(5):913–928

Qazi J (2016) Banana bunchy top virus and the bunchy top disease. J Gen Plant Pathol 82(1):2–11

Quénéhervé P (1989) Population of nematodes in soils under banana, cv. Poyo, in the Ivory Coast. 3. Seasonal dynamics of populations in mineral soil. Rev Nématologie 12(2):149–160

Rejeb IB, Pastor V, Mauch-Mani B (2014) Plant responses to simultaneous biotic and abiotic stress: molecular mechanisms. Plants 3(4):458–475

Rivera N, Garcia A (1981) Distribution in Cuba of watery rot of the pseudostem of banana caused by *Erwinia chrysanthemi*. Cienc Tecnia Agric Protect Planta 4(2/3):89–96

Robinson JC, Manicom BQ (1991) Rhizomes of new banana cultivars are susceptible to *Erwinia* corm rot. Inligtings Bull Navorsings instituut-vir-sitrus-en subtropiese-Vruget 228:1–2

Roncero MIG, Hera C, Ruiz-Rubio M, Garcí FI, Madrid MP, Caracuel Z, Velasco C (2003) Fusarium as a model for studying virulence in soilborne plant pathogens. Physiol Mol Plant Pathol 62(2):87–98

Rossmann B, Müller H, Smalla K, Mpiira S, Tumuhairwe JB, Staver C, Berg G (2012) Banana-associated microbial communities in Uganda are highly diverse but dominated by Enterobacteriaceae. Appl Environ Microbiol 78(14):4933–4941

Santana MF, Silva JC, Batista AD, Ribeiro LE, da Silva GF, de Araújo EF, de Queiroz MV (2012) Abundance, distribution and potential impact of transposable elements in the genome of Mycosphaerella fijiensis. BMC Genomics 13(1):720

Sarah JL (1989) Nematological review: banana nematodes and their control in Africa. Nematropica 19(2):199–216

Selvarajan R, Sheeba MM, Balasubramanian V (2011) Simultaneous detection of episomal banana streak mysore virus and banana bunchy top virus using multiplex RT-PCR. Curr Sci 100(1):31–34

Sengooba T (1986) Survey of banana pest problem complex in Rakai and Masaka districts, August 1986: preliminary trip report. Namulonge Research Station Uganda. Unpublished Mss. p 10

Sergeant K, Jenny R (2010) Plant biotic stress and proteomics. Curr Proteomics 7(4):275–297

Shekhawat UK, Ganapathi TR, Hadapad AB (2012) Transgenic banana plants expressing small interfering RNAs targeted against viral replication initiation gene display high-level resistance to banana bunchy top virus infection. J Gen Virol 93:1804–1813

Snyder WC, Hansen HN (1940) The species concept in Fusarium. Am J Bot 27:64–67

Srinivasan R, Kulothungan S, Sundararaju P, Govindasamy C 2011 Biodiversity of plant parasitic nematodes associated with banana in thanjavur district of tamilnadu 1(2)

Ssekiwoko F, Tushemereirwe WK, Batte M, Ragama PE, Kumakech A (2006) Reaction of banana germplasm to inoculation with Xanthomonas campestris pv musacearum. Afr Crop Sci J 14(2):151–155

Ssekiwoko F, Turyagyenda LF, Mukasa H, Eden-Green S, Blomme G (2008) Spread of Xanthomonas campestris pv. musacearum in banana (Musa spp.) plants following infection of the male inflorescence. IV Int Symp Banana: Int Conf Banana Plantain Afr: Harnessing Int 879:349–356

Stainton D, Martin DP, Muhire BM, Lolohea S, Lepoint P, Blomme G, Walters M (2015) The global distribution of banana bunchy top virus reveals little evidence for frequent recent, human-mediated long distance dispersal events. Virus Evol 1(1):vev009

Stevens C (2010) Industrial applications of natural fibres: structure, properties and technical applications, vol 10. In: J Müssig (ed). Wiley, Chichester, 560 pages. ISBN: 978-0-470-69508-1

Studholme DJ, Kemen E, MacLean D, Schornack S, Aritua V, Thwaites R, Jones JD (2010) Genome-wide sequencing data reveals virulence factors implicated in banana Xanthomonas wilt. FEMS Microbiol Lett 310(2):182–192

Su YY, Noireung P, Liu F, Hyde KD, Moslem MA, Bahkali AH, …Cai L (2011) Epitypification of Colletotrichum musae, the causative agent of banana anthracnose. Mycoscience 52(6):376–382

Swarupa V, Ravishankar KV, Rekha A (2013) Characterization of tolerance to Fusarium oxysporum f. sp., cubense infection in banana using suppression subtractive hybridization and gene expression analysis. Physiol Mol Plant Pathol 83:1–7

Talwana HAL, Speijer PR, De Waele D, Swennen RL (2005) Nematode root infection affects nutrient concentrations in banana leaves. Afr Crop Sci Conf Proc 7:1527–1537

Thomas JE, Smith MK, Kessling AF, Hamill SD (1995) Inconsistent transmission of banana bunchy top virus in micropropagated bananas and its implication for germplasm screening. Crop Pasture Sci 46(3):663–671

Thwaites R, Eden-Green SJ, Black R (2000) Disease caused by bacteria. In: Jones DR (ed) Diseases of banana, abaca and euset. CABI Publishing, Wallingford, pp 213–239

Tomlinson DL, King GA, Ovia A (1987) Bacterial corm and rhizome rot of banana (Musa spp.) in Papua New Guinea caused by Erwinia chrysanthemi. Int J Pest Manag 33(3):196–199

Tripathi JN, Lorenzen J, Bahar O, Ronald P, Tripathi L (2014) Transgenic expression of the rice Xa21 pattern-recognition receptor in banana (Musa sp.) confers resistance to Xanthomonas campestris pv. musacearum. Plant Biotechnol J 12(6):663–673

Turner DW (1995) The response of the plant to the environment. In: Gowen S (ed) Bananas and plantains. Springer, Netherlands, pp 206–229

Tushemereirwe W, Kangire A, Ssekiwoko F, Offord LC, Crozier J, Boa E, Smith JJ (2004) First report of Xanthomonas campestris pv. musacearum on banana in Uganda. Plant Pathol 53(6):802–802

Vetten HJ, Chu PWG, Dale JL, Harding R, Hu J, Katul L, Thomas JE (2005) Nanoviridae

Vurro M, Bonciani B, Vannacci G (2010) Emerging infectious diseases of crop plants in developing countries: impact on agriculture and socio-economic consequences. Food Secur 2(2):113–132

Wang Z, Zhang J, Jia C, Liu J, Li Y, Yin X, Jin Z (2012) De novo characterization of the banana root transcriptome and analysis of gene expression under Fusarium oxysporum f. sp. Cubense tropical race 4 infection. BMC Genomics 13(1):650

Wasukira A, Tayebwa J, Thwaites R, Paszkiewicz K, Aritua V, Kubiriba J, …Studholme D. J (2012) Genome-wide sequencing reveals two major sublineages in the genetically monomorphic pathogen Xanthomonas campestris pathovar musacearum. Genes 3(3):361–377

Watanabe S, Greenwell AM, Bressan A (2013) Localization, concentration, and transmission efficiency of banana bunchy top virus in four asexual lineages of Pentalonia aphids. Viruses 5(2):758–776

Xu J, Zheng HJ, Liu L, Pan ZC, Prior P, Tang B, Feng J (2011) Complete genome sequence of the plant pathogen Ralstonia solanacearum strain Po82. J Bacteriol 193(16):4261–4262

Yirgou D, Bradbury JF (1968) Bacterial wilt of Enset (Ensete ventri-cosum) incited by Xanthomonas musacearum sp. n. Phytopathology 58(1):111–112

Yirgou D, Bradbury JF (1974) A note on wilt of banana caused by the enset wilt organism Xanthomonas musacearum. East Afr Agric For J 40(1):111–11

Zakaria L, Sahak S, Zakaria M, Salleh B (2009) Characterisation of Colletotrichum species associated with Anthracnose of banana. Trop Life Sci Res 20(2):119

Zheng K, Zhang XR, Liang D (2013) The design and finite element analysis of the installation to smash banana stem. Adv Mater Res 791:694–698

Abiotic Stress Tolerance Research Using-Omics Approaches

6

Ewaut Kissel and Sebastien C. Carpentier

Abstract

The effects of abiotic stress on banana production become increasingly important, but the molecular processes behind tolerance still remain largely unknown. As the genome sequence is now publically available and cutting-edge high-throughput -omics technologies emerge, there lay multiple opportunities in the offing to close the knowledge gap. This chapter gives an overview of the molecular work that has already been performed on abiotic stress tolerance in banana. This research is mostly oriented towards cold and drought, while that on nutrient deficiencies is still lagging behind. Promising results as well as important gaps are formulated, and recommendations for future abiotic stress tolerance research are proposed. We show how new -omics technologies and the integration thereof enable a holistic view on abiotic stress tolerance in banana and can enhance our knowledge. Additionally, special emphasis is placed on the ultimate importance of phenotyping and a good definition of tolerance in the context of decent molecular research in this area.

Keywords

Musa • Abiotic stress • Proteomics • Genomics • Transcriptomics

6.1 Introduction

Until now, the main focus of molecular research on banana was oriented towards biotic constraints, even though, in some regions, e.g. East Africa, it is reported that abiotic constraints already override biotic constraints as main limiting factors for yield, with an emphasis on nutrient deficiencies and drought (Wairegi et al. 2010). As abiotic stress research is already

E. Kissel
Department of Biosystems, University of Leuven, Leuven, Belgium

S.C. Carpentier (✉)
Department of Biosystems, University of Leuven, Leuven, Belgium

SYBIOMA facility for systems biology based mass spectrometry, University of Leuven, Leuven, Belgium
e-mail: sebastien.carpentier@biw.kuleuven.be

S. Mohandas, K.V. Ravishankar (eds.), *Banana: Genomics and Transgenic Approaches for Genetic Improvement*, DOI 10.1007/978-981-10-1585-4_6

underrepresented in molecular banana research, reports on the molecular mechanisms behind tolerance are even scarcer. Furthermore, molecular technologies become rapidly more advanced and relatively less expensive, and the banana genome sequence reveals more and more of its secrets. Therefore, high-throughput profiling studies are increasingly becoming the standard in molecular plant research, and already some core processes during abiotic stress could be well described in model species. For these reasons, we intend to give an overview of the current state of molecular research on abiotic stress tolerance in banana and formulate future opportunities in this field. A special emphasis is put on the importance of phenotyping and the use of a solid definition on tolerance as the basis for decent molecular research. Ultimately, the plant phenotype is driven by the operation of genes to regulate growth in coordination with environmental limitations. As such, gene and cell function must always be considered in the context of the whole plant. Biological research on gene and cell function has mainly focused in the past on model organisms, and most of the functional genomic studies in the field of plant sciences are still performed on models or reference species that are characterized to a great extent. The power of proteomic and transcriptomic methods, i.e. high-throughput identification of candidate gene products, tends to be lost in so-called orphan species such as banana due to the lack of genomic information, due to the complexity of the genome or due to the sequence divergence to a related sequenced reference cultivar or to a related model organism. This chapter reviews the challenges encountered when working on non-model crops that were not selected by the research community for extensive study. In this perspective, the most important realizations concerning molecular research on abiotic stress tolerance in banana are reported and reviewed, and the main gaps, pitfalls and future perspectives are addressed in each of the three main areas of -omics approaches (genomics, transcriptomics and proteomics). To conclude, the power of integrating those different areas is shortly addressed and put in the context of abiotic stress tolerance research in banana.

6.2 Abiotic Stress

Before digging deeper into tolerance and its underlying genetics, abiotic stress should be well defined as both terms, stress and tolerance, are closely linked. Taiz and Zeiger (2002) define stress as 'a (non-living, ed.) external factor that exerts a disadvantageous influence on the plant'. For every type of abiotic stress, the range of this disadvantage will vary greatly. The intensity and duration of the stress also greatly influence stress progression and ultimately plant response. Depending on the intensity, very different mechanisms can be activated to cope with the stressor. In *Arabidopsis*, Skirycz et al. (2011) have shown that several genes that enhance survival during severe drought experiments did not confer better growth performance during mild osmotic stress experiments. On the contrary, most of these genes resulted in decreased growth for the specific genotype under mild osmotic stress. This clearly demonstrates that diverse mechanisms are active to cope with different intensity levels of abiotic stress.

6.3 Tolerance

Tolerance alleles are only relevant in reference to specific phenotypic features that reflect a tolerant phenotype. In essence, scientists always depart from an envisioned phenotype and use the phenotypical variation between genotypes for allele discovery, be it through a segregating population or direct cultivar comparison. That's why allele discovery should always follow an appropriate definition of a tolerant phenotype. A tolerant phenotype is a phenotype that reflects specific features which can be linked to tolerance. Therefore, it is important to have a clear view on tolerance first before even trying to define what a tolerant phenotype should look like. Only a solid definition leads to a good choice of features. To conclude, informative plant assessment needs (a) a good tolerance definition and (b) adequate feature selection.

6.3.1 Definition

In the case of biotic stress tolerance, it is relatively easy to define the tolerant phenotype. It is the phenotype that shows no or a reduced infestation or infection, which can mostly be linked directly to pathogen occurrence. For abiotic stress tolerance, however, the definition becomes less obvious. The definition of tolerance is closely linked with the one of stress. Following the definition from Taiz and Zeiger (2002), abiotic stress tolerance could be described as restoring the disequilibrium or disadvantageous influence created by the stressor. The most difficult part in this description is to determine the 'disadvantageous influence'. In a plant central approach, the disadvantageous influence is any influence that threatens its existence. Therefore, tolerance is solely focused on withstanding the adverse period. This approach, however, mainly focuses on the survival of the plant and does not necessarily take into account the growth potential under less favourable conditions. With regard to agriculture, this definition is therefore less suitable and may even be counterproductive since the focus should lie on production and profit maximization under limiting conditions. Moreover, the stress intensity has major implications on the plant's molecular response, and thus tolerance. The intensity should therefore be taken into account as well. Indeed, better survival of lethal drought stress is mainly associated with water-saving strategies and growth arrest, which is the complete opposite of production maximization. In this chapter, we focus on the agricultural output. Therefore, we identify yield (Y) decrease as the ultimate disadvantageous influence. Another important factor is the onset of stress, since the effect will be different when stress occurs at either the vegetative or generative stage.

6.3.2 Feature Selection

Yield of harvestable product is the preferred feature if one is only interested in the classification of genotypes according to drought tolerance in an agricultural perspective. However, when the focus lies on uncovering the mechanisms behind yield increase during adverse conditions or if yield cannot be measured due to experimental constraints, there is a need for a proxy for yield. Consequently, a new question arises, which proxy for yield should be used? In other words, which other features have an influence on yield? Abiotic stress negatively influences growth during the vegetative stage, the number of hands per bunch, the fingers per hand as well as filling of the fingers during the generative stage. Ultimately, these changes decrease bunch weight (Robinson and Alberts 1986; Nyombi et al. 2010; Obiefuna 1984).

Therefore, yield conservation during periods of stress is possible by maintaining biomass accumulation during the vegetative stage (BM), and by ensuring the full potential of fruit setting and filling during the generative phase (HI), and is given by

$$Y = \mathrm{BM} \times \mathrm{HI} \qquad (6.1)$$

Banana is a crop that is continuously grown throughout the year, so it is also necessary to take time into account as an extra dimension in the equation, i.e. the duration of the growth cycle (Δt_{GC}). A plant that needs less time to deliver the same yield is preferred, so that a new growth cycle can be initiated faster. Therefore, we can adjust Eq. 6.1 as follows:

$$Y = \mathrm{BM} \times \mathrm{HI} \times \Delta t_{\mathrm{GC}}^{-1} \qquad (6.2)$$

The best *proxy for yield* is, according to Eq. 6.2, any process that positively influences one of the core features in the right term (BM, HI, Δt_{GC}) during stress. These proxies therefore contribute to stress tolerance. For banana, the vegetative growth in terms of dry mass accumulation directly influences the duration of the growth cycle (Robinson and Alberts 1986). Water shortage during the vegetative stage will slow down the biomass accumulation and will prolong the duration of the growth cycle. A bigger biomass accumulation during the vegetative stage is also advantageous since bigger and stronger plants can resist stronger winds and at bunch formation it can carry a heavier bunch and allocate more resources towards the bunch. For these reasons dry mass accumulation during the vegetative stage is an ideal proxy.

Use efficiency should be included when a limiting resource, such as water or a soil nutrient, is at the basis of the abiotic stress. It is a measure for the conversion of limiting resource into harvestable product. On the other hand, a major drawback of selection on, for example, water use efficiency is that it can guide phenotype selection towards slower growers. As was suggested by Blum (2009), following de Wit's equation (Eq. 6.3),

$$B = \frac{n \times Tr}{E_0} \quad (6.3)$$

with B the biomass or fresh weight accumulated, Tr the transpiration, n a transpiration-independent crop constant and E_0 the free water evaporation. Decreasing transpiration will thus increase transpiration efficiency but further decrease biomass accumulation. Therefore it is equally important to take growth into account. In this perspective, we proposed the use of quadrants to integrate both use efficiency and growth for drought-tolerant genotype selection (Kissel et al. 2015).

Non-stressed conditions are often disregarded when looking for abiotic stress tolerance. It must be noted that this situation is the most prevalent situation in agriculture. For that reason, the selected features should be tested and compared between contrasting phenotypes under those conditions as well. Better performing genotypes under stress conditions are not necessarily the best performers under standard conditions. A clear example is given by the overexpression of the *Arabidopsis thaliana* transcription factor, *AtCBF1*, in banana cv. Dajiao (ABB). The overexpressing lines showed less electrolyte leakage and better ROS scavenging under cold stress, indicating improved tolerance. However, the overexpressing lines also showed growth retardation under standard conditions, making them less ideal for agricultural purposes (Liu Kai et al. 2012; Yang et al. 2015).

In conclusion, a tolerant phenotype for abiotic stress (due to a limiting resource) is a phenotype that best maintains growth both under normal and stress conditions (and uses the resource most efficiently).

6.4 Unravelling the Molecular Landscape of Abiotic Stress Tolerance

As already indicated above, the main focus of molecular research was on biotic constraints, while far less attention was paid to abiotic stresses. Moreover, most molecular research on abiotic stress is focused on one gene, while tolerance to abiotic stresses usually is a very complex interplay of different genetic adjustments and thus a multigenic trait. Sometimes more genes are monitored during an experiment, but they mainly encompass the usual suspects to prove an abiotic constraint rather than focusing on how tolerance is manifested (Shekhawat and Ganapathi 2013). It is very important to distinguish between genes that are differentially regulated due to the general stress response or due to a tolerance response. The best way to do this and to screen beyond the usual suspects is via holistic -omics approaches, where the whole genotypic variation at a certain moment and at multiple molecular levels is taken into account. The approach is therefore not specific and less biased. It is a hypothesis-generating rather than hypothesis-driven approach that can further be tested to go beyond the usual suspects (Horgan and Kenny 2011).

6.4.1 Molecular Variation Between Genotypes

Two types of molecular variation between genotypes can clearly be distinguished. Firstly, there is *static variation* due to genomic differences and secondly *dynamic variation* due to genome × time × environment interaction, which is reflected by the transcriptome and proteome, but can also comprise the metabolome and epigenome.

With every increase in the degree of functionality (genome to transcriptome to proteome), there is an *increase in molecular interaction and complexity* (Zivy et al. 2015). Let's take the smallest possible genomic difference, a single nucleotide polymorphism (SNP). Neither the different growth stages nor the environment influence this genomic difference; therefore the molecular variation at genome level can directly be assessed by sequencing. When such a SNP translates into an altered protein, it can change the transcriptome in multiple ways, for example, when the altered protein is directly involved in the transcription (e.g. RNA polymerases, transcription factors, activators), the translocation of transcription-related proteins (e.g. nucleopores) or the turnover of transcripts (e.g. RNases). At transcriptome level, this modification therefore already creates more molecular variation than can be predicted from the initial genomic difference. Going a step further to the proteome level, the altered protein could alternatively control translation or post-translation. The protein could even be altered differently by other proteins or by the stimulus itself. These processes would inflate the genomic difference at proteome level even more. To conclude, a genomic difference, even a SNP, may be a trigger for a shift in the chemical equilibrium resulting in a variety of genotypic differences which become more complex as the functionality level increases. Every extra genomic difference increases the complexity of the molecular variation even more due to possible alternative interactions between them. This holds the main challenge when conducting research at an increased functionality level. The advantage of increased functionality (gain of information) goes hand in hand with the disadvantage (obscureness of information) due to the increased complexity.

Dynamic molecular phenotyping therefore poses extra opportunities as well as challenges. Dynamic means that one can explore a *range of interesting moments* during stress progression, from the onset of stress through molecular signalling and adaptation to homoeostasis. Also the effect of an abiotic stressor on different tissues can be assessed. However, the dynamic nature and huge biological variability in sample groups make it extra challenging to disentangle the molecular soup and uncover the components of interest. For that reason, *precise and high-throughput phenotyping becomes increasingly important* to cope with increasing complexity. The more precise the phenotype is described, the more precise a correlation to the genotypic variation can be made.

As discussed before, *the environmental stimulus should be well defined* and demarcated as well, so that results can be compared across different studies and no nonsense comparisons are made, or even wrong genes are employed to construct 'tolerant' phenotypes (Skirycz et al. 2011). In what follows, several molecular techniques concerning abiotic stress research on banana are discussed. The main advantages, disadvantages and pitfalls of those techniques are listed.

6.4.2 Challenges for Banana

Molecular work, and thus the assessment of genotypic variation, on banana poses several challenges. Two major challenges for banana are identified: sterility of cultivars and difficulty to annotate the genome.

Sterility of the most important cultivars poses a first major challenge for molecular research. Because most cultivars are highly sterile, it is almost impossible to generate crosses. Research based on segregating hybrid populations is therefore impossible unless executed by crossing two crop (wild) relatives (https://www.croptrust.org/crop/banana/). Mapping of quantitative trait loci (QTL) through transgressive segregation is therefore excluded for the sterile cultivars. Especially for abiotic stress tolerance, a multifactorial process, this becomes an issue as it is important to link certain traits, related to tolerance, to genetic markers or loci of clustered tolerance genes. Another approach for mapping those traits to genetic markers could be through genome-wide association studies (GWASs). But therefore enough genetically closely related and perfectly phenotyped individuals with a high resolution of genetic markers are necessary.

The banana genome is *poorly annotated*. Very recently, molecular banana research experienced a breakthrough with the sequence of the doubled-haploid A genome of DH-Pahang (*M. acuminata* ssp. *malaccensis*) being published in 2012 (D'Hont et al. 2012). Two years later, the B genome of the diploid *M. balbisiana* var. Pisang Klutuk Wulung was sequenced and structurally annotated using the A genome as a template. However, several drawbacks have to be overcome for molecular research on banana to advance even further.

Only the A genome of one specific ancestor species has been sequenced until now. Yet, there are multiple *M. acuminata* ancestor species, which all had their specific contribution to what is now referred to as the A genome. As hybrid cultivars evolved further, crossover events, indels and other genomic rearrangements took place (De Langhe et al. 2010). For non-biased comparative (-omics) research between cultivars, it is therefore necessary to have a rough but complete overview of the genomic building blocks of the cultivars. Next-generation sequencing and the available template made it in principle straightforward to sequence the other ancestor and cultivar genomes and to map and resolve the biggest gaps. The functional annotation process becomes difficult, as the closest related model crop is rice, of which the genome already differs substantially from that of banana (D'Hont et al. 2012). Therefore, the function of many banana genes remains unknown. The main reason for this gap in genomic information can be attributed to the status of banana as a non-model crop and a large genetical distance with the closest model crops. As *a non-model crop*, less resources are oriented towards research on banana. Resources are necessary to sequence other ancestor *M. acuminata* species or to build up knowledge facilitating functional annotation of the genome, especially for banana-specific genes.

6.4.3 Genomics

At the level of the genome, in silico analyses like homology searches can be performed to uncover banana tolerance genes using tolerance-related genes in other species (cross species annotation). Such studies do not necessitate information about the full sequence nor the structure in overwhelming detail. Nevertheless, homology searches are restricted by the translatability gap between species. Will the homolog's functional annotation hold true when comparing banana with thale cress? Will the tolerance gene be equally important for tolerance in banana as in the original species? How much banana-specific tolerance genes are missed? Additionally, in the light of a multifactorial trait as abiotic stress tolerance, this approach is fundamentally arbitrary. Which gene will be picked? The holistic view on tolerance is completely lost. Alternatively, genomic information can be used through direct comparison of different genotypes and relate uncovered genetic differences to tolerance. This method implies the need of substantial resources and detailed sequential and structural information.

6.4.3.1 In Silico Analysis

Until recently, progress of genomic research in banana was arduous as the genome was not sequenced and annotated yet. Therefore the genomic abiotic stress research mainly focused on homology searches in small genomic databases of banana for cross species tolerance genes or for gene families or domains related to tolerance. Because the small databases did not cover the full genome, it was still very hard to find specific tolerance genes or to perform extensive research on how certain gene families are distributed and how they diversified across the genome. Therefore, genomic research was focused on homology search for the usual suspects, i.e. genes from other species that are somehow related to tolerance. When a gene of interest is discovered in a banana database, the next step involves linking the gene to tolerance. For that purpose, overexpressed lines can be created and correlated to a tolerant phenotype. An example of this workflow is the discovery and characterization of MusaDHN-1, a SK3-type dehydrin, which was identified from a banana leaf cDNA library. Subsequently, it was characterized to confer drought and salt stress tolerance because of high proline accumulation in overexpressed lines subjected to 15 days of water deprivation in the greenhouse (Shekhawat et al. 2011). However, growth and yield performance under stress and normal conditions were not assessed. This is problematic since osmolyte accumulation, such as proline, has been reportedly correlated to a reduced growth under normal conditions, while those are the most prevalent conditions in plantations (Maggio et al. 2002). As part of a survival strategy, these genes can be defined as tolerant; from an agricultural perspective, it is less obvious to classify them likewise at this point.

Direct discovery by homology search is a good starting point but is far from ideal, since abiotic stress tolerance is a multifactorial trait. It is a very arbitrary and rather limited approach. So which gene is relevant to pick for abiotic stress tolerance in banana? It is not evident to use the knowledge about a single tolerance gene and translate it to systemic tolerance in a complete other species. Moreover, different species follow complete different paths to convert inputs into output, let alone to sustain growth under abiotic stress (i.e. optimal adaptation of the unique flow from input to output). It is reasonable to think that homologous genes would influence the same basic processes in banana as they do in the original species, but this does not mean they confer drought tolerance the same way. The other way around, however, homology search is still of utmost importance for functional annotation of banana sequences using cross species databases.

6.4.3.2 Comparative Genomics

As stated in 6.4.2, detailed knowledge of the genome is necessary for comparative genomic research in banana. Due to the lack of a large genomic database until recently, research at genome level was hampered. As far as we know, a decent QTL mapping has not been performed in banana. As more crop (wild) relative species are becoming sequenced, more cultivars can be sequenced or mapped for markers, and QTLs can be discovered, through, e.g. genome-wide association studies (GWASs). Mapping through GWAS is largely dependent on high-throughput phenotyping, as desirable traits have to be phenotyped over a vast array of cultivars so they can be mapped reliably to a set of genetic markers (SNPs). Even before sequenced populations or closely related relatives are available, high-throughput phenotyping studies on abiotic stress can be performed. Later, when a sequence or marker map of those cultivars becomes available, the discovered tolerance traits can still be linked to markers. This may be one of the advantages of the easy vegetative propagation capabilities of banana and the use of the static genome. Whenever marker maps or sequenced genomes become available for cultivars that were phenotyped, GWAS can still be performed, given that no somaclonal mutations occurred during the conservation of the cultivars.

As an alternative, functional genomic studies (transcriptomics, proteomics) can be addressed, as the need of a sequenced genome or marker map of the study object is less crucial (Carpentier et al. 2008a, b). Functional genomic studies go beyond the differences in static information between genotypes and take into account the interaction effect with abiotic stress. Indeed, those techniques are based on differential expression levels of the genes/proteins caused by the abiotic stress stimulus directly. For that reason, obtained sequences can be readily mapped to genomic (cross species) databases, which does not necessitate knowledge about the entire genome.

6.4.4 Transcriptomics

Compared to other functional genomic studies like proteomics, it is a relative simple strategy as the transcriptome reflects one of the first steps from static information to dynamic functionality, while the molecular structure is still closely related to that of the genome. One way to assess dynamic genetic information is through the use of transgenic lines. After altering the expression of a particular gene that is hypothesized to confer tolerance, scientists assess the altered phenotype in the context of tolerance. However, this approach only sheds light on a subprocess as abiotic stress tolerance is influenced by the interplay of multiple genes. Alternatively, a complete overview of the expression profile can be generated using high-throughput transcriptomic methods. Major tolerance pathways can be uncovered by comparing profiles of different genotypes.

6.4.4.1 Altering Gene Expression

As mentioned in 6.4.3, homology searches are not ideal because of the translatability gap. However, an interesting approach was provided by Sreedharan et al. (2015) using transcriptomic data in the form of expressed sequence tag (EST) datasets. They performed a homology search on the *OsPIP2;6* aquaporin gene of rice in two contrasting EST datasets, one from stressed and the other one from non-stressed banana tissue. Doing so, they directed the homology search of a well-described abiotic stress-related gene in a distant species towards stress-responsive genes in banana itself. The selection of *OsPIP2;6* was based on its reported relation to salt tolerance in rice (Liu et al. 2013). A homolog was found in the contrasting EST databases and identified as *MusaPIP2;6*. In order to link this stress-responsive gene to tolerance, they compared a constitutive and inducible overexpressed line to an untransformed line of cv. Karibale Monthan (ABB) by assessing photosynthetic activity and membrane damage during salt stress. Plants were grown in the greenhouse and were 3–4 months old at the onset of severe salt stress (250 mM NaCl, 10 days). Based on both overexpressing lines, they concluded that the tolerance gene was related to better photosynthetic efficiency and reduced damage to the cellular membrane (Sreedharan et al. 2015). They did not, however, provide data on the growth performance of the overexpressed lines under abiotic stress.

Although this approach is well suited to confirm tolerance genes in banana, it does not offer a holistic view on the full differential reaction but merely focuses on a link in the chain. It is a hypothesis-driven and biased approach that does not necessarily focus on the most important processes in banana. High-throughput profiling offers a solution to overcome this issue as it provides an overview of the main processes, which allows for relevant hypotheses to be generated.

6.4.4.2 High-Throughput Profiling

EST comparison is a tag-based approach and can be used for medium-throughput profiling. Contrasting EST datasets from stressed and non-stressed tissue are compared between multiple genotypes. Sequences that are differentially abundant between the most tolerant and sensitive genotype(s) are marked as tolerance related. Sequences can be functionally annotated by using them as a template for homology search in (cross species) databases. However, quantification of the tags forms the basis of comparison studies, which is relatively complex and largely facilitated when a sequenced genome is available. As the genome sequence was not available yet, this method was only partly employed to screen for genes expressed at low (5 °C and 15 °C) and high (35 °C and 45 °C) temperature stress at different time points between 1 and 21 h after stress induction on in vitro propagated plantlets (±30 cm high) of *M. acuminata* ssp. *burmannicoides* var. Calcutta 4 (Santos et al. 2005). A *cold* and *hot* library was generated using the corresponding EST samples. Over both libraries, 1019 sequences could be reassembled, which is only 2.79% of the roughly 36,542 protein-coding models that were identified in the sequenced A genome (D'Hont et al. 2012). From these sequences only 517 could be functionally annotated, hence the medium-throughput label for this technique. Those annotated genes were mainly involved in protein modifications, chaperone functions, protein turnover, translation, carbohydrate metabolism, transport and signal transduction. Most of these identified processes are equally important during normal conditions. An additional search for enriched functions in the *cold* and *hot* libraries compared to a control EST dataset would have been more informative about the stress response. EST libraries from only one cultivar were generated, and no quantitative measurements were possible as the sequences could only be mapped to the available EST libraries. Therefore only a rough overview of a cold and hot transcriptome was constructed, and the major processes are merely indicative for the stress response, let alone for tolerance response.

Microarrays are widely used for system-wide transcriptome studies. In contrast to the previously described tag-based approaches, microarrays are highly dependent on sequenced genomes, since the probe sets for cDNA hybridization are specifically designed to capture a specific gene set. So it is a closed approach as unknown genes will not be detected since they are not spotted on the chip, and therefore the system-wide profiling is biased. Consequently, it was very challenging to use microarrays for abiotic stress research in banana before the genome was published. In 2009, however, when the genome was not available yet, scientists accepted the challenge and performed cross hybridization of banana genomic DNA to an *Arabidopsis thaliana* ATH-1 and a rice gene chip (Davey et al. 2009). The cross hybridization resulted in the selection of 33,700 highly homolog probe sets which could be used for transcriptomic profiling in banana. The selected probes may not be the most interesting ones for tolerance research, as the approach focuses specifically on well-conserved genes throughout distant species. Indeed, the authors reported that the stress-responsive genes showed a significant overlap with studies in other (model) species and that many of the transcripts are part of processes that are typically involved in plant stress response (Davey et al. 2009). Those genes therefore tend to be more involved in a general stress response that is conserved throughout several (distant) species than truly being responsible for tolerance. A possibility to distinguish between the general stress response and the tolerance response could be to repeat the study on genotypes with contrasting tolerant phenotypes. It should be noted that the imposed stress was attained by withholding water for 5 weeks. In pots, this would mean that the stress reached an extreme intensity that does not occur in banana production systems. Up till now, this study is still one of the most extensive transcriptomic profiling studies on drought stress in banana. As the banana genome reveals its secrets step by step and next-generation sequencing offers great and relatively cheap possibilities, there are new and better technologies arising.

RNA-sequencing or RNA-seq is the newest technique for transcriptomic profiling. The full mRNA sample is sequenced, using next-generation sequencing technology, and reads are long enough to be easily assembled to contigs with or without information of a reference genome. For that reason this technique does not necessitate a sequenced genome for quantification and becomes very interesting for non-model crops. Additional advantages over other techniques are the improved dynamic range as well as sensitivity. For banana, this technique offers great possibilities. The availability of the transcriptome helps to improve the structural annotation of the genome. Data from RNA-seq experiments help to determine the exact location of genes, intron-exon structures, splice variants and map SNPs. The first comparative RNA-seq that is reported for abiotic stress on banana was on short-term effects of low-temperature stress (10 °C) in the leaves after 3 and 6 h between cv. Dajiao (ABB) and cv. Cavendish (AAA) (Yang et al. 2015). At 6 h after stress induction, 28 common stress-responsive genes were identified, which links them to a general stress response to the cold. The authors defined cv. Dajiao as cold tolerant and identified 17 unique early stress-responsive genes for this cultivar compared to the sensitive one. They related those genes, which were involved in signal transduction, copper ion equilibrium, photosynthesis, photorespiration and sugar stimulation, as early stress genes to cold tolerance. Twelve early stress-responsive genes were further assessed under extended stress (3, 6, 24 and 48 h) to confirm the RNA-seq results and link them to stress progression. Especially *ICE1* and *MYBS3* were put forward as showing different expression patterns between the tolerant and sensitive cultivar during the course of the cold stress. Because of the selective induction of those genes by the tolerant genotype during early stress stages, they are thought to specifically initiate the cold tolerance pathways in ABB compared to AAA cultivars. Additionally, the authors stated that *MYBS3* was already placed in the context of long-term tolerance response in rice (Su et al. 2010). For those reasons, both genes are believed to be the main contributing factors to the tolerant phenotype during early stress (Yang et al. 2015). An extensive drought research is in the pipeline. We performed a mild drought experiment (5 % PEG, 3 days) on three genetically different cultivars, cv. Grande Naine (AAA), cv. Mbwazirume (AAAh) and cv. Cachaco (ABB), to assess a general early stress response. RNA-seq was performed on the root tips. Seventy to eighty-four percent of the reads could be mapped to the reference genome, and 92 genes were significantly upregulated in all three cultivars. The most important altered processes were glycolysis and fermentation. RT-qPCR on 22 of the genes confirmed the importance of those alternative energy-related processes during the early stress response. Osmotic stress causes a shift in metabolism which requires much energy. This extra energy demand rapidly depletes the energy level of high energy-demanding tissues like root tips. As a result, alternative energy sources become increasingly important to cope with the stress in an early stage (Zorrilla et al. 2016).

RNA-seq analysis was also used to molecularly profile the reaction of banana on osmotic stress. We conducted a mild drought experiment (5% PEG, 3 days) on three genetically different cultivars, cv. Grande Naine (AAA), cv. Mbwazirume (AAAh) and cv. Cachaco (ABB), to assess a general early stress response. RNA-seq was performed on the root tips. Seventy to eighty-four percent of the reads could be mapped to the reference genome, and 92 genes were significantly upregulated in all three cultivars. The most important altered processes were glycolysis and fermentation. RT-qPCR on 22 of the genes confirmed the importance of those alternative energy-related processes during the early stress response. Osmotic stress causes a shift in metabolism which requires much energy. This extra energy demand rapidly depletes the energy level of high energy-demanding tissues like root tips. As a result, alternative energy sources become increasingly important to cope with the stress in an early stage (Zorrilla-Fontanesi et al. 2016). Both of these studies were the first in their field (cold, osmotic stress) to be conducted on banana. The detailed results of the studies indicate the strength of such high throughput molecular profiling techniques, which promise to greatly enhance knowledge in the field of abiotic stress on banana.

6.4.5 Proteomics

Proteins are the product of transcription and translation of the genome and are the main functional elements in the cell. During abiotic stress, protein activity can be switched on or off, and therefore protein concentration is not the only determinant for their cellular activity. Hence, it is also interesting to assess protein activities, apart from quantitative high-throughput proteomic profiling.

6.4.5.1 Protein Activity Assessment

As part of (low-throughput) proteomic phenotyping, the alteration of several processes can be followed by assessing the activity of some proteins involved. One of the main consequences of most abiotic stresses is the imbalance of reactive oxygen species (ROS) production and scavenging. That's why many stress-responsive proteins are involved in redox balances and ROS scavenging. In order to get an overview on how the balance of ROS is shifted, the activity of ROS scavenging enzymes can be measured, as a way of a molecular (low-throughput) proteomic tool. When comparing cv. Williams (AAA) and cv. Cachaco (ABB) under low-temperature conditions (7 °C, 120 h), the level of several ROS increased in the leaves (Zhang et al. 2011). It was clear that cv. Cachaco experienced less membrane leakage and photoinhibition than cv. Williams. The authors therefore postulated that cv. Cachaco is projected to lose less yield than cv. Williams and can therefore be described as the more tolerant one. There was a general decrease of catalase in both cultivars. The enzymatic activity of superoxide dismutase, ascorbate peroxidase and peroxidase all increased in cv. Cachaco, but decreased in cv. Williams, explaining the reduced ROS damage in cv. Cachaco (Zhang et al. 2011). This is only one of the processes that are influenced under abiotic stress. In order to obtain an overview on all the processes that are changed, scientists use high-throughput techniques to quantify protein levels and compare them between treatments and genotypes.

6.4.5.2 High-Throughput Proteomic Profiling

Besides the high level of information that proteins offer, high-throughput proteomic profiling becomes also very interesting for non-model crops, of which the genome has not been sequenced (Carpentier and America 2014; Carpentier et al. 2008b). Protein sequences are more conserved than genomic or transcriptomic sequences. Moreover, proteins contain functional or structural domains that are most conserved at this level and which shed a light on the main function or even the location of the protein in the cell. The uncertainty that comes from the prediction of the open reading frame as in both genomic and transcriptomic research is also solved using proteomics. Two techniques are generally used and are denoted as gel based and gel-free.

The *gel-based technique* in particular is very useful when the genome is poorly sequenced. This method ensures a two-dimensional separation of the proteins according to their isoelectric point and molecular weight, leaving them intact. Proteins of interest can be picked from the gel and analyzed separately using mass spectrometry to identify the protein-derived peptides. In 2007, when the banana genome was not available yet, we performed a gel-based proteome study on the meristem of banana after osmotic stress induction (0.4 M sucrose, 2 weeks) (Carpentier et al. 2007). A banana plant only has one meristem that initiates all the new leaves during the vegetative phase and the bunch and flowers during the generative phase. Therefore changes in the meristem during abiotic stress become extremely important as they directly influence plant growth. The proteome study revealed many stress-responsive proteins that were active in a variety of processes. The main processes included recycling of storage proteins as a new source of amino acids, changes in carbohydrate metabolism to cope with altered energy demands under stress (Carpentier et al. 2007), preservation of cell wall integrity under low osmotic conditions and upregulation of molecular chaperones. Comparison of proteomic profiles of the meristem between a tolerant and sensitive cultivar yielded several possible candidates for drought tolerance as they were uniquely or differentially expressed. The tolerance-specific proteins were involved in energy metabolism (phosphoglycerate kinase, phosphoglucomutase, UDP-glucose pyrophosphorylase) and were associated with stress adaptation (OSR40, abscisic stress-ripening protein) (Carpentier et al. 2007). The proteome-based method even enables scientists to assess different isoforms or post-translational events, proving the valuable information the proteome can offer over other molecular levels. More and more scientists take the whole gene family or possible variants into account when performing a detailed analysis on tolerance genes, since they are differentially employed during stress and tolerance. For banana, paralog expression levels of the stress-related and well-described HSP70 family were investigated during osmotic stress (Vanhove et al. 2015). Instead of identifying the whole family as a tolerance marker, our group uncovered that a specific paralog on chromosome 2, with an abscisic acid-responsive element in the promoter region, was mostly responsible for the differential expression profile of the HSP70 family under osmotic stress. The gel-based technique proved to be the technique of choice as the intact proteins are separated on molecular mass and isoelectric point, clustering very similar proteins, which can subsequently be analyzed in great detail using MS/MS.

The *gel-free technique* has become the preferred way of performing high-throughput proteomics in banana. The technique has a higher throughput compared to the former technique because different steps inherent to gel separation and staining can be skipped. The drawback is that a complex sample of peptides is analyzed and has to be disentangled to reconstruct the original proteins. The availability of a sequenced genome greatly empowers the reconstruction of proteins from the peptide soup. An interesting approach to avoid the need of a fully sequenced genome was proposed by Yang et al. (2012). They constructed a transcriptome template using RNA-seq. The transcriptome holds even more advantages over the genome because the coding sequence is used directly and protein prediction is therefore not dependent on exon-intron or other structural predictions. The transcriptome template facilitated the authors to link peptides, making protein reconstruction easier. Due to the template, the exact coding sequences of the proteins could be derived, and the template could additionally be used for functional annotation. The authors analyzed the proteome from leaf tissue of cv. Dajiao (ABB) after 6 and 24 h of cold stress at 8 °C. GO annotations categorized the majority of the stress-responsive proteins in oxidation-reduction, photosynthesis, photorespiration, cell wall integrity and carbohydrate metabolic processes. Validation of the proteomic data was performed using Western blot analysis on seven differentially expressed proteins. Additionally, the quantity and activity of several ROS scavenging enzymes were compared between cv. Dajiao (ABB) and cv. Brazil (AAA). Cv. Dajiao seemed most efficient in scavenging ROS through the catalase pathway, partly explaining its success to withstand short cold periods better (Yang et al. 2012). However, it is surprising that the increased content and activity of catalase were put forward as the main factor determining tolerance, since Zhang et al. (2011) observed a decrease of catalase activity in cv. Williams as well as in the tolerant cv. Cachaco under low-temperature stress. Both experiments were performed using the same stress intensity (8 °C versus 7 °C). The description of plant material indicates that the plants were sampled at more or less the same stage as well. The biggest differences were found to be the duration of the stress (24 h versus 120 h) and the leaf that was sampled (newest leaf versus a random leaf). This indicates that, most likely, duration (or even the physiological stage of the sampled leaf) is a very important factor to take into account and to report when defining tolerance. Another possibility is that a genotypic difference within the same genomic group (ABB) is at the basis of the different reactions. In that case, it is not advisable to directly infer results from the comparison between two genotypes to a statement about their corresponding genomic groups. For that reason care should be taken when defining a whole genomic group (i.e. the ABB cultivars) as tolerant based on data of just one representative.

6.5 Omics Integration

Since the exponential rise of molecular techniques at different levels, scientists dreamed of integrating all information to dig deeper in core plant processes. Systems biology is the branch where scientists try to integrate the information of different -omics levels, as every functionality level contains unique information over another. Therefore, it is clear that the different levels depend on each other to gain detailed information (Zivy et al. 2015). For example, uncovering the high-value information at proteome level is much more efficient when genomic information is available. Another example is that the transcriptome is not entirely one on one correlated to the proteome due to different half-lives of transcripts and proteins and due to post-translational events (Haider and Pal 2013; Carpentier et al. 2008a). As a consequence, protein levels can give extra perspectives on the importance of differential transcript levels. Therefore, if possible, all levels should be used complementing one another.

The first line of integration in abiotic stress research is the use of the phenotype to guide the molecular analysis. This integration should always be conducted meticulously, by defining a phenotype and coupling molecular markers to it. Thereby, scientists are able to uncover tolerance-related genes. The genome is used extensively to complement other -omics research. As already mentioned previously, another integrative technique employed in abiotic stress research in banana is the use of the genome to facilitate molecular identifications at transcriptome and proteome level. In this matter, another interesting approach was employed by Yang et al. (2012) using RNA-seq data to improve protein identification in gel-free proteomics to find proteins related to cold tolerance in banana. For banana, where the genome was not readily or extensively available at that time, this technique offers great possibilities. Another use of transcriptome information for integrative purposes is to improve the structural annotation of the genome. Data from RNA-seq experiments can help determine the exact location of genes, intron-exon structures or even splice variants.

These integrations have improved the power of -omics research substantially; however the next step still remains the integration of expression levels at transcriptome and proteome level and eventually couple them to the change in metabolome. However, until now this challenge remains one of the hardest in systems biology, and no such study was already reported on banana.

6.6 Conclusion

Whereas the first molecular experiments in the early 2000s were mainly focused on single proteins and genes that could enhance abiotic stress tolerance, very recently more extensive profiling methodologies appeared at the scene. The problem with single-gene and single-protein studies is that they were linked to tolerance processes in other species, which are not necessarily equally important for tolerance in banana. Additionally, single-gene studies are merely focused on subprocesses of tolerance. As the technologies become more advanced and relatively less expensive and the genome sequence reveals more and more of its secrets, high-throughput profiling studies are increasingly becoming the standard, and already some core processes during abiotic stress could be well described in model species. Although biotic constraints are still a priority for banana, there has already been effort to characterize the genetics behind several abiotic stresses as well, especially low-temperature stress and drought. While extensive nutrient deficiency studies at molecular level for banana are nonexisting, in general extensive research towards abiotic stress tolerance should come off the ground.

As in other crops, focus of the little research that was already oriented towards abiotic stress was sometimes very scattered. The main problem is the lack of a clear demarcated vision on what abiotic tolerance exactly is. Slightly or very different views on what a tolerant phenotype looks like predominate research. Therefore different markers, genes, proteins or processes are put forward to explain tolerance. Sometimes tolerance response is even mistakenly taken for stress response. Scientists should clearly define a common tolerant phenotype and relate contributing features to this definition of a tolerant phenotype. For cold stress research, there is already a more or less common definition on tolerance. ABB cultivars are viewed as tolerant, while AAA cultivars are sensitive, which reflects the way they are perceived on the field. Membrane leakage, photosynthetic efficiency and ROS scavenging are mostly used to build up the tolerant phenotype, i.e. less membrane leakage, higher photosynthetic efficiency and better ROS scavenging are contributing factors to the tolerant phenotype. Short periods of cold stress are mostly envisioned, as the focus lies on cultivation in the subtropics where temperatures can sometimes, especially at night, drop to damaging depths for the plant, but do not persist throughout the day. For drought, however, a clear common definition is more difficult to construct. Furthermore there is a scattered focus, concerning abiotic stress, in terms of onset, duration and intensity of the stress. Also a wide diversity of genotypes under a broad variety of experimental growing conditions is being used. There is a need to link groups working on the same abiotic stress type and define a common aim.

It is recommended that the community puts forward a selection of key genotypes to focus (abiotic stress) research on, so that the available resources are efficiently oriented towards a limited set of genotypes. As a result multidisciplinary knowledge would be thoroughly collected and focused, which makes integration possible in a later stage. The key genotypes preferably contain several representatives of genomic groups that span a wide genetic range in the *Musa* species. KU Leuven and Bioversity International initiated such effort and have condensed the collection of more than 1500 accessions to a core list of 32 genotypes. Ideally, this core set should be adjusted by the community, if necessary, so that it becomes widely supported. Of course, a too narrow use of genotypes, as is currently seen in cold tolerance research in banana, where mostly cv. Dajiao is used as the tolerant cultivar and a cv. Cavendish is used as sensitive cultivar, should also be avoided. The use of cv. Cachaco as the tolerant cultivar resulted in a decreased activity of catalase during cold stress (Zhang et al. 2011), while its activity rose using cv. Dajiao, which led to define them as tolerant (Yang et al. 2012). Results may therefore not be 1:1 inferred to the corresponding genomic groups when using a too narrow set of genotypes. It must be noted however that in the context of cold tolerance, promising results were recently published by Yang et al. (2015) using RNA-seq for comparative transcriptomics. They identified core mechanisms active in early stress and tolerance response and put forward two candidate genes that putatively initiate a tolerance pathway specific to the tolerant cultivar (Yang et al. 2015). It would be interesting to see the results confirmed in other ABB and AAA cultivars as well, to confirm whether the tolerance path is ABB specific or not. As it is not evident that all ABB cultivars confer tolerance the same way. Other promising results on the general early stress response in banana were obtained by Zorrilla-Fontanesi et al. (2016). We discovered that the employment of alternative energy sources becomes increasingly important in growing root tips during stress. We postulate it as a general response, as the processes involved were upregulated in different cultivars spanning a broad genetic range (Zorrilla-Fontanesi et al. 2016). These new emerging researches prove the power of -omics technologies to provide a holistic view on abiotic stress tolerance in banana. Those technologies will definitely prove their value in the future.

References

Blum A (2009) Effective use of water (EUW) and not water-use efficiency (WUE) is the target of crop yield improvement under drought stress. Field Crop Res 112(2):119–123

Carpentier S, America T (2014) Proteome analysis of orphan plant species, fact or fiction? In: Jorrin-Novo JV, Komatsu S, Weckwerth W, Wienkoop S (eds) Plant proteomics, vol 1072. Methods in molecular biology. Humana Press, New York City, pp 333–346. doi:10.1007/978-1-62703-631-3_24

Carpentier SC, Witters E, Laukens K, Van Onckelen H, Swennen R, Panis B (2007) Banana (Musa spp.) as a model to study the meristem proteome: acclimation to osmotic stress. Proteomics 7(1):92–105. doi:10.1002/pmic.200600533

Carpentier SC, Coemans B, Podevin N, Laukens K, Witters E, Matsumura H, Terauchi R, Swennen R, Panis B (2008a) Functional genomics in a non-model crop: transcriptomics or proteomics? Physiol Plant 133(2):117–130. doi:10.1111/j.1399-3054.2008.01069.x

Carpentier SC, Panis B, Vertommen A, Swennen R, Sergeant K, Renaut J, Laukens K, Witters E, Samyn B, Devreese B (2008b) Proteome analysis of non-model plants: a challenging but powerful approach. Mass Spectrom Rev 27(4):354–377. doi:10.1002/mas.20170

D'Hont A, Denoeud F, Aury J, Baurens F, Carreel F, Garsmeur O (2012) The banana (Musa acuminata) genome and the evolution of monocotyledonous plants. Nature 488:213–217. doi:10.1038/nature11241

Davey MW, Graham NS, Vanholme B, Swennen R, May ST, Keulemans J (2009) Heterologous oligonucleotide microarrays for transcriptomics in a non-model species; a proof-of-concept study of drought stress in Musa. BMC Genomics 10:436–436. doi:10.1186/1471-2164-10-436

De Langhe E, Hřibová E, Carpentier S, Doležel J, Swennen R (2010) Did backcrossing contribute to the origin of hybrid edible bananas? Ann Bot 106(6):849–857. doi:10.1093/aob/mcq187

Haider S, Pal R (2013) Integrated analysis of transcriptomic and proteomic data. Curr Genomics 14(2):91–110. doi:10.2174/1389202911314020003

Horgan RP, Kenny LC (2011) 'Omic' technologies: genomics, transcriptomics, proteomics and metabolomics. Obstet Gynaecol 13(3):189–195. doi:10.1576/toag.13.3.189.27672

Kissel E, van Asten P, Swennen R, Lorenzen J, Carpentier SC (2015) Transpiration efficiency versus growth: exploring the banana biodiversity for drought tolerance. Sci Hortic 185:175–182. doi:http://dx.doi.org/10.1016/j.scienta.2015.01.035

Liu Kai HC-H, Fa-Xiu D, Zhang Y-E, Yue-Rong W, Gan-Jun Y (2012) Over-expression of the Arabidopsis CBF1 gene in Dongguandajiao (Musa spp.ABB group) and detection of its cold resistance. Sci Agric Sin 45(8):1653–1660. doi:10.3864/j.issn.0578-1752.2012.08.022

Liu C, Fukumoto T, Matsumoto T, Gena P, Frascaria D, Kaneko T, Katsuhara M, Zhong S, Sun X, Zhu Y, Iwasaki I, Ding X, Calamita G, Kitagawa Y (2013) Aquaporin OsPIP1;1 promotes rice salt resistance and seed germination. Plant Physiol Biochem 63:151–158. doi:http://dx.doi.org/10.1016/j.plaphy.2012.11.018

Maggio A, Miyazaki S, Veronese P, Fujita T, Ibeas JI, Damsz B, Narasimhan ML, Hasegawa PM, Joly RJ, Bressan RA (2002) Does proline accumulation play an active role in stress-induced growth reduction? Plant J 31:699–712. doi:10.1046/j.1365-313X.2002.01389.x

Nyombi K, van Asten PJA, Corbeels M, Taulya G, Leffelaar PA, Giller KE (2010) Mineral fertilizer response and nutrient use efficiencies of East African highland banana (Musa spp., AAA-EAHB, cv.

Kisansa). Field Crop Res 117(1):38–50. doi:http://dx.doi.org/10.1016/j.fcr.2010.01.011

Obiefuna JC (1984) Effect of potassium application during the floral initiation stage of plantains (Musa AAB). Fertil Res 5(3):315–319. doi:10.1007/BF01051631

Robinson J, Alberts A (1986) Growth and yield responses of banana (cultivar 'Williams') to drip irrigation under drought and normal rainfall conditions in the subtropics. Sci Hortic 30(3):187–202

Santos CMR, Martins NF, Hörberg HM, de Almeida ERP, Coelho MCF, Togawa RC, da Silva FR, Caetano AR, Miller RNG, Souza MT Jr (2005) Analysis of expressed sequence tags from Musa acuminata ssp. burmannicoides, var. Calcutta 4 (AA) leaves submitted to temperature stresses. Theor Appl Genet 110(8):1517–1522. doi:10.1007/s00122-005-1989-5

Shekhawat UKS, Ganapathi TR (2013) *MusaWRKY71* overexpression in banana plants leads to altered abiotic and biotic stress responses. PLoS ONE 8(10):e75506. doi:10.1371/journal.pone.0075506

Shekhawat US, Srinivas L, Ganapathi T (2011) MusaDHN-1, a novel multiple stress-inducible SK3-type dehydrin gene, contributes affirmatively to drought- and salt-stress tolerance in banana. Planta 234(5):915–932. doi:10.1007/s00425-011-1455-3

Skirycz A, Vandenbroucke K, Clauw P, Maleux K, De Meyer B, Dhondt S, Pucci A, Gonzalez N, Hoeberichts F, Tognetti VB (2011) Survival and growth of Arabidopsis plants given limited water are not equal. Nat Biotechnol 29(3):212–214

Sreedharan S, Shekhawat US, Ganapathi T (2015) Constitutive and stress-inducible overexpression of a native aquaporin gene (MusaPIP2;6) in transgenic banana plants signals its pivotal role in salt tolerance. Plant Mol Biol 88(1–2):41–52. doi:10.1007/s11103-015-0305-2

Su CF, Wang YC, Hsieh TH, Lu CA, Tseng TH, Yu SM (2010) A novel MYBS3-dependent pathway confers cold tolerance in rice. Plant Physiol 153:145–158. doi:10.1104/pp.110.153015

Taiz L, Zeiger E (2002) Plant physiology, 3rd edn. Pub Sinauer, Sunderland

Vanhove A-C, Vermaelen W, Cenci A, Swennen R, Carpentier SC (2015) Data for the characterization of the HSP70 family during osmotic stress in banana, a non-model crop. Data Brief 3:78–84. doi:10.1016/j.dib.2015.01.008

Wairegi LWI, van Asten PJA, Tenywa MM, Bekunda MA (2010) Abiotic constraints override biotic constraints in East African highland banana systems. Field Crop Res 117(1):146–153. doi:http://dx.doi.org/10.1016/j.fcr.2010.02.010

Yang Q-S, Wu J-H, Li C-Y, Wei Y-R, Sheng O, Hu C-H, Kuang R-B, Huang Y-H, Peng X-X, McCardle JA, Chen W, Yang Y, Rose JKC, Zhang S, Yi G-J (2012) Quantitative proteomic analysis reveals that antioxidation mechanisms contribute to cold tolerance in plantain (Musa paradisiaca L.; ABB group) seedlings. Mol Cell Proteomics 11(12):1853–1869. doi:10.1074/mcp.M112.022079

Yang Q-S, Gao J, He W-D, Dou T-X, Ding L-J, Wu J-H, Li C-Y, Peng X-X, Zhang S, Yi G-J (2015) Comparative transcriptomics analysis reveals difference of key gene expression between banana and plantain in response to cold stress. BMC Genomics 16(1):1–18. doi:10.1186/s12864-015-1551-z

Zhang Q, Zhang JZ, Chow WS, Sun LL, Chen JW, Chen YJ, Peng CL (2011) The influence of low temperature on photosynthesis and antioxidant enzymes in sensitive banana and tolerant plantain (Musa sp.) cultivars. Photosynthetica 49(2):201–208. doi:10.1007/s11099-011-0012-4

Zivy M, Wienkoop S, Renaut J, Pinheiro C, Goulas E, Carpentier S (2015) The quest for tolerant varieties: the importance of integrating "omics" techniques to phenotyping. Front Plant Sci 6:448. doi:10.3389/fpls.2015.00448

Zorrilla J, Rouard M, Cenci A, Kissel E, Do H, Dubois E, Nidelet S, Roux N, Swennen R, Carpentier S (2016) Differential root transcriptomics in a polypoloid non-model crop: the importance of respiration during osmotic stress. Sci Rep 6, art.nr. 22583

Molecular Analysis of Fruit Ripening in Banana

7

Antara Ghosh, T.R. Ganapathi, and V.A. Bapat

Abstract

Banana is a climacteric fruit and has a very short postharvest life. Many varieties of banana fruits are ripened artificially by treating them with hydrocarbons. The current methods of postharvest management practices used for fruits are not enough to control the ripening in banana. Recent advances in recombinant DNA technology and genetic engineering have resulted in the modification of fruit ripening in banana. Towards this, many genes involved in ripening have been cloned and characterized. Ripening in banana is characterized by a biphasic ethylene production with a sharp early peak and a post climacteric small peak. During banana fruit ripening, ethylene production induces a developmental cascade which results in the conversion of starch into sugars, an associated burst of respiratory activity, and an increase in the protein synthesis. Other changes include fruit softening, flavor and aroma development, change in pigmentation, and increased susceptibility to pathogens; also, banana fruit softening is attributed to activities of various cell wall hydrolases. The participation of various cell wall hydrolases in banana softening during ripening has also been reported recently. The enhancing and suppressive effects of ABA and IAA on activities of different cell wall hydrolases have been noticed during ethylene-induced ripening in banana. Simultaneously, decrease in polyphenols, higher alcohol acetyl transferase activity, chlorophyll degradation etc., have been earlier reported during ripening in

A. Ghosh • V.A. Bapat (✉)
Department of Biotechnology, Shivaji University, Kolhapur 416 004, India
e-mail: vabapat@gmail.com

T.R. Ganapathi
Plant Cell Culture Technology Section, Nuclear Agriculture and Biotechnology Division, Bhabha Atomic Research Centre, Trombay, Mumbai 400 085, India

S. Mohandas, K.V. Ravishankar (eds.), *Banana: Genomics and Transgenic Approaches for Genetic Improvement*, DOI 10.1007/978-981-10-1585-4_7

banana. Recently efforts are made to delay the ripening by antisense suppression of a transcription factor-related ripening. This review summarizes the various advances made in the field of fruit ripening in banana.

Keywords
Banana • Ripening • Somatic embryogenesis • Promoter • Agrobacterium • Transgenic

7.1 Introduction

Bananas and plantains (*Musa* spp.) are the perennial monocotyledonous herbs that grow well in the humid tropical and subtropical regions. There are wide varieties of historic references in banana and the crop is mentioned in the ancient Hindu, Greek, and Roman texts. The earliest reference to banana dates back to about 500 BC. Banana originated from two wild diploid species with genomic compositions of AA and BB, respectively (Cheesman 1948). It is cultivated in more than 130 countries across the world in an area of 5.14 M ha producing 105.32 million tonnes of banana and plantain (FAO 2012).

In spite of huge production levels, bananas and plantains have been the subject of limited research compared with other staple food crops. Estimates indicate that the production of banana in 2006 was 97.37 million tonnes in the world and export value amounted to US $ 4 billion annually. Banana, being a climacteric fruit, undergoes unrestrained ripening behavior once initiated, thus losing its market viability considerably in a very short span of time. According to FAO reports, in 1997–1998, annual loss due to perished fruits raised up to US $ 20 billion in India of which banana amounts the most. Various conventional practices have been used to increase the shelf life of banana. These include cooling treatments and controlled atmosphere storage, the use of ethylene-absorbing agents, and surface coating and moist sawdust treatments, but these practices proved to be either too costly or cumbersome. These factors shifted the focus of the researchers to the molecular mechanism of ripening process, and several genes associated with ripening in banana have been characterized, isolated, and cloned (Bapat et al. 2010).

Propagation of banana cultivars with a prolonged shelf life and those that are disease-free or disease resistant, using classical breeding techniques, remains difficult and a time-consuming endeavor due to the specific properties of banana such as polyploidy, sterility, and long generation period of most of the edible cultivars. On the other hand, modern tissue culture, cell biology, and molecular biological tools provide excellent opportunities to develop new germplasm better adapted to changing demands.

7.2 Molecular Studies on Fruit Ripening-Related Genes

Fruit ripening is characterized by a number of biochemical and developmental mechanisms that result in alterations in tissue metabolism rendering the fruit attractive for consumption by organisms leading to seed release and dispersal. Specific biochemical and physiological attributes of fruit ripening vary in different species and include changes in color, texture, flavor, aroma, nutritional quality, and susceptibility to pathogens. Fruit ripening is genetically determined and many ripening-related genes have been characterized in a variety of fruits which are related to different aspects like ethylene biosynthesis, fruit softening, aroma synthesis, changes in color, or in some cases transcription factors. Though earlier studies were focused more on the roles of cell wall modifying and structural proteins (Rose et al. 2004) and ethylene synthesis (Theologis 1992), the recent studies on ripening were carried out to reveal insights into primary ripening control upstream of ethylene and ripening-related signal transduction and metabolic pathways. Tomato ripening mutants have been a great success in understanding molecular basis of fruit ripening. One of the well-studied, tomato pleiotropic ripening mutations includes colorless ripening inhibitor (rin). This mutant locus encodes putative transcription factor. The rin mutation is dominant in nature and effectively blocks the ripening process resulting in failure either to produce higher amount of ethylene or to respond to exogenous ethylene in ripening. Liu et al. (2009) reported the involvement of MADS-box transcription factor gene (MuMADS1) in ethylene-induced fruit ripening in banana. The study reports the induction of MuMADS1 by ethylene during postharvest ripening.

Ethylene has long been believed to be the main regulator of ripening in climacteric fruits (Pathak et al. 2003; Vendrell 2003). Earlier studies on the mechanism of banana ripening have predicted the role of ABA and IAA (Pathak and Sanwal 1999; Jiang et al. 2000). Later reports indicate the effect of methyl jasmonate (MJ) and methyl salicylate (MS) on ripening-related genes. The role of ABA in ripening of climacteric and non-climacteric fruits through ethylene interaction is known. In McIntosh apples ABA treatment led to increase in respiration and other ripening-related changes (Vendrell 2003). Further, Zhang et al. (2009) observed in tomato that ABA buildup was seen in seeds and fruit flesh preceded by ethylene production. The effect of ABA in differentially regulating alcohol dehydrogenase gene during mango ripening was observed by Singh et al. (2010). ABA has been reported to have inducing effects on cell wall hydrolases during banana ripening (Lohani et al. 2004). In some of our unpublished reports, we have recorded a twofold increase in expression by glucanase promoter with a combination of ABA and ethephon treatment as compared to ethephon alone. ABA alone could enhance expression of expansin promoter which could be compared to ethephon induction (Ghosh et al. 2012). The categorization of *rin* (Vrebalov et al. 2002), *nor* (Adams-Phillips et al. 2004), and *cnr* (Manning et al. 2006) mutants of tomato is predominantly relevant as it directly indicates the involvement of other developmental factors that act upstream of ethylene and direct the ripening process (Giovannoni 2004). Additionally, other categories of genes influencing ripening in tomato are those encoding for the expression of auxin/IAA (Jones et al. 2002) which directly indicates the involvement of IAA as a regulator of ripening in climacteric fruits. There are contradictory reports on the exact involvement of IAA in ripening. Elevated levels of IAA with ripening progression have been reported, but this does not essentially predict stimulatory effects of IAA on cell wall hydrolases. Contrastingly there are also reports that indicate a decline in IAA levels at the onset of ripening (Bottcher et al. 2010). This study reported IAA as an agent that could delay ripening. Lohani et al. (2004) reported the oppressive effects of IAA on cell wall hydrolases.

Along with ethylene, methyl jasmonate has been considered as an agent that could induce fruit ripening (Rahman et al. 2008). A few reports indicate the stimulatory effects of jasmonate and salicylate during ripening on the expression of pathogenesis-related genes (Kesari et al. 2010; Ankala et al. 2009). These elicitors supposedly act upstream of ethylene and control the ripening (Ankala et al. 2009). We have observed inducing effects of methyl jasmonate (MJ) which is more in expansin followed by glucanase and chitinase (Ghosh et al. 2012). Chitinase, IFR, and expansin did not induce significantly to methyl salicylate (MS) treatment (unpublished observation by the authors). Slight induction in glucanase was observed with methyl salicylate.

The role of expansin as intermediaries of cell wall softening and increase in fruit size during the fruit development in various species such as tomato, banana, and pear has been reported (Hiwasa et al. 2003; Trivedi and Nath 2004). The upstream regulatory sequences of expansin gene are characterized in banana (Trivedi and Nath 2004). It is highly fruit specific and strongly upregulated by ethylene. Yet another gene found to be upregulated during ripening in banana is chitinase. Clendennen and May 1997 reported the isolation and characterization of chitinase from unripe banana pulp. While chitinases are involved in pathogenesis-related functions, research has shown that in banana pulp this protein is associated with fruit ripening and it is a transient storage protein (Peumans et al. 2002). Another cell wall hydrolase upregulated during banana ripening is β-1,3-glucanase (Clendennen and May 1997; Peumans et al. 2000). Although the primary function of this hydrolase is related to defense, there are evidences to prove its role in banana fruit ripening.

The involvement of ACC synthase gene with the ethylene biosynthesis and ripening in banana has been reported (Liu et al. 1999). These initial findings were later confirmed by Huang et al. (2006), who described many isoforms of ACS in banana. Among various cell wall hydrolases, Clendennen and May (1997) reported numerous upregulated and downregulated genes during ripening. Class III chitinase has a storage role in banana pulp. It also supplies amino acids for the protein synthesis during ripening (Peumans et al. 2002). The involvement of expansin and polygalacturonase in fruit ripening has also been investigated (Trivedi and Nath 2004; Asha et al. 2007; Asif and Nath 2005). Liu et al. (2009) investigated MADS-box transcription factor gene (MuMADS1) in ethylene-induced fruit ripening in banana. In their studies, MuMADS1 was induced by ethylene during postharvest ripening. In naturally ripened banana, its expression was noticed after 6 days postharvest and increased further till it reached the maximal level on the same day when ethylene production peaked.

The isoflavone reductase (IFR) genes are involved in stress, defense, detoxification, and plant secondary product synthesis in plants. In legumes, IFR is involved in catalyzing the last but one step of the biosynthesis of isoflavonoid phytoalexins. IFR has also been reported in other plants such as tobacco, grapes, and rice. Kim et al. (2004) have shown that IFR genes respond to signals emanating from ethylene and jasmonic acid. The expression of the IFR gene in banana is indeed highly responsive to ethylene and being activated within 10 min of ethylene exposure (Gupta et al. 2006). It is likely that the process of ripening, which results in the conversion of acids to sugars as well as progressive softening of the cell wall, makes the fruit susceptible to pathogen attacks, and the expression of these defense-related genes is a response toward the prevention of such attacks.

Another aspect of fruit softening involves cell wall metabolism and one of the major genes involved in this process is expansin. Researches show that ripening-related expansin protein abundance is directly correlated with fruit softening and has supplementary indirect effects on pectin depolymerization, indicating the involvement of this protein in the softening process. It also may restrict or control the activities of other ripening-related enzymes essential for the softening of fruits (Brummell and Harpster 2001).

7.3 Somatic Embryogenesis and Plant Regeneration: A Prerequisite for Engineering-Delayed Fruit Ripening

One of the major requisites in banana genetic transformation is to establish a well reproducible protocol for incorporation of alien genes to banana genome. Among the various modules attempted (Strosse et al. 2006), embryogenic cell suspension cultures (ECS) have been appropriately suited for banana genetic transformation. Cultured young male flowers after 2–3 months showed a small quantity of yellow callus which was followed by a white and translucent callus formed on the previous callus followed by the appearance of somatic embryos (Fig. 7.1). Embryogenic cell clusters generally characterize embryogenic cell suspensions with potential for somatic embryo development and further conversion to plants. In our work, friable embryogenic callus of cv. Robusta upon transfer to suspension medium (Liq. M2) released embryogenic cells with dense cytoplasm. Addition of ascorbic acid to the cell suspension medium prevented cell blackening due to the phenolic oxidation. Suspension cultures consisted of a heterogeneous mixture of different kinds of cells and cell clusters. The embryogenic cells were small and more spherical in shape and had a dense yellow cytoplasm with very few small vacuoles, while the non-embryogenic cells were distinct with large

Fig. 7.1 Different stages of somatic embryogenesis in Banana cv. Robusta. (**a**) Green nodular callus observed after transfer to embryogenesis medium. (**b**) Distinct globular embryos observed after plating embryogenic callus. (**c**) Secondary embryogenesis and early stage of germination. (**d**) Germinating somatic embryos. (**e**) Plantlets emerging from germinating somatic embryos

vacuolated cells with meager cytoplasm and irregular shape. Initially, these cultures were very heterogeneous and contained large translucent cells as well as small dense cells. Upon frequent subculture at 3–4-day intervals and subsequently a 2-week interval, the suspension cultures comprised only of aggregates of small tightly packed cells with a dense cytoplasm. Secondary somatic embryogenesis was also observed occasionally. This was characterized by typical globular stage embryos directly from primary somatic embryos and occurred in clusters, initially appearing as small, hyaline protuberances on the surface of the clusters (Fig. 7.1). The embryos passed through a recognizable heart-shaped stage, a primary torpedo stage, a later torpedo stage, and finally green plumule emerged from these embryos (Fig. 7.1).

Associated with morphological development, there was a change in opacity from hyaline to translucent to opaque with a simultaneous change in the color from pale white to yellow green. The embryo development was achieved by plating 0.1 ml of PCV on the M3 medium. The success of initiating a good quality ECS depends largely on the quality of the selected embryos. In this study, compact embryogenic calli were not at all proper for suspension initiation, as they did not disintegrate into small embryogenic cell clusters. In banana, compact calli and large embryos either turned dark and stopped releasing embryogenic cell material or dedifferentiated into globules that only released non-embryogenic cells (Strosse et al. 2006). For embryo germination, well-developed embryos were transferred to germination medium. These embryos showed good germination characterized by formation of green plumules (Fig. 7.1). For the development of complete plantlets, these germinated embryos were further transferred to regeneration medium supplemented with required cytokinin (Fig. 7.1). The well-developed somatic embryo-derived plantlets ("emblings") grew normally during hardening in the greenhouse. The development of plantlets in the greenhouse could be attributed to their developmental route through somatic embryogenesis and good root and shoot system of the transferred plantlets.

7.3.1 Genetic Transformation of Banana for Delayed Fruit Ripening

We have adopted the previously established and optimized transformation protocol (Ghosh et al. 2009) to transform two commercially important banana cultivars Rasthali and Robusta with ripening-related genes. Expansin and MADSrin genes are earlier isolated from banana cv. Harichal and cloned in antisense orientation in binary vector pBI121. Promoters of two genes, expansin and IFR which are highly upregulated during ripening, were earlier isolated and cloned. Two IFR promoters were used which differ from each other by only 144 nucleotides (designated as IFR 1.2, 1,359 bp and IFR 1.4, 1,215 bp). Expansin promoter used has previously been isolated by Trivedi and Nath (2004).

MaMADS and *MaExpansin* genes used in our study have earlier been isolated from banana cv. Harichal and cloned in antisense orientation in appropriate vectors. The ECS for banana cultivars Robusta and Rasthali were transformed with the above constructs using the procedure optimized in our laboratory. Transformed plants were obtained for both the cultivars which were hardened in the greenhouse and later transferred to contained greenhouse for growing them till fruiting. Fruiting was obtained for Rasthali plant fruits of which were further analyzed with respect to ripening parameters as visual examination, softness, and gene expression profile. Considerable increase in shelf life of *MADSrin* and *MaExpansin* transgenic fruits as compared to control fruits was observed (unpublished observation by authors).

Ripening was evident in control non-transgenic fruits from day 2. By day 3 the fruits were ripe and fit for consumption. Black patches that started to appear on day 2 intensified by day 4. The fruits further deteriorated and could not be consumed by days 5 and 6. By day 10 the fruits had become very soft and black and showed fungal infection (Fig. 7.2). Fruits of transgenic line of MADS gene did not show any signs of ripening until day 2. By day 3 fruits had ripened as was evident by the color of the peel. In the following days, ripening became more evident. The fruits could be consumed until day 6, as compared to control where the fruits were unfit for consumption by day 5. Fruits of MADS transgenic line were fit for consumption even on day 10, as compared to control which had started showing signs of fungal infection by day 10 (Fig. 7.2). These initial observations were further confirmed by measuring firmness using a penetrometer (Fig. 7.3). Two different MADSrin transgenic lines were analyzed and compared to control fruits with respect to firmness. One set was treated with

Fig. 7.2 Delayed ripening in MADSrin transgenic fruits as compared to control fruits of Banana cv. Rasthali

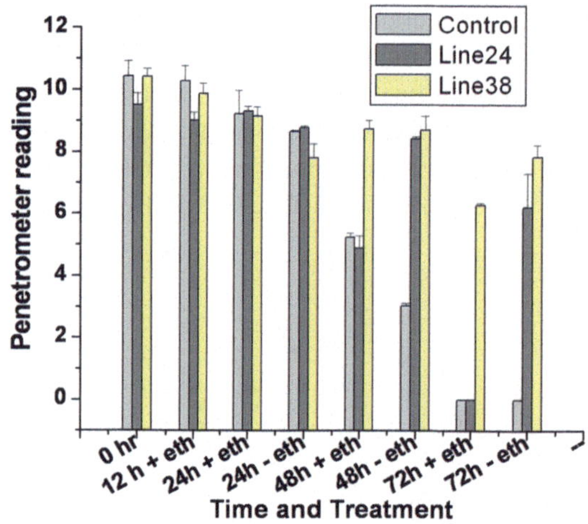

Fig. 7.3 Comparative analysis of fruit firmness of MADSrin transgenic fruits and control fruits of Banana cv. Rasthali

ethylene and another set was kept at room temperature without any ethylene treatment. At 0 h fruits of all the three lines were almost similar in firmness, with line 24 slightly less firm than the control and 38. No significant change in firmness was observed in all the three lines at 12 h after ethylene treatment. After 24 h both ethylene-treated and ethylene-untreated samples of all the three lines had almost similar firmness. Actual difference in firmness was clearly evident from 48 h. Ethylene-treated and ethylene-untreated samples of line 38 were much firmer than control at both 48 h and 72 h. Line 24 differed in this aspect. On ethylene treatment, the firmness of the fruits was same as that of control. But ethylene-untreated fruits were much firmer than control at both the above time points. Firmness of the transgenic fruits was maintained to a constant level until 24 h irrespective of ethylene treatment where as the control fruits showed a decreasing firmness trend until 24 h. There was a drastic decrease in firmness of both ethylene-treated and ethylene-untreated control fruits at 48 h. By 72 h control fruits had become so soft that penetrometer reading could not be taken. Ethylene-treated transgenic fruit was firmer than control at 48 h, but by 72 h, it had become too soft for reading to be taken. Ethylene-untreated transgenic fruits were much firmer than control at both above time points.

Expansin fruits did not show any signs of ripening until day 2. By day 3, ripening initiated as was evident by the color of the peel. In the following days, ripening became more evident. The fruits could be consumed until day 6, as compared to control where the fruits were unfit for consumption by day 5 (Fig. 7.4).

Firmness of the transgenic fruits was maintained to a constant level until 24 h irrespective of ethylene treatment, whereas the control fruits showed a decreasing firmness trend until 24 h. There was a drastic decrease in firmness of both ethylene-treated and ethylene-untreated control fruits at 48 h. By 72 h control fruits had become so soft that penetrometer reading could not be taken. Ethylene-treated transgenic fruit was firmer than control at 48 h, but by 72 h, it had become too soft for reading to be taken. Ethylene-untreated transgenic fruits were much firmer than control at both above time points (Fig. 7.5).

7.3.2 Promoters Useful for Engineering-Delayed Fruit Ripening

We have studied upstream regulatory sequences of cell wall hydrolases active during ripening. Expansin, chitinase, glucanase, and IFR were selected for their utility in delayed fruit ripening. The promoters were tagged with a reporter gene *gusA*. β-Glucuronidase expression was recorded fluorometrically. The data corresponding to the promoter strength under different conditions was studied, which can be utilized to identify the suitable promoter for fruit-specific expression to silence the target gene for delaying the ripening.

Chitinase promoter responded well to ethylene induction when it was treated with ethylene-releasing compound ethephon. Maximum expression level of 21,545.17 pmoles 4 MU/ h/ mg total protein was observed at 0.25 mM concentration of ethephon. This was approximately five times higher than un-induced cells. MJ in combination with ethephon (0.25 mM) further enhanced the level of gene expression which

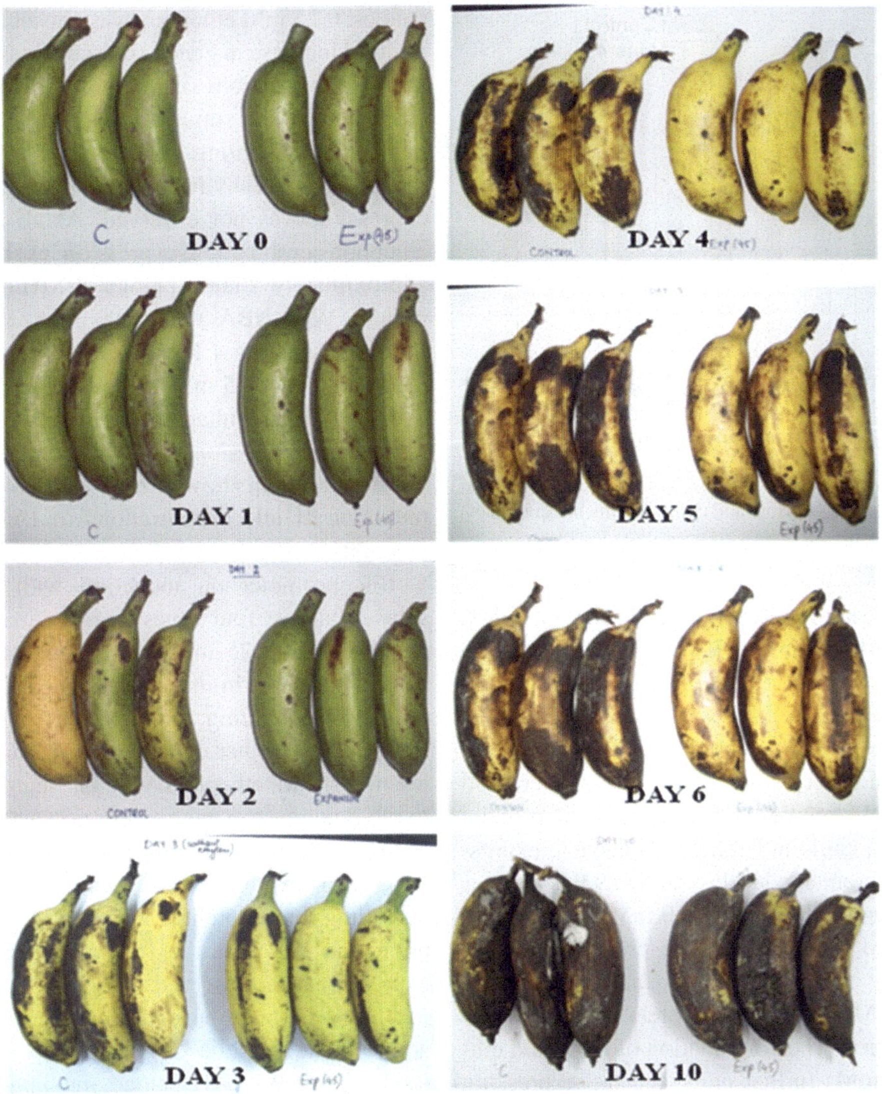

Fig. 7.4 Delayed ripening in Expansin transgenic fruits as compared to control fruits of Banana cv. Rasthali

reached up to 30,613.77 pmoles 4 MU/h/mg total protein. ABA in combination with ethephon treatment was not that effective with chitinase promoter as expression level dropped at all ABA concentration. However, ABA alone could induce an expression level that was at par with ethephon. At ABA 5 mM concentration, an expression level of 20,555 pmoles 4 MU/h/mg total protein was recorded. Significant induction was not noted with MS in combination with ethephon. At all concentrations of methyl salicylate, the expression level was lower than that of ethephon. IAA led to a significant reduction in gene expression at all concentrations. IAA 15 mM showed only 5,848.16 pmoles 4 MU/h/mg total protein (Ghosh et al. 2012).

The response of glucanase promoter with ethephon was even better than chitinase. Expression levels showed a significant increase of more than twofold with ethephon treatment. A maximum expression level of 30,173.95 pmoles 4 MU/h/mg total protein was recorded at 1 mM ethephon concentration. For further combination induction studies, 1 mM ethephon was chosen.

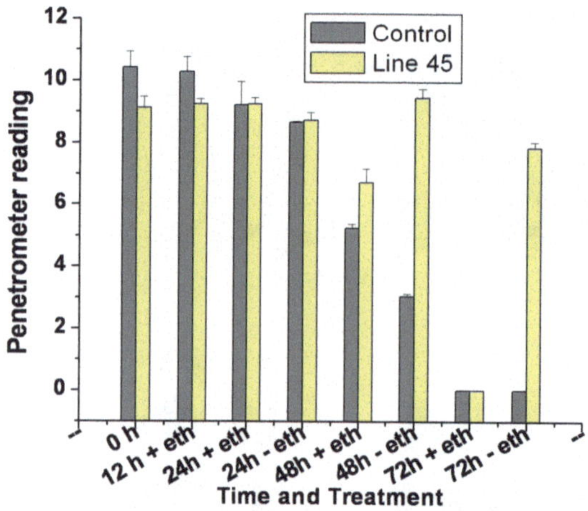

Fig. 7.5 Comparative analysis of fruit firmness of Expansin transgenic fruits and control fruits of Banana cv. Rasthali

Combinatorial treatment of MJ and ethephon showed further increase (two fold) in expression levels. Maximum expression with this combination was 49,222.97 pmoles 4 MU/h/mg total protein at 10 mM MJ concentration. The combination of ABA and ethephon (5 mM) further increased to 54,743.16 pmoles 4 MU/h/mg total protein. ABA alone could not enhance transgene expression which remained as low as 19,293.49 pmoles 4 MU/h/mg total protein at 10 mM ABA. MS (5 mM) could induce glucanase promoter as evidenced by the expression level of 39,043.77 pmoles 4 MU/h/mg total protein at 10 mM MJ concentration. The combination of ABA and ethephon (5 mM) further enhanced the expression level to 54,743.16 pmoles 4 MU/h/mg total protein. ABA alone could not elevate transgene expression. MS (5 mM) could induce glucanase promoter, and the transgene expression level which was 39,043.77 pmoles 4 MU/h/mg total protein was observed, which is comparatively higher compared to ethephon alone. IAA led to a significant reduction in the expression levels. Expression as low as 13,011.99 pmoles 4 MU/h/mg total protein was recorded with IAA 15 mM (Ghosh et al. 2012).

Expansin promoter on induction with ethephon showed almost three times increase in transgene expression at 0.25 mM as compared to untreated transformed cells with a maximum expression of 31,483.7 pmoles 4-MU/ h / mg total protein. For further combination induction studies, 0.25 mM ethephon was chosen. A further twofold increase in gene expression was recorded at all concentrations of MJ in combination with ethephon with maximum expression of 64,099.78 pmoles 4 MU/h/mg total protein at MJ 10 mM. With ABA any further enhancement of gene expression was not observed. At all ABA and ethephon combination, expression level remained approximately 20,000 pmoles 4 MU/h/mg total protein. With ABA alone, maximum expression of 15,987 pmoles 4 MU/h/mg total protein was recorded at ABA 5 mM. MS also could not produce any further enhancement in gene expression as highest expression was with ethephon alone. Gene expression showed a considerable twofold reduction at all concentrations of IAA (Ghosh et al. 2012).

IFR promoter on induction with ethephon showed almost four times increase in transgene expression at 0.25 mM as compared to control cells with a maximum expression of 23,987.84 pmoles 4 MU/h/mg total protein (unpublished results). For further combination induction studies, 0.25 mM ethephon was chosen. A further twofold increase in gene expression was recorded at MJ 5 mM in combination with ethephon with maximum expression of 43,465.25 pmoles 4 MU/h/mg total protein at MJ 10 mM. With ethephon and ABA, further enhancement of gene expression was observed. At ABA 15 mM and ethephon combination, expression level remained approximately 42,653.10 pmoles 4 MU/h/mg total protein. With ABA alone, maximum expression of 18,214 pmoles 4 MU/h/mg total protein was recorded at ABA 5 mM. A slight increase in gene expression was observed with MS. MS at 5 mM gave an expression level of 31,965 pmoles 4 MU/h/mg total protein. Gene expression showed a significant reduction at all concentrations of IAA (unpublished results by authors).

Beaudry et al. (1987) reported about climacteric nature and short shelf life of banana fruit. Ripening in banana is a complex developmental process involving different enzymes and genes (Brady 1987; Fischer and Bennett 1991). A wide range of hydrolases that regulate the process of fruit softening are activated at transcriptional and translational level. However, various studies suggested that these hydrolases might not be individually responsible for this process (Giovannoni et al. 1989; Smith et al. 1998; Brummell et al. 1999; Brummell and Harpster 2001).

Earlier work explained that expansins act as intermediaries of cell wall softening and increase the fruit size during the development of fruits reported in various species such as tomato, banana, and pear (Hiwasa et al. 2003; Kitagawa et al. 1995; Trivedi and Nath 2004). Trivedi and Nath (2004) reported the cloning and characterization of expansin promoter from banana which is strongly upregulated by ethylene and is ripening specific. Clendennen and May (1997) investigated the chitinase from unripe banana pulp; these chitinases are basically associated with pathogenesis-related functions; however, reports indicate that in banana pulp this abundant protein is closely associated with fruit ripening (Peumans et al. 2002). β-1,3-Glucanase is another cell wall hydrolase upregulated during banana ripening (Medina-Suarez et al. 1997; Clendennen and May 1997; Peumans et al. 2000). This hydrolase is involved in defense responses, but there are evidences to prove its role in banana fruit ripening. IFR genes in plants are related to stress, defense, detoxification, and plant secondary product synthesis. In legumes IFR is involved in catalyzing phytoalexins biosynthesis. IFR has also been reported in other plants such as tobacco, grapes, and rice. Recently Kim et al. (2004) described that IFR genes respond to ethylene, jasmonic acid, and wounding in rice. The expression of the IFR gene in banana is indeed highly responsive to ethylene and being activated within 10 min of ethylene exposure (Gupta et al. 2006). It is likely that the process of ripening results in the conversion of acids to sugars and softening of the cell wall, which makes the fruit susceptible to pathogen attack, and the expression of this defense-related genes is a response toward prevention of such an attack during ripening.

7.4 Future Directions

Proper revenue generation and maintenance of a superior quality of many fruit crops after harvest involve many economical considerations although conventional technologies have been standardized in some cases. Biotechnological strategies have evoked the resurgence of employing transgenic plants for integrating desirable genes or removal of genes from plant genome. Knowledge from traditional efficient practices and using modern techniques to speed up or reduce the fruit ripening process has now made clear elucidation of precise functioning of numerous genes responsible for ripening of the fruits. Work on transgenic plants during the last two decades has generated useful data for meaningful contributions to both the fundamental and applied aspects of fruit ripening. Establishment of a reproducible protocol of transgenic plant regeneration is a major challenging task which has to be accomplished for further research on ripening process of other cultivars of banana. In this connection, delayed ripening in tomato was one of earliest examples of successful demonstration of transgenic technology. In particular, extensive knowledge of the biochemistry of the metabolic steps in a ripening pathway is required for modifying and balancing the cellular metabolism. The complexity of such systems needs the competent expression of enzymes and the understanding of the effect of intermediates, end products, and elicitors on gene regulation and function. The development of these platforms can be combined with other strategies to acquire the desired end products or to develop new platforms. Rapid strides in genomics involving gene sequencing, gene annotation, gene editing, and phylogenetic analysis in conjunction with bioinformatics studies have considerably advanced research on fruit biotechnology. A cluster of gene-silencing agents such as DNA enzymes (DNAzymes), small interfering RNA (siRNA), antisense oligonucleotides, decoys, ribozymes, and aptamers are the other options needed to be explored for ripening process in fruits. *In planta* experiments on overexpression or suppression and mutants coupled with biochemical characterization of ripening pathway will successfully integrate genomics knowledge in fruit ripening process. Additionally, recent advances in the next-generation sequencing makes it possible to predict the targets for modification (Gapper et al. 2014) in a short time, and also the advances in information technology, the understanding of the fruit development, and modeling to predict the different factors responsible for fruit ripening can be achieved in a quick time. Recently, Bi et al. (2015) identified 26 novel miRNAs from the banana fruit, which are involved in ethylene-induced ripening. Further, functional analysis and experimental validation are required to pinpoint the exact miRNA, which can be targeted for successfully delaying the ripening of fruits in the near future.

Acknowledgment The authors thank Dr. PravendraNath, NBRI, Lucknow, for his guidance in the work as well as the National Botanical Research Institute, Lucknow, India, for providing laboratory facilities. VAB thanks the Indian National Science Academy, New Delhi, for senior scientist fellowship.

References

Adams-Phillips L, Barry C, Giovannoni J (2004) Signal transduction systems regulating fruit ripening. Trends Plant Sci 9(7):331–338. doi:10.1016/j.tplants.2004.05.004

Ankala A, Luthe DS, Williams WP, Wilkinson JR (2009) Integration of ethylene and jasmonic acid signaling pathways in the expression of maize defense protein Mir1-CP. Mol Plant Microbe Interact 22(12):1555–1564. doi:10.1094/Mpmi-22-12-1555

Asha, Sane VA, Sane AP, Nath P (2007) Multiple forms of alpha-expansin genes are expressed during banana fruit ripening and development. Postharvest Biol Technol 45(2):184–192. doi:10.1016/j.postharvbio.2007.03.003

Asif MH, Nath P (2005) Expression of multiple forms of polygalacturonase gene during ripening in banana fruit. Plant Physiol Biochem 43(2):177–184. doi:10.1016/j.plaphy.2005.01.011

Bapat VA, Trivedi PK, Ghosh A, Sane VA, Ganapathi TR, Nath P (2010) Ripening of fleshy fruit: molecular insight and the role of ethylene. Biotechnol Adv 28(1):94–107. doi:S0734-9750(09)00180-3 [pii] 10.1016/j.biotechadv.2009.10.002

Beaudry RM, Paz N, Black CC, Kays SJ (1987) Banana ripening – implications of changes in internal ethylene and Co2 concentrations, pulp fructose 2,6-bisphosphate concentration, and activity of some glycolytic-enzymes. Plant Physiol 85(1):277–282. doi:10.1104/Pp.85.1.277

Bi F, Meng X, Ma C, Yi G (2015) Identification of miRNAs involved in fruit ripening in Cavendish bananas by deep sequencing. BMC Genomics 16:776. doi:10.1186/s12864-015-1995-1

Bottcher C, Keyzers RA, Boss PK, Davies C (2010) Sequestration of auxin by the indole-3-acetic acid-amido synthetase GH3-1 in grape berry (Vitis vinifera L.) and the proposed role of auxin conjugation during ripening. J Exp Bot 61(13):3615–3625. doi:10.1093/jxb/erq174

Brady CJ (1987) Fruit ripening. Annu Rev Plant Physiol 38:155–178. doi:10.1146/annurev.pp.38.060187.001103

Brummell DA, Harpster MH (2001) Cell wall metabolism in fruit softening and quality and its manipulation in transgenic plants. Plant Mol Biol 47(1–2):311–340. doi:10.1023/A:1010656104304

Brummell DA, Hall BD, Bennett AB (1999) Antisense suppression of tomato endo-1,4-beta-glucanase Cel2 mRNA accumulation increases the force required to break fruit abscission zones but does not affect fruit softening. Plant Mol Biol 40(4):615–622. doi:10.1023/A:1006269031452

Cheesman EE (1948) Classification of the bananas. II: the genus Musa L. Kew Bull 2:106–117

Clendennen SK, May GD (1997) Differential gene expression in ripening banana fruit. Plant Physiol 115(2):463–469

FAO (2012) http://faostat3.fao.org/faostat-gateway/go/to/home/E; http://nhb.gov.in/area-pro/database-2012.pdf

Fischer RL, Bennett AB (1991) Role of cell-wall hydrolases in fruit ripening. Annu Rev Plant Physiol Plant Mol Biol 42:675–703. doi:10.1146/annurev.pp.42.060191.003331

Gapper NE, Giovannoni JJ, Watkins CB (2014) Understanding development and ripening of fruit crops in an 'omics' era. Hortic Res 1:14034. doi:10.1038/hortres.2014.34

Ghosh A, Ganapathi TR, Nath P, Bapat VA (2009) Establishment of embryogenic cell suspension cultures and Agrobacterium mediated transformation in an important Cavendish banana cv. Robusta (AAA). Plant Cell Tiss Org Cult 97:131–139

Ghosh A, Shekhawat UKS, Ganapathi TR, Bapat VA (2012) Analysis of banana fruit-specific promoters using transient expression in embryogenic cells of banana cultivar Robusta (AAA Group). J Plant Biochem Biotechnol 21(2):189–197. doi:10.1007/s13562-011-0070-5

Giovannoni JJ (2004) Genetic regulation of fruit development and ripening. Plant Cell 16:S170–S180. doi:10.1105/tpc.019158

Giovannoni JJ, Dellapenna D, Bennett AB, Fischer RL (1989) Expression of a chimeric polygalacturonase gene in transgenic rin (ripening inhibitor) tomato fruit results in polyuronide degradation but not fruit softening. Plant Cell 1(1):53–63

Gupta SM, Srivastava S, Sane AP, Nath P (2006) Differential expression of genes during banana fruit development, ripening and 1-MCP treatment: presence of distinct fruit specific, ethylene induced and ethylene repressed expression. Postharvest Biol Technol 42(1):16–22. doi:10.1016/j.postharvbio.2006.05.002

Hiwasa K, Rose JKC, Nakano R, Inaba A, Kubo Y (2003) Differential expression of seven alpha-expansin genes during growth and ripening of pear fruit. Physiol Plant 117(4):564–572. doi:10.1034/j.1399-3054.2003.00064.x

Huang FC, Do YY, Huang PL (2006) Genomic organization of a diverse ACC synthase gene family in banana and expression characteristics of the gene member involved in ripening of banana fruits. J Agric Food Chem 54(11):3859–3868. doi:10.1021/jf060001w

Jiang YM, Joyce DC, Macnish AJ (2000) Effect of abscisic acid on banana fruit ripening in relation to the role of ethylene. J Plant Growth Regul 19(1):106–111

Jones B, Frasse P, Olmos E, Zegzouti H, Li ZG, Latche A, Pech JC, Bouzayen M (2002) Down-regulation of DR12, an auxin-response-factor homolog, in the tomato results in a pleiotropic phenotype including dark green and blotchy ripening fruit. Plant J 32(4):603–613. doi:10.1046/j.1365-313X.2002.01450.x

Kesari R, Trivedi PK, Nath P (2010) Gene expression of pathogenesis-related protein during banana ripening and after treatment with 1-MCP. Postharvest Biol Technol 56(1):64–70. doi:10.1016/j.postharvbio.2009.11.012

Kim ST, Kim SG, Hwang DH, Kang SY, Koo SC, Cho MJ, Kang KY (2004) Expression of a salt-induced protein (SALT) in suspension-cultured cells and leaves of rice following exposure to fungal elicitor and phytohormones. Plant Cell Rep 23(4):256–262. doi:10.1007/s00299-004-0836-5

Kitagawa Y, Kanayama Y, Yamaki S (1995) Isolation of beta-galactosidase fractions from Japanese pear – Activity against native cell-wall polysaccharides. Physiol Plant 93(3):545–550. doi:10.1034/j.1399-3054.1995.930320.x

Liu XJ, Shiomi S, Nakatsuka A, Kubo Y, Nakamura R, Inaba A (1999) Characterization of ethylene biosynthesis associated with ripening in banana fruit. Plant Physiol 121(4):1257–1265. doi:10.1104/pp.121.4.1257

Liu JH, Xu BY, Hu LF, Li MY, Su W, Wu J, Yang JH, Jin ZQ (2009) Involvement of a banana MADS-box transcription factor gene in ethylene-induced fruit ripening. Plant Cell Rep 28(1):103–111. doi:10.1007/s00299-008-0613-y

Lohani S, Trivedi PK, Nath P (2004) Changes in activities of cell wall hydrolases during ethylene-induced ripening in banana: effect of 1-MCP, ABA and IAA. Postharvest Biol Technol 31(2):119–126. doi:10.1016/j.postharvbio.2003.08.001

Manning K, Tor M, Poole M, Hong Y, Thompson AJ, King GJ, Giovannoni JJ, Seymour GB (2006) A naturally occurring epigenetic mutation in a gene encoding an SBP-box transcription factor inhibits tomato fruit ripening. Nat Genet 38(8):948–952. doi:10.1038/ng1841

Medina-Suarez R, Manning K, Fletcher J, Aked J, Bird CR, Seymour GB (1997) Gene expression in the pulp of ripening bananas – Two-dimensional sodium dodecyl sulfate-polyacrylamide gel electrophoresis of in vitro translation products and cDNA cloning of 25 different ripening-related mRNAs. Plant Physiol 115(2):453–461. doi:10.1104/Pp.115.2.453

Pathak N, Sanwal GG (1999) Regulation of the ripening of banana (Musa acuminata) fruits by chemicals. Indian J Agric Sci 69(1):17–20

Pathak N, Asif MH, Dhawan P, Srivastava MK, Nath P (2003) Expression and activities of ethylene biosynthesis enzymes during ripening of banana fruits and effect of 1-MCP treatment. Plant Growth Regul 40(1):11–19. doi:10.1023/A:1023040812205

Peumans WJ, Barre A, Derycke V, Rouge P, Zhang WL, May GD, Delcour JA, Van Leuven F, Van Damme EJM (2000) Purification, characterization and structural analysis of an abundant beta-1,3-glucanase from banana fruit. Eur J Biochem 267(4):1188–1195. doi:10.1046/j.1432-1327.2000.01117.x

Peumans WJ, Proost P, Swennen RL, Van Damme EJM (2002) The abundant class III chitinase homolog in young developing banana fruits behaves as a transient vegetative storage protein and most probably serves as an important supply of amino acids for the synthesis of ripening-associated proteins. Plant Physiol 130(2):1063–1072. doi:10.1104/pp.006551

Rahman A, Chowdhury FR, Alam MB (2008) Artificial ripening: what we are eating. J Med 9:42–44

Rose JKC, Saladie M, Catala C (2004) The plot thickens: new perspectives of primary cell wall modification. Curr Opin Plant Biol 7(3):296–301. doi:10.1016/j.pbi.2004.03.013

Singh RK, Sane VA, Misra A, Ali SA, Nath P (2010) Differential expression of the mango alcohol dehydrogenase gene family during ripening. Phytochemistry 71(13):1485–1494. doi:10.1016/j.phytochem.2010.05.024

Smith DL, Starrett DA, Gross KC (1998) A gene coding for tomato fruit beta-galactosidase II is expressed during fruit ripening – Cloning, characterization, and expression pattern. Plant Physiol 117(2):417–423. doi:10.1104/Pp.117.2.417

Strosse H, Schoofs H, Panis B, Andre E, Reyniers K, Swennen R (2006) Development of embryogenic cell suspensions from shoot meristematic tissue in bananas and plantains (Musa spp.). Plant Sci 170(1):104–112. doi:10.1016/j.plantsci.2005.08.007

Theologis A (1992) One rotten apple spoils the whole bushel – the role of ethylene in fruit ripening. Cell 70(2):181–184. doi:10.1016/0092-8674(92)90093-R

Trivedi PK, Nath P (2004) MaExp1, an ethylene-induced expansin from ripening banana fruit. Plant Sci 167(6):1351–1358. doi:10.1016/j.plantsci.2004.07.005

Vendrell M (2003) Biology and biotechnology of the plant hormone ethylene III. IOS Press Book, Amsterdam, The Netherlands

Vrebalov J, Ruezinsky D, Padmanabhan V, White R, Medrano D, Drake R, Schuch W, Giovannoni J (2002) A MADS-box gene necessary for fruit ripening at the tomato ripening-inhibitor (Rin) locus. Science 296(5566):343–346. doi:10.1126/science.1068181

Zhang M, Yuan B, Leng P (2009) The role of ABA in triggering ethylene biosynthesis and ripening of tomato fruit. J Exp Bot 60(6):1579–1588. doi:10.1093/jxb/erp026

Metabolite Profiling in Banana

8

K.S. Shivashankara

Abstract

Metabolomics is relatively a new field of study used for identifying the metabolites involved in bringing about the phenotypic variations in terms of resistance to both biotic and abiotic stresses and also for fruit quality. Metabolomics also helps in identifying the proteins and genes responsible for the genotypic variations for the desired character. Comprehensive metabolomic studies on banana are not available in the literature. However the genotypic differences have been observed in banana for many phytochemicals related to disease and insect pest resistance. Wide variations for provitamin A carotenoids are available in banana genotypes. In addition to the carotenoids, catecholamines like dopamines are also reported to be associated with the antioxidant capacity of the bananas. Bananas are also good sources of phytosterols and B vitamins. Wide genotypic variability exists for sugar profiles in banana. Sucrose, fructose, and glucose are the major sugars in fruits. Esters are the major group of volatile aroma compounds in banana. Volatile compounds also include alcohols and aldehydes. Resistance to diseases has been associated with the higher phenolics especially cell wall-bound phenolics like lignins. Lignins have also been reported to be associated with nematode resistance in bananas. Banana stems, sap, and flowers have been used in many traditional medicines. Banana flowers and fruit peels are reported to have many antioxidant phytochemicals like anthocyanins, sterols, saponins, tannins, serotonins, and dopamines.

Keywords

Banana • Ripening • Metabolites • Carotenoids • Phenolics • Aroma volatiles • Diseases

K.S. Shivashankara (✉)
ICAR-Indian Institute of Horticultural Research,
Hessaraghatta Lake Post, Bengaluru 560089,
Karnataka, India
e-mail: shivaiihr@yahoo.com

S. Mohandas, K.V. Ravishankar (eds.), *Banana: Genomics and Transgenic Approaches for Genetic Improvement*, DOI 10.1007/978-981-10-1585-4_8

8.1 Introduction

Development of high-throughput technologies made significant advances in genomics, transcriptomics, and proteomics in plant science. Compared to these fields, metabolomics is relatively a young and growing area. Proteomics also gives information about the enzymes responsible for the production of various metabolites responsible for the phenotype. Metabolomics is the term used for high-throughput analyses of almost all the metabolites at a given point of time in plant tissues. This is made possible to some extent by the advances in mass spectrometry techniques in gas chromatography, liquid chromatography, capillary electrophoresis, etc. Since complete metabolite analysis is difficult in spite of the advances in the instrumentation, targeted metabolite profiling is preferred to explain the variation in phenotypes of the plants. The complexity of metabolomics is not only from the lack of high-throughput techniques but also from their large numbers. So far around 50,000 metabolites have been identified in plants and it is estimated that the number may exceed two lakh compounds in plants (De Luca and St. Pierre 2000; Pichersky and Gang 2000; Fiehn 2001, 2002). Thus, metabolomics represents a considerable challenge for plant scientists.

Metabolomics research will prove an invaluable tool for generating information and their use in many fields. Metabolomics information not only helps in identifying the complex interaction between many metabolic networks within the plants but also helps in elucidating the metabolic responses to the environmental challenges. The genetic variability in metabolic responses to the environments will help in explaining the biochemical responses through the metabolites. This will further provide more insights into the genetic and environmental interactions leading to the phenotypic differences between the genotypes. Plant resistance mechanisms, diversity in productivity mechanisms, and variations in the quality of the production can also be explained in better ways by the metabolomic studies. It also helps in identifying the metabolite pathways triggered for the development of tolerance or resistance against stresses. The major challenges are the nonavailability of reference compounds and the need for appropriate, dedicated bioinformatics tools, and these can be approached effectively with sufficient speed only through coordinated and collaborative efforts.

In banana such metabolomic studies are not yet available either for biotic and abiotic stresses or for establishing the genotypic diversity. However, some of the studies have established the genotypic variabilities in certain metabolites involved in fruit quality, and resistance for diseases and insect pests has been given here.

8.2 Medicinal Uses of Banana

Most parts of the banana plants have been reported to have medicinal properties (Amit and Shailandra 2006). The plant sap is used for curing hysteria, hemorrhages, leprosy, epilepsy, fevers and acute dysentery, and diarrhea; it is also applied for insect and other stings and bites. The medicinal effects of different plant parts and their constituents are summarized in Table 8.1.

Banana peel and pulp are also reported to have good antifungal and antibiotic principles against mycobacteria (Brooks 2008; Omojasola and Jilani 2009). Peel and pulp extracts of green fruits were reported to be active against a fungal disease of tomato plants (Ponnuswamy et al. 2011).

8.3 Metabolite Profiling of Banana Plant Parts

Metabolites reported in different plant parts are reported in Table 8.2. Banana plant parts are used in many culinary preparations and also as traditional medicine. Banana pseudo-stem and flowers are used in cooking and many different preparations are made. The stem and flower are known to have a good quantity of iron, potassium, and also lot of antioxidants.

Table 8.1 Various medicinal uses of banana plant

Diseases	Plant part	Effect	Constituent	References
Hysteria, leprosy, epilepsy, hemorrhages, fever, dysentery, diarrhea	Plant sap	Reduction	Not reported	Amit and Shailandra (2006)
Insect bites and stings	Plant sap	Reduction	Not reported	Amit and Shailandra (2006)
Malignant ulcers, dysentery, diarrhea	Ash of peels and leaves	Reduction	Not reported	Amit and Shailandra (2006)
Diabetes, bronchitis, and ulcers	Cooked flowers	Reduction	Not known	Amit and Shailandra (2006)
Burns and skin afflictions	Young leaves are placed	Curing	Not known	Girish and Satish (2008)
Digestive disorders	Roots and seed mucilages	Curing	Not known	Bhat et al. (2010b) and 2010a
Digestive disorders and ulcers	Ripe peel and pulp	Serotonin inhibits gastric secretion and stimulates smooth muscles of the intestine	Norepinephrine, dopamine, and serotonin	Anhwange et al. (2009) and Ratule et al. (2007)
Anemia	Fruit	Reduces anemia by stimulating hemoglobin	High iron content	Amit and Shailandra (2006)
Blood pressure, heart beats, body water balance	Fruit	Maintains BP, normalizes heart beats, and maintains water balance	High potassium content	Girish and Satish (2008) and Debabandya et al. (2010)
Stroke	Fruit	Reduces the risk by 40 %	Not known	Amit and Shailandra (2006)
Depression	Fruit	Reduces depression	High tryptophan and serotonin	Girish and Satish (2008)
Acidity, heartburn, and morning sickness	Fruit	Reduces acidity, heart burn and morning sickness	Natural antacid effect	Mokbel et al. (2005) and Amit and Shailandra (2006)
Hypertension	Fruit	Calms the nervous system	High B vitamins	Singh and Bhat (2003)
Temperature control	Fruit	Controls physical and emotional temperature in expectant mothers	Cooling effect	Mokbel et al. (2005)
Antimicrobial effect	Sap, peel, and leaves	Reduces microbial population	Phenols, tannins, saponins, alkaloids, steroids, flavonoids, and carbohydrates	Singh and Bhat (2003)

(continued)

Table 8.1 (continued)

Diseases	Plant part	Effect	Constituent	References
Antimicrobial activity	Ethanol extract of the peel	The least MIC was 16 mg/ml against *Salmonella typhi*, while *Bacillus subtilis* and *Staphylococcusaureus* showed the highest MIC of 512.5 mg/ml	Have glycosides, alkaloids, saponins, tannins, flavonoids, and volatile oils	Ehiowemwenguan et al. (2014)
		Salmonella typhi, *Micrococcus luteus*, and *Staphylococcus aureus* were not inhibited by the water extract		
Antifungal activity	Leaf methanol extract of *Musa sapientum*	MIC values against *Microsporum canis* were from 0.0625 mg/ml and 0.5 mg/ml, against *Trichophyton tonsurans*, no effect on *T. rubrum*	Activity was due to catecholamines, tryptophan, indole compounds, tannin, several triterpenes and carbohydrates, alkaloids, cardenolides, saponins, and flavonoids	Ige et al. (2015)

8.4 Varietal Variation in Metabolites

8.4.1 Carotenoids

Varietal variation in carotenoids and vitamin C exists in banana. Dwarf Brazilian ("apple") bananas recorded more vitamin C (12.7 mg/100 g fresh weight) than Williams fruit (4.5 mg/100 g). Dwarf Brazilian bananas also showed higher β-carotene (96.9 μg/100 g fresh weight) and α-carotene (104.9 μg /100 g) than Williams fruit (55.7 μg β-carotene and 84.0 μg α-carotene/100 g fresh weight). Bananas contain higher concentrations of lutein than that of provitamin A pigments, α- and β-carotene (Wall 2006). Germplasm diversity in carotenoid profile in banana pulp is given in Table 8.3. Ripe banana pulp is reported to be a good source of unsaturated fatty acids like linoleic and linolenic acids and also health-benefitting sterols.

A wide variation in the total carotenoid content was observed in banana germplasm. Major provitamin A carotenoid was *t*-β-carotene followed by *t*-α-carotene. *t*-β-Carotene content varied from 1 pmol/g dw to 59 mg/g dw indicating that certain genotypes of banana are excellent source of carotenoids and can be used for enhancing the vitamin A content in commercial varieties (Table 8.3). Wide genotypic variations in carotenoid content were reported by Mattos et al. (2010). Higher amount of carotenoids is associated with the B genome and was found to be more in triploids AAB and ABB genomic groups than AAA groups. Higher beta-carotene and vitamin A activity was observed in genotypes Aibwo, Fagufagu, and Ropa, and very low beta-carotene was seen in Grand Naine, Cavendish, Mpologoma, Pongani, Tereza, and Yangambi (Pereira and Marasch 2015).

Lipophilic constituents of the fruit other than carotenoids indicated that total fatty acids were more in cv. Williams, Zeig, and Gruesa (1281–1295 mg/kg dwt) and least fatty acids were seen in Dwarf Red (559 mg/kg dwt) (Table 8.4). Almost all the varieties except cv. Gruesa showed higher saturated fatty acids compared to unsaturated fatty acids. Palmitic acid was the major fatty acid (16:0) closely followed by linolenic (18:3), linoleic (18:2), and stearic acids (18:0). Wide variation was seen in the content of fatty acids among the genotypes. Higher unsaturated

Table 8.2 Metabolites in different plant parts of banana

Plant parts	Metabolites reported	References	Genotypes
Sap of dessert bananas	Flavonoids like apigenin, naringenin, myricetin, kaempferol, and quercetin, rich in phenolic compounds	Schieber et al. (2001) and Pothavorn et al. (2010)	Diverse species
Banana bracts	Anthocyanins like petunidin, malvidin, pelargonidin, delphinidin, cyanidin, and peonidin	Kitdamrongsont et al. (2008)	
Pulp of dessert banana and plantains	Condensed tannin like epigallocatechin, leucocyanidin	Lewis et al. (1999) and Santos et al. (2010)	Dessert and plantain cultivars
Pulp and peel of 13 varieties of banana	Caffeic acid-hexoside, ferulic acid-hexoside, sinapic acid-hexoside; ferulic acid-dihexoside; myricetin-deoxyhexose-hexoside (high in pulp), ferulic acid, sinapic acid, quercetin-deoxyhexose-hexoside (high in pulp), rutin, methylmyricetin-deoxyhexose-hexoside, quercetin-hexoside, kaempferol-deoxyhexose-hexoside, kaempferol-3-*O*-rutinoside, isorhamnetin-3-*O*-rutinoside. Pulp was dominated by hydroxycinnamic acids; peel was dominated by flavonoids	Tsamo et al. (2015)	Plantain and dessert varieties
Fruit pulp of Rasthali, Karpooravalli, Manjal Vazhaipazham (yellow), and Pachai Vazhaipazham (green)	Flavonoid content and metal-chelating activity were superior in Rasthali. Total phenolics, free radical scavenging activity, and reducing power high in green variety	Babu et al. (2012)	*Musa acuminata* sp.
Ripe pulp of *M. acuminata* and *M. balbisiana* species, namely 'Chinese Cavendish', 'Giant Cavendish', 'Dwarf Red', 'Grand Naine', 'Eilon', 'Gruesa', 'Silver', 'Ricasa', 'Williams' and 'Zelig'	Lipophilic fraction consisted mainly of fatty acids (579–1295 mg kg^{-1} of dry pulp) and sterols (155–316 mg kg^{-1} of dry pulp) and long-chain aliphatic alcohols (22–81 mg kg^{-1} of dry pulp). Hexadecanoic and octadeca- 9,12,15-trienoic acids and β-sitosterol were the major compounds found in all pulp samples	Vilela et al. (2014)	*M. acuminata* and *M. balbisiana* sp.
Peel and pulp of ripe dessert banana	Catecholamines, pyridoxine (vit. B6), phenolics and carotenoids, volatiles like esters and alcohols. Dopamine (80–560 mg per 100 g in the peel and 2.5–10 mg in the pulp) and palmitic, linoleic, linolenic, and oleic acids are the major fatty acids, and stearoyl glucosides (273–888 kg/kg) in ripe pulp	Kanazawa and Sakakibara (2000), Verde-Mendez et al. (2003), Van den Berg et al. (2000), Pérez et al. (1997), Nogueira et al. (2003) and Oliveira et al. (2008)	
Mauli banana stem extract	Tannins (67.59 %), saponin (14.49 %), alkaloid (0.347 %), flavonoid (0.253 %), ascorbic acid (mg/mL) (0.44)	Apriasari et al. (2014)	Mauli bananas

(continued)

Table 8.2 (continued)

Plant parts	Metabolites reported	References	Genotypes
Stem sap of banana accessions	Apigenin glycosides, myricetin glycoside, myricetin-3-*O*-rutinoside, naringenin glycosides, kaempferol-3-*O*-rutinoside, quercetin-3-*O*-rutinoside, dopamine, and *N*-acetylserotonin	Pothavorn et al. (2010)	*Musa balbisiana*, *Musa laterita*, *Musa ornata*, and *Musa acuminata*
Organic banana peel	High total phenols, among carotenoid lutein, zeaxanthin, β-cryptoxanthin, β- and α-carotene were present in decreasing amounts; catechins were also present	Pereira et al. (2012)	*Musa* spp. cv. Prata Anã
Banana pulp	90 % of carotenoids are trans-α-carotene and trans-β-carotene; 10 % are lutein and other compounds	Davey et al. (2006)	
Banana pulp	Total volatile aroma constituents in "Dwarf Cavendish," "Giant Cavendish," "Robusta," and "Williams" were 93.0, 116.5, 157.3, and 157.0 mg/kg, respectively. Major groups were esters, alcohols, aldehydes, ketones, and acids. 3-Methyl butyl butanoate ester was the major constituent. Dopamine levels were 205–10 mg per 100 g	Nogueira et al. (2003) and Kanazawa and Sakakibara (2000)	"Dwarf Cavendish," "Giant Cavendish," "Robusta," and "Williams"

fatty acids were observed in Gruesa, Williams, Zeig, and Ellon cultivars. Genotypes also exhibited very good content of sterols (155–316 mg/kg dwt) with sitosterol as the major constituent. Intake of phytosterols can reduce the absorption of cholesterols in the body (Vilela et al. 2014).

8.4.2 Phenolics

Banana peel and pulp showed genotypic variability in the phenolic content. The pulp of dessert and hybrid bananas showed higher myricetin content, and the content of ferulic acid and sinapic acids was very low compared to myricetin. On the other hand, some of the plantains had higher content of ferulic acid followed by myricetin. The major component in the peel was rutin closely followed by kaempferol, and the content of phenolics was significantly higher in the plantains when compared to the dessert and hybrid bananas.

Ferulic acid-hexoside was highest in genotype Mbeta 1, followed by Big Ebenga and Essang. It was higher in plantains. Lower ferulic and sinapic acids were observed in dessert and hybrid bananas. Myricetin-deoxyhexose-hexoside was the next higher compound and was found similar among the genotypes (Tsamo et al. 2015). Among the flavonoids, rutin was higher followed by kaempferol and myricetin. Among the genotypes, higher rutin was recorded in genotypes Columbia, Moto Ebanga, Mbeta 1, Red Yade, and Njombe, all belonging to plantains. Among dessert bananas, genotype Gros Michel had higher rutin and Grand Naine did not show any rutin (Tsamo et al. 2015). Kaempferol also showed a similar trend. Principal component analysis of the peel and pulp chemicals indicated that the dessert types, plantains, and hybrids grouped differently. Among the constituents, ferulic acid, caffeic acid, and sinapic acid, dihexosides, quercetin, and myricetins were more in the pulp, and rutin, kaempferol, and isorhamnetins were more in the peels (Tsamo et al. 2015).

The stem extracts of Mauli bananas contain many bioactive compounds with tannin and saponins as the major compounds amounting to 82 %

Table 8.3 Carotenoid variations in the banana genotypes

Genotype	*t*-AC	*t*-BC	*c*-BC	RAE	Country
Aibwo/Suria #1	23.58 mg/gww	59.45 mg/gww	nd	2.95 mg/g	Makira, Solomon Islands
Aibwo/Suria #2	15.17 mg/gww	25.72 mg/gww	nd	2.77 mg/g	Makira, Solomon Islands
Bantol Red	70 nmol/gdw	130 nmol/gdw	11 nmol/gdw	7.63 µg/g	Philippines
Batard	35 pmol/gdw	36 pmol/gdw	5 pmol/gdw	2.51 ng/g	Belgium
Batard	35 nmol/gdw	38 nmol/gdw	4 nmol/gdw	2.57 µg/g	Cameroon
Baubaunio	2.49 mg/gww	3.32 mg/gww	nd	0.37 mg/g	Makira, Solomon Islands
Cavendish	8 pmol/gdw	6 pmol/gdw	1 pmol/gdw	0.47 ng/g	Humid tropical Africa
Cavendish	8 nmol/gdw	5 nmol/gdw	1 nmol/gdw	0.42 µg/g	Cameroon
Cavendish	nd	4.6 µg/gww	nd	0.38 µg/g	Belgium
Chek Porng Moan	50 nmol/gdw	65 nmol/gdw	9 nmol/gdw	4.23 µg/g	Cambodia
Dimaemamosi	nd	24.17 µg/gww	nd	2.01 µg/g	Papua New Guinea
Duningi	nd	7.43 µg/gww	nd	0.62 µg/g	Papua New Guinea
Entukura	nd	4.90 µg/gww	nd	0.41 µg/g	Uganda
Enzirabahima	nd	3.19 µg/gww	nd	0.27 µg/g	Uganda
Fagufagu	15.24 mg/gww	34.28 mg/gww	nd	3.49 mg/g	Makira, Solomon Islands
Caico	nd	12.55 Hg/gww	nd	1.05 µg/g	Papua New Guinea
Gatagata/Vudito #1	0.79 mg/gww	6.95 mg/gww	nd	0.61 µg/g	Makira, Solomon Islands
Gatagata/Vudito #2	0.42 mg/gww	4.47 mg/gww	nd	0.39 mg/g	Makira, Solomon Islands
GCTV 215	nd	5.77 µg/gww	nd	0.48 µg/g	Belgium
Grand Naine	2 nmol/gdw	1 nmol/gdw	nd	0.09 µg/g	Cameroon
Grand Naine	nd	4.47 µg/gww	nd	0.37 µg/g	Belgium
Gunih	nd	14.27 µg/gww	nd	1.19 µg/g	Papua New Guinea
Hendemeyargh	44 nmol/gdw	124 nmol/gdw	9 nmol/gdw	6.73 µg/g	Philippines
Huld Matawa	2.93 mg/gww	2.96 mg/gww	nd	0.37 mg/g	Makira, Solomon Islands
IC2	nd	4.02 µg/gww	nd	0.33 µg/g	Belgium
Iholena Lele	78 nmol/gdw	107 nmol/gdw	6 nmol/gdw	6.67 µg/g	Hawaii
Kabucuragye	nd	1.41 µg/gww	nd	0.12 µg/g	Uganda
Catimor	74 nmol/gdw	84 nmol/gdw	7 nmol/gdw	5.57 µg/g	Philippines
Kibuzi	nd	4.28 µg/gww	nd	0.36 µg/g	Uganda
Kokopo	nd	11.42 µg/gww	nd	0.95 µg/g	Papua New Guinea
Musa paradisiaca	nd	0.72–122 µg/gww	nd	0.06–1.02 µg/g	Indonesia
Mbououkou-1	31 pmol/gdw	34 pmol/gdw	6 pmol/gdw	23.48 ng/g	Belgium
Mbououkou-1	29 nmol/gdw	34 nmol/gdw	5 nmol/gdw	2.28 µg/g	Cameroon
Mbououkou-3	28 pmol/gdw	26 pmol/gdw	3 pmol/gdw	18.56 ng/g	Belgium
Mbououkou-3	26 nmol/gdw	25 nmol/gdw	3 nmol/gdw	1.77 µg/g	Cameroon
Mbwazirume	nd	1.91 µg/gww	nd	0.16 µg/g	Uganda
Mpologoma	nd	1.46 µg/gww	nd	0.12 µg/g	Uganda
Musa spp. (Uht ipali)	5.46 µg/gww	11.81 µg/gww	nd	1.21 µg/g	Pohnpei, Micronesia

(continued)

Table 8.3 (continued)

Genotype	*t*-AC	*t*-BC	*c*-BC	RAE	Country
Musa spp. (Usr wac)	6.77 μg/gww	20.82 μg/gww	nd	2.00 μg/g	Kosrae, Micronesia
Musa troglodytarum (Uht en yap)	14.72 μg/gww	63.60 μg/gww	nd	5.91 μg/g	Pohnpei, Micronesia
Musa troglodytarum (Uht karat)	2.96 μg/gww	9.18 μg/gww	nd	0.89 μg/g	Pohnpei, Micronesia
Nakhaki	nd	4.62 μg/gww	nd	0.39 μg/g	Uganda
Nakitembe	nd	5.27 μg/gww	nd	0.44 μg/g	Uganda
Pagatau	nd	4.54 μg/gww	nd	0.38 μg/g	Papua New Guinea
Pisang Mas	nd	11.39 μg/gww	nd	0.95 μg/g	Belgium
Pitu	nd	11.27 μg/gww	nd	0.94 μg/g	Papua New Guinea
Pongani	nd	2.13 μg/gww	nd	0.18 μg/g	Papua New Guinea
Porapora	nd	7.88 μg/gww	nd	0.66 μg/g	Papua New Guinea
Pusit	99 nmol/gdw	101 nmol/gdw	9 nmol/gdw	6.93 μg/g	Philippines
Ropa	36.82 mg/gww	13.24 μg/gww	nd	2.64 mg/g	Makira, Solomon Islands
Saena	0.79 mg/gww	0.58 mg/gww	nd	0.08 mg/g	Makira, Solomon Islands
Tereza	nd	2.46 μg/gww	nd	0.21 μg/g	Uganda
Toraka Parao	2.50 mg/gww	5.26 mg/gww	nd	0.54 mg/g	Makira, Solomon Islands
Wambo	nd	19.04 Hg/gww	nd	1.59 μg/g	Papua New Guinea
Warowaro	<0.02 mg/gww	1.66 mg/gww	nd	0.14 μg/g	Makira, Solomon Islands
Williams	nd	6.20 μg/gww	nd	0.52 ng/g	Belgium
Yalim	nd	16.27 μg/gww	nd	1.35 μg/g	Papua New Guinea
Yangambi km5	3 pmol/gdw	1 pmol/gdw	0 pmol/gdw	0.11 ng/g	West and Central Africa
Yangambi km5	3 nmol/gdw	1 nmol/gdw	0 nmol/gdw	0.11 μg/g	Cameroon
Wasolay	172 μg/gdw	74 μg/gdw	28 μg/gdw	14.51 μg/g	Papua New Guinea
Jari Buaya	415 μg/gdw	525 μg/gdw	224 μg/gdw	70.37 μg/g	Malaysia
Malbut	145 μg/gdw	102 μg/gdw	34 μg/gdw	15.96 μg/g	Papua New Guinea
Jaran	127 μg/gdw	162 μg/gdw	nd	18.79 μg/g	Indonesia
Saba	9 μg/gdw	61 μg/gdw	13 μg/gdw	6.00 μg/g	Costa Rica
Caipira	2 μg/gdw	9 μg/gdw	nd	0.83 μg/g	Brazil
Bucaneiro	17 μg/gdw	5 μg/gdw	nd	1.13 μg/g	Saint Lucia
Nam	19 μg/gdw	22 μg/gdw	nd	2.62 μg/g	Thailand
Thap Maeo	135 μg/gdw	147 μg/gdw	36 μg/gdw	19.38 μg/g	Brazil

Pereira and Marasch (2015)

t-AC trans-alpha-carotene, *t-BC* trans-beta-carotene, *c-AC* is cis-alpha-carotene, *RAE* retinol activity equivalents

(continued)

(Table 8.2). Some of the other compounds are ascorbic acid, β-carotene, total flavonoid, lycopene, and alkaloid. The ferrous chelating effect of the stem extract had higher antioxidant activity, than hydroxyl scavenging activity and hydrogen peroxide scavenging activity. The data obtained that Mauli bananas' stem contained a lot of bioactive compounds and also has moderate to potent antioxidant and/or free radical scavenging activity (Apriasari et al. 2014). The sap of many banana accessions, namely, *Musa balbisiana*, *Musa laterita*, *Musa ornata*, and *Musa acuminata*, and some cultivars contained apigenin glycosides, myricetin glycoside, myricetin-3-*O*-rutinoside, naringenin glycosides, kaempferol-3-*O*-rutinoside, quercetin-3-*O*-rutinoside,

Table 8.4 Lipophilic constituents from the ripe pulp of banana genotypes

Rt (min)	Compound	'Chinese Cavendish'	'Giant Cavendish'	'Dwarf Red'	'Grand Naine'	'Eilon'	'Gruesa'	'Silver'	'Ricasa'	'Williams'	'Zelig'
	Fatty acids	1028	880	579	999	1055	1284	939	774	1295	1281
	Saturated	636	510	358	580	572	635	619	576	676	710
11.67	Dodecanoic acid	37	50	2	69	57	29	43	2	52	39
16.35	Tetradecanoic acid	20	26	9	24	24	15	24	6	23	21
18.57	Pentadecanoic acid	9	6	17	7	12	15	6	9	12	12
20.81	Hexadecanoic acid	396	251	245	269	256	402	276	257	378	456
22.73	Heptadecanoic acid	19	11	6	5	9	16	24	6	16	13
24.72	Octadecanoic acid	74	83	30	115	99	66	169	268	95	78
26.55	Nonadecanoic acid	34	50	10	38	54	30	42	4	49	37
28.36	Eicosanoic acid	11	6	9	20	17	16	6	6	14	11
30.10	Heneicosanoic acid	5	4	4	3	4	3	3	1	3	6
32.77	Docosanoic acid	7	6	7	15	16	11	6	4	15	10
33.38	Tricosanoic acid	4	3	4	3	4	6	3	2	3	5
34.94	Tetracosanoic acid	10	3	6	6	9	13	7	5	7	11
36.45	Pentacosanoic acid	5	2	2	2	4	6	4	3	3	5
37.92	Hexacosanoic acid	2	2	2	1	3	4	3	1	2	3
39.35	Heptacosanoic acid	1	2	2	1	1	1	1	1	1	nd
43.37	Triacontanoic acid	2	5	3	2	4	2	2	1	3	3
	Unsaturated	384	365	216	415	475	645	316	195	614	562
20.13	Hexadec-9-enoic acid	49	30	12	37	36	93	9	41	84	72
23.88	Octadeca-9,12-dienoic acid	99	65	12	90	119	52	198	72	87	178
24.11	Octadeca-9,12,15-trienoic acid	165	215	125	237	263	405	36	62	350	214
24.20	Octadec-9-enoic acid	71	55	67	51	57	95	73	20	93	98
	Diacids	6	1	2	1	2	3	1	1	4	5
15.36	Nonanedioic acid	6	1	2	1	2	3	1	1	4	5
	ω-Hydroxy acids	2	4	3	3	6	1	3	2	1	4
37.29	22-Hydroxy-docosanoic acid	2	4	3	3	6	1	3	2	1	4
	Long-chain aliphatic alcohols	45	57	49	58	81	22	52	35	50	62

(continued)

Table 8.4 (continued)

Rt (min)	Compound	'Chinese Cavendish'	'Giant Cavendish'	'Dwarf Red'	'Grand Naine'	'Eilon'	'Gruesa'	'Silver'	'Ricasa'	'Williams'	'Zelig'
18.95	Hexadecan-1-ol	17	21	19	26	42	7	19	15	20	26
22.41	(Z)-Octadec-9-en-1-ol	16	21	17	21	20	8	18	12	19	20
23.09	Octadecan-1-ol	6	7	7	7	11	4	9	5	7	8
30.39	Docosan-1-ol	2	2	2	2	3	1	1	1	2	2
33.65	Tetracosan-1-ol	nd	nd	1	nd	nd	nd	nd	nd	nd	nd
39.53	Octacosan-1-ol	4	6	3	2	5	2	5	2	2	6
	Sterols	155	263	176	233	296	275	182	316	237	169
40.39	Campesterol	26	27	18	37	59	44	30	43	46	28
40.86	Stigmasterol	23	46	26	30	49	43	29	43	37	26
41.75	β-Sitosterol	105	186	130	163	185	184	121	226	152	113
42.33	Cycloartenol Others	1	4	2	3	3	4	2	4	2	2
39.15	α-Tocopherol	nd	3	4	2	3	4	5	3	3	7

Reproduced with permission from Vilela et al. (2014)

dopamine, and N-acetylserotonin (Table 8.2). These compounds were reported to have biological activities. Moreover, the identities of these phytochemical compositions may be used as markers for the banana diet, the assessment of physiochemical status, or the classification of banana clones (Pothavorn et al. 2010).

8.4.3 Volatile Aroma of Fruits

Bananas have a large number of volatile aroma compounds, but the majority group is esters followed by alcohols. Esters like isoamyl acetate (Jordan et al. 2001), 2-pentyl acetate, 3-methylbutyl 3-acetate, 3-methylbutyl butanoate, 2-methyl 3-methylbutanoate, 3-methylbutyl 3-methylbutanoate, hexyl butanoate, Z-4-octen-1-yl 3-methylbutanoate, Z-4-octen-1-yl pentanoate, Z-5-octen-1-yl 3-methylbutanoate, and Z-4-decen-1-yl 3-methylbutanoate (Pino and Fables 2013) are reported to contribute for the banana odor. 2-Methylpropyl acetate, ethyl acetate, and butyl acetate are also important aroma contributors for the banana odor note.

Ethyl esters were found to comprise the largest chemical class accounting for 80.9 %, 86.5 %, 51.2 %, 90.1 %, and 6.1 % of the total peak area for Dwarf Cavendish, Prata, Ouro, Maçã, and Platano volatile fraction, respectively (Pontes et al. 2012). Around 80 volatile aromatic compounds were identified in Dwarf Cavendish banana grown in open field or under protected cultivation (Selli et al. 2012). The major compounds were esters followed by aldehydes, and banana odor-contributing compounds were isoamyl acetate, isoamyl butanoate, and 2-pentanol acetate. Open-field-grown bananas had higher volatile compounds than the bananas produced under protected cultivation. This indicated that the aroma quality changes with the environment in which the plant is growing. More sunshine hours and slightly higher temperatures may improve the aroma quality of fruits.

Aroma volatiles in plantains and dessert bananas were found to be varying. Esters like isoamyl acetate were higher in Dwarf Cavendish than in plantains. However, ethylbutanoic acid was more in plantains compared to dessert bananas (Aurore et al. 2011). Cavendish had the lowest overall concentration in glycosides (bound aroma) compared to the plantains (Aurore et al. 2011). The C13 norisoprenoids and shikimate derivatives were highest in plantains, while the shikimate derivatives were particularly low in Frayssinette variety. Plantains contain high levels of acetovanillone and 3-methylbutanoic acid.

Around 124 volatile compounds have been identified in banana cv. Giant Cavendish. The composition of banana fruit volatiles included 75 esters, 18 ketones, 14 phenols and derivatives, 13 alcohols, 7 aldehydes, 7 acids, and 12 miscellaneous compounds. The percentage of different groups of compounds was esters, 55.4 %; alcohols, 17.5 %; phenols and derivatives, 12.0 %; ketones, 6.9 %; aldehydes, 6.5 %; acids, 1.4 %; and miscellaneous compounds, 0.2 %. Major components (>8 mg kg^{-1}) were 3-methylbutyl acetate, 3-methylbutan-1-ol, eugenol, 3-methylbutyl 3-methylbutanoate, 2-methylpropyl acetate, and 2-pentyl acetate (Pino and Febles 2013).

Storage conditions also differentially affect the volatile constituents of the cultivars. Cold storage more strongly affects the cv. Nanicão than the Prata cultivar. Esters such as 2-pentanol acetate, 3-methyl-1-butanol acetate, 2-methylpropyl butanoate, 3-methylbutyl butanoate, 2-methylpropyl 3-methylbutanoate, and butyl butanoate were drastically reduced in the cold group of the Nanicão cultivar. This indicates that the metabolism responsible for the production of volatile compounds is related to the ability of the cultivar to tolerate low temperatures (Facundo et al. 2012).

8.4.4 Chilling Injury and Metabolites

The contents of acetaldehyde and ethanol increased in the peel and pulp of chilled fruits. There was an accumulation of α-keto acids in the peels of chilled fruits. Biosynthesis of citric and isocitric acids in chilled fruits was half of that of healthy fruits (Murata 1969). Hot water treatment has led to an induction of antioxidants (phenolics and flavonoids) in banana fruits as indicated by an increase of antioxidants and a decrease of H_2O_2 during ripening, all of which result in a delayed ripening of banana fruit (Ummarat et al. 2011). Considerable accumulation of the free polyamine, putrescine (Put), was found in both pulp and peel tissue of unchilled and rewarmed bananas (*Musa* AAA group) during normal fruit ripening with climacteric ethylene production, while free Put remained at the original level in continuously chilled fruit (Purwoko et al. 2002).

8.5 Metabolite Changes During Ripening

8.5.1 Metabolite Changes in the Pulp

The transcriptome studies in Dwarf Cavendish banana fruit during ripening indicated the genes involved in ethylene biosynthesis, perception and signaling processes, cell wall degradation, and the production of aromatic volatiles. There are a large number of metabolites involved in each of these processes. Changes in expansins and hemicelluloses were also observed during ripening using the transcriptome approach. These genes can be good candidates for future studies to establish their role in banana fruit ripening. The datasets developed in this study will help in developing strategies to manipulate banana fruit ripening and reduce postharvest losses.

In another study on the targeting, the carbohydrate metabolism during ripening indicated that starch and pectic fractions, xylose, decreased considerably. Some loss of mannose was also noticed. More than 80 % of the radioactivity of starch was incorporated into soluble sugars, namely, glucose, fructose, and sucrose (increased from 1.8 to 19 %) (Prabha and Bhagyalakshmi 1998).

Biochemical changes in the pulp of "Prata" bananas during storage at 16 °C indicated that starch decreased continuously throughout ripening and sugars increased. Sucrose increased initially up to 14 days afterward; fructose and glucose were the major sugars. Fructose and glucose content increased continuously throughout the storage period of 35 days with fructose being the major sugar. Starch loss followed a first-order reaction. The bioactive amines like putrescine, spermidine, and serotonin were also detected and found to decrease during ripening (Adao et al. 2005). In some of the plantains and bananas, sucrose was found to be around 70 % in the ripe fruits, indicating the differences in the sugar levels in different varieties. Plantains were found to have 9 % starch even in the ripe fruits when compared to bananas which had 1 % starch. Sugar content increased in overripe fruits with further reduction in the starch levels. The ratio of glucose-fructose was approximately a unity for bananas and plantains at all stages of ripeness (Marriott et al. 1981).

Indole acetic acid controls the degradation of starch in bananas. Starch breakdown is reported to begin only after the IAA levels were about 4 ng/g FW. The results herein suggest that IAA levels play a role during banana ripening in events like starch degradation with the consequence of banana sweetening (Purgatto et al. 2002).

8.5.2 Volatile Constituents During Ripening in the Pulp

Esters are the major group of volatiles in the ripe pulp but absent in the green pulp. Ketones are the other group which was present only in the ripe pulp. Alcohols are the major group in the green pulp, which was closely followed by aldehydes. Alcohols were the second major group in the ripe pulp (Facundo et al. 2013). Some of the major volatiles present in the ripe pulp are ethyl acetate, 2-pentanone, 2-methylpropyl acetate, 2-pentanol acetate, 3-methyl-1-butanol acetate, 2-methylpropyl butanoate, 2-methylpropyl 3-methylbutanoate, butyl butanoate, butyl pentanoate, 3-methylbutyl butanoate, 3-methylbutyl 3-methylbutanoate, 2-heptanol, 1-methylhexyl butanoate, hexyl 3-methylbutanoate, and 3-methylcyclohexanol.

In the green pulp, the most of the volatiles are fatty acid-derived hexanol metabolites. The major volatile constituents are α-pinene, hexanal, m-xylene, limonene, 2-hexenal, Z-ocimene, E-2-heptenal, 1-hexanol, 3-hexen-1-ol, E-2-hexen-1-ol, 2-ethyl-1-hexanol, E-2-nonenal, α-cedrene, β-cedrene, and 2,6-nonadienal (Facundo et al. 2013). Olfactometry of banana volatiles indicated that the esters contribute for the banana odor note. Green banana odor was dominated by the fatty acid derivative of C6 compounds. These volatiles contribute mainly to the green and grassy odor notes. Results indicate that most of the esters are synthesized during ripening by the action of alcohol acyl-CoA transferase enzyme.

8.5.3 Metabolites in Fruit Peels During Ripening

Peels of dessert (*Musa* AAA) banana, plantain (*Musa* AAB) cooking banana (*Musa* ABB), and hybrid (*Musa* AAAB) banana at three stages of ripeness were analyzed for fatty acids, amino acids, sterols, protein, sugars, and dietary fibers. Varietal variation in the content was not observed. The peel was rich in total dietary fiber (TDF) (40–50 %). The protein content in the peel of the banana and plantain was varied from 8 % to 11 %. Amino acids leucine, valine, phenylalanine, and threonine were present in significant quantities. Lipids varied from 2.2 % to 10.9 % and were found to have similar quantities of saturated and polyunsaturated fatty acids. Major fatty acids were palmitic, linoleic, and α-linolenic acid. Together they constitute more than 90 % of the fatty acids. Potassium was the most significant mineral element. During ripening, starch was converted to glucose and fructose and fructose was the major sugar in the ripe peel. Sucrose was not detectable (Emaga et al. 2007). Sterols represent about 49–71 % of the lipophilic extract with two triterpenic ketones (31-norcyclolaudenone and cycloeucalenone) as the major components. High amounts of stearoyl esters (469–24,405 mg/kg dry weight) and diacylglycerols (119–878 mg/kg dwt), in the banana peel extract, were also reported by Oliveira et al. (2008).

The peel dopamine levels decreased when ripening from 1290 to 500 mg/100 g dwt), and in the pulp, it decreased from 7 to 3.42 mg/100 g dwt (Kanazawa and Sakakibara 2000). Dopamine in the pulp was higher in the cultivar Prata (full yellow, 15 mg/100 g dry weight). Dopamine is a better antioxidant than other natural antioxidants like ascorbic acid, reduced glutathione, and several phenolic compounds, such as gallocatechin gallate (Kanazawa and Sakakibara 2000).

8.6 Diseases and Metabolites

8.6.1 Sigatoka Diseases

Resistant *Musa* genotypes exhibited activation of phenylalanine ammonia lyase and the subsequent accumulation of substances which block the fungal growth. The fungus of the sigatoka disease produces pentaketide metabolites flaviolin, 2-hydroxyjuglone, juglone, and 2,4,8-trihydroxytetralone (2,4,8-THT) chemicals which bring about the necrosis in the host plants. When inoculated host plants were treated with tricyclazole, extensive necrosis of both susceptible and resistant *Musa* cultivar leaves was observed. This indicates the importance of 2,4,8-THT for host-specific reactions, depending on its concentration at different stages of pathogenesis. Early activation of metabolism of fungal 2,4,8-THT by the resistant *Musa* cultivars caused necrotic micro-lesions and stimulation of postinfectional defense reactions leading to complete inhibition of fungal growth. Growth of the fungus on susceptible cultivars caused necrotizing doses of 2,4,8-THT (Hoss et al. 2000). Juglone a purified fungal metabolite showed a direct inhibitory effect of electron transfer of banana chloroplasts. Susceptibility of the chloroplasts of different cultivars to the chemical correlates with the sensitivity to the pathogen infection. Juglone also induces the oxidation of ascorbic acid. Chloroplast sensitivity can be used to screen the germplasm for the sensitivity to the disease. Antioxidant content may also give resistance against the pathogen. Recently using NMR and LCMS techniques, phenylphenalenones have been identified as the metabolites responsible for the sigatoka disease resistance (Hidalgo et al. 2016) using susceptible cultivar Williams and resistant cultivar "Khai

Thong Ruang." However the resistance was broken down by the more virulent strain of *M. fijiensis* which metabolized the phenylphenalenones into sulfate conjugates.

8.6.2 *Fusarium* and Phenolics

Banana cultivar 'GCTCV-218', which is more tolerant to *Fusarium*, produced a significantly higher total phenolic content when compared to 'Williams' at 0 and 6 h postinfection with *Fusarium*. Induction of total phenolics was observed after 6 h postinfection in GCTCV-218, but in Williams it occurred only after 24 h. Total phenolic productions in 'GCTCV-218' and 'Williams' were similar at 24 and 48 hpi. Free acids were significantly higher in both the varieties at 0 h but decreased after inoculation. 'GCTCV-218' also had a significantly higher level of free acids at 24 hpi. Increased induction of glycoside-bound phenolics was observed in GCTCV-218 and showed at 24 and 48 h, while in Williams at 6 hpi, an increased induction was visible at 24 and 48 h, while in 'Williams' induction was seen only after 48 h. Ester-bound phenolics also showed similar trend in both varieties. A significant increase in 'GCTCV-218' was seen in cell wall-bound phenolics at 24 and 48 h after infection when compared to Williams (van den Berg et al. 2009). Cell wall-bound phenolics were found to be the major metabolites giving resistance to *Fusarium*.

Strong deposition of lignin and cell wall strengthening of the roots was observed in the *Fusarium* wilt-resistant cultivar Goldfinger when compared to susceptible cv. Williams. No increase in callose content was observed for either clone (De Ascensao and Dubery 2000). Hydrogen peroxide and phenylalanine ammonia lyase are the most sensitive compound and enzyme detectable after inoculated with *Fusarium* spores and found to be the basis for the tolerance against *Fusarium* (Subramaniam et al. 2006). Goldfinger variety was inoculated with fungal elicitor and the variety responded quickly. Compositional and quantitative differences between induced phenolics (*p*-coumaric, ferulic, and sinapic acids) and those constitutively present (*p*-coumaric and ferulic acid). In addition, vanillic acid was found in the ester-bound fraction and protocatechuic acid in the cell wall-bound fraction of elicited tissue. There was an increase in the lignin synthesis. Elicitor stimulated the phenylpropanoid pathway resulting in the synthesis of cinnamic acid and benzoic acid derivatives that were esterified and incorporated into the cell wall fraction as part of the antimicrobial defenses activated in the root tissue (de Ascensao and Dubery 2003).

Proteomic studies indicated that there was an induction of proteins associated with pathogenesis-related (PR) response; isoflavonoid, flavonoid, and anthocyanin syntheses; cell wall strengthening; cell polarization; reactive oxygen species production and scavenging; jasmonic acid, abscisic acid, and auxin-mediated signaling conduction; molecular chaperones; energy; and primary metabolism. Comparison of resistant and susceptible genotypes indicated the distinct variation in protein profiles. Susceptible genotypes expressed proteins involved in the defense mechanisms; however, in resistant genotypes proteins related to PR response, cell wall strengthening, and antifungal compound synthesis were expressed in higher quantity (Li et al. 2013).

8.7 Metabolite Changes During Nematode Infection

Varietal variations in the phenylpropanoid pathway of secondary metabolism were observed in cultivars resistant to the banana nematode *Radopholus similis*, 'Yangambi Km5', 'Pisang Jari Buaya', and the susceptible 'Grand Naine'. Lignification of cells and the phenolic acids before infection were not associated with the resistance in some of the cultivars. Sinapic acid was found to accumulate in the thickened cell walls of the endodermis and vascular elements in both resistant and susceptible cultivars upon infection by nematode, while flavonoid-containing cells in the vascular cylinder were only found in resistant cultivars after infection. 'Grand Naine' roots which were susceptible cultivar contained the highest total phenolic content, which could be related to high amounts of proanthocyanidins. In 'Yangambi Km5' roots, the proanthocyanidin and total phenolic content were the lowest, but mean amounts of total phenols increased after infection, unlike in the other cultivars. Increased responses of the resistant cultivars in terms of total phenols particularly flavonoids were the major reasons for the resistance to the nematodes (Wuyts et al. 2005).

The *Musa* cultivars, Dwarf Cavendish, Yangambi Km5, and Kunnan, exhibit considerable differences in resistance to *Radopholus similis*. The infection increased condensed tannins and flavan-3,4-diols in roots significantly and is highest in the resistant cultivar Kunnan before and after infection. Tannins belong to mostly procyanidin group, but Kunnan also contained propelargonidins, suggesting that these tannins may be involved in the resistance mechanism (Collingborn et al. 2000).

Phenolic acid profile was estimated in the *Radopholus similis*-infected roots of *Musa* cv. Grand Naine and cv. Calcutta 4. Dopamine, cyanidin-derived compound, ferulic acid ester, ferulic acid, and hydrolysis product of ferulic acid were identified. All the compounds were significantly higher in Calcutta 4 which was tolerant to nematode infection (Wuyts et al. 2007). Resistant cultivars were found to induce the activities of phenylalanine ammonia lyase and cinnamyl-alcohol dehydrogenase several folds after infection in addition to the higher levels of cell wall-bound phenolics and lignins (Fogain and Gowen 1996). Significant increase in p-coumaric, ferulic, and sinapic acids was also observed (Vaganan et al. 2014). Yangambi Km5 resistance to penetration of the vascular bundle was attributed to the extensive presence of suberin in endodermal cell walls (Valette et al. 1997, 1998).

8.8 Conclusion

Varietal variation in banana exists for the tolerance against nematode infection and can be exploited for crop improvement programs. Even though most of the cultivated banana genotypes have low carotenoids, germplasm with very high provitamin A activity carotenoids has been reported and can be successfully used for breeding programs. Cell wall strengthening by the synthesis of lignin-forming phenolics has been associated with the tolerance for diseases also. From the available literature, it can be concluded that the wide variability exists in banana for phenolic acid content, carotenoids, volatiles, and phytosterols.

References

Adao RC, Beatriz M, Gloria A (2005) Bioactive amines and carbohydrate changes during ripening of 'Prata' banana (*Musa acuminata, M. balbisiana*). Food Chem 90:705–711

Amit R, Shailandra S (2006) Ethnomedicinal approach in biological and chemical investigation of phytochemicals as antimicrobials. Indian J Pharm Sci 41:1–13

Anhwange BA, Ugye TJ, Nyiaatagher TD (2009) Chemical composition of *Musa sapientum* (Banana) peels. EJEAF Che 8:437–442

Apriasari ML, Iskandar, Suhartono E (2014) Bioactive compound and antioxidant activity of methanol extract *Mauli bananas* (*Musa sp*) stem. Int J Bio Biochem Bioinforma 4:110–115

Aurore G, Ginies C, Ganou-parfait B, Catherine MGC, Fahrasmane RL (2011) Comparative study of free and glycoconjugated volatile compounds of three banana cultivars from French West Indies: Cavendish, Frayssinette and Plantain. Food Chem 129:28–34

Bhat MS, Prabhakar A, Rama KRR, Madhu GM, Rao GH (2010a) Statistical optimization and neural modeling of amylase production from banana peel using *Bacillus subtilis* MTCC 441. Int J Food Eng 56:34–45

Bhat MS, Prabhakar A, Rama KRR, Madhu GM, Rao GH (2010b) Optimization of process variables for amylase production using agro residual wastes by taguchi methodology. Chem Prod Process Model 5:4–7

Brooks AA (2008) Ethanol production potential of local yeast strains isolated from ripe banana peels. Afr J Biotechnol 7:3749–3752

Babu MA, Suriyakala MA, Gothandam KM (2012) Varietal impact on phytochemical contents and antioxidant properties of *Musa acuminata* (banana). J Pharm Sci Res 4(10):1950–1955

Collingborn FMB, Gowen SR, Mueller-Harvey I (2000) Investigations into the biochemical basis for nematode resistance in roots of three *Musa* cultivars in response to *Radopholus similis* infection. J Agric Food Chem 48:5297–5301

Davey MW, Keulemans J, Swennen R (2006) Methods for the efficient quantification of fruit pro-vitamin A contents. J Chromatogr A 1136:176–184

de Ascensao ARFDC, Dubery IA (2000) Panama disease: cell wall reinforcement in banana roots in response to elicitors from *Fusarium oxysporum* f.sp. *cubense* race four. Phytopathology 90:1173–1180 de Ascensao ARFDC, Dubery IA (2003) Soluble and wall-bound phenolics and phenolic polymers in *Musa acuminata* roots exposed to elicitors from *Fusarium oxysporum* f.sp. *cubense*. Phytochemistry 63:679–686

Debabandya M, Sabyasachi M, Namrata S (2010) Banana and its by-products utilization: an overview. J Sci Ind Res 69:323–329

De Luca V, St Pierre B (2000) The cell and developmental biology of alkaloid biosynthesis. Trends Plant Sci 5:168–173

Ehiowemwenguan G, Emoghene AO, Inetianbor JE (2014) Antibacterial and phytochemical analysis of banana fruit peel. J Pharm 4(8):18–25

Emaga TH, Andrianaivo RH, Wathelet B, Tchango JT, Paquot M (2007) Effects of the stage of maturation and varieties on the chemical composition of banana and plantain peels. Food Chem 103:590–600

Facundo HVV, Garruti DS, Carlos CTS, Cordenunsi BR, Lajolo FM (2012) Influence of different banana cultivars on volatile compounds during ripening in cold storage. Food Res Int 49:626–633

Facundo HV, Garruti DS, Cordenunsi BR, Lajolo FM (2013) Isolation of volatiles compounds in banana by HS-SPME: optimization for the whole fruit and pulp. Int J Bio Biochem Bioinforma 3:110–115

Fiehn O (2001) Combining genomics, metabolome analysis, and biochemical modelling to understand metabolic networks. Comp Func Genom 2:155–168

Fiehn O (2002) Metabolomics: the link between genotypes and phenotypes. Plant Mol Biol 48:155–171

Fogain R, Gowen SR (1996) Investigations on possible mechanisms of resistance to nematodes in *Musa*. Euphytica 92:375–381

Girish HV, Satish S (2008) Antibacterial activity of important medicinal plants on human pathogenic bacteria – a comparative analysis. World Appl Sci J 5(3):267–271

Hidalgo W, Chandran JN, Meneze RC, Otálvaro F, Schneider B (2016) Phenylphenalenones protect banana plants from infection by *Mycosphaerella fijiensis* and are deactivated by metabolic conversion. Plant Cell Environ 39:492–513. doi:10.1111/pce.12630

Hoss R, Helbig J, Bochow H (2000) Function of host and fungal metabolites in resistance response of banana and plantain in the black Sigatoka disease pathosystem (*Musa* spp. –*Mycosphaerella fijiensis*). J Phytopathol 148:387–394

Ige AO, Mebude OO, Adeniyi BA, Adewoye EO (2015) Antidermatophytic activities of *Musa sapientum* methanol leaf extract in-vitro. Br Microbiol Res J 10(1):1–7

Jordan MJ, Tandon K, Shaw PE, Goodner KL (2001) Aromatic profile of aqueous banana essence and banana fruit by gas chromatography–mass spectrometry (GC–MS) and gas chromatography–olfactometry (GC–O). J Agric Food Chem 49:4813–4817

Kanazawa K, Sakakibara H (2000) High content of dopamine, a strong antioxidant, in Cavendish banana. J Agric Food Chem 48(3):844–848

Kitdamrongsont K, Pothavorn P, Swangpol S, Wongniam S, Atawongsa K, Svasti J et al (2008) Anthocyanin composition of wild bananas in Thailand. J Agric Food Chem 56:10853–10857

Lewis DA, Fields WN, Shaw GP (1999) A natural flavonoid present in unripe plantain banana pulp (*Musa sapientum* L. var. paradisiaca) protects the gastric mucosa from aspirin-induced erosions. J Ethnopharmacol 65:283–288

Li X, Bai T, Li Y, Ruan X, Li H (2013) Proteomic analysis of *Fusarium oxysporum* f. sp. *cubense* tropical race 4-inoculated response to Fusarium wilts in the banana root cells. Proteome Sci 11:41–54

Marriott J, Robinson M, Karikari SK (1981) Starch and sugar transformation during the ripening of plantains and bananas. J Sci Food Agric 32:1021–1026

Mattos LA, Amorim EP, Amorim VBO, Cohen KO, Ledo CAS, Silva SO (2010) Agronomical and molecular characterization of banana germplasm. Pesq Agrop Brasileira 45:146–154

Mokbel M, Saif F, Hashinaga F (2005) Antibacterial and antioxidant activities of banana (*Musa*, AAA cv. *cavendish*) fruits peel. Asian J Biotechnol 2:1–4

Murata T (1969) Physiological and biochemical studies of chilling injury in bananas. Physiol Plant 22:401–411

Nogueira JMF, Fernandes PJP, Nascimento AMD (2003) Composition of volatiles of banana cultivars from Madeira Island. Phytochem Anal 14:87–90

Oliveira L, Freire CSR, Silvestre AJD, Cordeiro N (2008) Lipophilic extracts from banana fruit residues: a source of valuable phytosterols. J Agric Food Chem 56:9520–9524

Omojasola PF, Jilani OP (2009) Cellulase production by Trichoderma longi, Aspergillus niger and Saccharomyces cerevisae cultured on plantain peel. Res Microbiol 4:67–74

Pothavorn P, Kitdamrongsont K, Swangpol S, Wongniam S, Atawongsa K, Savasti J, Somana J (2010) Sap phytochemical compositions of some bananas in Thailand. J Agric Food Chem 58(15):8782–8787

Pereira A, Marasch M (2015) Banana (*Musa spp*) from peel to pulp: source of bioactive compounds and its relevance for human health. J Ethnopharmacol 160:149–163

Pereira A, Sônego M, Peruch LAM, Maraschin, M (2012) Qualitative and quantitative analysis of banana peel's secondary metabolites by spectrophotometric and chromatographic techniques. Simpósio de Integração das Pós-graduações do CCB/UFSC Florianópolis, 12 a 16 de agosto de

Pérez AG, Cert A, Rios JJ, Olias JM (1997) Free and glycosidically bound volatile compounds from two banana cultivars: valery and Pequena Enana. J Agric Food Chem 45:4393–4397

Pichersky E, Gang DR (2000) Genetics and biochemistry of secondary metabolites: an evolutionary perspective. Trends Plant Sci 5:439–445

Pino JA, Febles Y (2013) Odour-active compounds in banana fruit cv. Giant Cavendish. Food Chem 141:795–801

Ponnuswamy V, Lekshmi VD, Prakash SGV (2011) Bioprocessing of banana peel for amylase production by Sp. *Penicillium*. Asian J Exp Biol Sci 2:257–264

Pontes M, Pereira J, Câmara JS (2012) Dynamic headspace solid-phase microextraction combined with one-dimensional gas chromatography–mass spectrometry as a powerful tool to differentiate banana cultivars based on their volatile metabolite profile. Food Chem 134:2509–2520

Prabha TN, Bhagyalakshmi N (1998) Carbohydrate metabolism in ripening banana fruit. Phytochemistry 48:915–919

Purgatto E, do Nascimento JRO, Lajolo FM, Cordenunsi BR (2002) The onset of starch degradation during banana ripening is concomitant to changes in the content of free and conjugated forms of indole-3- acetic acid. J Plant Physiol 159:1105–1111

Purwoko BS, Susanto S, Novita T (2002) Studies on the physiology of polyamines and ethylene during ripening of banana and papaya fruits. Acta Hortic 575:651–657

Ratule MT, Osman A, Saari N, Ahmad SH (2007) Microstructure of peel cell wall and selected physic-chemical characteristics of Berangan banana (*Musa* cv. Berangan [AAA]) ripened at high temperature. Asia Pac J Mol Biol Biotechnol 15:8–13

Santos UJR, Bakry F, Brillouet JM (2010) A preliminary chemotaxonomic study on the condensed tannins of green banana flesh in the Musa genus. Biochem Syst Ecol 38:1010–1017

Schieber A, Stintzing F, Carle R (2001) By-products of plant food processing as a source of functional compounds – recent developments. Trends Food Sci Technol 12:401–413

Selli S, Gubbukb H, Kafkasc E, Gunesb E (2012) Comparison of aroma compounds in Dwarf Cavendish banana (Musa spp. AAA) grown from open-field and protected cultivation area. Sci Hortic 141:76–82

Singh B, Bhat TK (2003) Potential therapeutic applications of some antinutritional plant secondary metabolites. J Agric Food Chem 51:5579–5597

Subramaniam S, Maziah M, Sariah M, Puad MP, Xavier R (2006) Bioassay method for testing Fusarium wilt disease tolerance in transgenic banana. Sci Hortic 108:378–389

Tsamo CVP, Herent MF, Tomekpe K, Emaga TH, Leclercq JQ, Rogez H, Larondelle Y, Andre C (2015) Phenolic profiling in the pulp and peel of nine plantain cultivars (Musa sp.). Food Chem 167:197–204

Ummarat N, Matsumoto TK, Wall MM, Seraypheap K (2011) Changes in antioxidants and fruit quality in hot water-treated 'Hom Thong' banana fruit during storage. Sci Hortic 130:801–807

Vaganan MM, Ravi I, Nandakumar A, Sarumathi S, Sundararaju P, Mustaffa MM (2014) Phenylpropanoid enzymes, phenolic polymers and metabolites as chemical defenses to infection of *Pratylenchus coffeae* in roots of resistant and susceptible bananas (*Musa* spp.). Indian J Exp Biol 52:252–260

Valette C, Nicole M, Sarah JL, Boisseau M, Boher B, Fargette M, Geiger JP (1997) Ultrastructure and cytochem istry of interactions between banana and the nematode Radopholus similis. Fundam Appl Nematol 21:65–77

Valette C, Andary C, Geiger JP, Sarah JL, Nicole M (1998) Histochemical and cytochemical investigations of phenols in roots of banana infected by the burrowing nematode *Radopholus similis*. Phytopathology 88:1141–1148

van den Berg N, Berger DK, Hein I, Birch PRJ, Wingfield MJ, Viljoen A (2009) Genes up-regulated in tolerant Cavendish banana roots in response to *Fusarium oxysporum* f. sp. *cubense*. Infect Mol Plant Path 8(3):333–341

Van den Berg H, Faulks R, Granado HF, Hirschberg J, Olmedilla B, Sandmann G, Southon S, Stahl W (2000) The potential for the improvement of carotenoid levels in foods and likely systemic effects. J Sci Food Agric 80:880–912

Verde-Mendez CDM, Forster MP, Rodríguez-Delgado MA, Rodríguez- Rodríguez EM, Romer CD (2003) Content of free phenolic compounds in bananas from Tenerife (Canary Island) and Ecuador. Eur Food Res Technol 217:287–290

Vilela C, Santos SAO, Villaverde JJ, Oliveira L, Nunes A, Cordeiro N, Freire CSR, Silvestre AJD (2014) Lipophilic phytochemicals from banana fruits of several Musa species. Food Chem 162:247–252

Wuyts N, Lognay G, Sági L, De Waele D, Swennen R (2005) Secondary metabolites in roots and implications for nematode resistance in banana (*Musa* spp.). In: Turner DW, Rosales FE (eds) Banana root system: towards a better understanding for its productive management. Proceedings of an international symposium. The International Network for the Improvement of Banana and Plantain, Montpellier, pp 238–246

Wuyts N, Lognay G, Verscheure M, Marlier M, De Waele D, Swennen R (2007) Potential physical and chemical barriers to infection by the burrowing nematode *Radopholus similis* in roots of susceptible and resistant banana (*Musa* spp.). Plant Pathol 56:878–890

Wall MM (2006) Ascorbic acid, vitamin A, and mineral composition of banana (*Musa* sp.) and papaya (*Carica papaya*) cultivars grown in Hawaii. J Food Compos Anal 19:434–445

Part II

Genetic Engineering

Novel Gene Transfer Technologies 9

Harjeet Kaur Khanna and Pradeep Chand Deo

Abstract

Gene transfer is the process by which a gene from any source can be introduced into plant cells or tissues. Gene transfer technologies are used to manipulate plant cells for scientific research as well as for commercial purposes like production of transgenics. Transgenic plants generated using these technologies are used either for field deployment or for identifying and evaluating gene and promoter function. In the last few decades, significant developments have been made in gene transfer technology since the discovery of *Agrobacterium tumefaciens* as a natural tool for plant transformation, and there is a long list of transgenic crop varieties that have now been released for commercial production. These advances are related to major improvements in *Agrobacterium*-mediated and direct DNA delivery techniques, along with modifications in tissue culture techniques for regenerating transgenic plants from transformed cells or tissues. Bananas are not lagging too far behind in this race, with many laboratories engaged in field trials of transgenic bananas carrying genes of interest. With efficient gene transfer technologies available for banana, banana transformation research is now more focussed on the problems associated with generating cell suspensions of recalcitrant cultivars, identifying useful gene/trait associations, promoters and problems associated with stable integration and reliable expression of the DNA once it has been integrated. This chapter focuses on the gene transfer technologies currently available for generating transgenic banana cultivars and summarises the various traits of interest for which genes have been transformed into bananas.

H.K. Khanna (✉)
Sugar Research Australia, Brisbane, Australia
e-mail: harjeetk@hotmail.com

P.C. Deo
Centre for Tropical Crops and Biocommodities, Science and Engineering Faculty, Queensland University of Technology, Brisbane, Australia

S. Mohandas, K.V. Ravishankar (eds.), *Banana: Genomics and Transgenic Approaches for Genetic Improvement*, DOI 10.1007/978-981-10-1585-4_9

Keywords
Gene transfer • DNA delivery techniques • Electroporation • Microprojectile bombardment • Agrobacterium-mediated gene transfer • Centrifugation assisted somatic embryogenesis organogenesis

9.1 Introduction

Genes are the basic hereditary units of life. Gene transfer can be defined as a technique to stably introduce a foreign gene into the target cells. Genetic transformation is the introduction of foreign DNA into plant cells, tissues or organs either directly or indirectly (Alves et al. 1999). Genes carry the basic information necessary to produce all proteins in a living organism, and proteins ultimately perform the biological functions. Thus, when a new gene is transferred and stably integrated into a target cell, the protein encoded by the gene is produced and it results in a new phenotype.

Gene transfer technology is one of the major tools of biotechnology and was originally developed as a research tool for investigating gene expression and function. However, as new gene transfer technologies develop and as old technologies are refined, the potential applications keep expanding dramatically. At present, there are a number of gene transfer technologies available, and these vary greatly in the transfer efficiency and the types of cells they are capable of delivering genes into. The techniques for the transfer of DNA into organisms differ from organism to organisms. Generally, there are two approaches for DNA transfer, natural and artificial. The choice of method depends upon the target cells to be transformed. It also depends upon the objectives of the manipulation and it may be either stable or transient. Although the choice of DNA transfer method is very important, the other important considerations are selection of gene/trait, isolation of gene, method of preparation of recombinant DNA and selection of transformed cells. The ability to regenerate the transformed cell is also critically important, and all these factors guide the choice of the method used for transformation. Any successful plant transformation system requires the following: (1) strategy for transferring genes into plant cells, (2) a gene responsible for the useful new trait, (3) a promoter directing the appropriate level and pattern of expression, (4) an effective selection marker to inhibit the growth of non-transformed cells and (5) the ability to produce transgenic plants from transformed cells (Sharma et al. 2005).

Gene transfer can be done either indirectly by using a biological vector such as *Agrobacterium* or directly through physical or chemical processes such as electroporation, polyethylene glycol-mediated DNA uptake, microinjection, silicon carbide fibres and microprojectile bombardment (Taylor and Fauquet 2002). Apparently, any plant species can be transformed with any desirable trait found in nature. Further, with advancement in DNA database and synthesis technology, the entire transgene can be synthesised (Deo et al. 2010).

Plant transformation was first described in tobacco in 1984 (De Block et al. 1984; Horsch et al. 1984). Since that time, rapid developments in transformation technology have resulted in the genetic modification of many plant species. Genetic transformation is of great interest in banana because many pests, diseases and abiotic constraints limit banana cultivation causing significant trade losses and jeopardising food security. Although improvement of existing banana germplasm can technically be achieved by conventional breeding, the cultivated bananas are triploids and hence sterile, and therefore conventional breeding is very inefficient. Diploid bananas could be improved by conventional breeding and then triploids could be created from these improved varieties (Bakry et al. 1992, 2009), but conventional breeding of bananas is a very lengthy process and efficiency is very low and some traits are not easily available within existing germplasm. Unlike conventional breeding, genetic transformation provides a means of introducing only one or a few genes to an already accepted cultivar. The source of these genes might not necessarily be restricted to banana and can be derived from other plant species or completely different organisms. Therefore, production of improved banana via genetic transformation offers an attractive alternative to conventional breeding.

Success in transferring foreign genes into banana cells has been achieved in various laboratories as the protocols for electroporation of protoplasts derived from embryogenic cell suspensions (Sagi et al. 1998), particle bombardment of embryogenic cells (Sagi et al. 1995; Cote et al. 1997; Becker et al. 2000) and cocultivation of wounded meristems with *Agrobacterium* (May et al. 1995) are available for bananas and plantains. The *Agrobacterium*-mediated transformation method is also more widely used now as it offers several advantages over direct gene transfer methodologies like particle bombardment and electroporation (Gheysen et al. 1998; Hansen and Martha 1999; Shibata and Liu 2000).

Efficient plant transformation system requires a competent target tissue/cells for gene transfer, selection, proliferation and regeneration into plantlets. Generally, a large population of totipotent cells in the form of callus is multiplied prior to transformation (Koichi et al. 2002). Embryogenic cell suspension cultures are the most commonly used target tissue for the transformation and regeneration of transgenic plants. Such cultures (1) can be readily multiplied and thus provide ample target tissue, (2) consist of cytoplasmically dense small cell aggregates which provide high levels of both transient expression and transgene integration (Mahn et al. 1995), (3) when plated, form very thin large surface area thus enhancing the selection and identification of independent transformation events within the dispersed cell clusters and (4) allow the recovery of non-chimeric transgenics due to the unicellular origin of embryos (Aguado-Santacruz et al. 2002; Sahrawat et al. 2003; Ul-Haq 2005). In addition, these cells readily overcome the physical trauma induced by transformation (Santos et al. 2002). Further, thin cell layers on selection are efficient in minimising the frequency of escapes significantly. The protocol for banana embryogenic cell suspension and plant regeneration from immature male inflorescences (Cote et al. 1996) and scalps (Strosse et al. 2003) has been established. One of the difficulties in establishing banana cell suspension is the time it takes to form ideal callus for suspension. Once formed, the suspension has a window of 6–8 months for transformation before its regeneration potential and transformation competency decline dramatically. Consequently, fresh suspension cultures have to be established routinely. Such practise also eradicates somaclonal variations which could prevail in older cultures.

9.2 Gene Transfer Methods

9.2.1 Electroporation

Electroporation delivers DNA directly into viable and regenerative protoplasts using an electrical pulse. The electrical capacitance and field strength, duration and shape of electrical pulses, buffer composition, temperature and type of gene constructs and the concentration and form of DNA have been reported to affect the efficiency of transformation (Sagi et al. 1995). Consequently, these parameters have to be determined empirically for different cultivars of banana. The target material for transformation by electroporation in banana is the protoplasts. Protoplasts of monocotyledonous plants have generally been much more difficult to isolate and regenerate than those from dicot plants. In principle, each individual protoplast can reform a cell wall and later initiates either a callus through sustained divisions or an embryo, defined as a somatic embryo. In banana, protoplasts can be obtained from in vivo tissues or in vitro cultures (Cronauer and Krikorian 1986). Bakry (1984) reported the first successful isolation of viable banana protoplast from the inflorescence of Cavendish banana (AAA) and was confirmed by Da Silva Conceicao (1989). Banana is now amenable to in vitro culture of cultured protoplasts (Panis et al. 1993; Megia et al. 1993).

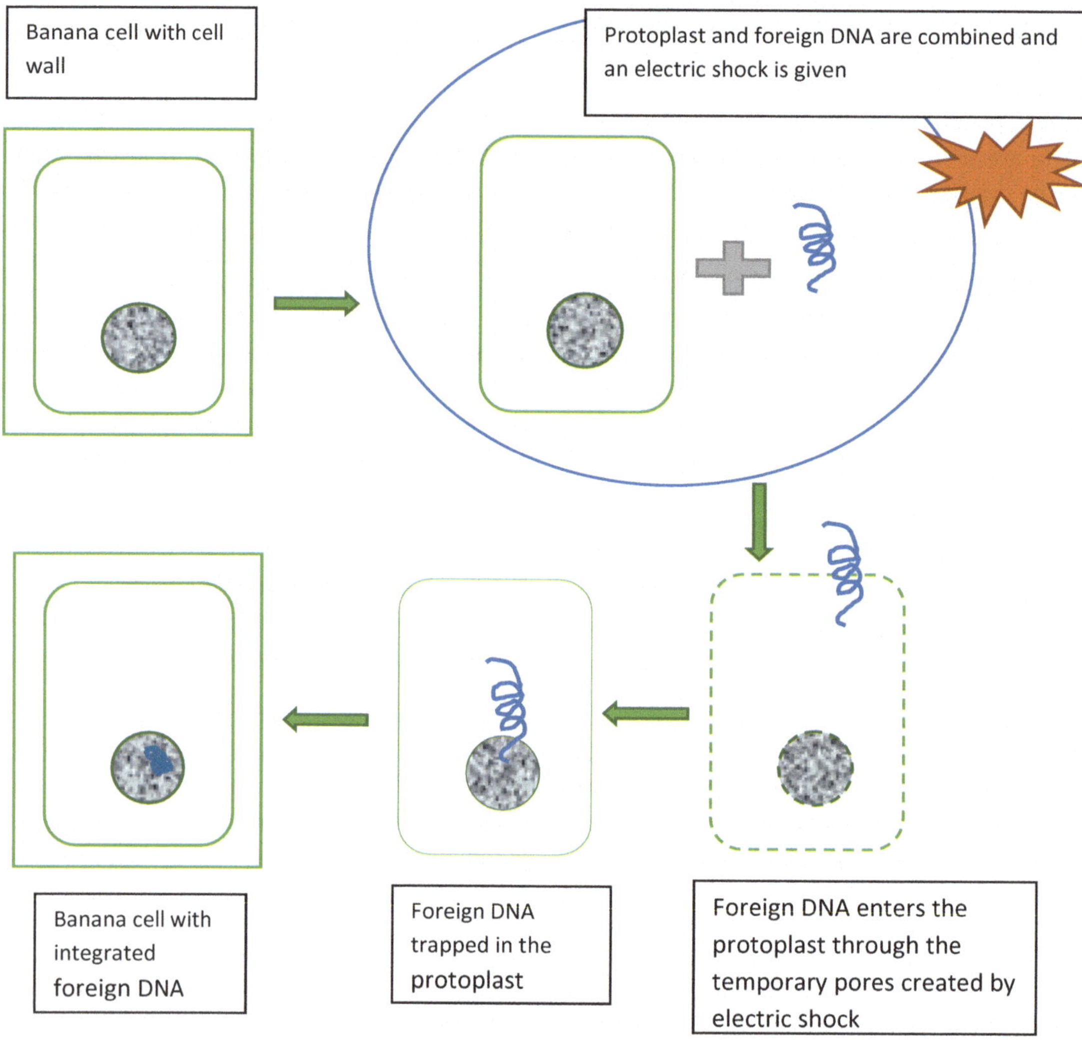

Fig. 9.1 Banana cell transformation using electroporation method

Electroporation of protoplasts has been attempted for transient GUS expression as a step towards creating transgenic banana cv. Bluggoe (ABB) (Sagi et al. 1994, 1995). Briefly, regenerable protoplasts are isolated and purified from cell suspensions. Gene cassettes are then introduced into the protoplast by electroporation. After electroporation, the protoplasts are diluted and incubated in the dark for recovery. The protoplasts are then allowed to develop into callus which could be induced to produce plants (Fig. 9.1).

However, using protoplast as a target tissue for banana transformation is not routine because regenerating plantlets back from transformed protoplasts are technically very challenging. Consequently, majority of banana transformations uses embryogenic cell suspensions with either *Agrobacterium* or microprojectile bombardment.

9.2.2 Microprojectile Bombardment

Microprojectile, biolistic or particle bombardment is another technique which can be used to introduce foreign DNA into a banana cell. Over the past several years, particle bombardment of

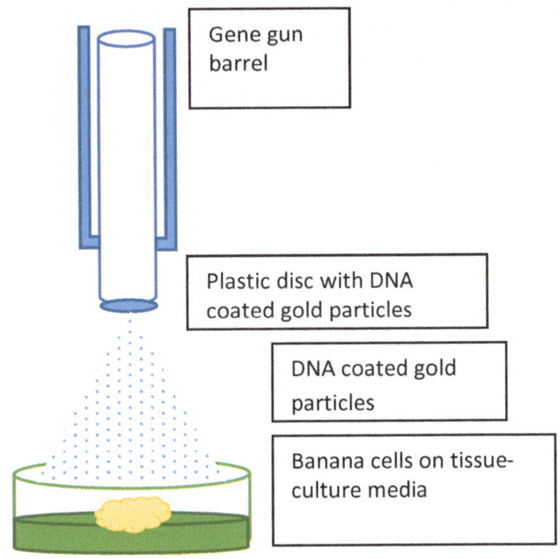

Fig. 9.2 Banana cell transformation using microprojectile bombardment method

banana has progressed into a valuable tool, permitting direct gene transfer to a range of cell and tissue types. In banana this method has been used to evaluate promoter function (Dugdale et al. 2001) and generating stable transgenics (Becker et al. 2000).

This method uses gold or tungsten microprojectile particles since they are inert and do not react with DNA. The DNA is coated onto these particles and then bombarded onto the cells using a helium pressure-driven gene gun. Bio-Rad PDS-1000/He™ (Hagio 1998) device and Particle Inflow Gun (PIG) (Finer et al. 1992) have both been popular choices for different reasons. The PDS-1000/He™ system, although more expensive, is superior to PIG in terms of uniformity in microprojectile distribution over target cells, consistency and reproducibility between successive bombardments and high frequency of transformation (Taylor and Fauquet 2002). One of the disadvantages of particle bombardment method is that it can result in higher frequency of high transgene copy number and transgene rearrangement, hence transgene silencing or co-suppression (O'Kennedy et al. 2001).

Stable transformation of plant cells using the biolistic method requires the penetration of cell wall hence nucleus by microprojectiles, insertion of the transgene into the host plant genome followed by integration and subsequent expression and, finally, continued growth of the transformed cells and regeneration of transgenic plantlets (Russell et al. 1992). Several factors reported to affect the efficiency of particle and DNA delivery into the plant cells and subsequent transient expression and stable integration of the transgene have been studied and optimised for different species (Quoirin et al. 1997; Marchant et al. 1998; Deroles et al. 2002; Janna et al. 2006) (Fig. 9.2).

Microcarriers (or microprojectiles) are made of different materials and vary in sizes (0.6 μm up to 1.6 μm). Both gold and tungsten are the most commonly used microcarriers. Being inert and uniform in size and shape, gold microprojectiles induce minimum damage to cells when compared with tungsten. Due to its surface oxidation, tungsten can catalytically degrade DNA, hence negatively affecting DNA binding (Sanford et al. 1993). In certain species, it can be toxic also (Russell et al. 1992). Consequently, gold is preferred over tungsten. However, for some applications, tungsten is perfectly satisfactory and cheaper to use than gold.

In general, the size of the microprojectiles can affect its momentum, the quantity of DNA it can carry and the degree of damage caused in plant cells, hence affecting cell survival (Klein et al. 1988; Häggman and Aronen 1998; Janna et al. 2006). For example, Tian and Seguin (2004) demonstrated that using 1.0 μm particles resulted in very high transient GUS expression when compared with 1.6 μm particles. Particles smaller than 1.0 μm are usually reserved for small cells such as microalgae, yeast and bacteria.

Other factors shown to affect transformation efficiency include helium pressure, macrocarrier flight distance to baffle screen, distance from baffle screen to target tissues and vacuum pressure. Further, multiple bombardments of the same target tissue has also been shown to both increase (Klein et al. 1988) and decrease (Janna et al. 2006) transient expression. Apparently excessive cell damage hence cell death resulted in lower transient expression.

The use of osmoticum, such as mannitol, sorbitol, high sucrose and myo-inositol in culture media, has also been reported to enhance microprojectile bombardment and subsequent reporter gene expression in some species. For example, treating embryogenic cell cultures of *Picea abies* with myo-inositol, pre- and post-bombardment, resulted in 5- to 12-fold increase in reporter gene expression (Clapham et al. 1995). Moreover, using sorbitol and mannitol resulted in a 20-fold increase in chloroplast transformation (Ye et al. 1990). Exposure to osmoticum treatment causes plasmolysis and reduces turgor pressure in cells. As such, extensive damage to cell membrane is minimised, and the leakage of cellular contents is prevented when the microcarriers perforate and penetrate the cells (Hagio 1998; Marchant et al. 1998; Santos et al. 2002). In addition, since plasmolysed cells are less rigid, particle entry may also be improved (Hagio 1998). Osmotic stress is not suitable for all species, and where it is used, the type and concentration are variables (Hagio 1998). It is apparent that optimal parameters for transformation efficiency must be arrived at empirically for each plant species.

In principle, biolistic transformation can be done using any regenerable target tissue that is actively growing. In case of banana, shoot apices and embryogenic cell suspensions are commonly used. Generic protocol for microprojectile-mediated transformation of banana cell suspensions involves using any actively growing, regenerable cells. Therefore, cells are harvested 4 days after media subculture and passed through a 500 μm mesh to harvest small cell clumps. At this stage the cells are in the log growth phase and hence are highly competent for transformation and regeneration. The filtrate is allowed to settle for a few minutes, and a 250 μL aliquot of settled cells, suspended in liquid medium ratio of 1:5, is dispensed onto a 70 mm diameter Whatman glass filter paper discs placed on agar-solidified medium in 90 × 15 mm Petri dishes. Cells are bombarded 4 days after plating. Reporter genes like *gus* or *gfp* are used for assessing the efficiency of transient or stable gene transfer. Preculturing of suspension cells on agar-solidified medium for 4 days prior to shooting allows the cells to adapt to the altered growing conditions and begin to proliferate thus increasing the number of actively dividing competent cells. Such actively dividing cells are reported to be ideal target tissue for particle bombardment as (1) they can easily overcome the stresses induced by microprojectiles (Santos et al. 2002) and (2) high frequency of transgene integration into the host genome, thus increasing the probability of obtaining stable transgenic plants (Yang et al. 1999).

9.2.3 *Agrobacterium*-Mediated Gene Transfer

Even though particle bombardment is used quite extensively for transient gene expression studies, *Agrobacterium*-mediated gene transfer is the preferred method for most practical applications where product development is involved. It is considerably less expensive (Lee et al. 2006), results in a lower transgene copy number and can transfer large segments of DNA with less transgene rearrangements (Jones 2005) thus reducing the probability of gene silencing. However, it is highly dependent on plant genotype. *Agrobacterium tumefaciens* is a well-known plant pathogens, but in the past couple of decades, *Agrobacterium* has been extensively used to transfer DNA to plant cells for the purposes of generating transgenic plants. This has involved modifying gene transfer methods and also manipulating this "natural genetic engineer" to extend the host range of the bacterium to economically important plant species and genotypes. In banana, major improvements involved modifications in cell tissue culture, bacterial pre-culture, coculture, post-transformation and selection and regeneration conditions because *Agrobacterium*-mediated plant transformation is highly complex and involves both the bacterium and the plant cell. The transfer of T-DNA and its stable integration into the plant genome is influenced by several factors. Plant tissues require an optimum density of *Agrobacterium* cells for a high frequency of transformation. In banana cell suspensions, although high inoculum densities (1.0 and 2.0) result in high transient gene expression, cell death would be also high (Khanna et al. 2004) indicating that densities giving high transient expression may not translate into a high number of stable transformants. Therefore, it is essential to optimise the inoculum levels of *Agrobacterium* so that cell necrosis is minimised while a high level of T-DNA transfer is maintained.

Incorporation of compounds like acetosyringone and Pluronic F68 in the infection and cocultivation media has been reported to increase the efficiency of transformation. Acetosyringone (a low molecular weight phenolic compound) is used as an exogenous stimulant for the induction of *vir* genes (Chakrabarty et al. 2002; Opabode 2006). Monocotyledonous plants producing only very low levels or none of these types of phenolic compounds are unable to activate the vir genes of *Agrobacterium* (Suzuki et al. 2001). Therefore, including acetosyringone in cocultivation medium has been suggested as one of the solutions to this problem (Fig. 9.3).

Preinduction of *Agrobacterium* in medium containing acetosyringone as well as the incorporation of acetosyringone in cocultivation medium increased T-DNA transfer into banana suspension cells (Khanna et al. 2004).

Agrobacterium causes cell necrosis after infecting them (Carvalho et al. 2004). Therefore, washing cocultured cells with bactericidal or bacteriostatic antibiotics such as Timentin and cefotaxime and incorporating it in the selection medium have been shown to be effective in eliminating *Agrobacterium* from plant cells (Carvalho et al. 2004). In addition, allowing explants to grow for a week without selection except against *Agrobacterium* after cocultivation could help them recover from infection and thus reduce cell necrosis (Carvalho et al. 2004; Bhalla and Singh 2008). Further, giving heat shock to plant cells before infection with *Agrobacterium* has been shown to improve cell survival and regeneration of large numbers of transgenic plants (Khanna et al. 2004). Plant cells produce heat shock proteins (Hsp) following pulses of heat which enables them to resist stress (Wang et al. 2004). The activation of heat shock genes and subsequent expression of proteins prior to *Agrobacterium* infection appear to enhance their resistance against the bacterium.

Modulating other cellular response such as impeding the programmed cell death (PCD) could also improve the frequency of stable transgenics (Khanna et al. 2007). Both biotic and abiotic stresses could induce PCD in plants which results in the expression of apoptotic genes to causing cell death, hence preventing the spread of infection (Dickman et al. 2001). Khanna et al. (2007) demonstrated that by expressing the animal antiapoptosis genes *Bcl-xL*, *Bcl-2* 3′ untranslated region and *CED-9* in banana suspension cells improved cell viability and transformation significantly.

9.2.3.1 Strategies for Enhancing *Agrobacterium*-Mediated Gene Transfer in Banana

9.2.3.1.1 Centrifugation-Assisted *Agrobacterium tumefaciens*-Mediated Transformation (CAAT)

Generating highly regenerable and transformation competent embryogenic cell suspensions are core to any form of banana transformation. Once established, the cell suspensions need to be screened for their competency for transformation and regeneration before any transformation could be carried out. A highly embryogenic regenerable cell suspension line will result in higher frequency of transient and hence stable transforma-

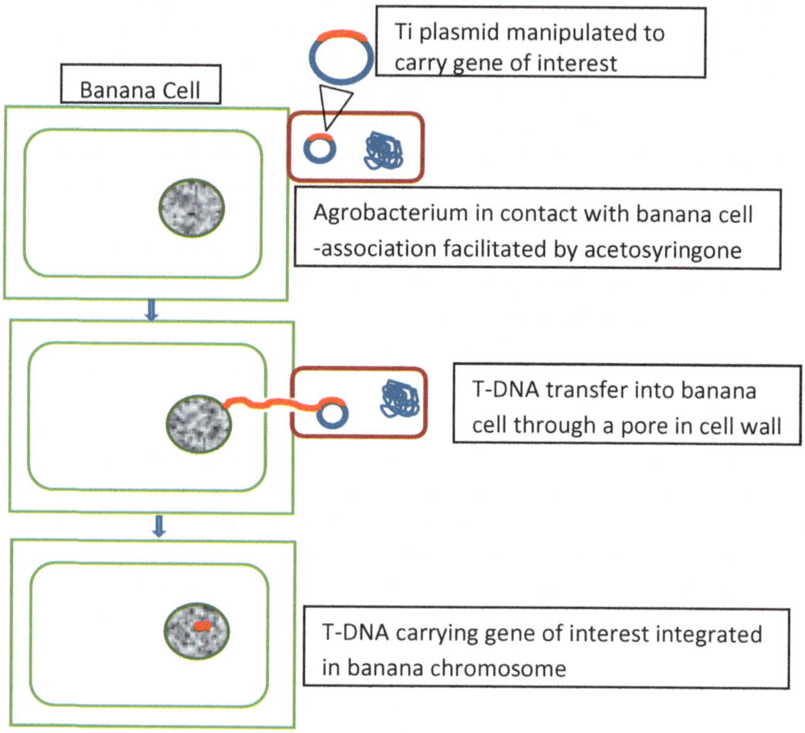

Fig. 9.3 Basic steps in *Agrobacterium*-mediated gene transfer

tion. But one process that can lead to a significant enhancement in transformation frequency is the introduction of a centrifugation step during infection. Stable transgenic banana plants have been generated at high efficiency from embryogenic cell suspensions using centrifugation to assist the gene transfer process (Khanna et al. 2004, 2008; Paul et al. 2011; Namuddu et al. 2013; Elayabalan et al. 2013).

In summary, the process involves harvesting cell suspensions 4 days after media subculture and filtering them through a 500 μm mesh and subjecting them to heat shock for 5 min at 42 °C. Acetosyringone pre-induced *Agrobacterium* culture at an optical density of 0.5 (OD_{600nm} = 0.5) is then forced into close contact with the heat-shocked cells by centrifugation for 10 min in presence of Pluronic F68. After resting the cells for 30 min at room temperature, the cells are plated on coculture medium containing acetosyringone and cocultivated at 23 °C for 3 days in the dark. After cocultivation, the cells are washed and processed for selection (Khanna et al. 2004).

9.2.3.1.2 Sonication and Vacuum Infiltration-Assisted *Agrobacterium tumefaciens*-Mediated Transformation

This method has been reported to enhance *Agrobacterium*-mediated transformation of juvenile banana suckers (Subramanyam et al. 2011). Basic micropropagation technique is used to generate large number of axenic plant material in vitro.

The suckers are initially pre-cultured for 2 days in liquid MS medium. Prior to the bacterial infection, the suckers are immersed in an *Agrobacterium* suspension containing acetosyringone and sonicated for 1–10 min using a bath sonicator to create micro wounds which become the access points for *Agrobacterium*. After sonication, the explants are vacuum infiltrated with the same *Agrobacterium* suspension for 1–10 min at 750 mm of Hg to allow the *Agrobacterium* to penetrate deep into the tissues. Following vacuum infiltration, the cultures are rested or incubated at 27 °C for 1 h. Then, the explants are blot dried to get rid of excess

Agrobacterium and then cocultivated in liquid medium for 3 days in the dark at 27 °C. Coculture media is changed every day to avoid *Agrobacterium* overgrowing and killing the explants (Subramanyam et al. 2011).

9.3 Post-transformation Selection and Regeneration

9.3.1 Regeneration of Transgenic Banana Plants via Somatic Embryogenesis

The success of gene transfer could be determined initially by checking the transient expression of reporter genes 72 h post-transformation for any of the methods used for transformation. This is the minimum time required by the cells to overcome the physical trauma induced by the process and express the genes. Therefore, depending on the reporter gene used, cells can be taken for fluorometric GFP or histochemical GUS assays 72 h post-bombardment. In general, a high level of transient expression is desirable since it is a useful indicator for the successful transformation. However, a high transient expression does not always translate into a high frequency of stable transformation (Tian and Seguin 2004).

The ability to select transformed plant tissue from non-transformed tissue is an important prerequisite in the generation of transgenic plants. In general, negative selection is used to select against non-transformed cells. The transformed cells are allowed to proliferate and form embryos.

The most commonly used selectable marker genes in plant transformation are antibiotic and herbicide resistance genes. The requirement for effective selection is the gradual cessation of growth of non-transformed cells rather than rapid death. Killing cells too rapidly is considered undesirable as dead and dying cells may release substances deleterious to neighbouring transgenic cells and reduce transformation efficiency. Therefore, the minimum effective concentration of the desired antibiotic preventing the growth of non-transformed cells needs to be determined by means of a kill curve.

One of the most popular selectable markers for selection of transgenic cells of banana is *nptII* gene. Kanamycin at 100 mg/L or Geneticin® at 50 mg/L can be used for negative selection, and Timentin® can be used for killing *Agrobacterium*, where the method involves this bacterium. Cells are normally maintained on selection for at least 4 months with monthly subculture on fresh medium. Within this period all untransformed cells die and the transformed cells proliferate and start producing putative transgenic embryos. The cultures are then transferred to embryogenesis medium and maintained in this medium for at least 3 months with monthly subculture. The prolonged culture on this media not only selects for transgenics but also allows embryos to mature so that they germinate readily upon transfer on germination medium. The putative transgenic plantlets are then transferred on higher selection dose (e.g. kanamycin 200 mg/L) where transgenics readily form roots. Escapes do not develop healthy proliferative roots and this can be used as a visual screen. A quick PCR can be performed on putative transgenic plants for the confirmation of the presence of the transgene. The plants generated using *Agrobacterium* method also need to be screened for *Agrobacterium*contamination to rule out false positives. Stable integration of transgene and transgene copy number in the banana genome can be confirmed by southern blot analysis or quantitative PCR methods.

To shorten the potentially lengthy selection/regeneration phase, transformed cell suspensions can be placed directly on regeneration media without a prior proliferation phase. Although potentially a more rapid procedure, there is the potential for escapes and lower transformation efficiency.

9.3.2 Regeneration of Transgenic Shoots via Organogenesis

Using meristem as a target tissue for *Agrobacterium*-mediated transformation of banana has been reported (May et al. 1995; Tripathi et al. 2008; Subramanyam et al. 2011). Initially adventitious shoots are generated de

Table 9.1 List of some traits and genes used in development of transgenic banana

Gene introduced	Phenotype	References
Bar	Resistance to herbicide Basta	Becker et al. (2000)
MSI-99	Resistance to *Fusarium oxysporum* and *Mycosphaerella musicola*	Chakrabarti et al. (2003)
Antimicrobial peptides	Resistance to *Fusarium oxysporum* and *Mycosphaerella fijiensis*	Remy et al. (2000) and Tripathi (2003)
Antimicrobial peptides	Resistance to *Verticillium theobromae* or *Trachysphaera fructigena*	Cary et al. (2000)
Antimicrobial peptide (Ace-AMP1)	Resistance to *Fusarium oxysporum* f.sp. cubense race 1	Mohandas et al. (2013)
Protein-engineered rice	Resistance to nematode*Radopholus similis*	Atkinson et al. (2004)
Cystatin (OcldeltaD86) (oryzacystatin-I)	Resistance to banana weevil (*Cosmopolites sordidus*)	Kiggundu et al. (2002)
Cysteine proteinase inhibitor		
Synthetic cercosporins	Resistance to bacterial wilt (*Xanthomonas* spp.)	Rajasekaran et al. (2001)
Plant ferredoxin-like protein (Pflp)	Resistance to bacterial wilt (*Xanthomonas campestris* pv. musacearum)	Namukwaya et al. (2012)
Rice pattern recognition receptor (PRR), XA21	Resistance to bacterial wilt (*Xanthomonas campestris* pv. musacearum)	Tripathi et al. (2014)
Antiretroviral genes (adefovir, tenofovir)	Resistance to banana streak virus	Helliot et al. (2003)
Human lysozyme gene	Resistance to *Fusarium oxysporum*	Pei et al. (2005)
Antiapoptosis gene (Bcl-xL, Ced9, Bcl-2 3′ UTR)	Resistance to *Fusarium oxysporum Race 1*	Paul et al. (2011) and Magambo (2012)
Carica papaya cystatin (CpCYS)	Resistance to nematode	Namuddu et al. (2013)
RNAi	Resistance to BBTV	Elayabalan et al. (2013) and Shekhawat et al. (2012)
RNAi	Resistance to *Fusarium*	Ghag et al. (2015)
Aquaporins – membrane intrinsic protein gene (MusaPIP1;2)	Abiotic stress	Sreedharan et al. (2013)
WRKY transcription factors	Modified abiotic and biotic stress response	Shekhawat and Ganapathi (2013)
Phytoene synthase (psy), phytoene desaturase (CrtI), ferritin (FER), nicotianamine synthase (NAS), ferric reductase oxidase (FRO), iron-regulated transporter (IRT), yellow stripe-like (YSL)	Provitamin A biofortification Iron biofortification	Khanna et al. (2011) and Dale et al. (2013)

novo from intercalary meristematic tissues from the corms. The shoots are micro wounded using gold microprojectiles and infected with *Agrobacterium* containing the gene construct. After cocultivation, the infected shoots are proliferated under selection to regenerate transgenic plants.

After cocultivation, the explants are washed in liquid shoot induction medium containing cefotaxime (500 mg/L) or Timentin (200 mg/L) and then cultured in shoot induction medium containing cefotaxime (500 mg/L) and hygromycin (30 mg/L) for 3 weeks to induce shoot buds. Once the shoot buds are induced, they are transferred into multiplication medium with selection for further 3 weeks. The shoots are elongated for 3 weeks before being transferred to rooting medium. Putative transgenic shoots root profusely in the presence of the selective agent (e.g. hygromycin).

This system has been reported to produce transgenic banana plants at a frequency of 34.9 %. Although it is lower than the frequency of getting transgenic plants from embryogenic cell suspensions, this method is promising for those cultivars which are difficult to establish in cell suspensions. Also, this ensures continuous supply of in vitro axenic target tissue, whereas banana flowers have to be sourced from mature field plants which need to be decontaminated. The initiation of embryonic cultures from flowers follows seasonal cycle; hence, there are inconsistencies in the efficiency of embryogenic callus and cell suspension formation. However, meristem transformation could produce chimeric plants due to its multicellular nature.

9.4 Traits and Genes of Interest for Transgenic Banana

Following development of efficient protocols for banana transformation and regeneration of stable transgenics, there has been many reports on generation of banana transgenics. A summary of these is provided below (Table 9.1).

9.5 Conclusion

Thus, there are many different ways by which the genes can be introduced into the banana cells. Genetic transformation is a powerful tool and an important technique, not only for generating transgenics but for the study of banana functional genomics, i.e. gene discovery and gene function. Molecular tools and technologies have now made it much easier to identify new genes and promoters and introduce them into banana cells to generate transgenics. Gene transfer through different gene transfer technologies has led to the generation of transgenics with many traits of interest. Improvements in gene transfer are now required in terms of transgene cassette designs and integration technology to permit gene targeting. Further developments in this field will make it possible to target transgenes to specific locations in the banana genome.

References

Aguado-Santacruz GA, Rascón-Cruz Q, Cabrera-Ponce JL, Martínez-Hernández A, Olalde-Portugal V et al (2002) Transgenic plants of blue grama grass, Bouteloua gracilis (H.B.K.) Lag. ex. Steud. from microprojectile bombardment of highly chlorophyllous embryogenic cells. Theor Appl Genet 104:763–771

Alves AC, Quecini VM, Vieira MLC (1999) Plant transformation: advances and perspectives. Sci Agric 56:1–8

Atkinson HJ, Grimwood S, Johnston K, Green J (2004) Prototype demonstration of transgenic resistance to the nematode *Radopholus similis* conferred on banana by a cystatin. Transgenic Res 13:135–142

Bakry F (1984) Application des techniques de culture In vitro pour l'amélioration du bananier (Musa sp.). Thesis Paris-Sud University

Bakry F, Haïcour R, Horry JP et al (1992) Applications of biotechnologies to banana breeding: haplogenesis, plant regeneration from protoplasts and transformation. Applications of Biotechnologies to Banana Breeding. pp 52–62

Bakry F, Carreel F, Jenny C et al (2009) Genetic improvement of banana. In: Breeding plantation tree crops: tropical species. Jain, S.M. and P.M. Priyadarshan (Eds.). Springer, New York, pp 3–50

Becker DK, Dugdale B, Smith MK et al (2000) Genetic transformation of Cavendish banana (Musa spp. AAA group) cv. "Grand Nain" via microprojectile bombardment. Plant Cell Rep 19:229–234

Bhalla PL, Singh MB (2008) *Agrobacterium*-mediated transformation of *Brassica napus* and *Brassica oleracea*. Nat Protoc 3:181–189

Carvalho CHS, Zehr UB, Gunaratna N et al (2004) *Agrobacterium*-mediated transformation of sorghum: factors that affect transformation efficiency. Genet Mol Biol 27:259–269

Cary JW, Rajasekaran K, Jaynes JM et al (2000) Transgenic expression of a gene encoding a synthetic antimicrobial peptide results in inhibition of fungal growth in vitro and in planta. Plant Sci 154:171–181

Chakrabarti A, Ganapathi TR, Mukherjee PK et al (2003) MSI-99, a magainin analogue, imparts enhanced disease resistance in transgenic tobacco and banana. Planta 216:587–596

Chakrabarty R, Viswakarma N, Bhat SR et al (2002) *Agrobacterium*-mediated transformation of cauliflower: optimization of protocol and development of Bt-transgenic cauliflower. J Biosci 27:495–502

Clapham D, Manders G, Yibrah HS et al (1995) Enhancement of short and medium term expression of transgenes in embryogenic suspensions of Picea abies (L.) Karst. J Exp Bot 46:655–662

Cote FX, Domergue R, Monmarson S et al (1996) Embryogenic cell suspensions from the male flower of Musa AAA cv. Grand Nain. Physiol Plant 97:285–290

Cote FX, Legavre T, Grapin A et al (1997) Genetic transformation of embryogenic cell suspension in plantain (Musa AAB) using particle bombardment. In: Proceedings of the international symposium on biotechnology of tropical and subtropical species: part 1. Acta Horticult 460:126–129

Cronauer SS, Krikorian AD (1986) Banana (Musa spp.). In: Bajaj YPS (ed) Biotechnology in agriculture and forestry, vol 1. Trees 1. Springer, Berlin Heidelberg New York pp 233–252

Da Silva Conceicao (1989) Isolement et culture de protoplastes de bananiers (Musa sp.). Etude de divers facteurs. DEA génétique et sélection Animale et Végétale, Université Paris Sud Orsay

Dale J, Harding R, Tushemeirwe W et al (2013) Use of transgene expression to elevate micronutrients in banana. Paper presented at plant and animal genome, Asia. 17–19 March

De Block M, Herrera-Estrella L, van Montagu M et al (1984) Expression of foreign genes in regenerated plants and their progeny. EMBO J 3:1681–1689

Deo PC, Tyagi AP, Taylor M et al (2010) Factors affecting somatic embryogenesis and transformation in modern plant breeding. South Pac J Nat Appl Sci 28:27–40

Deroles S, Smith MAL, Lee C (2002) Factors affecting transformation of cell cultures from three dicotyledonous pigment-producing species using microprojectile bombardment. Plant Cell Tissue Org Cult 70:69–76

Dickman MB, Park YK, Oltersdorf T et al (2001) Abrogation of disease development in plants expressing animal antiapoptotic genes. Proc Natl Acad Sci U S A 98:6957–6962

Dugdale B, Becker D, Harding RM et al (2001) Intron-mediated enhancement of the banana bunchy top virus DNA-6 promoter in banana (Musa spp) embryogenic cells and plants. Plant Cell Rep 20(3):220–226

Elayabalan S, Kalaiponmani K, Subramaniam S et al (2013) Development of *Agrobacterium*-mediated transformation of highly valued hill banana cultivar Virupakshi (AAB) for resistance to BBTV disease. World J Microbiol Biotechnol 29:589–596

Finer JJ, Vain P, Jones MW et al (1992) Development of a particle inflow gun for DNA delivery into plant cells. Plant Cell Rep 11:323–328

Ghag SB, Shekhawat UKS, Hadapad AB et al (2015) Stacking of host-induced gene silencing mediated resistance to banana bunchy top virus and fusarium wilt disease in transgenic banana plants. Curr Trends Biotechnol Pharm 9(3):212–221

Gheysen G, Angenon G, Montague MV (1998) *Agrobacterium* mediated plant transformation: a scientifically intriguing story with significant application. In: Lindsey k (ed) Transgenic plant research. Harwood Academic Press, Amsterdam, pp 1–33

Häggman H, Aronen T (1998) Transgene expression in regenerating cotyledons and embryogenic cultures of Scots pine. J Exp Bot 49:1147–1156

Hagio T (1998) Optimizing the particle bombardment method for efficient genetic transformation. Japan Agric Res Q 32:239–247

Hansen G, Martha SW (1999) Recent advances in the transformation of plants. Trends Plant Sci 4:226–231

Helliot B, Panis B, Frison E et al (2003) The acyclic nucleoside phosphonate analogues, adefovir, tenofovir and PMEDAP, efficiently eliminate banana streak virus from banana (Musa sp.). Antivir Res 59:121–126

Horsch RB, Fraley RT, Rogers SG et al (1984) Inheritance of functional foreign genes in plants. Science 223(4635):496–498

Janna OA, Maziah M, Ahmad PGKA et al (2006) Factors affecting delivery and transient expression of β-glucuronidase gene in Dendrobium sonia protocorm-like-body. Afr J Biotechnol 5:88–94

Jones HD (2005) Wheat transformation: current technology and applications to grain development and composition. J Cereal Sci 41:137–147

Khanna H, Becker D, Kleidon J et al (2004) Centrifugation assisted *Agrobacterium tumefaciens*-mediated transformation (CAAT) of embryogenic cell suspensions of banana (Musa spp. Cavendish AAA and Lady finger AAB). Mol Breed 14:239–252

Khanna HK, Paul J-Y, Harding RM et al (2007) Inhibition of *Agrobacterium*-induced cell death by antiapoptotic gene expression leads to very high transformation efficiency of banana. Mol Plant-Microbe Interact 20:1048–1054

Khanna H, Becker D, Whitelaw E et al (2008) Biofortification of Bananas. In: Proceedings of the tropical fruits in human nutrition and health conference, Couran Cove Island Resort, Gold Coast, Australia. 08–11 November 2008. pp 9–18

Khanna H, Harding R, Becker D et al (2011) Metabolic engineering for provitamin A biofortified bananas: a sustainable strategy to combat vitamin A deficiency. Gordon conference on plant metabolic engineering Waterville Valley, New Hampshire. 24–29 July 2011

Kiggundu A, Kunert K, Viljoen A et al (2002) Designing proteinase inhibitors for banana weevil control. In: 3rd international symposium on molecular and cellular biology of bananas, Luven

Klein TM, Fromm M, Weissinger A (1988) Transfer of foreign genes into intact maize cells with high-velocity microprojectiles. Proc Natl Acad Sci U S A 85:4305–4309

Koichi T, Bae C-H, Seo MS et al (2002) Overcoming of barriers to transformation in monocot plants. J Plant Biotechnol 4:135–141

Lee S-H, Lee D-G, Woo H-S et al (2006) Production of transgenic orchardgrass via *Agrobacterium*-mediated transformation of seed derived callus tissues. Plant Sci 171:408–414

Magambo B (2012) Generating transgenic banana (cv. Sukali Ndizi) resistant to Fusarium wilt. Master's thesis, Queensland University of Technology, Brisbane. eprints.qut.edu.au/61024/1/Betty_Magambo_Thesis.pdf

Mahn A, Matzk A, Sautter C et al (1995) Transient gene expressions in shoot apical meristems of sugarbeet seedlings after particle bombardment. J Exp Bot 46:1625–1628

Marchant R, Power J-B, Lucas JA (1998) Biolistic transformation of rose (Rosa hybrida L.). Ann Bot 81:109–114

May GD, Afza R, Mason HS et al (1995) Generation of transgenic banana (Musa acuminata) plants via *Agrobacterium*-mediated transformation. Biotechnology 13:486–492

Megia RR, Haicour S, Tizroutine V et al (1993) Plant regeneration from cultured protoplasts of the cooking banana cv. Bluggoe (Musa ssp., ABB group). Plant Cell Rep 13:41–44

Mohandas S, Sowmya HD, Saxena AK et al (2013) Transgenic banana cv. Rasthali (AAB, Silk gp) harboring Ace-AMP1 gene imparts enhanced resistance to Fusarium oxysporum f.sp. cubense race 1. Sci Hortic 164:392–399

Namuddu A, Kiggundu A, Mukasa SB et al (2013) *Agrobacterium*-mediated transformation of banana (Musa sp.) cv. Sukali Ndiizi (AAB) with a modified Carica papaya cystatin (CpCYS) gene. Afr J Biotechnol 12:1811–1819

Namukwaya B, Tripathi L, Tripathi JN, Arinaitwe G, Mukasa SB, Tushemereirwe WK (2012) Transgenic banana expressing Pflp gene confers enhanced resistance toXanthomonas wilt disease. Transgenic Res 21:855–865

O'Kennedy MM, Burger JT, Berger DK (2001) Transformation of elite white maize using particle inflow gun and detailed analysis of a low copy integration event. Plant Cell Rep 20:721–730

Opabode JT (2006) *Agrobacterium*-mediated transformation of plants: emerging factors that influence efficiency. Biotechnol Mol Biol Rev 1:12–20

Panis B, Wauwe AV, Swennen R (1993) Plant regeneration through direct somatic embryogenesis from protoplasts of banana (Musa spp.). Plant Cell Rep 12:403–407

Paul J-Y, Becker DK, Dickman MB et al (2011) Apoptosis-related genes confer resistance to Fusarium wilt in transgenic 'Lady Finger' bananas. Plant Biotechnol J 9(9):1141–1148

Pei XW, Chen SK, Wen RM et al (2005) Creation of transgenic bananas expressing human lysozyme gene for panama wilt resistance. J Integr Plant Biol 47:971–977

Quoirin M, Aragão F, Rech E et al (1997) Transient expression of a reporter gene introduced by bioballistic bombardment into Racosperma mangium (Leguminosae family) tissues. Braz J Genet 20: 507–510

Rajasekaran K, Stromberg KD, Cary JW et al (2001) Broad-spectrum antimicrobial activity in vitro of the synthetic peptide D4E1. J Agric Food Chem 49:2799–2803

Remy S, Buyens A, Cammue BPA et al (2000) Production of transgenic banana plants expressing antifungal proteins. International Symposium on Banana in the Subtropics. Acta Hortic 490:219–277

Russell JA, Ray MK, Sanford JC (1992) Physical trauma and tungsten toxicity reduce the efficiency of biolistic transformation. Plant Physiol 98:1050–1056

Sagi L, Remy S, Panis B et al (1994) Transient gene expression in electroporated banana (Musa spp., cv. 'Bluggoe', ABB group) protoplasts isolated from regenerable embryogenetic cell suspension. Plant Cell Rep 13:262–266

Sagi L, Remy S, Verelst B et al (1995) Transient gene expression in transformed banana (Musa cv. Bluggoe) protoplasts and embryogenic cell suspension. Euphytica 85:89–95

Sagi L, Remy S, Swennen R (1998) Genetic transformation for the improvement of bananas – a critical assessment. In: INIBAP annual report. FRA, Montpellier, pp 33–36

Sahrawat AK, Becker D, Lütticke S et al (2003) Genetic improvement of wheat via alien gene transfer, an assessment. Plant Sci 165:1147–1168

Sanford JC, Smith FD, Russell JA (1993) Optimizing the biolistic process for different biological applications. Methods Enzymol 217:483–509

Santos MO, Barros EVSA, Tinoco MLP et al (2002) Repetitive somatic embryogenesis in cacao and optimization of gene expression by particle bombardment. J Plant Biotechnol 4:71–76

Sharma KK, Bhatnagar-Mathur P, Thorpe TA (2005) Genetic transformation technology: status and problems. In Vitro Cell Dev Biol Plant 41:102–112

Shekhawat UKS, Ganapathi TR (2013) Musa WRKY71 overexpression in banana plants leads to altered abiotic and biotic stress responses. PLoS ONE 8:75506

Shekhawat UKS, Ganapathi TR, Hadapad AB (2012) Transgenic banana plants expressing small interfering RNAs targeted against viral replication initiation gene display high-level resistance to banana bunchy top virus infection. J Gen Virol 93:1804–1813

Shibata D, Liu YG (2000) *Agrobacterium*-mediated plant transformation with large DNA fragments. Trends Plant Sci 5:354–357

Sreedharan S, Shekhawat UKS, Ganapathi TR (2013) Transgenic banana plants overexpressing a native plasma membrane aquaporin MusaPIP1;2 display high tolerance levels to different abiotic stresses. Plant Biotechnol J 11(8):942–952

Strosse H, Domergue R, Panis B et al (2003) Banana and plantain embryogenic cell suspensions. In: Vézina A, Picq C (eds) INIBAP technical guidelines 8. The International Network for the Improvement of Banana and Plantain, Montpellier

Subramanyam K, Subramanyam K, Sailaja KV et al (2011) Highly efficient *Agrobacterium*-mediated transformation of banana cv. Rasthali (AAB) via sonication and vacuum infiltration. Plant Cell Rep 30:425–436

Suzuki S, Supaibulwatana K, Mii M et al (2001) Production of transgenic plants of the liliaceous ornamental plant Agapanthus praecox ssp. orientalis (Leighton) via *Agrobacterium*-mediated transformation of embryogenic calli. Plant Sci 161:89–97

Taylor NJ, Fauquet CM (2002) Microprojectile bombardment as a tool in plant science and agricultural biotechnology. DNA Cell Biol 21:963–977

Tian L, Seguin A (2004) Microprojectile particle effect on stable transformation of black spruce via bombardment. Plant Mol Biol Report 22:199a–199f

Tripathi L (2003) Genetic engineering for improvement of Musa production in Africa. Afr J Biotechnol 2:503–508

Tripathi L, Tripathi JN, Tushemereirwe WK (2008) Rapid and efficient production of transgenic East African Highland Banana (Musa spp.) using intercalary meristematic tissues. Afr J Biotechnol 7:1438–1445

Tripathi JN, Lorenzen J, Bahar O et al (2014) Transgenic expression of the rice Xa21pattern-recognition receptor in banana (Musa sp.) confers resistance to Xanthomonas campestris pv. musacearum. Plant Biotechnol J 12:663–673

Ul-Haq I (2005) Callus proliferation and somatic embryogenesis in cotton (Gossypium hirsutum L.). Afr J Biotechnol 4:206–209

Wang J, Seliskar DM, Gallagher JL (2004) Plant regeneration via somatic embryogenesis in the brackish wetland monocot Scirpus robustus. Aquat Bot 79:163–174

Yang J, Lee H-J, Shin DH (1999) Genetic transformation of Cymbidium orchid by particle bombardment. Plant Cell Rep 18:978–984

Ye GN, Daniell H, Sanford JC (1990) Optimization of delivery of foreign DNA into higher plant chloroplasts. Plant Mol Biol (Hist Arch) 15:809–819

Somatic Embryogenesis as a Tool in Genetic Transformation

10

H.D. Sowmya, T.R. Usharani, and Sukhada Mohandas

Abstract

Bananas and plantains are the most important staple food crop in the world. The production of banana is hampered by various stresses. Genetic transformation has become an important potential tool in developing improved banana with desired agronomic traits, which is highly difficult to achieve through conventional breeding. To be successful, genetic engineering technique requires reliable, efficient in vitro regeneration protocol through tissue culture. Plant regeneration through somatic embryogenesis has become an important tool due to high proliferation potential, minimal genetic instability, and single-cell origin which in turn reduces the formation of chimera. However, banana is highly recalcitrant toward the development of somatic embryogenesis. Considerable progress has been achieved in the regeneration of banana through somatic embryogenesis, but there are still many factors to overcome. In present review key factors such as age of the explant, genotype, and plant growth regulators affecting the induction and regeneration of plants by embryogenic callus are discussed. This review also provides special focus on methods being applied in plant transformation through somatic embryogenesis, different factors which affect somatic embryogenesis, and various strategies to improve the transformation efficiency using somatic embryogenesis.

Keywords

Agrobacterium • Embryogenic cell suspension • Genetic transformation • Particle bombardment • Somatic embryogenesis

H.D. Sowmya • T.R. Usharani
Indian Institute of Horticultural Research, Hessaraghatta, Bangalore 560 089, India

S. Mohandas (✉)
Division of Biotechnology, ICAR-Indian Institute of Horticultural Research, Hessaraghatta Lake Post, Bangalore 560 089, India

SM Plant Production Technologies Pvt Ltd., 138 Kodi Halli, Madure Hobli, Doddaballapura Taluk, Bangalore 562103, India
e-mail: sukhada.mohandas@gmail.com

S. Mohandas, K.V. Ravishankar (eds.), *Banana: Genomics and Transgenic Approaches for Genetic Improvement*, DOI 10.1007/978-981-10-1585-4_10

10.1 Introduction

Bananas and plantains are perennial, monocotyledonous crop cultivated in tropical and subtropical region of the world. It is considered as world's fourth most important staple food crop after rice, wheat, and corn due to its richness in providing a balanced diet for millions of people. Presently, banana is cultivated approximately in around 150 countries across the world in an area of more than 10 million hectares with a total production of 144 million tonnes (FAOSTAT 2013). The production is primarily hampered by several abiotic and biotic factors. The application of the classical breeding approach for the improvement of crop has limitation due to their polyploidy, long generation time, narrow genetic variability, and high sterility (Sasson 1997). The application of genetic engineering technology provides a powerful tool for the agronomic improvement of banana within a short period. However, to be successful, genetic engineering technique requires reliable, efficient regeneration protocol (Hansen and Martha 1999). Presently, in vitro micro-propagation of banana has been achieved by using different exploits such as apical meristem (Arinaitwe et al. 2000), multiple bud clumps (Yip et al. 2011), leaf sheath (Venkatachalam et al. 2006), and anthers (Assani et al. 2003) in different genotypes. However, development of transgenic banana by these explants often produces chimeric plants and the transformation efficiency is also very low. (Ramírez-Villalobos and de García 2008). Plant regeneration through somatic embryogenesis becomes an important tool due to the high proliferation potential and minimal genetic instability (Sharp et al. 1980; Ahloowalia 1991; Henry 1998). Many workers have emphasized somatic embryogenesis as a preferred method for genetic improvement, since somatic embryos arise from a single cell which reduces the formation of chimeras (Dhed'a et al. 1991; Toonen and De Vries 1996; Côte et al. 1996; Becker et al. 2000). Other than genetic improvement, somatic embryogenesis is also considered as a valuable tool for the production of secondary metabolites, polyploid, somatic hybridization, and virus-free plant development (Kamle et al. 2011).

So far, a number of researchers have utilized embryogenic cells as target explants in the agronomic improvement of banana for different traits (Mohandas et al. 2013b: Shekhawat et al. 2012; Namukwaya et al. 2012).

10.2 Somatic Embryogenesis

Somatic embryogenesis is a process where somatic cells are reprogrammed to embryogenic pathway by differentiation, resulting in the development of embryos that resemble morphologically and physiologically to zygotic embryos. They are bipolar in nature, bear typically embryogenic organ, and are not connected with the original tissue by vascular part (Solís-Ramos et al. 2012). The process is feasible because plants possess cellular totipotency whereby individual somatic cells can regenerate into a whole plant. The somatic embryos develop either directly without intermediated callus (direct) or through the callus (indirect) or directly or indirectly from the cultured somatic embryos (secondary somatic embryos) (Gaj 2004). These embryos are developed either by unicellular or multicellular pathway (Haccius 1978; Raghavan 1976). The unicellular origin somatic embryos have coordinated cell division and embryos may connect to the maternal tissue through suspensor-like structure, whereas multicellular origin embryos lack coordinated cell division, initially observed as a protuberance, and those embryos in contact with the basal area are typically fused to the maternal tissue (Williams and Maheswaran 1986; Quiroz-Figueroa et al. 2006).

10.3 Explants

The regeneration of banana plants via somatic embryogenesis has been achieved using different explants like zygotic embryos, proliferating meristems, leaf sheath, rhizome, and female and male flowers (Côte et al. 1996; Escalant and Teisson 1989; Escalant et al. 1994; Grapin et al. 2000; Ma 1991; Mohandas et al. 2011; Novak et al. 1989; Strosse et al. 2006). Although several protocols have been established using different explants for the development of somatic embryos in different genotypes of banana, female and male flower and rhizome have been considered as a good source (Elayabalan et al. 2013a; Chang and Shu 2013). However, utilization of female flower as an initial explant material causes revenue loss to farmer, as it is required for banana fruit development (Chang and Shu 2013). Several researchers have opted male inflorescence as a much better choice for somatic embryo development in different genotypes than scalp method, perhaps due to more availability of male flowers and extensive material preparation phase with corms (Schoofs et al. 1999), associated with a high load of microbial flora.

10.4 Different Stages of Somatic Embryo Development

The morphogenic sequential stage of somatic embryo development can be divided into different stages: formation of proembryogenic mass, pre-maturation of somatic embryos, maturation, and plant regeneration.

10.4.1 Induction of Pro-embryogenic Mass

Banana is a highly recalcitrant crop toward embryogenic callus response (Xu et al. 2008). Escalant et al. (1994) obtained a variable percentage of embryogenic callus response (0–7 %) in five different banana genotypes. Navarro et al. (1997) found a variable 2–6 % of embryogenic callus in banana cv. Grand Nain from male flowers. A 21.75 % embryogenic callus response was found in banana cv Maa (AA) by Jalali et al. (2003). Strosse et al. (2006) reported 1.8 %, 3.3 %, and 6.0 % embryogenic callus response from the scalp in Cavendish-type banana (AAA), AAB plantains, and cooking banana (ABB), respectively. Schoofs et al. (1999) reported the embryogenic callus response is lower than 1 % in Cavendish group banana, whereas Mohandas et al. (2013a) found 6.27 % embryogenic callus induction from male flowers of banana cv. Elakki Bale (AB).

10.4.1.1 Factors Influencing Induction of Embryogenic Calli

The development of somatic embryos are influenced by various factors, such as the type of explants, genotype, age, position of the explant, and plant growth regulators.

10.4.1.1.1 Genotype and Age of the Explant

The genotype of plant plays a key factor during the induction of embryogenic calli. The research carried out with different genotypes has shown the variation in the development of somatic embryos. Strosse et al. (2003) found that the frequency of embryogenic callus induction varied between genotypes and also differed within genotypes. Escalant et al. (1994) found the variable rates of callus induction from 0 % to 7 % in AAA, AAB, and ABB genome when tested under similar callus induction medium conditions. Similar types of results were also reported by Ganapathi et al. (1999) where cv. Rasthali (AAB) was found to be more responsive compared to AAA genotypes such as Basrai, Lokhandi, and Trikoni. Variation in the embryogenic callus induction within genotype has been reported in different cultivars. Navarro et al. (1997) obtained 2–6 % embryogenic calli in cv. Grand Nain (AAA), Strosse et al. (2003) 8 %, and Youssef et al. (2010) reported 7.5 % and 10 % response in cv. Grand Nain and Williams, respectively. Elayabalan et al. (2013a) observed 57.78 % embryogenic callus response in banana cv. Virupakshi which belongs to the AAB group. Morais-Lino et al. (2008) found 20 % embryogenic callus induction from the male inflorescence of genome AAB (Brazilian cultivar Terra) while Mohandas et al. (2011) reported 6.25 % embryogenic callus induction from banana cv. Rasthali (AAB).

The age of the explant also has great influence on induction of embryogenic callus (George et al. 2008). Youssef et al. (2010) investigated the effect of age of explant on embryogenic callus induction by taking 1- and 2-week-old male inflorescence in banana cv. Williams. Maximum number of embryogenic callus was observed in a 2-week-old than in a 1-week-old male inflorescence.

10.4.1.1.2 Plant Growth Regulators

Among different external factors, plant growth regulators such as auxins and cytokinins play a crucial role in the induction of embryogenic callus. In particular, the presence of auxin promotes callus proliferation and inhibits differentiation. Ganapathi et al. (1999) and Navarro et al. (1997) reported that induction of embryogenic calli from male flowers requires three different types of auxins such as NAA, IAA, and 2,4-D, while cytokinin and auxins are required for embryogenic callus multiplication and embryo development (Ganapathi et al. 1999). Daniels et al. (2002) suggested that 4 mg l^{-1} 2,4-D was a preferable concentration for the induction of embryogenic callus. A similar type of result was reported by other research groups (Dai et al. 2010; Mohandas et al. 2011, 2013a). Kitamiya et al. (2000) reported the influence of 2,4-D on the induction of somatic embryos, where 2,4-D triggers the stress-related genes of somatic embryogenic pathway; the reprogrammed somatic cells undergo differentiation with the expression of stress-related genes and form somatic embryos. Xu et al. (2004) observed maximum number death of the explant at higher concentration of 2,4-D (8 mg l^{-1}). The induction of good scalp was achieved at lower concentration of thidiazuron (TDZ), and the combination of 6-benzylaminopurine (BAP) and TDZ at 2.30/1.25 and 2.80/1.00 mg l^{-1}, respectively (Sadik et al. 2007).

10.4.1.1.3 Other Factors

The position of the male floral hands plays an important role in the response of embryogenic calli. Various authors reported different responses from different floral hand positions for the induction of embryogenic calli in cultivars. Generally, the male floral hands between 8 and 16 positions were more responsive in embryogenic callus induction (Jalil et al. 2003; Chong et al. 2005), whereas Wei et al. (2005) found a floral position from 6 to 12 was more responsive (5.79). Dai et al. (2010) reported 9.2 % embryogenic callus induction in banana cv. Da Jiao in the floral cluster ranked between 4 and 12. Mohandas et al. (2013a, b) reported the highest frequency of embryogenic callus induction in banana cv Elakki Bale (Ney Poovan) from floral position 8–16. Other than positional effect, seasonal factor also influences embryogenic callus induction. Escalant et al. (1994) observed that the response of embryogenic callus induction was more (74 %) in male flowers harvested between September and October, while less in December to January.

10.4.2 Establishment of Embryogenic Cell Suspension

Initiation and multiplication of good quality embryogenic cell suspension is mainly dependent on the nature of the embryogenic calli selected. Embryogenic callus cluster often contains both embryogenic and non-embryogenic cells. Georget and coworkers (2000) studied five different types of cell aggregates in embryogenic cell suspension of banana cv. Grand Nain (AAA). Histological studies revealed that type I cells correspond to small aggregates with non-joining cells (Fig. 10.1A, B), and type II cell aggregates are closely attached in nature and characterized by high nucleoplasmic ratio, small vacuole, and starch reserve (Fig. 10.1C, D), whereas type III aggregates at peripheral region contain cellular proliferating zone and at center embryogenic cells (Fig. 10.1E, F). These cells can be distinguished from type II cells by embryonic zone organized around peripheral zone. Type IV cells in the peripheral region contain protodermal cell masses (Fig. 10.1G, H). These masses are characterized by a nucleus with small nucleoli and vacuoles. Type V cells are nodular in nature characterized by an external layer of starchy cells with a center region composed of meristematic cells with large nuclei (Fig. 10.1I, J). Among these cell aggregates, types II and III are highly embryogenic in

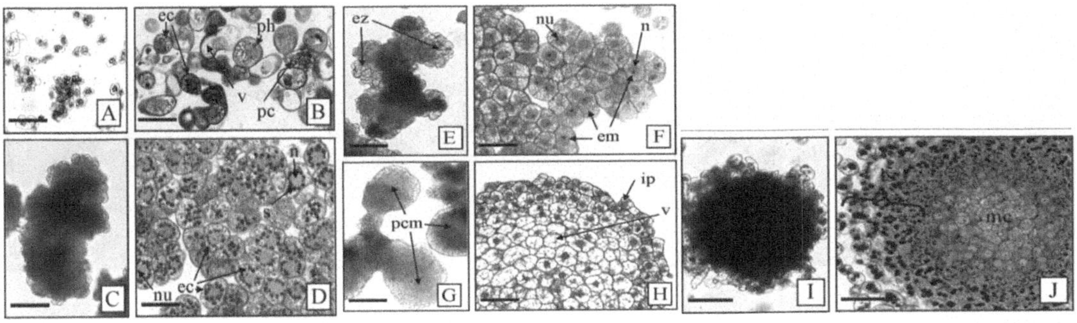

Fig. 10.1 Morphological aspects (**A**, **C**, **E**, **G**, **J**) and histological section (**B**, **D**, **F**, **H**, and **J**) of banana embryogenic cell aggregates (Source: Georget et al. 2000)

nature and germination of somatic embryos into plantlets from these cell aggregates is high. Similar type of opinion was also reported by Dhed'a et al. (1991).

10.4.3 Plantlet Germination via Somatic Embryogenesis

The regeneration of embryogenic cells through somatic embryos in banana is highly variable and depends on the quality of the ECS. Georget et al. (2000) investigated germination efficiency of five different types of embryogenic cells aggregates. Maximum percentage of germination was observed in type II (90 %) and III (80 %) followed by type I (48 %) cells, while no plant germination could be observed in type V cell aggregates.

Besides cell aggregates, other factors such as media phase (semisolid/liquid), incubation time, size, and density have also influenced germination of somatic embryos into plantlets. There are two types of media phases that are followed for the regeneration of somatic embryos such as temporary immersion system and semisolid system. Husina et al. (2014) attained regeneration rate of 10.6-fold in banana cv. Berangan in a short period of time in liquid medium compared to semisolid medium. Wong et al. (2006) reported that approximately 31,835 plants regenerated from 1 ml of settled cell volume (SCV) of *Musa acuminata* cv. Mas (AA) in liquid-based embryo development medium. Gomez et al. (2002) reported a very high germination rate in a temporary immersion system of bioreactor than in semisolid system in banana cv. FHIA-18 (AAAB). Escalant et al. (1994) found much higher germination of about 60–70 % in the temporary immersion system than from semisolid system. This could be due to the less oxidation of somatic embryos and easy uptake of nutrient by cells in liquid culture medium (Gomez et al. 2002; Husina et al. 2014) (Fig. 10.2).

Germination frequency of somatic embryos was also affected by the duration of incubation on M3 medium (somatic embryo development medium). Grapin et al. (1996) achieved maximum number of somatic embryo germination by the prolonged incubation of embryogenic cells on M3 medium. Côte et al. (1996) investigated the role of size of the embryos on regeneration. The result suggested that cells 800–1000 μm in size germinated around 20 %, while 100–250 μm size germinated approximately 3 % on M4 medium. Daniels et al. (2002) found that other than cell size, cell density also impacts frequency of somatic embryo development and its germination.

10.5 Genetic Transformation

Plant genetic transformation is the technique of introduction of desired foreign DNA into plant cells using biological or non-biological tool. It offers a significant advancement in banana breed-

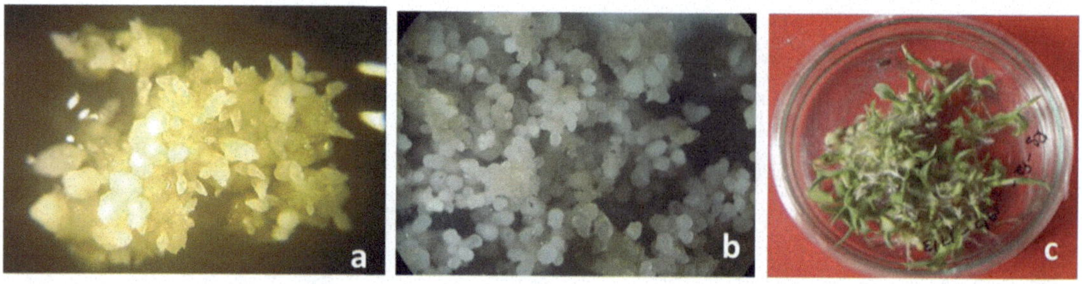

Fig. 10.2 Different stages of somatic embryo development. (**a**) Embryogenic callus. (**b**) Somatic embryos. (**c**) Germinating somatic embryos (Source: Mohandas et al. 2013a)

ing programs for the production of desired agronomical important trait without altering the native characters (Singh et al. 2004). Various sources of banana explants have been used for the transgenic development; however, the choice of high competent explants is prerequisite for the development of the transgenic banana; otherwise, it ends with poor transformation efficiency because of the recalcitrant nature of the explant. The explants that are employed for transformation always do not yield an expected frequency of all transformed cells. Besides, it depends on various parameters that influence the transformation and regeneration efficiency. The plants that are regenerated from transformed tissues via organogenesis often resulted in the production of chimera. Hence, it is emphasized to select a single cell for transformation. Plant genetic transformation using somatic embryos as explant is an extremely valuable tool in plant biotechnology due to its unicellular origin, which reduces the formation chimera (Sagi et al. 1995; Taylor and Fauquet 2002; Roux et al. 2001). In addition, cell culture contains cytoplasmically rich, actively dividing cells, which provide high frequency of transgene integration (Mahn et al. 1995). It is also believed that these cells can tolerate the stress induced by transformation and produces a lower number of somaclonal variants (Santos et al. 2002; Kamle et al. 2011). Transformation techniques that have been employed for the banana improvement through somatic embryogenesis are discussed below.

10.5.1 *Agrobacterium*-Mediated Transformation

Currently, *Agrobacterium*-mediated transformation is considered as the most amenable and routinely applied technique for genetic manipulation of plant system, and it has been successfully applied to the transformation of various economically and horticultural important crops (Ishida et al. 1996; Cheng et al. 1997; Herrera-Estrella et al. 2005). *Agrobacterium*-mediated transformation has been reported to be the most preferred method compared to other methods due to its efficient and stable transformation of foreign DNA into the plant genome (Arinaitwe et al. 2004; Gheysen et al. 1998: Shibata and Liu 2000; Ghosh et al. 2009). Further, it allows the transfer of large DNA fragments with minimal rearrangement with one or few copies. However, *Agrobacterium*-mediated transformation has certain limitation due its limited host range, integration of extra vector sequence other than T-DNA sequence, and its persistence on host tissue for long periods (Christou 1996).

The *Agrobacterium*-mediated transformation is well established for dicots, whereas monocots are recalcitrant. To date, successful genetic modification has been achieved in monocots by standardizing the various parameters. In case of banana, to reduce the recalcitrant nature toward *Agrobacterium*, several parameters have been optimized such as *Agrobacterium*strain used, method of transformation, duration of cocultivation, and age of the explant (Khanna et al. 2004; Arinaitwe 2008; Esuola et al. 2011) and are discussed below. So far, several researchers have used embryonic cells for the development of transgenic banana for different agronomic traits that are summarized in Table 10.1. *Agrobacterium*-mediated genetic transformation using embryogenic cells derived from the scalp was first reported by Ganapathi et al. (2001).

10.5.1.1 *Agrobacterium* Cell Strain, Density, and Duration

Types of *Agrobacterium* strain and cell density play an important role for efficient delivery of T-DNA into plant cells and recovery of the transformed cells. Various authors have used different *Agrobacterium* strains and cell density for cocultivation of embryogenic cell suspension in different genotypes. Khanna et al. (2004) tested two different *Agrobacterium* strains such as LBA4404 and highly virulent strain AGL1 with different cell density (0.1–2 OD_{600nm}) on ECS of banana cv. Grand Nain (AAA) and Lady Finger (AAB). It was reported that OD_{600nm} at 0.5 and 1.0 is found to be best for AGL1 and LBA4404, respectively, depending on the extent of cell death and level of reporter gene expression in both genotypes. Ghosh and coworkers (2009) have used EHA105, a highly virulent strain, at 0.2 OD_{600nm} to develop the transformants through embryogenic cell suspension of Cavendish banana cv. Robusta (AAA). Huang et al. (2007) demonstrated the development of transgenic banana cv. Mas (diploid AA) using EHA105 cells at 0.8 OD_{600nm} using the liquid cocultivation method. The difference in the use of *Agrobacterium* cell density by the various researchers may be due to the virulent nature of the *Agrobacterium* strain and cultivar employed for the research (Huang et al. 2007; Arinaitwe 2008; Esuola et al. 2011; Yip et al. 2011).

10.5.1.2 Age of Embryogenic Cell Suspension

Age of the embryogenic cell suspension (ECS) is another important factor which influences T-DNA integration into cells. Various researchers have been using different age of ECS for cocultivation. Arinaitwe (2008) investigated the effect of age of ECS (1, 3, 5, 7, and 9 days) in banana cv. Obino l'Ewai for efficient integration of T-DNA using the GUS reporter gene. It was found that with the increase in age of ECS, number of GUS signal spots also increased significantly but decreased in 7-day-old ECS. Ganapathi et al. (2001) found maximum transformation efficiency when 7-day-old ECS cells of banana cv. Rasthali were used, whereas 4-day-old ECS was utilized for the cocultivation of ECS in banana cv. Lady Finger and Grand Nain (AAA) (Khanna et al. 2004).

10.5.1.3 Genotype

The frequency of T-DNA transformation by *Agrobacterium* is genotype dependent. In order to investigate transformation frequency, Arinaitwe (2008) cocultivated embryogenic cells of different genotypes with *Agrobacterium* strain EHA101 harboring pFAJ3000*uidA*. It was found that consistent and higher level of *uidA* expression was *observed in* Grand Nain and Obino l'Ewai (95 %) followed by Three Hand Planty (83 %) and Orishele (81 %). A similar type of variability in transformation frequency has been reported in other crops such as wheat and barley (Harwood 2011). The differential necrosis, the presence of inhibitors of the *Agrobacterium* sensory machinery, hypersensitive response, and subsequent cell death of host cells are attributed to be the cause of varied level of transformation efficiency in different genotypes (Hansen 2000; Khanna et al. 2004).

10.5.1.4 Other Factors

Certain phenolic compounds such as acetosyringone and surfactants aid in the efficiency of *Agrobacterium*-mediated transformation. Acetosyringone is an exogenous stimulant known to trigger the *Agrobacterium* virulence genes required for transfer and incorporation of T-DNA into the host (McCullen and Binns 2006; Opabode 2006). In general, monocotyledons produce insufficient level of phenolic compounds which failed to activate the *Agrobacterium* vir genes (Suzuki et al. 2001). To improve the transformation efficiency in banana, several researchers

Table 10.1 Summary of *Agrobacterium*-mediated transformation of banana embryogenic cells for different agronomical important traits

Sl. No	Genotype	Cultivar	Source of explant	Explant	Gene	*Agrobacterium* strain	Trait	References
1.	AAB	Rasthali	Male inflorescence	ECS	MusaDAD1	EHA105	Fusarium wilt	Ghag et al. (2014b)
					MusaBAG1			
					MusaBI1			
2.	AAB	Rasthali	Male inflorescence	ECS	Defensin (Sm-AMP-D1)	EHA105	Fusarium wilt	Ghag et al. (2014a)
3.	AAB	Rasthali	Male inflorescence	ECS	Petunia floral defensin (PhDef1/PhDef2)	EHA105	Fusarium wilt	Ghag et al. (2012)
4.	AAB	Rasthali	Male inflorescence	ECS	Velvet (vel)	EHA105	Fusarium wilt	Ghag et al. (2014c)
					Fusarium transcription factor 1 (Ftf1)			
5.	AAB	Lady Finger	Male inflorescence	ECS	Bcl-xL	AGL1	Fusarium wilt	Paul et al. (2011)
					Ced-9			
					Bcl-2 3′ UTR			
6.	AA	Furenzhi	Male inflorescence	ECS	chit42	EHA105	Fusarium wilt	Hu et al. (2013)
7.	AAB	Rasthali	Male inflorescence	ECS	Ace-AMP1	LBA4404	Fusarium wilt	Mohandas et al. (2013b)
8.	AAB	Sukali Ndizi	Male inflorescence	ECS	Mced9	AGL1	Fusarium wilt	Magambo (2012)
9.	AAB	Rasthali	–	ECS	MSI-99	EHA105	Fusarium wilt	Chakrabarti et al. (2003)
10.	AAA	Gros Michel	Male inflorescence	ECS	rcc2	EHA105	Black sigatoka	Kovács et al. (2013)
					Rcg3			
11.	ABB	Sukali Ndiizi	Male inflorescence	ECS	SAP1	EHA105	Bacterial wilt	Betty (2011)
	EA-AAA	Nakinyika						
12.	AAA-EAHB	Mpologoma	–	ECS	Hrap	AGL1	Bacterial wilt	Tripathi et al. (2010)
	ABB	Sukali Ndiizi						
13.	EA – AAA	Nakinyika	–	ECS	Plant ferredoxin-like protein (Pflp) gene	EHA105	Bacterial wilt	Namukwaya et al. (2012)
	ABB		Sukali Ndizi					

14.	AAB	Virupakshi	Male inflorescence	ECS	Rep gene	LBA4404	Virus (BBTV)	Elayabalan et al. (2013b)
15.	AAB	Rasthali	Male inflorescence	ECS	Rep gene Pro Rep	EHA105	Virus (BBTV)	Shekhawat et al. (2012)
16.	ABB	Sukali Ndiizi	Male inflorescence	ECS	Carica papaya cystatin (CpCYS)	LBA4404	Nematode (*Radopholussimilis*)	Namuddu et al. (2013)
17.	AAB	Gonjamanjaya	Tiny multiple bud clumps	ECS	Maize cystatin	–	Nematode (*Radopholussimilis*)	Roderick et al. (2012)
18.	AAB	Rasthali	Male inflorescence	ECS	MusaWRKY71	EHA105	Abiotic stress	Shekhawat and Ganapathi (2013)
19	AAB	Rasthali	Male inflorescence	ECS	MusaPIP1;2	MusaPIP1;2	Abiotic stress	Sreedharan et al. (2013)
20.	AAB	Rasthali	Male inflorescence	ECS	MusaDHN-1	EHA105	Abiotic stress	Shekhawat et al. (2011)
21.	AAB	Rasthali	Male inflorescence	ECS	MusaNAC68	EHA105	Abiotic stress	Negi et al. (2016)

have used acetosyringone in preinduction and cocultivation medium from 100 to 200 μM (Khanna et al. 2004; Ghosh et al. 2009; Mohandas et al. 2013a, b).

The surfactant may also improve the T-DNA transfer by aiding the elimination of certain substances that inhibit *Agrobacterium* attachment and also facilitate the *Agrobacterium* to attach host cells (Opabode, 2006). Cheng et al. (1997) suggested that addition of Pluronic F-68 (0.01–0.2 % w/v) and Silwet L-77 (0.01–0.075 % v/v) in medium enhanced the T-DNA delivery in immature embryos of wheat. The incorporation of F-68 to the inoculation medium significantly improved the GUS expression up to 100 times in sorghum (Carvalho et al. 2004). Khanna et al. (2004) reported that addition of surfactant F-68 improved the T-DNA delivery into embryogenic cells of banana.

Inclusion of thiol compounds in the cocultivation medium further enhanced transformation efficiency by reducing the oxidative burst and also improved the T-DNA delivery and led to stable expression of gene. In order to reduce the necrosis and to improve transformation efficiency, several authors have used L-cysteine in the cocultivation medium ranging from 40 to 400 mg l^{-1} (Ganapathi et al. 2001; Khanna et al. 2004).

Providing a heat shock before the cocultivation of the embryogenic cells has proven to increase the viability of the embryogenic cells, which further recovered the large number of transgenic plants (Khanna et al. 2004). This indicates that the release of heat shock proteins from explant increased its immunity against *Agrobacterium.*

10.5.2 Biolistic-Mediated Transformation of Somatic Embryos

Particle bombardment is a type of direct DNA transfer method, where foreign DNA-coated micro carriers or projectiles accelerated to adequate kinetic energy to deliver the DNA fragment into cells or tissue. These micro carriers are able to penetrate the membranes of the cell and to the nucleus, where a DNA fragment is delivered (Trick and Finer 1997). The foreign DNA then integrates into the chromosomal DNA by illegitimate or homologous recombination (Sanford 1990). This technique has been widely applied to plant species that appeared to be recalcitrant to *Agrobacterium* infection. It is independent of genotype, which can be applied to any cell or tissue type. Unlike *Agrobacterium*-mediated transformation, it does not transfer the gene to a specific region in the plant genome. However, transformation with this technique inserts the foreign gene randomly in multiple copies into plant system. But the high cost of equipment limits the use in genetic transformation (Finer et al. 1992; Christou 1996; Taylor and Fauquet 2002). Application of particle bombardment method for transgenic development has been reported in several crops including barley, soybean, wheat, and rice (Finer and Mc Mullen 1991; Li et al. 1993; Hadi et al. 1996; Yao et al. 2006; Tanasienko et al. 2011). This method of transformation was first utilized in the transgenic banana development by Sagi et al. (1995a, b).

Several factors have been reported to influence the transformation efficiency and subsequent stable integration and expression using particle bombardment method. Chee and coworkers (2005) studied the effect of different target distance and helium gas pressure on stable integration of the GUS gene in immature somatic embryo using biolistic method. The histological study found that maximum expression of GUS was observed in immature embryos when the helium pressure of 1350 psi and at a target distance of 6 cm used to bombard. The strong GUS activity was observed when higher acceleration, pressure, and shorter target distance were used. Weak GUS staining was observed when higher target distance and lower acceleration pressure were used. Similarly, Houllou-Kido et al. (2005) used target distance of 9 cm and helium pressure 1100 psi to develop the transgenic banana cultivar Maçã (AAB) with three different constructs containing GUS gene. Becker et al. (2000) developed an effective method of microprojectile bombardment for stable transformation and regeneration of Cavendish banana cv. Grand Nain using the embryogenic cell suspension. To bombard the embryogenic cells, target distance of 7.5 cm and helium pressure of 550 KPa were used.

10.6 Conclusion

Somatic embryogenesis-mediated transformation is at present widely used to transform banana to obtain transformants as it reduces formation of chimera. A large number of plants can be produced through this method if good embryogenic cells are identified. However, development of embryogenic cells and regeneration of plants are a laborious process. The time taken to produce embryogenic cells varies between cultivars. Some cultivars take as long as 7–8 months for producing embryogenic cells. If we can standardize methods to obtain embryogenic cells within 3–4 months and proliferate them in a short time, somatic embryogenesis would be the most preferred method for plant transformation.

References

Ahloowalia BS (1991) Somatic embryos in monocots. Their genesis and genetic stability. Rev Cytol Biol Veget Bot 14:223–235

Arinaitwe G (2008) Improved Agrobacterium-mediated transformation methods for banana and plantain. Catholic University of Leuven, Belgium. Ph.D. thesis

Arinaitwe G, Rubaihayo PR, Magambo MJS (2000) Proliferation rate effects of cytokinins on banana (Musa spp.) cultivars. Sci Hortic 86:13–21

Arinaitwe G, Remy S, Strosse H, Swennen R, Sági L (2004) Agrobacterium- and particle bombardment-mediated transformation of a wide range of banana cultivars. In: Mohan, Jain S, Swennen R (eds) Banana improvement: cellular, molecular biology, and induced mutations. Science Publishers Inc., Enfield, pp 351–357

Assani A, Bakry F, Kerbellec F, Haicour R, Wenzel G, Fehr W (2003) Production of haploids from anther culture of banana (Musa balbisiana (BB)). Plant Cell Rep 21:511–516

Becker DK, Dugdale B, Smith MK, Harding RM, Dale JL (2000) Genetic transformation of Cavendish banana (Musa spp. AAA group) cv. Grand Nain'via microprojectile bombardment. Plant Cell Rep 19:229–234

Betty N (2011) Expression of amphipathic protein (sap1) from sweet pepper for induction of resistance to xanthomonas campestris pv. musacearum in banana. B.Schn. Makerere university

Carvalho CHS, Zehr UB, Nilupa G, Anderson J, Kononowicz HH, Hodges TK, Axtell JD (2004) Agrobacterium-mediated transformation of sorghum: factors that affect transformation efficiency. Genet Mol Biol 27:259–269

Chakrabarti A, Ganapathi TR, Mukherjee PK, Bapat VA (2003) MSI-99, a magainin analogue, imparts enhanced disease resistance in transgenic tobacco and banana. Planta 216:587–596

Chang S, Shu H (2013) A method to suppress the browning in banana (Musa- AAA) embryogenic callus induced. Res J Biotech 8:63–69

Chee WW, Jalil M, Abdullah Meilina OR, Othman Y, Khalid N (2005) Comparison of β-glucuronidase expression and anatomical localization in bombarded immature embryos of banana cultivar Mas via biolistic transformation. As Pac J Mol Biol Biotechnol 13:15–22

Cheng M, Fry JE, Pang S, Zhou H, Hironaka CM, Duncan DR, Conner TW, Wan Y (1997) Genetic transformation of wheat mediated by Agrobacterium tumefaciens. Plant Physiol 115:971–980

Chong B, Gómez R, Reyes M, Bermúdez I, Gallardo J, Freire M, Posada L, Herrera I, Swennen R (2005) New methodology for the establishment of cell suspensions of Grande Naine (AAA). InfoMusa 14:13–17

Christou P (1996) Transformation technology. Trends Plant Sci 1:423–431

Côte FX, Domergue R, Monmarson S, Schwendiman J, Teisson C, Escalant JV (1996) Embryogenic cell suspensions from male flower of Musa AAA cv. Grand nain. PhysiolPlantarum 97:285–290

Dai XM, Xiao W, Huang X, Zhao JT, Chen YF, Huang XL (2010) Plant regeneration from embryogenic cell suspensions and protoplasts of dessert banana cv. Da Jiao' (Musa paradisiacal ABB Linn.) via somatic embryogenesis. In Vitro Cell Dev Biol Plant 46:403–410

Daniels D, Gomez KR, Reyes VM (2002) plant regeneration system via somatic embryogenesis in the hybrid cultivar FHIA-21 group). In Vitro Cell Dev Biol Plant 38:330–333

Dhed'a D, Dumortier F, Panis B, Vuylsteke D, De Langhe E (1991) Plant regeneration in cell suspension cultures of the cooking banana cv. Bluggoe' (Musa spp. ABB group). Fruits 46:125–135

Elayabalan S, Kalaiponmani K, Michael Pillay A, Chandrasekar R, Selvarajan KK, Kumar, Balasubramanian P (2013a) Efficient regeneration of the endangered banana cultivar Virupakshi' (AAB) via embryogenic cell suspension from immature male flowers. Afr J Biotechnol 12:563–569

Elayabalan S, Kalaiponmani K, Subramaniam S, Selvarajan R, Panchanathan R, Muthuvelayoutham R, Kumar KK, Balasubramanian P (2013b) Development of Agrobacterium-mediated transformation of highly valued hill banana cultivar Virupakshi (AAB) for resistance to BBTV disease. World J Microbiol Biotechnol 29:589–596

Escalant JV, Teisson C (1989) Somatic embryogenesis and plants from immature zygotic embryos of the species Musa acuminate and Musa balbisiana. Plant Cell Rep 7:665–668

Escalant JV, Teisson C, Cote FX (1994) Amplified somatic embryogenesis from male flowers of triploid banana and plantain cultivars (Musa sp.). In Vitro Cell Dev Biol Plant 30:181–186

Esuola CO, Tripathi L, Fawole I (2011) Effect of various virulent strains of *Agrobacterium tumefaciens* on genetic transformation of banana (Musa sp.) cultivar Williams. Afr J Hortic Sci 5:84–91

FAOSTAT Agriculture Data (2013) - http://faostat3.fao.org/download/Q/QC/E

Finer JJ, Mc Mullen MD (1991) Transformation of soybean via particle bombardment of embryogenic suspension culture tissue. In Vitro Cell Dev Biol Plant 27:175–182

Finer JJ, Vain P, Jones MW, McMullen MD (1992) Development of the particle inflow gun for the DNA delivery to plant cells. Plant Cell Rep 11:323–328

Gaj MD (2004) Factors influencing somatic embryogenesis induction and plant regeneration with particular reference to Arabidopsis thaliana (L.) Heynh. Plant Growth Reg 43:27–47

Ganapathi TR, Suprasanna P, Bapat VA, Kulakarni VM, Rao PS (1999) Somatic embryogenesis and plant regeneration from male flower buds in banana. Curr Sci 76:1229–1231

Ganapathi TR, Higgs NS, Balint-Kurti PJ, Arntzen CJ, May GD, Van Eck JM (2001) Agrobacterium-mediated transformation of the embryogenic cell suspensions of the banana cultivar Rasthali (AAB). Plant Cell Rep 20:157–162

George EF, Hall MA, De Klerk G-J (2008) Plant propagation by tissue culture, vol 1, 3rd edn, The background. Springer, Dordrecht, p 551

Georget F, Domergue R, Ferrier N, Cote FX (2000) Morpho-histological study of the different constituents of a banana (Musa AAA, cv. Grand Naine) embryogenic cell suspension. Plant Cell Rep 19:748–754

Ghag SB, Shekhawat UKS, Ganapathi TR (2012) Petunia floral defensins with unique prodomains as novel candidates for development of Fusarium wilt resistance in transgenic banana plants. PLoS One 7:e39557

Ghag SB, Shekhawat UKS, Ganapathi TR (2014a) Transgenic banana plants expressing a Stellaria media defensin gene (Sm-AMP-D1) demonstrate improved resistance to *Fusarium oxysporum*. Plant Cell Tissue Organ Cult 119:247–255

Ghag SB, Shekhawat UKS, Ganapathi TR (2014b) Native cell-death genes as candidates for developing wilt resistance in transgenic banana plants. *AoB PLANTS* plu037; doi:10.1093/aobpla/plu037

Ghag SB, Shekhawat UKS, Ganapathi TR (2014b) Host induced posttranscriptional hairpin RNA-mediated gene silencing of vital fungal genes confers efficient resistance against Fusarium wilt in banana. Plant Biotechnol J12:541–553

Gheysen G, Angenon G, Van Montague M (1998) Agrobacterium mediated plant transformation: a scientifically intriguing story with significant application. In: Lindsey K (ed) Transgenic plant research. Harwood Academic Press, Amsterdam, pp 1–33

Ghosh A, Ganapathi TR, Nath P, Bapat VA (2009) Establishment of embryogenic cell suspension cultures and Agrobacterium mediated transformation in an important Cavendish banana cv. Robusta (AAA). Plant Cell Tissue Organ Cult 97:131–139

Gomez KR, De Feria MS, Posada LP, Gilliard T, Bernal FM, Reyes MV, Chavez MM, Quiala EM (2002) Somatic embryogenesis of the banana hybrid cultivar FHIA-18 (AAA-B) in liquid medium and scaled –up in a bioreactor. Plant Cell Tissue Organ Cult 68:21–26

Grapin A, Schwendiman J, Teisson C (1996) Somatic embryogenesis in plantain banana. In Vitro Cell Dev Biol Plant 32:66–71

Grapin A, Ortı'z JL, Lescot T, Ferrie're N, Côte FX (2000) Recovery and regeneration of embryogenic cultures from female flowers of False Horn Plantain. Plant Cell Tissue Organ Cult 61:237–244

Haccius B (1978) Question of unicellular origin of non-zygotic embryos in callus cultures. Phytomorphology 28:74–81

Hadi MZ, McMullen MD, Finer JJ (1996) Transformation of 12 different plasmids into soybean via particle bombardment. Plant Cell Rep 15:500–505

Hansen G (2000) Evidence for Agrobacterium-induced apoptosis in maize cells. Mol Plant Microbe Interact 13:649–657

Hansen G, Martha SW (1999) Recent advances in the transformation of plants. Trends Plant Sci 4:226–231

Harwood WA (2011) Advances and remaining challenges in the transformation of barley and wheat. J Exp Bot 63:1791–1798

Henry RJ (1998) Molecular and biochemical characterization of somaclonal variation. In: Jain SM, Brar DS, Ahloowalia BS (eds) Somaclonal variation and induced mutations in crop improvement. Kluwer Academic Publishers, Dordrecht, 485–499

Herrera-Estrella L, Simpson J, Martinez-Trujillo M (2005) Transgenic plants: an historical perspective. Methods Mol Biol 286:3–32

Houllou-Kido LM, Kido EA, Falco MC, Silva Filho MCS, Figueira AVO, Nogueira NL, Rossi ML, Neto AT (2005) Somatic embryogenesis and the effect of particle bombardment on banana 'Maçã' regeneration. Pesqagropec bras Brasília 40:1081–1086

Hu CH, Wei YR, Huang YH, Yi GJ (2013) An efficient protocol for the production of chit42 transgenic Furenzhi banana (Musa spp. AA group) resistant to

Fusarium oxysporum. In Vitro Cell Dev Biol Plant 49:584–592

Huang X, Huang XL, Xiao W, Zhao JT, Di X, Chen YF, Li XJ (2007) Highly efficient Agrobacterium-mediated transformation of embryogenic cell suspensions of Musa acuminata cv. Mas (AA)via a liquid co-cultivation system. Plant Cell Rep 26:1755–1762

Husina N, Jalil M, Othman RY, Khalid N (2014) Enhancement of regeneration efficiency in banana (Musa acuminate cv. Berangan) by using proline and glutamine. Sci Hortic 168:33–37

Ishida Y, Saito H, Ohta S, Hiei Y, Komari T, Kumashiro T (1996) High efficiency transformation of maize (*Zea mays* L.) mediated by Agrobacterium tumefaciens. Nat Biotechnol 14:745–750

Jalil M, Khalid N, Othman RY (2003) Plant regeneration from embryogenic suspension cultures of Musa acuminata cv. Mas (AA). Plant Cell Tissue Organ Cult 75:209–214

Kamle M, Bajpai A, Chandra R, Kalim S, Kumar R (2011) Somatic embryogenesis for crop improvement. GERF Bull Biosci 2:54–59

Khanna H, Becker D, Kleidon J, Dale J (2004) Centrifugation Assisted Agrobacterium tumefaciens-mediated Transformation (CAAT) of embryogenic cell suspensions of banana (Musa spp. Cavendish AAA and Lady finger AAB). Mol Breed 14:239–252

Kitamiya E, Suzuki S, Sano T, Nagata T (2000) Isolation of two genes that were induced upon initiation of somatic embryogenesis on carrot hypocotyls by high concentrations of 2,4-D. Plant Cell Rep 19:551–557

Kovács G, Sági L, Jacon G, Arinaitwe G, Busogoro JP, Thiry E, Strosse H, Swennen R, Remy S (2013) Expression of a rice chitinase gene in transgenic banana ('Gros Michel', AAA genome group) confers resistance to black leaf streak disease. Transgenic Res 22:117–130

Li L, Qu R, de Kochko A, Fauquet C, Beachy RN (1993) An improved rice transformation system using the biolistic method. Plant Cell Rep 10:401–405

Ma SS (1991) Somatic embryogenesis and plant regeneration from- cell suspension culture of banana. In: Department of Agriculture, National Taiwan University (ed) Proceedings of the symposium on tissue culture and horticultural crops. National Taiwan University, Taipei, 181–188

Magambo B (2012) Generation of transgenic banana (cv. SukaliNdizi) resistance to Fusarium wilt, B.Schn. Queensland Universisty of Technology. Australia

Mahn A, Matzk A, Sautter C, Schiemann J (1995) Transient gene expressions in shoot apical meristems of sugarbeet seedlings after particle bombardment. J Exp Bot 46:1625–1628

McCullen CA, Binns AN (2006) Agrobacterium tumefaciens and plant cell interactions and activities required for interkingdom macromolecular transfer. Annu Rev Cell DevBiol 22:101–127

Mohandas S, Sowmya HD, Manjula R, Pratibha KY, Meenakshi S, Ajay KM (2011) Development of highly regenerative embryogenic cell suspensions of 'NanjangudRasbale' (syn. 'Rasthali', Musa, AAB, Silk Subgroup) and transformants with AMP gene. Acta Hort 897:245–249

Mohandas S, Sowmya HD, Meenakshi S (2013a) Somatic Embryogenesis and plant regeneration through cell suspension in Diploid (AB) banana cultivar Elakki Bale (synNeypoovan). J Plant Biochem Biotechnol 22:245–249

Mohandas S, Sowmya HD, Saxena AK, Meenakshi S, Rani RT, Mahmood R (2013b) Transgenic banana cv. Rasthali (AAB, Silk gp) harboring Ace-AMP1 gene imparts enhanced resistance to Fusarium oxysporumf. sp. cubense race. Sci Hortic 164:392–399

Morais-Lino LS, Santos-Serejo JA, Oliveira eSilva S, Ferreira de Santana JR, Kobayashi AK (2008) Cell suspension culture and plant regeneration of a Brazilian plantain, cultivar Terra. Pesqagropec bras Brasília 43:1325–1330

Namuddu A, Kiggundu A, Mukasa SB, Kurnet K, Karamura E, Tushemereirwe W (2013) Agrobacterium mediated transformation of banana (Musa sp.) cv. Sukali Ndiizi (ABB) with a modified Carica papaya cystatin (CpCYS) gene. Afr J Biotechnol 12:1811–1819

Namukwaya B, Tripathi L, Tripathi JN, Arinaitwe G, Mukasa SB, Tushemereirwe WK (2012) Transgenic banana expressing Pflp gene confers enhanced resistance to Xanthomonas wilt disease. Transgenic Res 21:855–865

Navarro C, Escobedo RM, Mayo A (1997) In vitro plant regeneration from embryogenic cultures of a diploid and a triploid, Cavendish banana. Plant Cell Tissue Organ Cult 51:17–25

Negi S, Tak H, Ganapathi TR (2016) Expression analysis of MusaNAC68 transcription factor and its functional analysis by overexpression in transgenic banana plants. Plant Cell Tissue Organ Cult 125:59–70

Novak FJ, Afza R, Van Duren M, Perea-Dallos M, Conger BV, Xiaolang T (1989) Somatic embryogenesis and plant regeneration in suspension cultures of dessert (AA and AAA) and cooking (ABB) bananas (Musa spp.). Biol Technol 7:154–159

Opabode JT (2006) *Agrobacterium*-mediated transformation of plants: emerging factors that influence efficiency. Biotechnol Mol Biol Rev 1:12–2

Paul JY, Becker DK, Dickman MB, Harding RM, Khanna HK, Dale JL (2011) Apoptosis-related genes confer resistance to Fusarium wilt in transgenic Lady Finger' bananas. Plant Biotechnol J 9:1141–1148

Quiroz-Figueroa FR, Rojas-Herrera R, Galaz-Avalos RM, Loyola- Vargas VM (2006) Embryo production through somatic embryogenesis can be used to study cell differentiation in plants. Plant Cell Tissue Org Cult 86:285–301

Raghavan V (1976) Adventive embryogenesis: induction of diploid embryoids. In: Sutchliffe JF (ed) Experimental embryogenesis in vascular plants. Academic, London, pp 349–381

Ramírez-Villalobos M, de García E (2008) Obtainment of embryogenic cell suspensions from scalps of the banana CIEN-BTA-03 (Musa sp., AAAA) and regeneration of the plants. J Biotechnol 5:11

Roderick H, Tripathi L, Babirye A, Wang D, Tripathi J, Urwin PE, Atkinson HJ (2012) Generation of transgenic plantain (Musa spp.) with resistance to plant pathogenic nematodes. Mol Plant Pathol 13:842–851

Roux N, Dolezel J, Swennen R, Zapata-Arias FJ (2001) Effectiveness of three micropropagation techniques to dissociate cytochimeras in Musa spp. Plant Cell Tiss Org 66:189–197

Sadik K, Rubaihayo PR, Magambo MJS, Pillay M (2007) Generation of cell suspensions of East African highland bananas through scalps. Afr J Biotechnol 6:1352–1357

Sagi L, Panis B, Remy S, Schoofs H, De Smit K, Swennen R, Cammue BPA (1995a) Genetic transformation of banana and plantain (Musa spp.) by particle bombardment. Biol Technol 13:481–485

Sagi L, Panis B, Remy S, Verelst B, Cammue BPA, Volckaert G, Swennen R (1995b) Transient gene expression in transformed banana (Musa spp., cv. Bluggoe') protoplasts and embryogenic cell suspensions. Euphytica 85:89–95

Sanford JC (1990) Biolistic plant transformation. Physiol Plantarum 79:206–209

Santos M, Albuquerque de Barros E, Tinoco M, Brasileiro A, Aragão F (2002) Repetitive somatic embryogenesis in cacao and optimisation of gene expression by particle bombardment. J Plant Biotechnol 4:71–76

Sasson A (1997) Importance of tropical and subtropical horticulture, future prospects of biotechnology in tropical and subtropical horticulture species. Acta Hortic 460:12–26

Schoofs H, Panis B, Strosse H, Mosqueda AM, Torres JL (1999) Bottlenecks in the generation and maintenance of morphogenic banana cell suspensions and plant regeneration via somatic embryogenesis therefrom. Info Musa 8:3–7

Sharp WR, Sondahl MR, Caldas LS, Maraffa SB (1980) The physiology of in vitro asexual embryogenesis. Hortic Rev 2:268–310

Shekhawat UKS, Ganapathi TR (2013) MusaWRKY71 overexpression in banana plants leads to altered abiotic and biotic stress responses. PLoS One 8:e75506

Shekhawat UKS, Srinivas L, Ganapathi TR (2011) MusaDHN-1, a novel multiple stress-inducible SK3-type dehydrin gene, contributes affirmatively to drought- and salt-stress tolerance in banana. Planta 234:915–932

Shekhawat UKS, Ganapathi TR, Hadapad AB (2012) Transgenic banana plants expressing small interfering RNAs targeted against viral replication initiation gene display high-level resistance to banana bunchy top virus infection. J Gen Virol 93:1804–1813

Shibata D, Liu YG (2000) *Agrobacterium*-mediated plant transformation with large DNA fragments. Trends Plant Sci 5:354–357

Singh M, Jaiswal U, Jaiswal VS (2004) In vitro regeneration and improvement in tropical fruit trees: an assessment. In: Srivastava PS, Narula A, Srivastava S (eds) Plant biotechnology and molecular markers. Anamanya Publishers, New Delhi, pp 228–243

Solís-Ramos LY, Andrade-Torres A, Carbonell LAS, Carlos M, Salfn O, Castaño de la Serna E (2012) Somatic embryogenesis in recalcitrant plants, embryogenesis. In: Sato K-I (ed) ISBN: 978-953-51-0466-7, InTech, pp 597–618

Sreedharan S, Shekhawat UKS, Ganapathi TR (2013) Transgenic banana plants overexpressing a native plasma membrane aquaporin MusaPIP1;2 display high tolerance levels to different abiotic stresses. Plant Biotechnol J 11:942–952

Strosse H, Domergue R, Panis B, Escalent JV, Côte F (2003) Banana and plantain embryogenic cell suspensions. INIBAP technical guidelines 8. INIBAP, In: Vézina A, Picq C (Eds). International Network for the Improvement of Banana and Plantain, Montpellier, France.

Strosse H, Schoofs H, Panis B, Andre E, Reyniers K, Swennen R (2006) Development of embryogenic cell suspension from shoot meristematic tissue in bananas and plantains (Musa spp.). Plant Sci 170:104–112

Suzuki S, Supaibulwatana K, Mii M, Nakano M (2001) Production of transgenic plants of the Liliaceous ornamental plant Agapanthus praecox ssp. Orientalis (Leighton) Leighton via Agrobacterium-mediated transformation of embryogenic calli. Plant Sci 161:89–97

Tanasienko IV, Emets AI, Ia VP, Korkhovoĭ VI, Abumkhadi N, IaB B (2011) Developing transgenic barley lines producing human lactoferrin using mutant alfa-tubulin gene as selective marker gene. Tsitol Genet J 45:3–10

Taylor NJ, Fauquet CM (2002) Microprojectile bombardment as a tool in plant science and agricultural biotechnology. DNA Cell Biol 21:963–977

Toonen MAJ, De Vries SC (1996) Initiation of somatic embryos from single cells. In: Wang TL, Cuming A (eds) Embryogenesis: the generation of a plant. Bios Scientific, Oxford, pp 173–189

Trick NH, Finer JJ (1997) SAAT: sonication-assisted *Agrobacterium*-mediated transformation. Transgenic Res 6:329–336

Tripathi L, Mwaka H, Tripathi JN, Tushemereirw WK (2010) Expression of sweet pepper Hrap gene in banana enhances resistance to *Xanthomonas campestris* pv. Musacearum. Mol Plant Pathol 11:721–731

Venkatachalam L, Thimmaraju R, Sreedhar RV, Bhagyalakshmi N (2006) Direct shoot and cormlet regeneration from leaf explants of 'silk' banana (AAB). In Vitro Cell Dev Biol Plant 42:262–269

Visarada KBRS, Sarma NP (2004) Transformation of indica rice through particle bombardment: factors influencing transient expression and selection. Biol Plantarum 48:25–31

Wei YR, Huang XL, Li J, Huang X, Li Z, Li XJ (2005) Establishment of embryogenic cell suspension culture and plant regeneration of edible banana Musa acuminate cv. Mas (AA). Chin J Biotechnol 21:58–65

Williams EG, Maheswaran G (1986) Somatic embryogenesis: factors influencing coordinated behavior of cells as an embryogenic group. Ann Bot 57:443–462

Wong WC, Jalil M, Ong-Abdullah M, Othman RY, Khalid N (2006) Enhancement of banana plant regeneration by incorporating a liquid-based embryo development medium for embryogenic cell suspension. J Hortic Sci Biotech 81:385–390

Xu CX, Liang QR, Peng GK, Chen HB (2004) Factors affecting induction of callus from immature male flowers of bananas(Musa AAA). Adv Hortic 6:96–99

Xu C, Zou R, Pan X, Chen H (2008) Somatic embryogenesis in banana (Musa spp.). Int J Plant Dev Biol 2:52–58

Yao Q, Cong L, Chang JL, Li KX, Yang GX, He GY (2006) Low copy number gene transfer and stable expression in a commercial wheat cultivar via particle bombardment. J Exp Bot 57:3737–3746

Yip MK, Lee SW, Su KC, Lin YH, Chen TY, Feng TY (2011) An easy and efficient protocol in the production of pflp transgenic banana against Fusarium wilt. Plant Biotechnol Rep 5:245–254

Youssef M, James A, Mayo-Mosqueda A, Ku-Cauich JR, Grijalva-Arango R, Escobedo RM (2010) Influence of genotype and age of explant source on the capacity for somatic embryogenesis of two Cavendish banana cultivars (*Musa acuminata* Colla, AAA). Afr J Biotechnol 9:2216–2223

Promoter Analysis in Banana

11

Efrén Santos, Ricardo Pacheco, Liliana Villao, Luis Galarza, Daniel Ochoa, Carlos Jordán, and José Flores

Abstract

Currently, several research laboratories are developing genetically modified bananas and plantains for different purposes, while new transgenic crops are being developed and released in several countries. Therefore, a more regulated transgene expression is needed. Furthermore, native banana promoters should imply a more public acceptance than heterologous sequences. Promoter isolation could be performed by an insertional mutagenesis approach or indirectly by analyzing expression patterns of related genes. Advances in genomics led to the genome sequencing and gene annotation of banana, as a source to identify candidate promoters for specific expression patterns. The challenge in computational approaches for promoter characterization is to precisely identify pattern of expression. Actually, promoter analysis for activity characterization has been confirmed through experimentation with different techniques, including reporter genes, bioinformatics analysis for candidate *cis*-acting element and promoter prediction, and expression analysis of related genes. A review of characterization of different banana promoters is summarized, and the analysis of available banana promoter sequences in the GenBank was performed with available bioinformatics tools and a novel method to

E. Santos (✉) • L. Galarza • J. Flores
Escuela Superior Politécnica del Litoral, ESPOL, Facultad de Ciencias de la Vida, Campus Gustavo Galindo Km 30.5 vía Perimetral, 09-01-5863, Guayaquil, Ecuador

Escuela Superior Politécnica del Litoral, ESPOL, Centro de Investigaciones Biotecnológicas del Ecuador, CIBE, Campus Gustavo Galindo, Km 30.5 vía Perimetral, 09-01-5863, Guayaquil, Ecuador
e-mail: gsantos@espol.edu.e

R. Pacheco • L. Villao
Escuela Superior Politécnica del Litoral, ESPOL, Centro de Investigaciones Biotecnológicas del Ecuador, CIBE, Campus Gustavo Galindo, Km 30.5 vía Perimetral, 09-01-5863, Guayaquil, Ecuador

D. Ochoa • C. Jordán
Escuela Superior Politécnica del Litoral, ESPOL, Centro de Visión y Robótica, Facultad de Ingeniería en Electricidad y Computación, Campus Gustavo Galindo Km 30.5 vía Perimetral, 09-01-5863, Guayaquil, Ecuador

S. Mohandas, K.V. Ravishankar (eds.), *Banana: Genomics and Transgenic Approaches for Genetic Improvement*, DOI 10.1007/978-981-10-1585-4_11

identify motif sequences. Furthermore, a list of promoters used in the development of genetically modified banana is presented, indicating an increase of plant and native banana promoters used to drive expression of related gene in a specific pattern.

Keywords
Regulatory sequence • Reporter genes • *In silico* • Bioinformatics • *Musa* • Gene expression • Promoter tagging

11.1 Introduction

Regulatory sequence analysis is necessary for expected expression of heterologous genes in different organisms. The second and third generation of transgenic plants required a more accurate transgene expression, leading to the search of novel regulatory sequences. Heterologous protein expression and localization in temporarily manner and/or tissue specificity implies the use of regulatory regions called promoters. Promoters are the first step of gene regulations, which involve a sequential process leading to transcription initiation. However, gene regulation could imply other (un)related processes. The use of native regulatory sequences, in the development of genetic modified banana plants, should provide more public acceptance than the use of heterologous regulatory sequences.

Methods for promoter isolation include mRNA differential expression analysis or the use of a T-DNA tagging approach which uses a promoter-less reporter or selectable marker gene integrated in plant cell genomes (Teeri et al. 1986; André et al. 1986). Furthermore, using T-DNA tagging, cryptic promoters were discovered in plants and could be tissue specific (tobacco: seed coat, Fobert et al. 1994; *Arabidopsis*: callus and roots, Ökrész et al. 1998; guard cells, Plesch et al. 2000; and roots, Mollier et al. 2000; Sivanandan et al. 2005).

As the banana genome has been sequenced (Droc et al. 2013), selection of specific genes could be performed for promoter identification where gene expression is already known. Additionally, promoter sizes range typically from 0.5 kb up to 1.7 kb (average length of stress-responsive promoters in *Arabidopsis*, Kristiansson et al. 2009). For instance, the constitutive promoter 35S from cauliflower mosaic virus (CaMV) is only 538 nt. The majority of transgenic crops approved uses this promoter.[1,2] The PlantProm DB contains sequences of 200 bp plus 50 bp after the transcription start site (TSS), indicating that basal promoter activity is within this range[3].

Analysis of promoters requires experimental assays mainly by fusion of partial sequences at the 5′ or 3′ region to reporter genes to identify relevant elements (*cis*-acting elements), which are necessary for a specific gene expression pattern. *In silico* analysis of promoters involves the use of computational methods to identify already known *cis*-acting elements or the promoter architecture itself. However, any *cis*-acting element could be in any DNA sequence as they are mainly short; therefore, some elements might be false positives (*cis*-acting elements which are not functional), leading to a final experimental assay to ensure promoter activity (reviewed by Radhamony et al. 2005). This chapter briefly explains the basic concepts of promoters, current techniques for promoter analysis in banana, and an in silico comparison of the known banana promoters with a novel bioinformatics tool.

[1] www.cera-gmc.org/GMCropDatabase
[2] www.isaaa.org/
[3] www.softberry.com/berry.phtml?topic=plantprom&group=data&subgroup=plantprom

11.2 How Gene Expression Starts: Promoter Structure

A series of multiple events occurs in eukaryotic gene expression, which includes *locus* decondensation, remodeling of the nucleosome, histone modifications, and interaction of transcriptional activators to promoters leading to the recruitment of the basal machinery to the core promoter to start transcription (reviewed by Smale and Kadonaga 2003). DNA is packaged in the chromatin (DNA and histones); therefore, promoter sequences could be available after chromatin remodeling to start transcription (reviewed by Mellor 2006).

The promoter could be identified as the DNA sequence located at the 5′ region of a gene, containing specific sequences where the transcription machinery interacts. Regulatory sequences within promoters are also denominated *cis*-acting elements, which direct the rate or pattern of transcription of the gene located at the 3′ region (Praz et al. 2002). *Cis*-acting elements, also known as transcription factor (TF) binding sites (Venter and Botha 2004), could have a length from 5 to 20 nucleotides (nt) (Rombauts et al. 2003). Therefore, promoter sequences, located upstream of the TSS, contain sites for recognition of transcriptional activators or repressors (Wu et al. 2001).

The main region in the promoter is the core, which is in direct contact with the basal machinery of transcription (Lee and Young 2000). The RNA polymerase II interacts with TFs including the TFIID, which contains a TATA box-binding protein, and, in combination with other TFs, locates the RNA polymerase II to start transcription after the preinitiation complex is formed (reviewed by Venter and Botha 2004). A tridimensional configuration is formed due to the bending of the promoter and enhancer sequences to activate transcription (Berk 1999; Struhl 2001). Rombauts et al. (2003) suggest that *cis*-acting elements are grouped into modules and could be arranged in four different *cis*-regulatory modules (CRMs) including (1) a compact arrangement of clustered *cis*-acting elements denominated "typical," (2) a "diffuse" CRM where several *cis*-elements are distributed in a long stretch of DNA, (3) clustered CRMs distributed over a large DNA fragment denominated "composite," and (4) the "simple compact" CRM showing no dense cluster of binding sites. The CRMs are also considered as enhancers (Halfon 2006; Janssens et al. 2006).

Promoters from eukaryotic organisms consist of a TATA box located 30 nt at the 5′ region of the TSS (Breathnach and Chambon 1981), an initiator element located at the TSS (Smale 1994), and downstream promoter elements (DPE), located 30 nt. at the 3′ region of the TSS (Kutach and Kadonaga 2000). Shahmuradov et al. (2005) indicated that approximately 30–50 % of known plant promoters contains a TATA box between 45 and 25 nt at the 5′ region of the TSS. Furthermore, some housekeeping and photosynthesis-related genes are TATA-less (Dynan 1986; Nakamura et al. 2002). Recently, Kumari and Ware (2013) revealed through computational analysis of the genome-wide core promoter elements from four monocot and four dicot plants that only 17 % of promoters contain TATA box, while 80 % are TATA-less.

DNA free energy profiles revealed that structural properties of promoter regions are different from non regulatory genome sequences. TATA-less promoters might be activated by interaction of transcription machinery with sequences surrounding the TSS (Smale 1997) or the DPE (Shahmuradov et al. 2005). Other structures, like the "YR rule," might play an important role in transcription initiation. The "YR rule" corresponds to a pyrimidine (C/T) and a purine (A/G) around the TSS at the position −1 and +1, respectively, which is found in 77 % (10,806) of *Arabidopsis* promoters (Yamamoto et al. 2007a, b; Yamamoto and Obokata 2007). The "YR rule" is considered to be recognized by the TFIID as a less stringent initiator region (Yamamoto et al. 2007a). Elements at the 3′ region of the TSS and located in gene regions, including the 5′ untranslated region (UTR), could be involved in higher gene expression regulations. For instance, high promoter activity was obtained with 5′ UTR and first intron together with the maize ubiquitin promoter (Christensen et al. 1992).

11.3 Promoter Analysis

11.3.1 T-DNA Tagging for Promoter Discovery

Disruption of gene function could be accomplished by insertional mutagenesis leading to a recognizable phenotype (Springer 2000). For insertional mutagenesis, transposon (Sundaresan et al. 1995; Martienssen 1998) or T-DNA (Feldmann 1991; Krysan et al. 1999) could be used. The resulting phenotype of a specific line is further characterized to identify the disrupted gene by the presence of the tag (inserted DNA fragment) which greatly facilitates the isolation of the disrupted gene (reviewed by Springer 2000). One major drawback of these techniques occurs when several copies of the inserted tag are present in the genome. Therefore, the interpretation of a phenotype is not straightforward, and further isolation of flanking sequences is required (Springer 2000; Santos et al. 2009). On the other hand, insertional mutagenesis is an important tool for functional genomics in different plant species, including *Arabidopsis thaliana*, *Oryza sativa*, and *Medicago truncatula*.

For the discovery of novel promoters, Teeri et al. (1986) and André et al. (1986) developed a technique in which selectable marker genes without a promoter placed near a T-DNA border were used for transformation of plants. As T-DNA integration is random, only a portion of transgenic lines will show activity of the selectable marker gene if the T-DNA is integrated downstream of a promoter sequence. In *Nicotiana* spp., an estimated 0.1–5 and 5–19 % of screened transgenic lines survived under kanamycin pressure (Teeri et al. 1986; André et al. 1986, respectively). Major improvement of the techniques was the use of an alternative selectable marker gene for selection of transgenic lines (Koncz et al. 1989) and the use of a promoter-less reporter gene instead of the selectable marker gene (Kertbundit et al. 1991; Fobert et al. 1991). Different plant species have been used for promoter tagging and T-DNA activation frequency could vary from 0.06 % (Remy et al. 2005) and up to 78 % (Topping et al. 1991). Different factors may affect activation frequency, including explant/plant species, stress/condition, tissue/development screened, and selectable/reporter gene used. Further improvement in the promoter tagging technique to increase the of tagging frequency is the addition of one or multiple stop codons in-frame upstream of the start codon of the reporter gene (Koncz et al. 1989; Mudge and Birch 1998). The addition of splice acceptor or donor sequences upstream of the reporter gene may allow expression if T-DNA is inserted in an intron (Sundaresan et al. 1995; Springer 2000; Calderon-Villalobos et al. 2006). Additionally, the absence of the start codon of the reporter gene allows translational fusion with the tagged gene (Koncz et al. 1989).

Three main reporter genes have been used in promoter tagging: the β-glucuronidase (*uid*A; *gus*) of *E. coli* (Jefferson et al. 1987), the green fluorescent protein (*gfp*; GFP) of the Pacific jellyfish *Aequorea victoria* (Heim et al. 1995; Chiu et al. 1996), and the firefly luciferase (*luc*) of the American firefly *Photinus pyralis* (Ow et al. 1986).

11.3.2 Reporter Genes in Promoter Characterization

Indirectly, promoter activity could be inferred by analyzing the gene expression of the corresponding gene with different techniques including RT-qPCR, Northern blot, digital gene expression, and microarray, among others. Alternatively, to confirm level and pattern of expression, promoters could be characterized by fusion with reporter genes and analyzing the activity of the protein derived from the reporter gene. The main advantages of using reporter genes are the analysis of tissues for which isolation of transcripts is difficult, in low abundance of transcripts, and if promoter characterization is needed in several time points, avoiding the extraction of mRNA in different plant tissues. Functional *cis*-acting elements could be discovered by fusion of deleted variants of promoters to reporter genes or by point mutation of the elements.

Several factors influence the use of a specific reporter gene. GUS is the reporter gene most used for promoter tagging and promoter characterization (reviewed by Santos Ordoñez 2008). Spatial analysis is often performed when using the GUS reporter system due to the long half-life (~50 h in living mesophyll protoplasts, Jefferson et al. 1987). However, GUS activity assays should be carefully performed especially when screening at different developmental stages or under stress conditions due to the long half-life. For instance, analysis of circadian gene expression using chloramphenicol acetyltransferase (*cat*) gene was not detected; on the other hand, when using LUC, the pattern activity was detected (Wood 1995). The main reason of different results obtained with different reporter genes was the long half-life of the protein in CAT (50 h) than in LUC (3 h, Wood 1995; Thompson et al. 1991). Analysis of downregulated gene expression using GUS is difficult and 5 days are needed to perform characterization (Castle et al. 2005). Kinetic studies require sampling at several time points, because GUS assays are detrimental (Mandal et al. 1995). In summary, when temporal gene expression is expected, the *uid*A reporter gene is not the best choice (Table 11.1).

The GFP reporter system avoids the use of exogenous substrates and is not invasive. Gene expression in tissues and under specific stimuli could be accomplished with GFP (Chiu et al. 1996). However, plant tissue could generate background fluorescence (Hraska et al. 2006) including in 2-days-old *Arabidopsis* seedlings (Castle et al. 2005). The background fluorescence could be avoided by using appropriate filters (Maximova et al. 1998; Elliott et al. 1999; Hraska et al. 2006). The estimated half-life of the modified eGFP is 1 day (Verkhusha et al. 2003); thus temporal gene expression could be performed after 1 day.

The LUC reporter gene system allows real-time *in planta* gene expression. However, the substrate luciferin has to be added to the samples for photon generation reaction. LUC enzyme detection activity is noninvasive and nondestructive (Ow et al. 1986). Promoter characterization of inducible or developmental regulated activity is suitable when using LUC due to its short half-life (~15.3 min after luciferin is applied, Van Leeuwen et al. 2000). For instance, a low-temperature responsive promoter was identified in banana using a modified-improved luciferase (*luc*$^+$) reporter gene (Santos et al. 2009). However, once luciferin is added, LUC activity emits photons that need to be acquired with a sophisticated equipment such as a charge-coupled device (CCD) digital camera (Remy et al. 2004).

A summary of characteristics of the three most used reporter gene systems is described (Table 11.1). Furthermore, promoter analysis in banana could be performed with the three reporter genes (Fig. 11.1).

11.4 Promoter Prediction

Available gene expression data could be used to identify promoters indirectly. Genome annotation of open reading frames could be used to

Table 11.1 Reporter genes used for promoter characterization in plants

Reporter	Half-life protein	Specialized equipment	Use substrate	Sensitivity	Routinely in vivo analysis	Best suited
GUS	50 h	Fluorometer (for quantification)	Yes	++	No	Spatial
LUC	15 min–3 h	CCD camera, luminometer	Yes	++++	Yes	Temporal expression, low activity
GFP	24 h	Fluorometer, fluorescence stereo/microscope	No	+++	Yes	Spatial

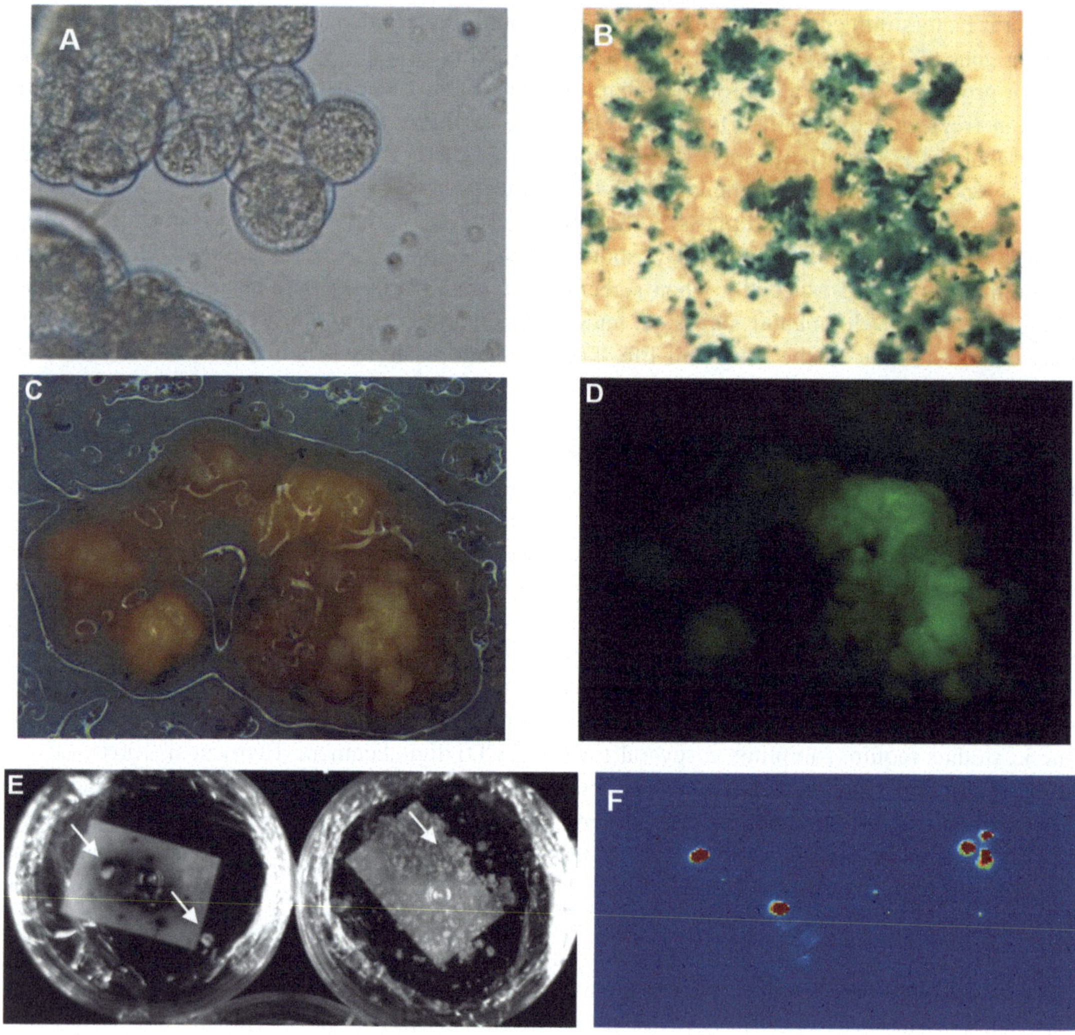

Fig. 11.1 Reporter gene systems in banana. (**a**) Embryogenic cell suspensions of the banana cultivar 'Williams' (AAA) used for *Agrobacterium*-mediated transformation. (**b**) Histochemical GUS analysis of the plantain 'Barraganete' (AAB) callus after 3 months of transformation with the pCAMBIA1301 T-DNA. (**c**, **d**) Banana colonies after 3 months of transformation with the pCAMBIA1304. GFP was detected (**d**). (**e**, **f**) Banana colonies after 3 months of transformation with the pLVCIBE1 (banana promoter fused to the LUC2 reporter gene). Live image (**e**) and in complete darkness (**f**) image was processed in pseudocolor. *White arrows* (**e**) indicate banana colonies with LUC activity. *Red color* indicates high LUC activity (**f**)

retrieve DNA sequences in the 5′ region of the ORF (Seki et al. 2002). However, determination of promoter activity prior to experimental data is not a simple task. Plant *cis*-acting elements databases including PLACE (Higo et al. 1999) and PlantCARE (Lescot et al. 2002) are mainly used to identify candidate elements (Table 11.2). However, due to the short length of *cis*-acting elements, any given DNA sequence may contain elements with no biological function, indicating that the sequences in silico might be indicative, but not sufficient to prove their function (Radhamony et al. 2005; reviewed by Hernandez-Garcia and Finer 2014). Therefore, other factors need to be considered while analyzing sequences in silico for promoter prediction such as CpG islands (Rombauts et al. 2003) and DNA bending and curvature (Marilley and Pasero 1996; Schatz and Langowski 1997). The distance between a TATA box and TSS could be critical as indicated by Zhu et al. (1995) when changing the distance in the rice phenylalanine ammonia-lyase promoter. The plant promoter analysis could be performed by measuring base composition, relative entropy, and periodicity and curvature properties (Pandey and Krishnamachari 2006). Differences in DNA curvature were observed between promoter and non-promoter sequences, while the peaks of relative entropy values were detected in TATA and TATA-less promoters. Therefore, promoter prediction software should take into considerations different properties to efficiently identify promoter sequences and to further elucidate promoter activity patterns.

For banana promoter analysis, different techniques are used, including bioinformatics tools through PlantCARE and/or PLACE, promoter activity with reporter genes, and gene expression analysis (Table 11.2). Most of the reports refer to the characterization of promoters active during fruit ripening. Few reports used specific techniques to confirm the interaction of transcription factors with promoter sequences, including yeast one-hybrid and EMSA (Table 11.2).

The TSSP plant promoter prediction program (www.softberry.com, Shahmuradov et al. 2003) is based on discriminant analysis of sequence features and plant regulatory motifs (a sequence pattern found in several promoters). This program contains a database of 576 experimentally verified promoters with TSS identified and a dataset of 3503 and 4220 promoters with predicted TSS based on mapping of full-length DNAs from *Arabidopsis* and rice. The collected data correspond to promoter sequences, including position −200 to +50 where the TSS is +1.[4] Most of banana promoter sequences queried from GenBank were confirmed for promoter identity using the TSSP (Table 11.3).

One way to identify *cis*-acting elements resulting in a specific gene expression is to use transcription profile data and analyzed promoters of selected genes to look for over-representative elements (Maruyama et al. 2012). These approaches might reveal new *cis*-acting elements or confirm those already known. On the other hand, dealing with bioinformatic software like PLACE and/or PlantCARE might be indicative of regulatory sequence functionality, and their interaction with transcription factors should be confirmed through experimentation. The problem of finding the motif sequences where a transcription factor binds could be modeled as an optimized search for a nucleotide pattern. Well-known approaches to find motifs in promoter regions from biota are BioProspector (Liu et al. 2001) and MEME (Bailey et al. 2006), which are based on Gibbs sampling and expectation-maximization modulation, respectively. For analysis of *cis*-acting regulatory elements in plants, besides PLACE and PlantCARE, POCO (Kankainen and Holm 2005) is a motif finding tool available for research. In our case, the banana promoters were analyzed using an evolutionary method named MBMEDA (Jordan and Jordan 2015; Table 11.4). MBMEDA is a motif search algorithm based on the estimation of distribution algorithm. It estimates a probabilistic model of the best individuals in the population, beach individual being a candidate motif for the solution. New candidate motifs are sampled from the model to a population pool with a better probability of finding the real motif sequences in the subsequent generations.

The results of applying MBMEDA for 15 promoters of *Musa* regions are shown (Table 11.4). The number of columns represents the number of motif instances found in the promoter regions. The motif window column is the size of the corresponding motif instances. Both parameters were obtained from the constitution of the promoter base and the criteria of the researcher. Sequence consensus analysis of selected promoter regions is indicated (Figs. 11.2a, 11.2b, and 11.2c).

The location of the motif sequences on each of promoter regions is denoted using uppercase letters (Table 11.5). Some motif sequences are close, as in AF119096.1 and BD268975.1. In JF320825.1 and JQ957542.1, the nucleotide sequences are located far from each other. This particular property could affect significantly the binding of the transcription factor and consequently the corresponding gene expression.

[4]http://www.softberry.com/berry.phtml?topic=plantprom&group=data&subgroup=plantprom

Table 11.2 Reports of banana promoter analysis since 2008

Type of promoter	Promoter	References	Analysis 1	2	3	4	5
Abiotic stress inducibility	MusaDHN-1	Shekhawat et al. (2011)	PLACE	PlantCARE	GUS	RT-qPCR	
Inducible: ethylene and ripening	MaBSD1	Ba et al. (2014b)	PLACE	PlantCARE	Dual luciferase assay	RT-qPCR	
Three promoters were not related to ripening response. Eight promoters were reduced in postethylene treatment	25 promoters from MPK	Asif et al. (2014)	PlantCARE	Transcriptome			
Inducible to ethylene and ripening: MaLBD1, MaLBD2, and MaLBD3	Four MaLBD promoters	Ba et al. (2014a)	PLACE	PlantCARE	Dual luciferase assay	RT-qPCR	
MabHLH1, MabHLH4, and MabHLH2 cold inducible and MeJa inducible	MabHLH1/MabHLH2/MabHLH4	Peng et al. (2013)	PLACE	PlantCARE	GFP	RT-qPCR	
Peel and pulp tissues were ethylene induced	MaACS1, MaACO1	Xiao et al. (2013)	Yeast one-hybrid	GUS	RT-qPCR		
Stress- and hormone-responsive TF that might be involved in chilling and pathogen stress responses	MabZIP3	He et al. (2013)	PLACE	PlantCARE	RT-qPCR		

Osmotic stress response	Ten HSP70 promoters	Vanhove et al. (2015)	PLACE	PlantCARE	Proteomics		
Response to ethylene	MaACO1	Liu et al. (2015)	Yeast one-hybrid	GUS	RT-qPCR		
Cold and ethylene inducible	MaNAC1	Shan et al. (2014)	Yeast one-hybrid	GFP, dual luciferase assay	EMSA	PLACE	PlantCARE
MaEBF2 expressed constitutively. MaEBF2 induced to ethylene	MaEBF2	Kuang et al. (2013)	PLACE	GFP	RT-qPCR		
SA and MeJa induced	MaPR1, MaPR2, MaCHIL1, and MaPR10c	Tang et al. (2013)	Yeast one-hybrid	PlantCARE	RT-qPCR		
Fruit ripening	MaNAC1, MaNAC2, MaNAC3, MaNAC4, MaNAC5, MaNAC6	Shan et al. (2012)	PlantCARE	GFP	RT-qPCR		
Banana fruit-specific promoters	Chitinase, expansin, glucanase	Ghosh et al. (2012)	GUS				
Fruit ripening, regulation of ethylene production	MA-ACS1	Roy Choudhury et al. (2011)	PLACE	PlantCARE			
Regulating the expression of fruit ripening	Sucrose-phosphate synthase (SPS)	Roy Choudhury et al. (2010)	PLACE	PlantCARE	Southwestern		
Inducible: ethylene and fruit ripening	SPS	Roy Choudhury et al. (2009)	PLACE	PlantCARE	RT-PCR		
Inducible: ethylene and fruit ripening, response to ethylene, auxin, wounding, low temperature	SPS	Roy Choudhury et al. (2008a)	PLACE	EMSA	GUS	RT-PCR	
Inducible: ethylene and fruit ripening	MA-ACS1, MA-ACO1	Roy Choudhury et al. (2008b)	PLACE	RT-PCR			

Table 11.3 TSSP analysis of promoter sequences found in the GenBank

Cultivar	Size (nt)	Banana promoter	GenBank accession	TSSP SoftBerry		
				Promoters/enhancers	Position	TATA
Musa ABB group "KaribaleMonthan"	986	SK3-type dehydrin (DHN-1)	JF320825.1	Enhancer	914	–
				Promoter	853	825
Musa acuminata, cultivar *Robusta*	1173	1-Aminocyclopropane-1-carboxylate synthase and promoter region (acs)	AJ622943.1	Promoter	172	158
Musa acuminata AAA group, Cavendish	2505	1-Aminocyclopropane-1-carboxylate synthase and promoter region (acs)	AF119096.1	Promoter	1229	1216
Musa acuminata, "TuuGuia"	752	Thaumatin-like protein (Thau)	JQ957542.1	Promoter	585	561
Musa AAB group, Three Hand Planty	1232	Unknown promoter region	EU161097.1	Promoter	431	NO
Musa acuminata AAA group, AAA group, Cavendish	290	1-Aminocyclopropane-1-carboxylate synthase (BACS1)	U88062.1	NO		
Musa acuminata AAA group, Cavendish	1000	Lectin	JX628603.1	Promoter	867	840
				Promoter	443	427
Musa AB group	325	Chitinase (ChiI2)	GU391235.2	NO		
Musa AB group	846	Chitinase (ChiI1)	GU391234.2	Promoter	800	NO
Musa acuminata	1523	1-Aminocyclopropane-1-carboxylate oxidase (ACO)	AF221107.1	Promoter	1473	1439
Musa ABB group, "KaribaleMonthan"	1254	NAC domain-containing transcription factor NAC68 (NAC68)	KP861890.1	Promoter	744	711
Musa acuminata	660	Sucrose-phosphate synthase (SPS)	AY850374.1	Promoter	509	NO
Musa ABB group "Bluggoe" *Musa acuminata* x *Musa balbisiana* (ABB genome)	1162	ACT1	AF285176.1	NO		
Musa sp.	2453	Banana promoter and melon promoter	BD268975.1	Promoter	914	880

Table 11.4 Motif consensus founded by MBMEDA for *Musa* promoter regions

Name	GenBank accession	Number of variables	Motif window	Motif consensus
SK3-type dehydrin (DHN-1)	JF320825.1	7	6	CATTGG
1-Aminocyclopropane-1-carboxylate synthase and promoter region	AJ622943.1	8	6	ATCGAC
1-Aminocyclopropane-1-carboxylate synthase (ACS) gene, promoter	AF119096.1	17	6	ATCGAA
Thaumatin-like protein (Thau)	JQ957542.1	8	6	TGAACA
Unknown promoter region	EU161097.1	9	6	GGGATC
1-Aminocyclopropane-1-carboxylate synthase (BACS1)	U88062.1	2	6	ACGACA
Lectin	JX628603.1	7	6	CGAGAG
Chitinase (ChiI2)	GU391235.2	7	6	ATTTCC
Chitinase (ChiI1)	GU391234.2	6	6	GGAGAT
1-Aminocyclopropane-1-carboxylate oxidase (ACO)	AF221107.1	10	6	AAACAT
Lectin	DQ979391.1	7	6	TGAGAG
NAC domain-containing transcription factor NAC68 (NAC68)	KP861890.1	9	6	ACACGT
Sucrose-phosphate synthase (SPS)	AY850374.1	5	6	TTGTTG
ACT1	AF285176.1	8	6	GCAGCT
Banana promoter and melon promoter	BD268975.1	16	6	AAAGGA

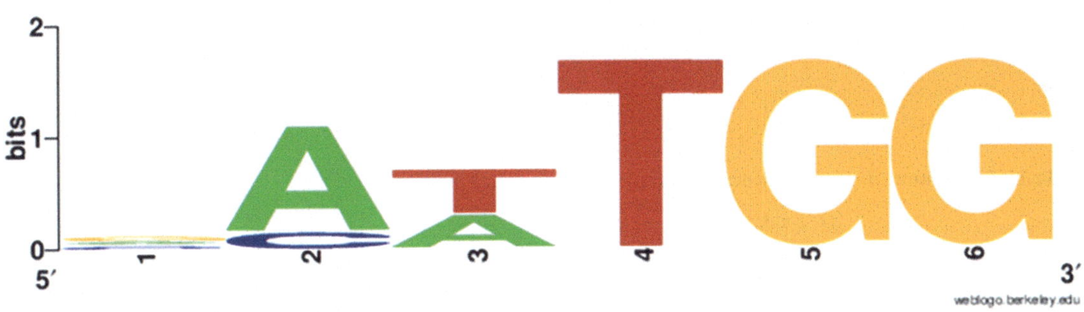

Fig. 11.2a Sequence logo for JF320825.1

Sequence logos (Schneider and Stephens 1990) are a graphical representation of amino acid sequence alignment; it provides a graphical representation of a consensus on the sequences of motifs. The height, expressed in bits as the maximum sequence conservation (Crooks et al. 2004), of a certain nucleotide represents its conservation on the corresponding column. The representation using sequence logos for the motif of selected promoter sequences is shown (Figs. 11.2a, 11.2b, and 11.2c).

11.5 Promoters Used in Genetically Modified Bananas

Since the first report of banana transgenic lines (Sági et al. 1992) and the first transgenic banana plants generated (Sági et al. 1995; May et al. 1995), several promoters have been used to drive (trans)gene expression. At least 35 different promoters have been used (Table 11.6), mainly to drive highly and constitutive expression, of which the CaMV35S promoter derivate (and derivates) is the most used (reviewed by Swennen et al. 2003; Santos Ordoñez 2008). A different tissue specificity and/or inducible promoter activity was also obtained, mainly in native banana promoters specific for ethylene and/or fruit-specific expression.

Fig. 11.2b Sequence logo for AJ622943.1

Fig. 11.2c Sequence logo for AF119096.1

Table 11.5 Location of motif sequences found for *Musa* promoter regions

Name	GenBank accession	Motif sequences on promoter regions
SK3-type dehydrin (DHN-1)	JF320825.1	GATTGGattcctgcctcctccccccggaatgcgccaagatcactagccatagtcaattccaca aggtaccaaaagcctgtcatgcatggtacgagttgggacgtgcgaccatgtgAATTGGcacag gactgaagtacttgcgtacgagaaaagatggcaatgtcaccgcgacccccggCAATGGcagc gagatagctcctgggagagccctccttgcccggccgcttctccttcgccatatttgtgatgatgtatgta gtactcctcccccaaccactgcactacgaggctaggactaagcAATTGGctcaacatctactc gacatgagatgacctcctcgactctatacgtagttctctttgaacgcgcttaaagacgaaacgaagaaa gcaaaaaagcaccggcactcaaccgacactctttatcccttcgaggctcctccgccaataatagccat aacaaatctcgcactcccgcacacgactcgttGAATGGtagagggaacgtgagttgcgttccac cgctccttccggaccacgatgatcgacgaaagaGATTGGctctagggtcatcgaagcgttacac gcgtcgctcaacgaaaccatgacacggcgcggtgtagcttactcgcgaggcatcgcacgcgacccg taaatatttcgaggtcgacaggccacgtgtcgcgaccgaccttccggatacgtccaccggctgctgac aGAATGGagggcccgcgatacgtgacggtttctGAATGGaggtctgtgagggccgacca ttccgctcccatcttttccgccgtcctctgttgtcacttctttactctttaaatttcccatggttttctttaattc gaatccttcgacgcgtaggtaacgcgtcaagcatctgacccttccactccaccccgccccacccca ccgccctctacccaacgcaaccggctaaggttctatataagcccgcggctctCCTTGGagtcc

(continued)

Table 11.5 (continued)

Name	GenBank accession	Motif sequences on promoter regions
1-Aminocyclopropane-1-carboxylate synthase and promoter region	AJ622943.1	actatagggcacgcgtgGTCGACggcccgggctggtaaatagccagtcatttcaaacatga agatcatgcttttattgtttctttttggtagaagagtgaatgagatccacattttccttaagaggatgc atataagccaattaaattaacatccatatgtataaaatccaattaggcaaaggcaaATCGAAg tagttagatacgtttctatctgatccgattaatattctcttttacgattgataaagaatacatgctaattg atatgggaggcaattcccgacgtatcatactcatccgATCGATgcgtaatatcgttaggtcagt gatgaagatgttaatttgtctcacgagATCGAAatatataattagaaaagttgacatgacaagtc agttattctataaccatagttttgattcatgactcatcgtgacattattatcgtcttaattaatatcatatatta taatataaaaggtggtctagcgggagtttgacttaatacatattaatttatgctgatgaatctcaaccttt actaacctaatatgactaaataagtatgCTCGACagattttctacgtgataatcttgactcttaaca cctcactgagtcaactcggaaCTCGAAgtggactccgtcagactccttaactctgtaacaactg atctaattggcaccaatcaaatcagatgcattgagctaaatggatctaatattagaaaacattgtgatttt ccttctttcgcaaatagcatgtcATCGATgagattgaggtttacaaagagcggcgcacattttgtt tgggataaataattctgtgcttacaatatagtagagttcgagtcgaaagcgactcccgagttcggaac acgtcattgttgccgccaacactgaagcttcctatttggcgtcacctGTCGATgttacggcgcat ccatcgccaatcacgtccatgatttacacgctgccggATCGACtcggttttcatgtccttcttttcc agcctgaagtcctctttgttgacctctttggatgtttgaatggtctcgggatttgcctattaatggtcatc ggaATCGACtcttgcaaactgcagcagctgcttctccttcttcttctctgctcgcttcagccttttc cggtacgtacctgagataacgggtcac
1-Aminocyclopropane-1-carboxylate synthase (ACS) gene, promoter	AF119096.1	ccattggctttgtgCTCGGAgatgacggcatggctaagctgcttccccgagcacagtatcagc cgcaaggagaccagtcgcacttgcagcaagcacggcctcacatttgcacgagggccaaatgc aggagcggctccgatccacgaaggtctcagtgtttgatcacaacttctctatctctGTCAAA gctttgacctccccactacaatattagcactttgttctgttctccctgcaactgccatcactctggcccg gccaggaagcgtttgattttgatacgcctttgatctttgattccttgctatccattaacaactcagtacac aatctggtgaacccccaccaccgctggtggttctcATCCAAatctcggcagatacaacatgat agtatcttctgtcagcaatagccacatctccacagcttaacaaaatcacgcgactccgtgttcgcgcg tataacATCGAGaaaaaggaagagagaggaagaaagaaggaatgttttctgcgaagaacac ggccatgaaagctttctcgtggacgtcgcgtcaccgtccgtgaggaacaaccacagctccgtagtt tagcaggaatgggtttcacggatccgacgatggcaccgagaaccaagaaagaaatggggatgt ccaATCTAAaacgttgccgacatcttaatcgtgcctgagatcatatattacctacaaacacacg ttatttattctggctttcagtggaataattactctttactttctttgatggattgctggaaatggattgtttgg attgcgaacgagaatataagaacaacattctacTTCAAAtattaatttcacttttgattCTCGGA aactacaccccataaatttgatcTTCGAAgcctaagaattaaatctagataatggtagaagcataa tttatttcaggtaggtggatggcttacatgctcacctaataatttatttaagtgggacccatgaataggcca cgttacaTTCAAAatttgatgattagtgtaccatgatgagctgtgtaatgcgtgtcttgattgcgac ttcccgtagaccccatcgatattagattgagacggaaatagcgtcggtcggtagacaacatagtac gtagatcgtgatagaacaaTTCGAAataataagacataccaaaaaatttggtatacgactcgat atataaaagttaaagttatttctttcatctctatatatatacacaatttgtaagtttcctaaaagatattagat tttgggtcaactataattgccTTCCAAcaaagtagaatgcctctATCAGAatctcgcgccta caaacaacaccaatcctaatcattcacatgatcttaacgttaatgaagtagcaaaTTCCAAatc acatttaaatagccagtcatTTCAAAcatgaagatcatgcttttattgtttctttttggtagaagagt gaatgagatccacattttccttaagaggatgcatataagccaattaaattaacatccatatgtataaa ATCCAAttaggcaaaggcaaATCGAAgtagtcagatacgtttctatctgatccgattaatat tctcttttacgattgataaagaatacatgctaattgatatgggaggcaattcccgacgtatcatactcat ccgatcgatacgtaatatcgttaggtcagtgatgaagatgttaatttgtctcacgagATCGAAata tataattagaaaagttgacatgacaagtcagttattctataaccatagttttgatttcatgactcatcgtg acatttattatcgtcttaatttaattatcatatgttataatataaaaggtgttctagtgggggttttgacttaat acattatttaattttaatgcttgattgaattcTTCAAActtttactaacctaaatattgactaaataag tatgcTTCGACaagatttttctacgtgattaaatctttgacctcttaaacacctcaactgga GTCAAACTCGGAaaCTCGAAagtggacctccgtcagactcctttaacatctgttaa acaacttgaATCTAATTCGACactaatcaaaATCAGActgcattgagcctaaatgga tctaaatattaagaaaaccattgttgaattttccttctttcgcaaacagcatgtcatcgatgagattaagg tttacaaagagcggcgcacaattttgttttgggataaataattctgtgcttacaatatagaagagttcga GTCGAAagcgactcccgagTTCGGAacacgtcattgttgccgccaacactgaagcttc ctatttggcgtcacctgtcgatgttacggcgcatccatcgccaatcacgtccatgatttacacgctg ccggatcgactcggttttcatgtccttcttttccagcctgaagtcctctttgttgacctctttggatgtttg aatggtctcgggatttgcctattaatggtcATCGGAatcgactcttgcaaact

(continued)

Table 11.5 (continued)

Name	GenBank accession	Motif sequences on promoter regions
Thaumatin-like protein (Thau)	JQ957542.1	aattcttagatgtttggatggaaccatctatcattctgatagcagatcatgagagggaggacatgcttatt tatttgtacatgagataattgcgaaaatttTGAACAcgttattggtaataaagcaatgtcctttagtta ttaaatagtacctataatcctcaaatggaggatgaatcgataaatcagcatcacctctggttggta AGAACCccatgtttcatcttaacgaatgtctgctgttgttcatcttgacaactcatgtatgcattac tccgttactgtttctttttaagctgtcattgatgcacctactcTGAACAggctggatattttagggt tttgtagttgaattctgtttgatgacatgcttttcagagtaaatcatggccataattgatgtggtagaaac aaagtccagatgaagtttcctttgtatcctcctgaagtccatgttattgTCAACAcacacaaacg ataatgtttgacaatttgacagagagttgacctacagaaatgaagagctcaatatatatctatatctat ctatatatatatatatatatatataaagcttagcttttacccctaaaaatttacgtgtttatatataagatgt gtaaattttagttgatattaatgatctttaaaaatttaatcatgatgatatTGAATAcatttgtttttattt tataaaaatattccgacaatcaaattgatgggtattgtcacgaatgttgatggaatttata
Unknown promoter region	EU161097.1	ttccgtagagatgatccatgagcaattttgaagagatcgccaaatatgaaccgacagaagatatatag tgacagaaccaaccttccgctcccagctTGGATCGggccctgtggtcgttcaagatcgacagg ccgtgcttgaggaccttgttGGATCGccgggccttgagcttgcgcacgaggcagaggtactcc cacgcccccccgccgccgttctcgacctggctctcgagccgctccatccgcgctgcctccgacgc ccccaccatcggcctgcgcttgccggcgaaaacctccttgcgggcgagatgtagaaggaagaga gaggaaacagagcaacgccggccgccgtgattccaatagcgatcgtcctcgaaagcaatggga gatccaaggaaaacgtaccccagaacagatcctctccagcacgtgtCGGATCCaatagctcc tctccacagcccagccgtccacagtGGGATCCacgtgcatcaatgataccaatggagggttac gatagaagaatgagcggatggagggtggagatcgagagaggttaatcatcggacaggttcgatttg gagaggatcacataaaccttcgtccaaggCGGATCCgatgacatgactggagttgggttattgg acttggggctcgatctagactcgcaacacatgagatatgaactcgggtgcaaatgagtccaatccatct gaaaacagttcaaatttattcgttaatattaatggtgtttaattaattatttatttatcttaaccatgattgcttg tattggagaccgaggcagaaaaattaatagattaaaaataaaatgaaataacaaaaaatagaaatctt aaatcaattaaatggaatccggacccgattagagCGGGTCgcatatgataaatccgacccgctg ccatccgctatCGCATCGGGTCagcgtgagactgacgtggcacgcgcgccactgttacttg gtacaccaagcgtcacagcatcacacaaagggagtgcttcgcgcactcgcagaaacgcccgtcatc tcctcgattaaaatttatatgcggacaaggtcggtccttcgattccaaccgttcgatgatctcgcata GGGATCGcttccggacgtacgccgtcggatgttatcgttggttaaatgacgttaatacccttgcc ctcggtttcttgtgcctccctcacgcctctcgcatgtgatcagttgctaatccaactggtcatgacgt ACGATCgatcgaaaaacccgtagctagctagctagctagctagctcgatcgctagctcgctag ctcgatcgctagctcagagctaagctcgctagctagctcgatcgctagat
1-Aminocyclopropane-1-carboxylate synthase (BACS1)	U88062.1	ggtggctgtgggtgcgtgctacatgccatgcacggatgtcctaaaagattgcggggcggtctctcg ctggctcggtcaatgtttccactcgatacaaggtagttaagaaggtcgtcctGACACAacggt aaaccgtcagattaaggcggtttgctagcttataatggtcaaagagacgcctACGACAataaa tgacctgtaaagtagaagaacgattgctgagcgttgatggcgattgatattttgatgatggagcaaa aatgaatctaaatctaaatatcagtatcttc
Lectin	JX628603.1	cccaaaataacatcgttAGAGAGgaggatagttgacctccttcattctctcccataaatatgta ggaactcagggatatacgatttccctagataacacaatctcctttgtacgcgcggttttctattttacga gtttttacatgtcaagcttcgcaagatgacgaacaccttttagaaaattTGAGATttttattttctat ttttccactgcatatgtgatgtcgcccttagatttccctatggttgtgtcatattgatacttatacataaga acttttTGAGATgatatattaccttggattccttgttaCGAGATcatttgtggcttacaaaatt atctttgtattatgaattgtccatcaagtatttttaatttatttttataaattcttaaaagatcctagaacgtta atccctttggatcaatatataaaagtgacatacgacttagtaaaaatcaaataaaattaaatataaatat atatttatgaagaataaaataaaattttgtattacttaatcatatcttacagatctacgatctgcaccatca aacccaaaaataAAAGGGgcatatctctcctacgtagccgaagtattgaactctcaagtctc cccaacacgtttccatgtgtcgaccgccacgccatatccttatccgctcctctttaattattgaccacac caccatccgtgatcgtcgccagtcacgtcttcgtcgtCGAGAGgctctacacgaccctcgtca cgtgatacacgctttcctcgaggaagccacggaCGAGGGggtgagcggtacacctccattcc accccgacatcagagccatggcgatccgctataagtagcctctctccctcttccccaacattatcttc tgcagaagAAAGGGaggagtttggtgcacacaacacgatggTGAGAGgctctctatac acacagttttctgtctgtggaacagtttggttacctccgtgcgttatgtgttgacgtcga
Chitinase (ChiI2)	GU391235.2	attagaggagaagcataggccgaagaaacTTTTCCtattggtgaaagagacacacacaag aaaaagaggagcgagagagagaaggaagcgccactcttgactggtcttgtccatgagcaattaa ttatttgtcgactaatgagctactaacattaatgtcacagagagcaatagatgagaagctgtagactt ggtcgcaggacttatccgaagacttgcctctgcgATTTCCcacatgcacctcatttggtccaaa ggaagcttcacagcgggcaggaatccatttctctatataagcaccacctcccacccacaccaccacc

(continued)

Table 11.5 (continued)

Name	GenBank accession	Motif sequences on promoter regions
Chitinase (ChiI1)	GU391234.2	cttcttggtatggattcaattcagctttttctgaagctggaggacaaccacacactggcgaccagacta aaagtgataactctgtcccAGAAATaagttcttaagtgttgatttctctggaagaatctgag AGAGATcactgtcatgtggatccatctagtttcgactccagcAGAGATcatagacaacgt atcgaacaagaacaaaaggagaaggtagtcaatctacagacttggtttttgtttttacatcttaaatag aagtatatggctctcaccaatatgcatcttGGAGATggtcaatgcacacaccattatttgc AGAAATggaatgattgtttctatggcatgaaccattgacgtcctgaccttagggaagaagcat aaacttctcttaaaaatcaaaggaaaacctctcaacaccgtgatttatcaatttttgctctaaattgag tttaatttagaTGTGATgtgaaaaccttataactatcagttcttcatcatttctaagatggtttatgag caatttattttaaaaaaaactctTGTGATaagcttttataccctcatatttaattttatcaagagtgctc cttttatttaaaattatgataataccccttgaattgattgatttggtccaatCGAGATgtcttaatccaa atcagtttgattcggatcagatcctttcgaatcaagtcaaatcagatcccttccactaactcataaaccat ctccacttcctccttctccctctctcctttggcttccttctcttctccctcctctcactcctcctactccacca cctcccacccacaccaccaccaccaccactgctaaggagg
1-Aminocyclopropane-1-carboxylate oxidase (ACO)	AF221107.1	gaCAACAGggtgaattggcatcatcatcctttacctgtccctgatttgagctgtagtgctgcgctc agccagcgttttatcgaagaagatAAACATgcgtgagcctacatgcgccaaagcttgccggca agtcatgcatagctgcAAACATgtgacaggcaccgaaccaacaattgaagaagatacgat AAACATgcgtgagcctacatgcaccaaagcttgccgacaagttatgtttgggtgcacaatgt gtcctcatcttacttgcacatctgctgttgcaCAACAGcagattgcatggaggtgtttttccgg caatgcaatctttgatgttggttctcttttctcccttcttgcattgtttatagctctgtttcttgtgctcttctt tcacgtagattcatagcgtagcttaagttgttatagattacctgctttactgggcAAACTTgtgca acccaggaatattcccatgtgcaccttcttcctgttttcctctgtcaaactgttctgttcatgatgaggca gcaccgaatctaagagaaatatcctaatgttgattgatttaacctcataAAACTTgaagcagaa tatgcttgccgctttcatgtgatcaattgaattgtttgcttgcttcacgagaacaccacattctgaacc cattgctttcttgtggccaccaaccggagaaagggagtctataTAACTAgccgagcgaggat tttcccatgacctgttcatctcacgtagagatggtgatttggttatagttatagcgatctatgatcgaag aatgagaaaatacccagataacggagatccatgcgtcaccagatggaacctcggccgagtgaca ctgtttgcacaccggatacttcatgttcacggcaatggcCGACATgccgaacagccatcgagc gttgaatgataggcaggaatggcccatttctcacatatgagagggatacgagtggaaagggcgc tctaatgagctgtgaatcgaaacaatttctacctatcgatccctgttcttttgatatgaagtatagc CAACAGgtcaagagaagacgagtacacacgcatcgccgatgctgtgacgttactttctgaggt tggcaatttgtcactacaatccaagcggaagccatgcacgcgaggcgtcgccatggaagaac tcaaCAACATgatgccttcccgggtctcctcaaaggggagagaccgatggaagcagcc AAACTTggtccccgatcgtgatgggacgcgagaggtggaagcaaggagggtggagaa caaggccaaaggtggtggggctgagagatggccaactgggtcaccttatggaatcggctccgt tacgtcttccactgctgttgctctcgtcgatagatccttctcCAACTTtgcttcttcattcatttcg tccctcgacgtcaagaacgcctataaattgcctggtaatcagcagcacctagcacactccagata gaaaggacaagtgcaatcagggaagaaagagcgtgtc
Lectin	DQ979391.1	cccaaaataacatcgttAGAGAGgaggatagttgacctccttcattctctcccataaatatgta ggaactcagggatatacgatttccctagataacacaatctcctttgtacgcgcggttttctattttac gagtttttacatgtcaagcttcgcaagatgacgaacaccttttagaaaattTGAGATttttatttt ctatttttccactgcatatgtgatgtcgcccttagatttccctatggttgtgtcatattgatacttataca taagaacttttTGAGATgatatattaccttggattccttgttaCGAGATcatttgtggctta caaaattcatctttgtattatgaattgtccatcaagtatttttaatttatttttataaattcttaaaagatc ctagaacgttaatccctttggatcaatatataaaagtgacatacgacttagtaaaaatcaaataaa attaaatataaatatatatttatgaagaataaaataaaattttgtattacttaatcatatcttacagatct acgatctgcaccatcaaacccaaaaataAAAGGGgcatatctctcctacgtagccgaagt attgaactctcaagtctccccaacacgtttccatgtgtcgaccgccacgccatatccttatccgct cctctttaattattgaccacaccaccatccgtgatcgtcgccagtcacgtcttcgtcgtCGAGAG gctctacacgaccctcgtcacgtgatacacgctttcctcgaggaagccacggaCGAGGG ggtgagcggtacacctccattccaccccgacatcagagccatggcgatccgctataagtagcct ctctccctcttccccaacattatcttctgcagaagAAAGGGaggagtttggtgcacacaac acgatggTGAGAGgctctatacacacagttttctgtctgtggaacagtttggttacctcc gtgcgttatgtgttgacgtcga

(continued)

Table 11.5 (continued)

Name	GenBank accession	Motif sequences on promoter regions
NAC domain-containing transcription factor NAC68 (NAC68)	KP861890.1	taaaaaaaactctccaagtacaacgttggcttctttttcctttaagaaagagaaatgggagaccttttact ttcttccgtacttcggaggACACGTgacaaacaagtaatttgctgtccgatgttctaaaatgaata aaaaagtggtgtgaatactgtttgtcaaaaccaacccACACGTgacagaaatgaaagggtcaa cgaaaaggactcgactgctccgcggtcatgcccttttaccgaaggtttgctcgcatcagaaaccttg gcgtcggctgagaattttctaaagccgccgcattaatccccagtcgtatgactttgtgttcctttcg ACACGActgactgggcGCATATggattcatcatctaaagtacctaattaatcaaagaattat ccgatgcataatgctcaattaaggactcaactcggatcaggcgtcaataggacaagaatggcttgc ttggtggacaagaaagagactcCTACGTgcggaaatgcaaggcgcccagatgcaatgcac gggcacggcgatggggcgtctcaacagtgaccatccagaaaccctaacgggtgatggcaccg cctcccaaggttttaactggggacgtactcatcacggcccccgtaactgtgtcaccacctttcctc ttccttgtctcttttcgccatctccaggtccaatggatccatgtgtcatcattccatgttctctatgaaac cctagctcgctgtcagtcctcatcccTCACGTgagctgaagggttctgagcaagggagacg tacatattcatgtcgtgaatccccctccttGCACAAcggctgcgtaattaacagtgctagtccc tccgactccaaattccatctcgtcttcatgcaattcatACATGTgttcggtatgcgtacactaagc aataatgagaattcataccaaatgtaGCACAAgaaacaaaattgatatgaaacgatgaaaac acagaaatgaagtaaaaaaaatatatagagcattaaggtgaattatgcataaataaatatcatataat atgaattatgcattaatatgtaaatttacctagactgaacggtttggggttccagcatgggtaggtgt ctgcgggcaccagcgcgttggtcttcgccaccttgaccaatgcggtcggccgcctcacgcgaag cttcccggatcccACACGTggttgctgcgactcatccattcccctatgtatccaACACGA tgggagttcaaagtatctc
Sucrose-phosphate synthase (SPS)	AY850374.1	ggatttaattaattagtaataattattcTTGTTGtatttatttgtttatttcttattataacctcagtgtcc tctcagttacagaaacgaaaaggtgcttactgtaaggtgcgggcgacctgacgcttaacgtacgc ggataagcgtatcgccgttggattcttataggtcacgaatcttagagccaaacaccgccggaggat gcgaacgacacacgaatcttcaagccgaaaggctgatctccaatgctGTGTTGtctccgacg agcaccggcagattcccaagagcgacgaacctcgaatcgtcgtcgttgattctgatggtttgcttat ctcgaagcaacagtgttcatatccaacggtcgagaATGTTGttcgcacgaccccattcgtgcg gtctggtcacgccgtgtcgttgtgggggacgcttctcttccccgtgtctctcgcttctttcttacatcgc cgcacatatattgattgcctccacccgagagagagagagaaagaaagagagagagagagagaac ttctcctcctccctctccTTGTTGtctcttctgttcttgtcagttcgatcatggcggggaacgattgg atcaacagctaTCTGGAggcgatcctggacgccgggccctccatcgacgcggccaagtcct
ACT1	AF285176.1	accttagtctgAGAGCTctgatgcactagatcgtaaacccactcttaggattatccgagaatacc caccctacaaatcgagcctacctttccacacttacatcacatatctaccgccgaggctgatcttttctc ttattaataaggtgctgaaagcccactaccgacaagaactcaataagtacaagcaaggcagcaacc agtgcttccaatacggatagaggaagttcagactcaagccggccagttgcagcctGCAGCT caggtacccttagactcaaaccggGCAGCTcgagccttgaactcaccaattacagactgaaa ccggGCAGCTcaagccttggatcgatcggtcaaagactaaaacgagcttccccaaggccct aacccggcagcaacagaattaacccaatgagtataagaattgaaccggtaagtataaggctagttc cagtccaaaagtaagaattgcactaatgggtctcaggcttAGAGCTagtggcaacagaatgg agccggtaaatcaaagcctgtagtggtaggaactcggacaaagactcagaattacacctatgagt cgaagccaacaagcagacaagccgacaagacacggatagacaagcagacaaaccacacttta gtcaagtactctttcgacatttgcatcgagccaagtcactactaatgtgacgcggcagcgttgttgca cgggaaaacggaaaaaatcttagaaatctggTCAGCTattaggggaaatctggtcactgcc aaaaaacccgggttttgtcaagaaatcctcgaaattgacCCAGCTattagcggaaatccggtt gctaccaaaaaacacgggttttggtggacgcccaccccgcctcaaccttccaacaccgcacgtca tgccggacctgatccgacggcctagatgaccctgatccgagctccatcggacgattggcgttccct ctcgagACAGCTcattgacgaacacaaaaggcatattcgacttccgtgtatagctcaagcgac aaaacactgggtttaatctttttattcgacattctctgtgggatccgcaacacgtatcctaaatcccata agtcgagttaatagaaagaccatttaagctgtctcgtcgctgggtagagacacaggagctgcttgtt

(continued)

Table 11.5 (continued)

Name	GenBank accession	Motif sequences on promoter regions
Banana promoter and melon promoter	BD268975.1	gctattaaggcttgatcccaagaccctcgctctctatcttcgcaacaactcacgtcattgttattggttcc cttttgccttcgtctcaaatgtctcatattgtataatcttcaagaagggtattcacgacctcctacgat GAAGGAgtgcaccctcgtctgctttcgtttcttatcattgcttcgttagggaaacataataactc ggGAAGGAgacacaaacaatgtttatagtgatgagtcatgtaagGAAGGAga GAAGAAagttgtgtcgtgattgcctccttccccttaacctttgttggatgaaaaagatcattag gactcgaaattttaaaaggtggaGAAGGAgacccaagataccctcctcatagcaagat aagagatatccgagatgaatgtgaGGAAGAAaacgatagcaaacgatgtaagttatcatg aaaaTAAAGAgaaaatatgagaacctcatgatgaggctttagtgtcacctcgataat TAAAGACGAGGAtaacaacgtgacaacaataaccaacaagggacataaacgataaa ggcgttgattgacgagaccaaagtcgaacataataatattttttaagataaaaaaaaaagta AAAGGAtgtattttaGAAGAAAAGAAataaaagattataattttttgagaatttgtc cgaatacgaatatatattattttgaatattaattaaaTAAAGAtaccaacgcgtcgctttggttc atcgtctttctttaacgcggcggacgggaacgtgaggccgacaaaggtttcatgattcctagtgg cgtctttatgatttccactctgatgctgatggaaacgtgagcggcgAAAGAAgcgccaca attgatcgaagcgctcctctataaatggcgagtaccgggagggagcctcaagcagtgccttgt cccggttgattcgagtcccgtcctccgatttcgtgcaagaagagaagggaatcgagcgagat ggcggagAAAGGAtcggtgatcgggtgccacaccatcgcccagtggaaccggcagct ccaactcgccagcgagtccgggaagctggtaaggttccgacttgctcgaccctttgtgtttgctt ctanggttttgggggtaaatgtccattgcgacctcgtcatggaaacaccaactcttttcttgcacta aaaacccngcattggcAGAGGAgctttatatgccgagcttagaattttagactcagggtt ttgagtcattctttATAGGAttttatgagttgatattattttctatgctctatctgttgatttatattgg tctaagatcattgttattgtgcaagaattttacAGAGGGaaaggtaatgacagaaaaagaaa ggagagtgccatcaaatgctatttgggtgccaagggacaaatttgttctgatgccatgttccatt ctttgaccctaatgtctaatatcaatggtatctgaattaaacgatgtcattgtgacggtggatgtcta atgtctaagttacagaaaacAAAGGAgggtgtcatcaaatgctattatctggtaattggatg ccaaggggcaaattatcatggtgtctaatcaatgatatctaatttaaaggctactgcctaaagccca accttttcttcctttatttattttaattgttaactctttatttgaacttttatattctgtgtgtactccacttaatt gagcatatcaagttctgtcaaactaagtctatcaaacttttggtggtgaagctgatgattccaagaga tcaGGAAGAtctttctattcaacatccatgtttttcttataccattgtttacaacttagttacatctttt cttccgttcctgtttcataaatgcatatttattttctgaaaaatgtaaagctaaacagagtcgaaatctatc aaactgctgcatctcttgttggctcTAAAGAtttgtctggtttacaggtggtcgtagatttcactt cttcatggtgtggtccttgccgtatgattgccccgttcttcgctgagctagctaataagttcaccgat gccatcttccTAAGGGtggacgtcaatgagctgaaggtaaaaactatcaacatgcagattaa actcgtgccgctgttgtttgcgaggtgaaattatgtttctaattctccaacctgtgtcaccaaacggc AGAGGGttgccctggactgtgcgatcgagacactgccaaccttcatcttcctgaggcaggg aaacattgtggatcgcgttgttggtgctcgtaaagatctgttgccGAAGAAgattgagctcca catgaggaactgaatgctcgcttgcagtattagtgtcggtgttgggtcgtgaactcggagattttgt ggggttagaataaacatatgtactgaattccatcggttcatgattatattaactattgaataa actagtctttcatcc

Table 11.6 Promoters used in genetically modified bananas

Promoters (may include derivates)	Type	Source	First report
35S	Viral	Cauliflower mosaic virus (CaMV)	Sági et al. (1992)
Sc	Viral	Sugarcane bacilliform badnavirus (ScBV)	Schenk et al. (1999)
My	Viral	Banana streak badnavirus (BSV)	Schenk et al. (2001)
BT	Viral	Banana bunchy top virus (BBTV)	Dugdale et al. (1998)
T1200	Viral	Taro bacilliform badnavirus (TaBV)	Yang et al. (2003)
Nos	Bacterial	*Agrobacterium tumefaciens*	May et al. (1995)
Mas	Bacterial	*Agrobacterium tumefaciens*	Remy (2000)
Emu	Recombinant	Maize, *Agrobacterium*	Sági et al. (1995)
Gelvin super-promoter	Recombinant	*Agrobacterium tumefaciens*	Remy et al. (1998)
Ubi1	Plant	Maize	Sági et al. (1995)
Adh1	Plant	Maize	Dugdale et al. (2001)
Act1	Plant	Rice	May et al. (1995)
REG-2	Plant	Rice	Chong-Pérez et al. (2013)
TUB-1	Plant	*Arabidopsis*	Atkinson et al. (2004)
ubq3	Plant	*Arabidopsis*	Chakrabarti et al. (2003)
Ahas	Plant	*Arabidopsis*	Matsumoto et al. (2002)
act2p	Plant	*Arabidopsis*	Matsumoto et al. (2002)
HSP18.2	Plant	*Arabidopsis*	Chong-Pérez et al. (2012)
Gmhsp17.6-L	Plant	Soybean	Chong-Pérez et al. (2012)
PR-10	Plant	Alfalfa	Vishnevetsky et al. (2011)
Act1	Plant	Banana	Hermann et al. (2001)
EFE	Plant	Banana	Sunil Kumar et al. (2005)
ACC	Plant	Banana	Wang and Peng (2001a)
ACO1	Plant	Banana	Wang and Peng (2001b)
MusaDHN-1	Plant	Banana	Sreedharan et al. (2015)
17-1	Plant	Banana	Santos et al. (2009)
RGC2	Plant	Banana	[a]
Exp4	Plant	Banana	[b]
Exp1	Plant	Banana	[b]
Ext	Plant	Banana	[b]
MT2A	Plant	Banana	[b]
ACS	Plant	Banana	[b]
BAC	Plant	Banana	Ghosh et al. (2012)
EXP	Plant	Banana	Ghosh et al. (2012)
GLU	Plant	Banana	Ghosh et al. (2012)

[a] www.ogtr.gov.au/internet/ogtr/publishing.nsf/Content/dir107-3/$FILE/dir107rarmp.pdf
[b] www.ogtr.gov.au/internet/ogtr/publishing.nsf/content/dir109-3/$FILE/dir109rarmp1.pdf

11.6 Conclusion

Promoter analysis in banana has been performed using different methods, including in silico and experimental. In silico approaches still need to fully simulate functionality and promoter activity patterns in plants. Novel banana promoters should be discovered and used in the heterologous expression of genes. Furthermore, the generation of intra- or cisgenic banana plants should lead to a better acceptance for consumptions and facilitate risk analysis for granting permission for cultivation.

Acknowledgments Promoter activity in banana transgenic lines was performed in the framework of the SENESCYT project: PIC-08-0000300.

References

André D, Colau D, Schell J, Montagu M, Hernalsteens JP (1986) Gene tagging in plants by a T-DNA insertion mutagen that generates APH(3′)II-plant gene fusions. Mol Gen Genet 204:512–518

Asif M, Lakhwani D, Pathak S et al (2014) Genome-wide identification and expression analysis of the mitogen-activated protein kinase gene family from banana suggest involvement of specific members in different stages of fruit ripening. Funct Integr Genomics 14(1):161–175. doi:10.1007/s10142-013-0349-9

Atkinson HJ, Grimwood S, Johnston K, Green J (2004) Prototype demonstration of transgenic resistance to the nematode *Radopholus similis* conferred on banana by a cystatin. Transgenic Res 13:135–142

Ba L, Shan W, Kuang J et al (2014a) The banana MaLBD (lateral organ boundaries domain) transcription factors regulate EXPANSIN expression and are involved in fruit ripening. Plant Mol Biol Report 32(6):1103–1113. doi:10.1007/s11105-014-0720-6

Ba L, Shan W, Xiao Y et al (2014b) A ripening-induced transcription factor MaBSD1 interacts with promoters of MaEXP1/2 from banana fruit. Plant Cell Rep 33(11):1913–1920. doi:10.1007/s00299-014-1668-6

Bailey TL, Williams N, Misleh C, Li WW (2006) MEME: discovering and analyzing DNA and protein sequence motifs. Nucleic Acids Res 34:W369–W373

Berk AJ (1999) Activation of RNA polymerase II transcription. Curr Opin Cell Biol 11:330–335

Breathnach R, Chambon P (1981) Organization and expression of eukaryotic split genes-coding for proteins. Annu Rev Biochem 50:349–383

Calderon-Villalobos LIA, Kuhnle C, Li HB, Rosso M, Weisshaar B, Schwechheimer C (2006) LucTrap vectors are tools to generate luciferase fusions for the quantification of transcript and protein abundance *in vivo*. Plant Physiol 141:3–14

Castle J, Szekeres M, Jenkins G, Bishop GJ (2005) Unique and overlapping expression patterns of Arabidopsis *CYP85* genes involved in brassinosteroid C-6 oxidation. Plant Mol Biol 57:129–140

Chakrabarti A, Ganapathi TR, Mukherjee PK, Bapat VA (2003) MSI-99, a magainin analogue, imparts enhanced disease resistance in transgenic tobacco and banana. Planta 216:587–596

Chiu WL, Niwa Y, Zeng W, Hirano T, Kobayashi H, Sheen J (1996) Engineered GFP as a vital reporter in plants. Curr Biol 6:325–330

Chong-Pérez B, Kosky RG, Reyes M et al (2012) Heat shock induced excision of selectable marker genes in transgenic banana by the Cre-lox site-specific recombination system. J Biotechnol 159(4):265–273. doi:10.1016/j.jbiotec.2011.07.031

Chong-Pérez B, Reyes M, Rojas L, Ocaña B et al (2013) Excision of a selectable marker gene in transgenic banana using a Cre/lox system controlled by an embryo specific promoter. Plant Mol Biol 83:143–152. doi:10.1007/s11103-013-0058-8

Christensen AH, Sharrock RA, Quail PH (1992) Maize polyubiquitin genes: structure, thermal perturbation of expression and transcript splicing, and promoter activity following transfer to protoplasts by electroporation. Plant Mol Biol 18:675–689

Crooks GE, Hon G, Chandonia JM, Brenner SE (2004) WebLogo: a sequence logo generator. Genome Res 14:1188–1190

Droc G, Lariviere D, Guignon V et al (2013) The banana genome hub. Database 2013:bat035. doi:10.1093/database/bat035

Dugdale B, Beetham PR, Becker DK, Harding RM, Dale JL (1998) Promoter activity associated with the intergenic regions of banana bunchy top virus DNA-1 to-6 in transgenic tobacco and banana cells. J Gen Virol 79:2301–2311

Dugdale B, Becker DK, Harding RM, Dale JL (2001) Intron-mediated enhancement of the banana bunchy top virus DNA-6 promoter in banana (*Musa* spp.) embryogenic cells and plants. Plant Cell Rep 20:220–226

Dynan WS (1986) Promoters for housekeeping genes. Trends Genet 2:196–197

Elliott AR, Campbell JA, Dugdale B, Brettell RIS, Grof CPL (1999) Green fluorescent protein facilitates rapid in vivo detection of genetically transformed plant cells. Plant Cell Rep 18:707–714

Feldmann KA (1991) T-DNA insertion mutagenesis in *Arabidopsis*: mutational spectrum. Plant J 1:71–82

Fobert PR, Miki BL, Iyer VN (1991) Detection of gene regulatory signals in plants revealed by T-DNA-mediated fusions. Plant Mol Biol 17:837–851

Fobert PR, Labbe H, Cosmopoulos J et al (1994) T-DNA tagging of a seed coat-specific cryptic promoter in tobacco. Plant J 6:567–577

Ghosh A, Shekhawat U, Ganapathi T et al (2012) Analysis of banana fruit-specific promoters using transient expression in embryogenic cells of banana cultivar Robusta (AAA Group). J Plant Biochem Biotechnol 21(2):189–197. doi:10.1007/s13562-011-0070-5

Halfon MS (2006) (Re)modeling the transcriptional enhancer. Nat Genet 38:1102–1103

He S, Shan W, Kuang J et al (2013) Molecular characterization of a stress-response bZIP transcription factor in banana. Plant Cell Tiss Org Cult 113(2):173–187. doi:10.1007/s11240-012-0258-y

Heim R, Cubitt AB, Tsien RY (1995) Improved green fluorescence. Nature 373:663–664

Hermann SR, Harding RM, Dale JL (2001) The banana actin 1 promoter drives near-constitutive transgene expression in vegetative tissues of banana (*Musa* spp.). Plant Cell Rep 20:525–530

Hernandez-Garcia CM, Finer JJ (2014) Identification and validation of promoters and cis-acting regulatory elements. Plant Sci 217–218:109–119

Higo K, Ugawa Y, Iwamoto M, Korenaga T (1999) Plant cis-acting regulatory DNA elements (PLACE) database: 1999. Nucleic Acids Res 27:297–300

Hraska M, Rakousky S, Curn V (2006) Green fluorescent protein as a vital marker for non-destructive detection of transformation events in transgenic plants. Plant Cell Tiss Org Cult 86:303–318

Janssens H, Hou SL, Jaeger J et al (2006) Quantitative and predictive model of transcriptional control of the *Drosophila melanogaster even skipped* gene. Nat Genet 38:1159–1165

Jefferson RA, Kavanagh TA, Bevan MW (1987) GUS fusions: β-glucuronidase as a sensitive and versatile gene fusion marker in higher plants. EMBO J 6:3901–3907

Jordan CI, Jordan CJ (2015) MBMEDA: an application of estimation of distribution algorithms to the problem of finding biological motifs. Lecture notes in computer science. Artificial computation in biology and medicine, vol 9107. Springer International Publishing, Cham, pp 39–46

Kankainen M, Holm L (2005) POCO: discovery of regulatory patterns from promoters of oppositely expressed gene sets. Nucleic Acids Res 33(Web Server issue):W427–W431

Kertbundit S, De Greve H, Deboeck F, Van Montagu M, Hernalsteens JP (1991) *In vivo* random β-glucuronidase gene fusions in *Arabidopsis thaliana*. Proc Natl Acad Sci U S A 88:5212–5216

Koncz C, Martini N, Mayerhofer R et al (1989) High-frequency T-DNA-mediated gene tagging in plants. Proc Natl Acad Sci U S A 86:8467–8471

Kristiansson E, Thorsen M, Tamás MJ, Nerman O (2009) Evolutionary forces act on promoter length: identification of enriched *Cis*-regulatory elements. Mol Biol Evol 26(6):1299–1307. doi:10.1093/molbev/msp040

Krysan PJ, Young JC, Sussman MR (1999) T-DNA as an insertional mutagen in Arabidopsis. Plant Cell 11:2283–2290

Kuang J, Chen L, Shan W et al (2013) Molecular characterization of two banana ethylene signaling component MaEBFs during fruit ripening. Postharvest Biol Technol 85:94–101. doi:10.1016/j.postharvbio.2013.05.004

Kumari S, Ware D (2013) Genome-wide computational prediction and analysis of core promoter elements across plant monocots and dicots. PLoS One 8(10):e79011

Kutach AK, Kadonaga JT (2000) The downstream promoter element DPE appears to be as widely used as the TATA box in *Drosophila* core promoters. Mol Cell Biol 20:4754–4764

Lee TI, Young RA (2000) Transcription of eukaryotic protein-coding genes. Annu Rev Genet 34:77–137

Lescot M, Dehais P, Thijs G et al (2002) PlantCARE, a database of plant *cis*-acting regulatory elements and a portal to tools for *in silico* analysis of promoter sequences. Nucleic Acids Res 30:325–327

Liu X, Brutlag DL, Liu JS (2001) Bioprospector: discovering conserved DNa motifs in upstream regulatory regions of co-expressed genes. Pac Symp Biocomput 6:127–138

Liu J, Liu L, Li Y et al (2015) Role for the banana AGAMOUS-like gene MaMADS7 in regulation of fruit ripening and quality. Physiol Plant 155(3):217–231

Mandal A, Sandgren M, Holmstrom KO, Gallois P, Palva ET (1995) Identification of *Arabidopsis thaliana* sequences responsive to low temperature and abscisic acid by T-DNA tagging and *in-vivo* gene fusion. Plant Mol Biol Report 13:243–254

Marilley M, Pasero P (1996) Common DNA structural features exhibited by eukaryotic ribosomal gene promoters. Nucleic Acids Res 24:2204–2211

Martienssen RA (1998) Functional genomics: probing plant gene function and expression with transposons. Proc Natl Acad Sci U S A 95:2021–2026

Maruyama K, Todaka D, Mizoi J et al (2012) Identification of Cis-acting promoter elements in cold- and dehydration-induced transcriptional pathways in Arabidopsis, rice, and soybean. DNA Res 19(1):37–49. doi:10.1093/dnares/dsr040

Matsumoto K, Morais LS, Vianna GR, Aragão FJL, Rech EL (2002) Genetic transformation of banana embryogenic cells through particle bombardment using a herbicide resistance gene as selectable marker. Acta Horticult 575:61–67

Maximova SN, Dandekar AM, Guiltinan MJ (1998) Investigation of *Agrobacterium*-mediated transformation of apple using green fluorescent protein: high transient expression and low stable transformation suggest that factors other than T-DNA transfer are rate-limiting. Plant Mol Biol 37:549–559

May GD, Afza R, Mason HS et al (1995) Generation of transgenic banana (*Musa acuminata*) plants via *Agrobacterium*-mediated transformation. Biotechnology 13:486–492

Mellor J (2006) Dynamic nucleosomes and gene transcription. Trends Genet 22:320–329

Mollier P, Hoffmann B, Orsel M, Pelletier G (2000) Tagging of a cryptic promoter that confers root-specific *gus* expression in *Arabidopsis thaliana*. Plant Cell Rep 19:1076–1083

Mudge SR, Birch RG (1998) T-DNA tagging and characterisation of a novel meristem-specific promoter from tobacco. Aust J Plant Physiol 25:637–643

Nakamura M, Tsunoda T, Obokata J (2002) Photosynthesis nuclear genes generally lack TATA-boxes: a tobacco photosystem I gene responds to light through an initiator. Plant J 29:1–10

Ökrész L, Mathe C, Horvath E et al (1998) T-DNA trapping of a cryptic promoter identifies an ortholog of highly conserved SNZ growth arrest response genes in Arabidopsis. Plant Sci 138:217–228

Ow DW, Wood KW, DeLuca M et al (1986) Transient and stable expression of the firefly luciferase gene in plant cells and transgenic plants and transgenic plants. Science 234:856–859

Pandey SP, Krishnamachari A (2006) Computational analysis of plant RNA Pol-II promoters. Biosystems 83:38–50

Peng H, Shan W, Kuang J et al (2013) Molecular characterization of cold-responsive basic helix-loop-helix transcription factors MabHLHs that interact with MaICE1 in banana fruit. Planta 238(5):937–953. doi:10.1007/s00425-013-1944-7

Plesch G, Kamann E, Mueller-Roeber B (2000) Cloning of regulatory sequences mediating guard-cell-specific gene expression. Gene 249:83–89

Praz V, Perier R, Bonnard C, Bucher P (2002) The eukaryotic promoter database, EPD: new entry types and links to gene expression data. Nucleic Acids Res 30:322–324

Radhamony RN, Prasad AM, Srinivasan R (2005) T-DNA insertional mutagenesis in *Arabidopsis*: a tool for functional genomics. Electron J Biotechnol 8:82–106

Remy S (2000) Genetic transformation of banana (*Musa* spp.) for disease resistance by expression of antimicrobial proteins. Dissertationes de agricultura. Ph.D. thesis 420.KatholiekeUniversiteit Leuven, Belgium. Faculteit Bio-ingenieurswetenschappen. 341 p

Remy S, Buyens A, Cammue BPA, Swennen R, Sági L (1998) Production of transgenic banana plants expressing antifungal proteins. Acta Horticult 490:433–436

Remy S, De Weerdt G, Deconinck I, Swennen R, Sági L (2004) An ultrasensitive luminescent detection system in banana biotechnology: from promoter tagging to southern hybridization. In: Mohan Jain S, Swennen R (eds) Banana improvement: cellular, molecular biology, and induced mutations. Science Publishers, Enfield, pp 307–319

Remy S, Thiry E, Coemans B et al (2005) Improved T-DNA vector for tagging plant promoters via high-throughput luciferase screening. Biotechniques 38:763–770

Rombauts S, Florquin K, Lescot M et al (2003) Computational approaches to identify promoters and cis-regulatory elements in plant genomes. Plant Physiol 132:1162–1176

Roy Choudhury S, Roy S, Das R, Sengupta DN (2008a) Differential transcriptional regulation of banana sucrose phosphate synthase gene in response to ethylene, auxin, wounding, low temperature and different photoperiods during fruit ripening and functional analysis of banana SPS gene promoter. Planta 229:207–223. doi:10.1007/s00425-008-0821-2

Roy Choudhury S, Roy S, ParamitaSaha P, Kumar Singh S, Sengupta DN (2008b) Characterization of differential ripening pattern in association with ethylene biosynthesis in the fruits of five naturally occurring banana cultivars and detection of a GCC-box-specific DNA-binding protein. Plant Cell Rep 27:1235–1249. doi:10.1007/s00299-008-0547-4

Roy Choudhury S, Roy S, Sengupta DN (2009) A comparative study of cultivar differences in sucrose phosphate synthase gene expression and sucrose formation during banana fruit ripening. Postharvest Biol Technol 54(1):15–24. doi:10.1016/j.postharvbio.2009.05.003

Roy Choudhury S, Roy S, Singh S, Sengupta D (2010) Understanding the molecular mechanism of transcriptional regulation of banana Sucrose phosphate synthase (SPS) gene during fruit ripening: an insight into the functions of various cis-acting regulatory elements. Plant Signal Behav 5(5):553–557. doi:10.4161/psb.11092

Roy Choudhury S, Roy S, Singh K, Sengupta DN (2011) Ma-ACS1: a key operator in ethylene biosynthesis in banana – its role and regulation during fruit ripening. Acta Horticult 897:187–194

Sági L, Remy S, Panis B, Volckaert G, Swennen R (1992) Transient gene expression in banana protoplasts. Banana Newslett 15:42

Sági L, Panis B, Remy S et al (1995) Genetic transformation of banana and plantain (*Musa* spp.) via particle bombardment. Biotechnology 13:481–485

Santos Ordóñez EG (2008) Characterization and isolation of T-DNA tagged banana promoters active during *in vitro* regeneration and low temperature stress. Dissertationes de agricultura. Ph.D. thesis 787. KatholiekeUniversiteit Leuven, Belgium. Faculteit Bio-Ingenieurswetenschappen, 188 p

Santos E, Remy R, Thiry E et al (2009) Characterization and isolation of a T-DNA tagged banana promoter active during in vitro culture and low temperature stress. BMC Plant Biol 9:77. doi:10.1186/1471-2229-9-77

Schatz T, Langowski J (1997) Curvature and sequence analysis of eukaryotic promoters. J Biomol Struct Dyn 15:265–275

Schenk PM, Sági L, Remans T et al (1999) A promoter from sugarcane bacilliform badnavirus drives transgene expression in banana and other monocot and dicot plants. Plant Mol Biol 39:1221–1230

Schenk PM, Remans T, Sági L et al (2001) Promoters for pregenomic RNA of banana streak badnavirus are active for transgene expression in monocot and dicot plants. Plant Mol Biol 47:399–412

Schneider TD, Stephens RM (1990) Sequence logos: a new way to display consensus sequences. Nucleic Acids Res 18:6097–6100

Seki M, Narusaka M, Kamiya A et al (2002) Functional annotation of a full-length Arabidopsis cDNA collection. Science 296:141–145

Shahmuradov IA, Gammerman AJ, Hancock JM, Bramley PM, Solovyev VV (2003) PlantProm: a database of plant promoter sequences. Nucleic Acids Res 31:114–117

Shahmuradov IA, Solovyev VV, Gammerman AJ (2005) Plant promoter prediction with confidence estimation. Nucleic Acids Res 33:1069–1076

Shan W, Kuang J, Chen L et al (2012) Molecular characterization of banana NAC transcription factors and their interactions with ethylene signalling component EIL during fruit ripening. J Exp Bot 63(14):5171–5187. doi:10.1093/jxb/ers178

Shan W, Kuang J, Lu W et al (2014) Banana fruit NAC transcription factor MaNAC1 is a direct target of MaICE1 and involved in cold stress through interacting with MaCBF1. Plant Cell Environ 37(9):2116–2127. doi:10.1111/pce.12303

Shekhawat U, Srinivas L, Ganapathi T (2011) MusaDHN-1, a novel multiple stress-inducible SK3-type dehydrin gene, contributes affirmatively to drought- and salt-stress tolerance in banana. Planta 234(5):915–932. doi:10.1007/s00425-011-1455-3

Sivanandan C, Sujatha TP, Prasad AM et al (2005) T-DNA tagging and characterization of a cryptic root-specific promoter in Arabidopsis. Biochim Biophys Acta-Gene Struct Expr 1731:202–208

Smale ST (1994) Core promoter architecture for eucaryotic protein-coding genes. In: Conaway RC, Conaway JW (eds) Transcription: mechanisms and regulation. Raven Press, Ltd., New York, pp 63–81

Smale ST (1997) Transcription initiation from TATA-less promoters within eukaryotic protein-coding genes. Biochim Biophys Acta-Gene Struct Expr 1351:73–88

Smale ST, Kadonaga JT (2003) The RNA polymerase II core promoter. Annu Rev Biochem 72:449–479

Springer PS (2000) Gene traps: tools for plant development and genomics. Plant Cell 12:1007–1020

Sreedharan S, Singh Shekhawat UK, Ganapathi TR (2015) Constitutive and stress-inducible overexpression of a native aquaporin gene (MusaPIP2;6) in transgenic banana plants signals its pivotal role in salt tolerance. Plant Mol Biol 88:41–52. doi:10.1007/s11103-015-0305-2

Struhl K (2001) Gene regulation – a paradigm for precision. Science 293:1054–1055

Sundaresan V, Springer P, Volpe T et al (1995) Patterns of gene action in plant development revealed by enhancer trap and gene trap transposable elements. Genes Dev 9:1797–1810

Sunil Kumar GB, Ganapathi TR, Revathi CJ, Srinivas L, Bapat VA (2005) Expression of hepatitis B surface antigen in transgenic banana plants. Planta 222:484–493

Swennen R, Arinaitwe G, Cammue BPA et al (2003) Transgenic approaches for resistance to *Mycosphaerella*leaf spot diseases in *Musa* spp. In: Mycosphaerella leaf spot diseases of bananas: present status and outlook. Proceedings of the 2nd international workshop on *Mycosphaerella* leaf spot diseases of bananas. L. Jacome, P. Lepoivre, D. Marin, R. Ortiz, R. Romero and J.V. Escalant, editors. San José, Costa Rica, 20–23 May 2002. INIBAP, Montpellier, France. pp. 209–238

Tang Y, Kuang J, Wang F et al (2013) Molecular characterization of PR and WRKY genes during SA- and MeJA-induced resistance against *Colletotrichum musae* in banana fruit. Postharvest Biol Technol 79:62–68. doi:10.1016/j.postharvbio.2013.01.004

Teeri TH, Herrera-Estrella L, Depicker A, Van Montagu M, Palva ET (1986) Identification of plant promoters *in situ* by T-DNA-mediated transcriptional fusions to the *npt*-II gene. EMBO J 5:1755–1760

Thompson JF, Hayes LS, Lloyd DB (1991) Modulation of firefly luciferase stability and impact on studies of gene regulation. Gene 103:171–177

Topping JF, Wei WB, Lindsey K (1991) Functional tagging of regulatory elements in the plant genome. Development 112:1009–1019

Van Leeuwen W, Hagendoorn MJM, Ruttink T et al (2000) The use of the luciferase reporter system for *in planta* gene expression studies. Plant Mol Biol Report 18:143a–143t

Vanhove AC, Vermaelen W, Swennen R, Carpentier SC (2015) A look behind the screens: characterization of the HSP70 family during osmotic stress in a non-model crop. J Proteomics 119:10–20. doi:10.1016/j.jprot.2015.01.014

Venter M, Botha FC (2004) Promoter analysis and transcription profiling: integration of genetic data enhances understanding of gene expression. Physiol Plant 120:74–83

Verkhusha VV, Kuznetsova IM, Stepanenko OV et al (2003) High stability of Discosoma DsRed as compared to Aequorea EGFP. Biochemistry 42:7879–7884

Vishnevetsky J, White TL Jr, Palmateer AJ et al (2011) Improved tolerance toward fungal diseases in transgenic Cavendish banana (Musa spp. AAA group) cv. Grand Nain. Transgenic Res 20:61–72. doi:10.1007/s11248-010-9392-7

Wang XL, Peng XX (2001a) Cloning of promoter of banana fruit-specific ACC synthase gene and primary study on its function. Chin J Biotechnol 17:293–296

Wang XL, Peng XX (2001b) Cloning of promoter of banana fruit ripening-related ACO1 and primary study on its function. Chin J Biotechnol 17:428–431

Wood KV (1995) Marker proteins for gene expression. Curr Opin Biotechnol 6:50–58

Wu CH, Madabusi L, Nishioka H et al (2001) Analysis of core promoter sequences located downstream from the

TATA element in the hsp70 promoter from *Drosophila melanogaster*. Mol Cell Biol 21:1593–1602

Xiao Y, Chen J, Kuang J et al (2013) Banana ethylene response factors are involved in fruit ripening through their interactions with ethylene biosynthesis genes. J Exp Bot 64(8):2499–2510. doi:10.1093/jxb/ert108

Yamamoto YY, Obokata J (2007) PPDB: a plant promoter database. Nucleic Acids Res 36:D977–D981

Yamamoto YY, Ichida H, Matsui M et al (2007a) Identification of plant promoter constituents by analysis of local distribution of short sequences. BMC Genomics 8:67

Yamamoto YY, Ichida H, Abe T et al (2007b) Differentiation of core promoter architecture between plants and mammals revealed by LDSS analysis. Nucleic Acids Res 35:6219–6226

Yang IC, Iommarini JP, Becker DK et al (2003) A promoter derived from taro bacilliform badnavirus drives strong expression in transgenic banana and tobacco plants. Plant Cell Rep 21:1199–1206

Zhu Q, Dabi T, Lamb C (1995) TATA box and initiator functions in the accurate transcription of a plant minimal promoter *in vitro*. Plant Cell 7:1681–1689

Enhancing Abiotic Stress Tolerance 12

Anjana Rustagi, Shashi Shekhar, Shalu Jain,
Deepak Kumar, and Neera Bhalla Sarin

Abstract

The banana is one of the most important staple fruit crop which feed millions of people around the globe. In the era of climate change, banana cultivation and production is seriously hampered by different biotic and abiotic constraints. Abiotic stresses arise from adverse environmental conditions which can hinder yield stability every cropping season, thus reducing food security adversely. Improved genetic material and breeding methods can help in solving the abiotic stress, challenges such as drought, salinity and temperature stress. Scientific advancements in the twenty-first century through advanced molecular technology and approaches have been made for a robust banana improvement programme. The development of climate resilient crops can be achieved by combining traditional breeding, tissue culture, and biotechnological approaches. This chapter reviews some of the basic aspects related to genetic transformation of banana along with some advanced techniques for further improvement.

A. Rustagi
Department of Botany, Gargi College, University of Delhi, New Delhi 110049, India

S. Shekhar • N.B. Sarin (✉)
School of Life Sciences, Jawaharlal Nehru University, New Delhi 110067, India
e-mail: neerasarin@rediffmail.com

S. Jain
Department of Plant Pathology, North Dakota State University, Fargo, ND 58102, USA

D. Kumar
Department of Bioscience and Biotechnology, Banasthali University, Rajasthan 304022, India

S. Mohandas, K.V. Ravishankar (eds.), *Banana: Genomics and Transgenic Approaches for Genetic Improvement*, DOI 10.1007/978-981-10-1585-4_12

Embracing modern biotechnological tools combined with tissue culture and breeding techniques is imperative for developing abiotic stress-tolerant/stress-resistant banana.

Keywords
Banana • Genetic engineering • Transformation • *Agrobacterium* • Abiotic stress • Molecular markers

12.1 Introduction

The present scenario of global climate change seriously hampers the agroecosystems through changes in the key variables, such as temperature, pattern of rainfall, drought and water logging. These changes not only affect crop productivity but are also responsible for the decline in the food quality. Wheeler and von Braun (2013) suggested that, in the coming decades, the climate change will result into shrinkage of food availability along with reduced accessibility and stability of food supply around the globe. Banana and plantains are important fruit crops for food security, and they may prove to be a useful resource for livelihood and economic stability under poor environmental conditions. Bananas and plantains (*Musa spp.*) are the perennial herbs which were grown on 10.6 million ha in 2011 with an average fruit yield of 13.6 tha^{-1} (FAO 2013). There are about 1000 dessert and cooking cultivars of banana and plantain reported which have been developed through inter and intraspecific hybridization of wild diploid (*Musa acuminata* and *Musa balbisiana*) ancestors. Yield is a prime trait which breeders and farmers use for the selection of new varieties. However, various environmental stresses damage banana crop year after year and reduce the yield. Abiotic stress, in particular drought, salinity and low temperature, can cause major yield reductions. From the ancient times, moulding and manipulation of available natural resources have been going on to fulfil the field demands, and it is the hallmark of different civilizations. Genomics research related to *Musa* species has developed rapidly since the year 2000. 'The Global Musa Genomics Consortium' coordinates the research in various aspects of genomics and domestication of the banana. The major aim of the consortium is to assure the sustainable production of banana by integrating genetic and genomic approaches. In the current century, the advancement in genetic concepts immensely helped to look into the crop design and enable breeders and genomic scientists to find and evaluate those genes which are responsible for agronomic traits. Towards this pursuit some of the research groups utilize the advanced genetic technologies to develop cultivars with beneficial agronomic traits which encompass biotic and abiotic stress tolerance/resistance. On the other hand, the germplasm collections provide valuable resources for search for agronomically important traits along with conservation of biodiversity. The combination of scientific knowledge and skill to exploit the gene pool opened up a new vista of opportunities in the development of *Musa* genomics for the future. This chapter focuses on the progress towards the development of stress-tolerant banana cultivars through genetic manipulation and tissue culture based on *in vitro* approaches for sustainable agriculture and food security.

12.2 Breeding in Banana: Present and Future

Conventional breeding is time consuming and tedious for banana owing to inherent parthenocarpy, polyploidy, sterility, long life cycle, etc. The oldest breeding programme in banana is the Fundacion Hondureña de Investigacion Agricola (FHIA, Honduras). The FHIA initially focused on breeding dessert bananas but later on also started breeding in plantains (Ortiz and Swennen

2014). Ortiz (2013) also reported that the cultivars (mostly triploids or $2n=3\times=33$) show fruit parthenocarpy and sterility, which makes the crossbreeding of this crop challenging. According to MusaNet (2012), the genetic base of triploid banana and plantain cultivars appears narrow due to their developmental programme which involves 20–25 meiosis events which make them very fragile in comparison to wild species and diploid cultivars of banana. Though a few reports on the use of breeding in the development of biotic stress-tolerant banana plants are available, there are no reports noteworthy available for the generation of abiotic stress-tolerant banana through crossbreeding. Today, breeding in diploid banana remains an important activity for genetic enhancement of banana (Amorim et al. 2011). Some of the advanced techniques, such as marker-assisted breeding, doubled haploid production and next-generation sequencing, are likely to accelerate the improvement of the diploid cultigen pool. Intermediate diploid-breeding sources may enhance genetic gains in different cultigens of banana and plantain. The chromosome doubling of diploid species and further use of their derivative(s) to produce triploids (by 4×–2× crossing) could be another approach for incorporating diversity in the breeding pool (Goigoux et al. 2013; Tomepke and Sadon 2013). Moreover, the sterility in the breeding lines is a problem, but Jenny et al. (2013) indicated that fertility could be restored in the breeding lines (4×) by colchicine treatment to their respective ancestors. Based on these reports, it can be concluded that a multidisciplinary approach can lead to overcoming different hurdles of development of abiotic stress-tolerant plants.

12.3 Role of Cell/Tissue Culture in Improvement of Banana

Cell/tissue culture is not only used in mass multiplication of banana and plantain but also plays a crucial role in crop improvement through the production of desirable variants. Together cell/tissue culture and genetic engineering can successfully be used to incorporate valuable agronomic traits in banana and plantain. The biotechnological research along with cell/tissue culture and omics approaches could be utilized simultaneously for the production of abiotic stress-tolerant banana and plantains. To incorporate a valuable and important agronomic trait in the *Musa* genome, development of explants is a prerequisite which is dependent on the nature and type of source tissue or cells. Several research findings reported different types of sources of explant along with their respective merits and demerits. In vitro regeneration of banana has the potential to provide genetically uniform and disease-free planting material apart from being the prerequisite for the improvement of crop by genetic transformation. Regeneration in banana has been well documented in literature by organogenesis, embryogenesis and through protoplast cultures as discussed in this chapter.

12.3.1 Organogenesis

Different types of explants have been used for initiating in vitro cultures through organogenesis in banana, namely, shoot tips, zygotic embryos, immature male flowers, leaf sheaths and corms. Each of the varieties of explants used in tissue culture provides its own advantages and disadvantages. Shoot tips are widely used as starting material because they are simple and easy to culture from a wide range of *Musa* genotypes (Banerjee and De Langhe1985; Vuylsteke 1989). Shoot tips do not require access to the field and are independent of flowering season, unlike other methods. Leaf sheaths and corms also have the advantage that the methodologies can be applied to many banana varieties irrespective of the genotype. Regeneration from leaves and corm slices has been reported by Okole and Schulz (1996) and Venkatchalam et al. (2006). Male inflorescence, when utilized as an explant, reduces the rate of contamination compared to suckers and produces more number of shoots in culture conditions. Okole and Schultz (1996) used micro-cross sections of leaf segments and reported the formation of adventitious shoot and callus.

12.3.2 Somatic Embryogenesis

In bananas and plantains, an embryogenic callus can be induced from a suitable explant, in a semi-solid medium. Embryogenic cell suspension (ECS) with a high regeneration capacity is prepared by transferring embryogenic calli into liquid medium (Fig. 12.1). The ECS can be used for mass clonal propagation and to regenerate protoplasts in banana (Strosse et al. (2004). More importantly being single celled, they are the explants of choice for inducing mutations and for genetic transformation experiments. The ECS could be initiated from different explant sources, e.g. zygotic embryos, rhizome slices and leaf sheaths, immature male/female flowers, multiple meristem cultures (scalps) and bracts. Recently Nurain et al. (2014) developed the most efficient banana regeneration system for ECS derived from immature male-flower clusters by including L-proline and L-glutamine in embryo development media.

The use of highly differentiated rhizome and leaf tissues for induction of embryogenic calli has not been explored effectively. The Male-flower technology uses the starting material from flowering plants (Escalant et al. 1994). However, this method has its own limitations of poor embryogenic response of male flowers soon after harvest, seasonal dependence, field work and is not applicable in cultivars which do not produce male flowers. Female flowers can be used in place of male flowers; however, it is not useful on a large scale. On the other hand, scalps, which are highly compact meristem cultures derived from proliferating meristem cultures, are very promising explants (Fig. 12.2).

Establishing ECS in banana has several limitations. An elaborate material preparation phase is required for obtaining scalps, which are generally used as the explants for induction of embryogenesis. Embryogenic responses are often very low and vary for different varieties and cultivars. The response of the embryogenic calli to give rise to regenerable cell suspensions is also very low, which is about 20–50 %. Consequently, the synchronized cell suspensions have to be maintained for long periods. The longer generation and maintenance time of these cultures in liquid medium proportionately increases their chance of contamination. Thus the ECS cultures need to be checked periodically for bacterial contamination, if any. The regeneration potential of cells in suspension decreases with time. The cell suspension-derived plants may show somaclonal variation, which may range from 2 % to 99 %. Since the establishment of *Musa* ECS is labour intensive and time consuming and the embryogenic capac-

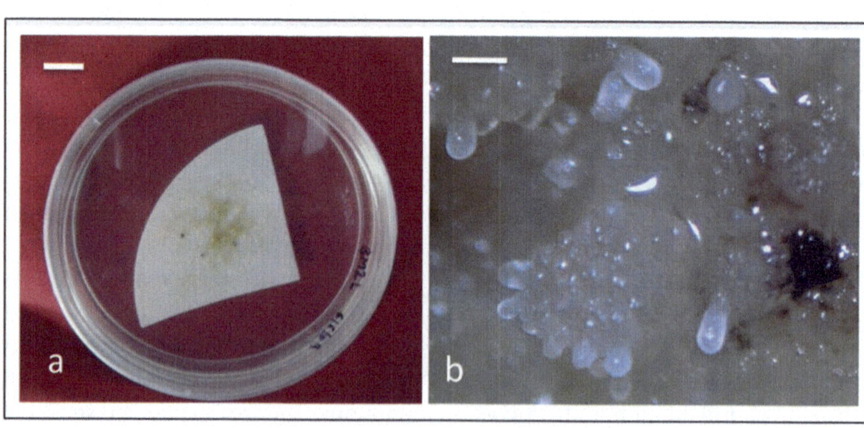

Fig. 12.1 (**a**) Embryogenic cell suspension (ECS) of *Musa* sp. var. Matti cultured on regeneration medium (*bar* = 1 cm) and (**b**) torpedo-stage embryos regenerating from ECS (*bar* = 2 mm)

Fig. 12.2 Different stages of regeneration of scalps of *Musa* sp. var. Matti: (**a**) scalp formation (*bar*=1 cm), (**b**) regeneration of multiple shoots from scalps (*bar*=1.2 cm), (**c**) rooting of shoots (*bar*=1,2 cm), (**d**) hardening of regenerated plantlets (*bar*=8 cm), and (**e**) hardened plants in the glasshouse (*bar*=10 cm)

ity is lost over time, cryopreservation of a small amount of the ECS will be helpful (Panis et al. 2004).

12.3.3 Protoplast Culture

Protoplasts represent a single cell system which has potential to regenerate plants. This method has been used for the development of transgenic banana plants by the incorporation of transgene using electroporation (Sagi et al. 1994). There are different research groups who isolated protoplasts from explants such as inflorescence, basal tissue of the youngest leaf, etc. The best results were obtained when embryogenic cells and cell suspensions were used as a source of explant (Matsumoto and Oka 1998).

12.3.4 Somaclonal Variation

A genetic change in the plants that occurs naturally during conventional as well as *in vitro* propagation is referred to as somaclonal variation. While the incidence of somatic mutations is low in bananas propagated conventionally, it is frequent in micropropagated material and often constrains regeneration by somatic embryogenesis (Strosse et al. 2006). Somatic mutations are limited to nonreproductive cells, particularly when the regeneration follows the callus phase. These variations are associated with a mutation like point mutation, gene duplication and rearrangement of chromosomes along with changes in chromosomal complements (Babita et al. 2013). Stress in tissue culture is the most common cause of this variation in banana and plantains, and it is more frequent when the plants are propagated by meristematic cells. Several improved banana cultivars have originated through somaclonal variation (Heslop-Harrison and Schwarzacher 2007). Therefore, it can be concluded that somaclonal variation is important for genetic improvement of banana. Somaclonal variants of banana might be able to withstand different types of abiotic stresses.

12.3.5 Mutagenesis

Mutagenesis is a traditional method for inducing novel genetic variation in both sexually and asexually propagated species. Mutation-derived varieties have been released in a wide range of plant species including banana (Ahloowalia et al. 2004). It can be a non-transgenic alternative to create genetic modification in crops. The most commonly reported chemical mutagen is ethyl methanesulphonate (EMS). It induces primarily GC-AT base-pair transition mutations (Greene et al. 2003). To develop a platform suitable for vegetatively propagated species, Jankowicz-Cieslak et al. (2012) treated banana shoot apical meristems with EMS. The plantlets were recovered and screening for induced mutations was performed. A large number of GC-AT transition mutations were identified and were stably inherited in subsequent generations. Therefore, genotypic insights into the fate of totipotent cells after mutagenesis were investigated, and banana improvement through induced mutations was suggested.

12.4 Development of Abiotic Stress-Tolerant Banana Through Tissue Culture

Tissue culture is a reliable technique, which has enabled the evaluation of tolerance to environmental stresses because it allows manipulation of plants *in vitro*. Said et al. (2015) used tissue culture-based *in vitro* selection method to develop drought stress-tolerant banana plantlets. This study concluded that pretreating the plantlets with low concentration of trehalose is most effective in combating drought stress. They artificially induced polyethylene glycol (PEG)-mediated drought stress by increasing the strength of PEG (by 1, 2 and 3 %). To combat the PEG-mediated drought stress, the plantlets were pretreated with different concentrations (20, 60 and 100 mM) of trehalose to test the best suited concentration which showed the most prominent effect in resisting the drought stress. Several research findings suggested that trehalose is a putative signalling molecule while others assumed its direct role in stress tolerance by their physical interaction with the membrane lipids, which maintain the structural integrity of the cell wall during stress (Iondachescu and Imai 2008). In 2010, Stolker hypothesized in his thesis report that trehalose performs the protective function during stress by scavenging of reactive oxygen species (ROS).

In another tissue culture-based *in vitro* study, Bidabadi et al. (2012) attempted to select and characterize the water stress-tolerant lines of *Musa* spp., with AAA genome through ethyl methane sulphonate (EMS) based mutagenesis. In this study, PEG at three different concentrations (10, 20 and 30 g L^{-1}) was used to induce water stress followed by mutagenesis of shoot tips. They demonstrated that the tolerant lines showed higher fresh weight and proliferation rate as compared to the control plants under water stress condition. The tolerant lines also showed significant increase in the biomass as compared to the control plants with higher proline content. Significant decrease in the MDA, H_2O_2 content and water loss from leaves of mutated plantlets was observed as compared to non-mutated lines after PEG treatment.

Mahmood and his co-workers (2012) reported the enhancement of water stress tolerance in the shoot tip cultures of banana (*Musa acuminata* cv. 'Berangan', AAA). This was achieved by the exogenous application of methyl jasmonate (MeJA) prior to imposition of PEG-induced water stress. They used various concentrations (5–160 μM) of MeJA before and during the PEG treatment. Significant improvement in the proliferation rate was achieved with increasing level of MeJA up to 80 μM in comparison to the untreated plants. These findings suggest that MeJA treatment leads to the restriction of PEG-mediated water stress in banana possibly through the suppression of oxidative damage which was evident from the significant reduction in H_2O_2 and MDA content of banana plantlets. Therefore, manipulation of pathways involving MeJA may result in enhanced drought tolerance in banana as indicated in this study.

12.5 Development of Abiotic Stress-Tolerant Banana Through Genetic Manipulation

There are a few reports on the development of abiotic stress-tolerant banana through genetic manipulation. The genes used for this purpose have been listed in Table 12.1. However, with rapid advances in the genomics for resistance gene identification and cloning in banana and other species, more useful genes in banana will be successfully transferred. One of the plant-specific hydrophilic transcriptional regulator proteins (abscisic acid stress and ripening-induced protein, ASR) is involved in sucrose stress, salt stress and wounding in banana (Miao et al. 2014; Henry et al. 2011). Henry et al. (2011) reported four members of *Musa Asr* gene, and all these exhibited two structural patterns, i.e. they had two exons separated by one intron and have 'ABA/WDS' (abscisic acid/water-deficit stress) domain. Two members of *Asr* gene family possessed two versions which corresponded to the two sub-genomes of *Musa acuminata* and *Musa balbisiana*. However, their role in salt tolerance in banana is not yet known. Miao et al. (2014) evaluated the role of banana ASR1 in salt stress tolerance by overexpressing it in *Arabidopsis*. The results of this study showed that *MaASR1* was preferentially expressed in roots and leaves at 4 h of stress and the expression gradually decreased later. There was a low expression of *ASR* which was evident in fruits, rhizomes and flowers as compared to the root and leaves. Overexpression of *MaASR1* induces salt stress tolerance in *Arabidopsis* by a reduction in the expression of the ABA/stress-responsive genes without affecting the expression of either ABA-independent pathway or biosynthesis pathway genes. Therefore, the *MaASR1* could be a promising candidate gene, the overexpression of which led to enhanced salinity tolerance as well as tolerance to drought and/or abscisic acid-mediated stresses in banana and plantain. As an important gene, it can be further validated and intensively characterized by identification of its targets. Functional characterization of *MaASR1* will improve our understanding of the molecular mechanisms of interaction of *MaASR1* in enhancement of salt stress tolerance.

The plants manifest various cellular defence mechanisms during stress conditions. One of the mechanisms to resist the stress is reduction of cellular water potential. The cell produces certain proteins, enzymes, metabolites, etc. to tide over the unfavourable conditions. Dehydrin (DHN) is a multifamily of hydrophilic proteins present in plants that is produced in response to water and cold stress. Shekhawat et al. (2011) identified and characterized a SK-3-type dehydrin (*MusaDHN-1*) gene and further overexpressed it in banana to check its efficacy towards abiotic stress tolerance. By the study of expression profile, they found that *MusaDHN-1* was highly expressed in leaves under several abiotic stresses such as drought, salinity, cold, oxidative and heavy metal stress. Signalling molecules like abscisic acid, ethylene and methyl jasmonate treatment also induced the expression of *MusaDHN-1*. The accumulation of proline was higher, while the level of malondialdehyde was lower under stress condition in transgenic lines in comparison to the control plants. Therefore, it can be concluded that MusaDHN-1 affirmatively helps in salinity and drought stress tolerance in banana. This can help in harnessing the production of stress-tolerant banana and plantains.

Another group of proteins called the aquaporins are membrane channel proteins and central players in plant water relations which facilitate transport of water and small neutral solutes (Maurel et al. 2008). In response to different environmental stimuli, aquaporins regulate water transport. To elucidate the role of aquaporins in

Table 12.1 List of genes used in genetic engineering for abiotic stress tolerance in banana

S.N.	Gene	Source	Gene's characteristics	Transformation method	Stress targeted and result	References
1	Asr	*Musa paradisiaca*	Abscisic acid, stress, ripening induced	Floral dip/*Agrobacterium* mediated	Salinity tolerance in *Arabidopsis*	Miao et al. (2014)
2	MusaPIP1;2	Banana	Water channel proteins	Embryogenic cells/*Agrobacterium* mediated	Cold, drought and salinity tolerance	Sreedharan et al. (2013)
3	Musa SAP1	Banana	Stress-associated protein	Embryogenic cells/*Agrobacterium* mediated	Drought and salinity tolerance	Sreedharan et al. (2012)
4	MusaDHN-1	Banana	Dehydrin protein	Embryogenic cells/*Agrobacterium* mediated	Drought and salinity tolerance	Shekhawat et al. (2011)
5	PR10	Groundnut	Pathogenesis-related protein	Scalp/*Agrobacterium* mediated	Drought and salinity tolerance	Rustagi et al. (2015)
6	WRKY	Banana	Transcription factor	Embryogenic cells/*Agrobacterium* mediated	Oxidative and salinity tolerance	Shekhawat and Ganapathi (2013)
7	MusaPIP2;6	Banana	Aquaporin gene	Embryogenic cells/*Agrobacterium* mediated	Salinity tolerance	Sreedharan et al. (2015)
8	MaPIP1;1	*Musa acuminata*	Aquaporin gene	Floral dip-mediated infiltration/*Agrobacterium* mediated	Drought and salinity tolerance	Xu et al. (2014)
9	MusabZIP53	Banana	bZIP transcription factors	*Agrobacterium* mediated	Enhanced drought, salt and cold tolerance with stunted plants	Shekhawat and Ganapathi (2014)
10	P5CS	Banana	Δ1-pyrroline-5-carboxylate synthetase for proline accumulation	Microprojectile bombardment	Salt stress tolerance	Ismail et al. (2005)
11	tps-tpp	*Saccharomyces cerevisiae*	trehalose-6-phosphate synthase (tps) and trehalose-6-phosphate phosphatase (tpp)	*Agrobacterium* mediated	Salt stress tolerance	Santamaria et al. (2009)

abiotic stress, Sreedharan et al. (2013) cloned and characterized a plasma membrane intrinsic protein gene (*MusaPIP1;2*) from banana. Their finding showed that MusaPIP1;2 protein is localized in plasma membrane of transformed banana cells. The constitutive expression led to better abiotic stress tolerance by lowering of malondialdehyde levels but elevating the level of proline, relative water content and photosynthetic efficiency as compared to their control counterparts under imposed abiotic stress treatments. Similar stress tolerance levels were observed in *in vitro* and in *in vivo* assays when *MusaPIP1;2* was overexpressed under dehydrin protein promoter (Sreedharan et al. 2013). In a similar study, PIP1 subfamily *AQP* (*MaPIP1; 1*) gene from banana was cloned and overexpressed in *Arabidopsis* plants (Xu et al. 2014). The localization study of *MaPIP1;1* was done through its fusion with green fluorescent protein (GFP), and it was confirmed that MaPIP1;1 was localized in the plasma membrane in transformed plants. The transgenic *Arabidopsis* plants showed an increment in primary root elongation, root hair numbers and survival rate under imposed salinity and drought stresses. The cytosolic ratio of K^+/Na^+ was also high in the transgenic plants as compared to the control plants which showed that *MaPIP1;1* is directly responsible for imparting stress tolerance by adjustment of cellular osmotic potential in the transgenic lines. In continuation of the previous study on aquaporin, Sreedharan et al. (2015) identified another aquaporin gene, *MusaPIP2;6*, by comparing EST data sets of stressed and non-stressed banana tissue. This gene has putative function in salt stress signalling in banana. It was further characterized by overexpressing in banana plants. It was evident that the overexpression of *MusaPIP2;6* either under constitutive promoter or under inducible promoter resulted in higher salinity tolerance in transgenic banana (Sreedharan et al. 2015). They also supported the findings of Xu and co-workers (2014) about the cellular localization of aquaporins in the plasma membrane. The banana plants overexpressing *MusaPIP2;6* had better photosynthetic efficiency and lower membrane damage under imposed saline stress. The aquaporins are also involved in the regulation of complex metabolic processes as well as in certain physiological and developmental processes, including seed germination, stomatal movement and leaf water transport (Maurel et al. 2008). In continuation of the previously reported role of aquaporin (AQP) in water stress tolerance, recently Hu and co-workers (2015) identified 47 banana *AQP* genes (*MaAQPs*) based on the banana genome sequence which were clustered into four subfamilies. AAQP-like or major intrinsic protein (MIP) domains were found conserved in all banana AQPs. The MaAQPs had highly specific intronic sequences of two to four introns for each subfamily. They found that *MaAQP* genes were also expressed during fruit development and postharvest ripening; a number of *MaAQP* genes were strongly induced under stress conditions.

The pathogenesis-related class 10 (PR-10) genes are one of the integral components of the plant defence system and have been shown to be transcriptionally responsive across a large range of abiotic stress conditions such as drought, salinity, heat and cold stresses, heavy metals, wounding and UV exposure (Xie et al. 2010; Takeuchiet al. 2011; Jain et al. 2012). Several proteins with similarities to the PR-10 family members were shown to be up-regulated in peanut callus cultures exposed to salt stress conditions, in a study undertaken by Jain et al. (2006). The overexpression of a peanut salinity-induced PR-10 gene (*AhSIPR10*) in tobacco exhibited enhanced tolerance to salt, heavy metal ($ZnCl_2$) and mannitol-induced drought stress (Jain et al. 2012). Xie et al. (2010) reported that two PR10 proteins showed higher abundance in maize, which resulted in tolerance to multiple abiotic stresses including wounding and biotic stresses. Towards development of stress-tolerant banana, Rustagi et al. (2015) overexpressed salinity-induced pathogenesis-related class 10 protein gene from *Arachis hypogaea* (*AhSIPR10*) in indigenous banana cv. Matti (AA). The role of *AhSIPR10* in abiotic stress tolerance was evident from experiments where transgenic banana over-expressing *AhSIPR10* performed better when exposed to NaCl-induced salinity and mannitol-induced drought stress (Rustagi et al. 2015). The transgenic plants showed higher membrane stability under stress conditions with less accumulation of electrolytes and higher chlorophyll retention than the control plants. Despite the ubiquitous presence of PR10 proteins across several plant species, their involvement in combating both abiotic and biotic stress conditions sensed by plants and their functional mechanism is still unclear. Some of the PR-10 proteins exhibiting RNase activity inhibit the growth of pathogens through direct cytotoxic impact on pathogens possibly by participating in the induction of plant cell apoptosis and development of hypersensitive reactions (Filipenko et al. 2013). This needs to be explored and understood in diverse stress conditions and stress defence signalling network.

In addition Goulas et al. (2007) demonstrated that vegetative storage protein (VSP), a homolog of PR-10 protein, exhibited *in vitro* cryoprotective role in autumn and winter conditions which may endow chilling stress tolerance in plants.

Plant stress-associated proteins (SAP) containing A20/AN1 zinc finger domains are associated with different stress response pathways. In their study, Sreedharan et al. (2012) identified a *SAP* gene (*MusaSAP1*) from banana and characterized it by overexpressing it in banana plants. They concluded that the expression profile of *MusaSAP1* gene was up-regulated in salinity, drought, thermal and oxidative stress as well as by treatment with abscisic acid (ABA). Cellular localization study conclusively emphasized that MusaSAP1 protein is incompletely translocated to the nucleus. The stress endurance characteristic of transgenic banana plants was better as indicated by reduced malondialdehyde levels in transgenic leaves subjected to drought and salt as compared to the control counterparts in both *in vitro* and *in vivo* assays. The transgenic banana plants also displayed strong up-regulation of a polyphenol oxidase (PPO) transcript by wounding and methyl jasmonate treatment. The outcome of this study showed a positive role of *MusaSAP*1 in stress amelioration pathways of banana.

Apart from plant-associated proteins, several transcription factors have been identified, the overexpression of which in heterologous systems resulted in abiotic stress tolerance in several plants. The WRKY transcription factors are involved in the transcriptional reprogramming of various key processes, including abiotic or biotic stress conditions in plants (Rushton et al. 2010). Shekhawat and Ganapathi (2013) expressed *MusaWRKY71* in banana and found that the transgenic banana had enhanced tolerance against oxidative and salt stress, which was evident from better photosynthetic efficiency and higher membrane stability. The quantitative RT-PCR analysis for expression of putative downstream genes of transgenic expressing *MusaWRKY71* showed differential expression patterns of WRKY family, PR proteins family and chitinase family genes. These findings suggest that *MusaWRKY71* is involved in the transcriptional reprogramming of diverse stress responses in banana.

There are other factors involved in regulating gene expression. These include microRNAs or miRNAs which are noncoding RNAs and are involved in post-transcriptional regulation of gene expression. These fascinating molecules are fine-tuned with plant growth and development and also perform a pivotal role under diverse biotic and abiotic stresses. Although plant miRNAs have been extensively studied in certain model and crop plant systems, there are hardly reports available on their role in fruit crops such as banana. To explore the horizon of miRNAs in gene regulation, recently Ghag et al.(2015) investigated the small RNA expression profiles of two economically important banana cultivars (Grand Naine and Rasthali). A total of 170 and 244 miRNAs were identified in cv. Grand Naine and cv. Rasthali, respectively. To validate the role of miRNAs, they overexpressed *MusamiRNA156* (a common miRNA in both cultivars) in banana. The transgenic plants overexpressing the *MusamiRNA156* showed stunted growth with alteration in leaf anatomy. This research provides a foundation for further investigation at the omics level to search for factors which might be operational in the development of this peculiar morphology.

12.6 Modern Technologies for Improvement of Banana Under Abiotic Stress: Future Perspective

Increased production of bananas and plantains under abiotic stresses is a major challenge. In the last few decades, scientists have successfully developed new molecular technologies to help in improving crop varieties at a faster pace. Recent developments in genomics, high-throughput sequencing and phenotyping platforms have changed conventional breeding to molecular breeding. The progress in breeding of plantain and banana has been hindered due to complex genetic background of *Musa* species. However, the new technologies have emerged as having a great promise for banana improvement.

12.6.1 Association Mapping and Marker-Assisted Breeding

Genetics and genomics studies help in identification of genes related to useful traits. Genetic mapping in *Musa* has been restricted due to lack of large segregating, rendering poor male and female fertility and low level of seed viability. Therefore, quantitative trait loci (QTLs) analysis with large mapping populations of F_2 or recombinant inbred lines, backcross or doubled haploid plants is not feasible. However, association mapping (also called linkage disequilibrium mapping) may be an alternative approach for identification of genes related to the trait of interest and valuable alleles in *Musa*. In association mapping, a set of accessions is genotyped with large number of genetic markers, and phenotypic data are collected for different traits. Genes which are associated with important traits are identified by marker-trait association analysis. Germplasm banks a diversity of *Musa* accessions which can be used in marker-trait association analysis.

The *Musa* Germplasm Information System (MGIS) (http://www.crop-diversity.org/mgis/) is a key source of *Musa* germplasm diversity containing 2330 accessions managed in 6 collections around the world. It contains valuable information on passport data, botanical classification, morpho-taxonomic descriptors, molecular studies, plant photographs and GIS information on these banana genetic resources. Over 1200 *Musa* accessions are deposited and maintained in the genebank at the Biodiversity International *Musa* Germplasm Transit Centre (ITC) in Belgium. Screening of banana germplasm for abiotic stress resistance needs to be conducted on priority basis so that accessions with better gene pool can be used for developing improved varieties of banana. Data collected from multiple environments will provide an opportunity for modelling 'stress impacts' on crops (Kole et al. 2015). The collections held in MGIS may play a role in candidate gene discovery. Once genes controlling stress resistance have been mapped, identification of different alleles available at that locus can be achieved. Genetic variability present in the sequence of the gene may help in the identification of the best alleles for stress resistance that can be used for future banana improvement programmes.

Marker-assisted breeding in banana will be advantageous over conventional breeding due to relatively long life cycle of banana. The key to successful integration of marker-assisted breeding into conventional breeding programmes will depend on identifying markers which are closely associated with agronomic traits (Muthamilarasan et al. 2013). Molecular markers are a powerful tool to reveal genetic variability through DNA analysis and have been widely used in different species for assessing genetic diversity (Mondini et al. 2009). Different PCR-based genetic markers, including amplified fragment length polymorphism (AFLP), random amplified polymorphic DNA (RAPD) and simple sequence repeats (SSR), have been successfully applied in genetic diversity and phylogenetic studies of banana (Opara et al. 2010; Carreel et al. 2002; Creste et al. 2003). Youssef et al. (2011) used the sequence-related amplified polymorphism (SRAP) technique for accessing genetic diversity and phylogenetic relationships among 40 *Musa* accessions. Cheung and Town (2007) identified 352 SSR loci in *M. acuminata* Calcutta 4 bacterial artificial chromosome (BAC) end sequences. Approximately 1500 diversity array technology (DArT) markers were developed in *Musa* species by the Generation Challenge Program, which is applicable to develop dense genetic maps (Kilian 2007). A genetic map with 11 linkage groups consisting 167 SSRs and 322 DArT markers spanning 1197 cM was developed in an F_1 progeny developed from a cross between 'Borneo' and 'Pisang Lilin', two genetically distinct accessions of *M. acuminata* (Hippolyte et al. 2010). High-throughput technologies based on single nucleotide polymorphism (SNP) or insertion-deletion (indel) offer advantages over other marker systems such as RFLP, RAPD or AFLP because of high abundance and polymorphism (Dillon et al. 2007). Due to advancement of next-generation sequencing (NGS) technologies, genome sequences of many plants are available in GenBank. The banana reference genome sequence (*Musa acuminata*) has been released (D'Hont et al. 2012), and new results were generated from sequence analyses. Droc et al. (2013) developed putative SNPs from the RNA-Seq data of Pahang and DH-Pahang genotypes of *Musa acuminata*. Overall, 2 689 intra-genotype SNPs were generated in Pahang, covering 1 SNP every 1 394 bp of physical map. All the genetic and phenotypic information related to molecular markers and maps, quantitative trait loci, genetic diversity and a short description of genetic resources of banana are stored in the TropGene DB (Hamelin et al. 2013).

Once closely associated markers are identified, they can be used in marker-assisted selection (MAS). The advantage of MAS over conventional plant breeding techniques is that large number of advanced breeding lines can be screened in shorter time period as large field trials and difficult phenotyping procedures can be eliminated if closely associated marker is available. Currently, many MAS projects have been successfully carried out in food crops to increase the resilience to changing climate and various diseases (Varshney et al. 2013; Kole et al. 2015) and will continue in the future to ensure food security. The banana being a new crop for genomic research is lagging behind but hopefully in the future will benefit from MAS.

12.6.2 Tilling

Targeted induced local lesions in genomes (TILLING) is a variant of mutation method which help in creating changes in the gene of interest (Comai and Henikoff 2006). Ecotilling was used for the identification and characterization of SNPs and small insertions/deletions (indels) to find polymorphism in *Musa* species (Till et al. 2010). This method is high throughput and has high accuracy. Gene target regions of about 700–1600 bp size are amplified using fluorescently labelled gene-specific primers for ecotilling using enzymatic mismatch cleavage. Hybridization of polymorphic amplicons after denaturing and annealing generates heteroduplexed molecules. Single-strand-specific nuclease CEL I present in crude extract of celery juice is used to cleave mismatched regions in otherwise double-stranded duplex. Fluorescent-labelled digested products are resolved on denaturing polyacrylamide gel electrophoresis (PAGE) as shown by Till et al. (2006). Eighty banana and plantain accessions were used to discover and characterize nucleotide polymorphisms using the ecotilling approach. Over 800 novel alleles were obtained by tilling 14 gene targets from these accessions.

12.6.3 Genome Editing

Another recent form of mutagenesis, called genome editing or targeted genome engineering is predicted to have great success in generating new varieties in the near future (Voytas and Gao 2014). Targeted double-strand DNA breaks are generated in the genome at or close to place of gene target using sequence-specific nucleases such as zinc finger nucleases (ZFNs) and transcription activator-like effector nucleases (TALENs). The DNA break can be exploited to introduce desirable changes in the specific DNA such as insertion or deletion and repaired by non-homologous end joining (NHEJ) or homologous recombination (HR). Genome editing has potential of genetic engineering into elite varieties of crops in an efficient and precise manner. Recently, additional method of genome editing based on the bacterial type II CRISPR (clustered regularly interspaced short palindromic repeats)/Cas (CRISPR-associated) immune system has been identified (Belhaj et al. 2013). A customizable small noncoding RNA guides the targeted breakage of genomic DNA in CRISPR/Cas system. This results in DNA modifications by both non-homologous end joining (NHEJ) and homology-directed repair (HDR) mechanisms. Although this method is still in infancy, the plants generated using genome editing technology may have more social acceptance than genetically engineered crops (Ronald 2014).

12.7 Conclusions

Despite being a staple crop, banana cultivation faces many challenges such as sterile nature of cultivated banana varieties, very few sources of resistance and high polyploidy levels. Modern agricultural techniques such as genetic manipulation of plants can be practised for developing abiotic stress-tolerant banana varieties by transferring resistant genes into the banana genome without having an effect on other useful characteristics of a given cultivar. The successful transformation of several genes conferring abiotic stress resistance and high transformation efficiency with improved tissue culture methods has been achieved in various cultivars of banana. From a practical application perspective, further evaluation of transgenic banana possessing abiotic stress tolerance is required. Multilocation field trials of these genetically modified plants will be the next phase of transgenic research. Although there are large collection of accessions of banana, exploration of the factors responsible for conferring abiotic stress tolerance in these is still awaited. Molecular breeding which includes genetic engineering and conventional breeding using DNA markers has potential to breed for stress-tolerant plants in the future. In conclusion, genetically improved varieties of banana can be derived from transgenic technology, chemically induced mutation or recent technologies such as genome editing. These techniques can help in reducing the impact of abiotic stresses in plants and enhancing sustainable agriculture and food security.

Acknowledgement The authors thank Dr. Nishakant Pandey, Dr. Arun Vincent Kisku and Neelam Prabha Negi for their help in preparing the manuscript.

References

Ahloowalia BS, Maluszynski M, Nichterlein K (2004) Global impact of mutation-derived varieties. Euphytica 135:187–204

Amorim EP, Amorim VBO, Silva SO, Pillay M (2011) Quality improvement of cultivated *Musa*. In: Pillay M, Tenkouano A (eds) Banana breeding: progress and challenges. CRC, New York, pp 252–280

Babita, Rattanpal HS, Thakur KK (2013) Recent advances in banana improvement through biotechnological approaches. J Chem Biol Phys Sci 3:1853–1862

Banerjee N, De-Langhe E (1985) A tissue culture technique for rapid clonal propagation and storage under minimal growth conditions of *Musa* (Banana and Plantain). Plant Cell Rep 4:351–354

Belhaj K, Chaparro-Garcia A, Kamoun S, Nekrasov V (2013) Plant genome editing made easy: targeted mutagenesis in model and crop plants using the CRISPR/Cas system. Plant Methods 9:39–49

Bidabadi SS, Meon S, Wahab Z, Subramaniam S, Mahmood M (2012) *In vitro* selection and characterization of water stress tolerant lines among ethyl methane sulphonate (EMS) induced variants of banana (Musa spp., with AAA genome). Aust J Crop Sci 6:567–575

Carreel F, Gonzalez de Leon D, Lagoda P, Lanaud C, Jenny C, Horry JP, Tezenas du Montcel H (2002) Ascertaining maternal and paternal lineage within *Musa* by chloroplast and mitochondrial DNA RFLP analyses. Genome 45:679–692

Cheung F, Town CD (2007) A BAC end view of the *Musa acuminate* genome. BMC Plant Biol 7:29–35

Comai L, Henikoff S (2006) TILLING: practical single-nucleotide mutation discovery. Plant J 45:684–694

Creste S, Neto TA, Silva SO, Figueira A (2003) Genetic characterization of banana cultivars (*Musa* spp.) from Brazil using microsatellite markers. Euphytica 132:259–268

D'Hont A, Denoeud F, Aury JM, Baurens FC, Carreel F, Garsmeur O, Noel B, Bocs S, Droc G, Rouard M, Da Silva C, Jabbari K, Cardi C, Poulain J, Souquet M, Labadie K, Jourda C, Lengellé J, Rodier-Goud M,

Alberti A, Bernard M, Correa M, Ayyampalayam S, Mckain MR, Leebens-Mack J, Burgess D, Freeling M, Mbéguié-A-Mbéguié D, Chabannes M, Wicker T, Panaud O, Barbosa J, Hribova E, Heslop-Harrison P, Habas R, Rivallan R, Francois P, Poiron C, Kilian A, Burthia D, Jenny C, Bakry F, Brown S, Guignon V, Kema G, Dita M, Waalwijk C, Joseph S, Dievart A, Jaillon O, Leclercq J, Argout X, Lyons E, Almeida A, Jeridi M, Dolezel J, Roux N, Risterucci AM, Weissenbach J, Ruiz M, Glaszmann JC, Quétier F, Yahiaoui N, Wincker P (2012) The banana (*Musaacuminata*) genome and the evolution of monocotyledonous plants. Nature 488(7410):213–217. doi: 10.1038/nature11241

Dillon SL, Shapter FM, Henry RJ, Cordeiro G, Izquierdo L, Lee LS (2007) Domestication to crop improvement: genetic resources for *Sorghum* and *Saccharum* (Andropogoneae). Ann Bot 100:975–989

Droc G, Larivière D, Guignon V, Yahiaoui N, This D et al. (2013) The banana genome hub. Database: The Journal of Biological Databases and Curation: bat035. http://doi.org/10.1093/database/bat035. Accessed on 14 Sept 2015

Escalant JV, Teisson C, Cote F (1994) Amplified somatic embryogenesis from male flowers of triploid banana and plantain cultivars (*Musa* spp.). In Vitro Cell Dev Biol Anim 30:181–186

FAO (2013) FAOSTAT: core production data. Food and Agriculture Organization of the United Nations, Rome. Available online at http://faostat.fao.org/site/340/default.aspx. Accessed on 15 Sept 2015

Filipenko EA, Kochetov AV, Kanayama Y, Malinovsky VI, Shumny VK (2013) PR-proteins with ribonuclease activity and plant resistance against pathogenic fungi. Russ J Genet 3:474–480

Ghag SB, Shekhawat UKS, Ganapathi TR (2015) Small RNA profiling of two important cultivars of banana and over-expression of miRNA156 in transgenic banana plants. PLoS ONE 10:e0127179. doi:10.1371/journal.pone.0127179

Goigoux S, Salmon F, Bakry F (2013) Evaluation of pollen fertility of diploid and doubled-diploid clones of Mlali and their potential use for banana breeding. Acta Horticult 986:195–204

Goulas E, Richard-Molard C, Le Dily F, Le Dantec C, Ozouf J, Ourry A (2007) A cytosolic vegetative storage protein (TrVSP) from white clover is encoded by a cold-inducible gene. Physiol Plant 129:567–577

Greene EA, Codomo CA, Taylor NE, Henikoff JG, Till BJ et al (2003) Spectrum of chemically induced mutations from a large-scale reverse-genetic screen in Arabidopsis. Genetics 164:731–740

Hamelin C, Sempere G, Jouffe V, Ruiz M (2013) Trop Gene DB, the multi-tropical crop information system updated and extended. Nucleic Acids Res 41:1172–1175

Henry IM, Carpentier SC, Pampurova S, Hoylandt AV, Panis B, Swennen R (2011) Structure and regulation of the Asr gene family in banana. Planta 234:785–798

Heslop-Harrison JS, Schwarzacher T (2007) Domestication, genomics and the future of the banana. Ann Bot 100:1073–1084

Hippolyte I, Bakry F, Seguin M, Gardes L, Rivallan R (2010) A saturated SSR/DArT linkage map of *Musa acuminata* addressing genome rearrangements among bananas. BMC Plant Biol 10:65–65

Hu W, Hou X, Huang C, Yan Y, Tie W (2015) Genome-wide identification and expression analyses of aquaporin gene family during development and abiotic stress in banana. Int J Mol Sci 16:19728–19751

Iondachescu M, Imai R (2008) Trehalose biosynthesis in response to abiotic stresses. J Integr Plant Biol 50:1223–1229

Ismail IA, Salama MI, Hamid NA, Sadik AS (2005) Production of transgenic banana plants conferring tolerance to salt stress. Ann Agric Sci (Cairo) 50:263–279

Jain S, Srivastava S, Sarin NB, Kav NNV (2006) Proteomics reveals elevated levels of PR10 proteins in saline-tolerant peanut (*Arachis hypogaea*) calli. Plant Physiol Biochem 44:253–259

Jain S, Kumar D, Jain M, Chaudhary P, Deswal R, Sarin NB (2012) Ectopic over-expression of a salt stress-induced pathogenesis related class 10 protein (PR10) gene from peanut (*Arachishypogaea* L.) affords broad spectrum abiotic stress tolerance in transgenic tobacco. Plant Cell Tissue Organ Cult 109:19–31

Jankowicz-Cieslak J, Huynh OA, Brozynska M, Nakitandwe J, Till BJ (2012) Induction, rapid fixation and retention of mutations in vegetatively propagated banana. Plant Biotechnol J 10:1056–1066

Jenny C, Holtz Y, Horry JP, Bakry F (2013) Synthesis of new interspecific triploid hybrid from natural AB germplasm of banana (*Musa* sp.). Acta Horticult 986:209–218

Kilian A (2007) Towards effective deployment of diversity arrays technology (DArT) in banana genomics and sequencing. In: Plant and animal genomes XV conference, San Diego. 13–17 January 2007

Kole C, Muthamilarasan M, Henry R, Edwards D, Sharma R (2015) Application of genomics-assisted breeding for generation of climate resilient crops: progress and prospects. Front Plant Sci 6:563. doi:10.3389/fpls.2015.00563

Mahmood M, Bidabadi SS, Ghobadi C, Gray DJ (2012) Effect of methyl jasmonate treatments on alleviation of polyethylene glycol mediated water stress in banana (*Musa acuminata* cv. 'Berangan', AAA) shoot tip cultures. Plant Growth Regul 68:161–169

Matsumoto K, Oka S (1998) Plant regeneration from protoplasts of a Brazilian dessert banana (*Musa* spp., AAB group). Acta Hortic 490:455–462

Maurel C, Verdoucq L, Luu D, Santoni V (2008) Plant aquaporins: membrane channels with multiple integrated functions. Annu Rev Plant Biol 59:595–624

Miao H, Wang Y, Liu J, Jia C, Hua W, Sunb P, Jin Z, Xu B (2014) Molecular cloning and expression analysis of the MaASR1 gene in banana. Electron J Biotechnol 17:287–295

Mondini L, Noorani A, Pagnotta MA (2009) Assessing plant genetic diversity by molecular tools. Diversity 1:19–35

MusaNet (2012) http://www.musanet.org/ Accessed on 23 Sept 2015

Muthamilarasan M, Theriappan P, Prasad M (2013) Recent advances in crop genomics for ensuring food security. Curr Sci 105:155–158

Nurain H, Mahanom J, Rofina YO, Norzulaani K (2014) Enhancement of regeneration efficiency in banana (*Musa acuminata* cv. Berangan) by using proline and glutamine. Sci Horticult 168:33–37

Okole BN, Schulz FA (1996) Micro-cross sections of banana and plantains (Musa spp.): morphogenesis and regeneration of callus and shoot buds. Plant Sci 116:185–195

Opara UL, Jacobson D, Al-Saady NA (2010) Analysis of genetic diversity in banana cultivars (*Musa* cvs.) from the South of Oman using AFLP markers and classification by phylogenetic, hierarchical clustering and principal component analyses. J Zhejiang Univ (Sci) 11:332–341

Ortiz R (2013) Conventional banana and plantain breeding. Acta Horticult 986:77–194

Ortiz R, Swennen R (2014) From crossbreeding to biotechnology-facilitated improvement of banana and plantain. Biotechnol Adv 32:158–169

Panis B, Strosse H, Remy S, Sagi L, Swennen R (2004) Cryopreservation of banana tissues: support for germplasm conservation and banana improvement. In: Jain SM, Swennen R (eds) Banana improvement: cellular, molecular biology, and induced mutations. Science, Enfield, pp 3–16

Ronald PC (2014) Lab to farm: applying research on plant genetics and genomics to crop improvement. PLoS Biol 12(6):e1001878. doi:10.1371/journal.pbio.1001878

Rushton PJ, Somssich IE, Ringler P, Shen QJ (2010) WRKY transcription factors. Trends Plant Sci 15:247–258

Rustagi A, Jain S, Kumar D, Shekhar S, Jain M, Bhat V, Sarin NB (2015) High efficiency transformation of banana [*Musa acuminata* L. cv. Matti (AA)] for enhanced tolerance to salt and drought stress through over-expression of a peanut salinity-induced pathogenesis-related class 10 protein. Mol Biotechnol 57:27–35

Sagi L, Remy S, Panis B, Swennen R, Volckaert G (1994) Transient gene expression in electroporated banana (*Musa* spp, cv Bluggoe, ABB group) protoplasts isolated from regenerable embryogenetic cell suspensions. Plant Cell Rep 13:262–266

Said EM, Mahmoud RA, Al-Akshar R, Safwat G (2015) Drought stress tolerance and enhancement of banana plantlets *In Vitro*. Austin J Biotech Bioeng 2:1040–1046

Santamaria JM, Hernandez-Portilla D, Chi-Manzanero B, Espadas F, Castano E, Iturriaga G, Rodriguez-Zapata LC (2009) Incorporation of two trehalose biosynthetic genes in banana increases trehalose levels and protects the photosynthetic apparatus from salt-stress damage. J Hortic Sci Biotechnol 84:665–671

Shekhawat UKS, Ganapathi TR (2013) *MusaWRKY71* overexpression in banana plants leads to altered abiotic and biotic stress responses. PLoS One 8(10):e75506. doi:10.1371/journal.pone.0075506

Shekhawat UKS, Ganapathi TR (2014) Transgenic banana plants overexpressing MusabZIP53 display severe growth retardation with enhanced sucrose and polyphenol oxidase activity. Plant Cell Tissue Organ Cult 116:387–402

Shekhawat UKS, Srinivas L, Ganapathi TR (2011) MusaDHN-1, a novel multiple stress-inducible SK3-type dehydrin gene, contributes affirmatively to drought- and salt-stress tolerance in banana. Planta 234:915–932

Sreedharan S, Shekhawat UKS, Ganapathi TR (2012) MusaSAP1, a A20/AN1 zinc finger gene from banana functions as a positive regulator in different stress responses. Plant Mol Biol 80:503–517. doi:10.1007/s11103-012-9964-4

Sreedharan S, Shekhawat UKS, Ganapathi TR (2013) Transgenic banana plants overexpressing a native plasma membrane aquaporin MusaPIP1;2 display high tolerance levels to different abiotic stresses. Plant Biotechnol J 11:942–952

Sreedharan S, Shekhawat UKS, Ganapathi TR (2015) Constitutive and stress-inducible overexpression of a native aquaporin gene (MusaPIP2;6) in transgenic banana plants signals its pivotal role in salt tolerance. Plant Mol Biol 88:41–52

Strosse HI, Houwe VD, Panis B (2004) Cell and tissue culture, and mutation induction. In: Mohan Jain S, Rony S (eds) Banana improvement: cellular, molecular biology, and induced mutations. Science Publishers, Enfield, pp 3–16

Stolker R (2010) Combating abiotic stress using trehalose. Thesis report, Wageningen University & Research Centre 42

Strosse H, Schoofs H, Panis B, Andre E, Reyniers K, Swennen R (2006) Development of embryogenic cell suspensions from shoot meristematic tissue in bananas and plantains (Musa spp.). Plant Sci 170:104–112

Takeuchi K, Gyohda A, Tominaga M, Kawakatsu M, Hatakeyama A (2011) RSOsPR10 expression in response to environmental stresses is regulated antagonistically by jasmonate/ethylene and salicylic acid signaling pathways in rice roots. Plant Cell Physiol 52:1686–1696

Till BJ, Zerr T, Comai L, Henikoff S (2006) A protocol for TILLING and Ecotilling in plants and animals. Nat Protoc 1:2465–2477

Till BJ, Jankowicz-Cieslak J, Sagi L, Huynh OA, Utsushi H, Swennen R, Terauchi R, Mba C (2010) Discovery of nucleotide polymorphisms in the Musa gene pool by Ecotilling. Theor Appl Genet 121:1381–1389

Tomepke K, Sadon L (2013) Ploidy manipulations by conventional and mutation breeding for developing new bananas. International conference on plant genetics and breeding technologies, Vienna, Austria, 18–20 February 2013

Varshney RK, Mohan SM, Gaur PM, Gangarao NV, Pandey MK (2013) Achievements and prospects of genomics-assisted breeding in three legume crops of the semi-arid tropics. Biotechnol Adv 31:1120–1134

Venkatachalam L, Thimmaraju R, Sreedhar RV, Bhagyalakshmi N (2006) Direct shoot and cormlet regeneration from leaf explants of 'silk'banana. In Vitro Cell Dev Biol Anim 42:262–269

Voytas DF, Gao C (2014) Precision genome engineering and agriculture: opportunities and regulatory challenges. PLoS Biol 12:e1001877. doi:10.1371/journal.pbio.1001877

Vuylsteke D (1989) Shoot-tip culture for the propagation, conservation and exchange of *Musa* germplasm. International Board of Plant Genetic Resources, Rome

Wheeler T, von Braun J (2013) Climate change impacts on global food security. Science 341:508–513

Xie YR, Chen ZY, Brown RL, Bhatnagar D (2010) Expression and functional characterization of two pathogenesis-related protein 10 genes from *Zea mays*. J Plant Physiol 167:121–130

Xu Y, Hu W, Liu J, Zhang J, Jia C, Miao H, Xu B, Jin Z (2014) Banana aquaporin gene, *MaPIP1;* involved in tolerance to salt and drought stress. BMC Plant Biol 14:59

Youssef M, James AC, Rivera-Madrid R, Ortiz R, Escobedo-Graciamedrano RM (2011) *Musa* genetic diversity revealed by SRAP and AFLP. Mol Biotechnol 47:189–199

13 Transgenic Technologies for Bacterial Wilt Resistance

Leena Tripathi, Jaindra Nath Tripathi, and Jerome Kubiriba

Abstract

Banana production is severely affected by bacterial diseases jeopardizing the food security of millions of inhabitants in countries where farmers depend upon banana as staple food. Bacterial diseases like *Xanthomonas* wilt, Moko, blood, and Bugtok are the most important diseases threatening banana cultivation in several tropical and subtropical countries. Genetic improvement of banana through classical breeding is difficult due to the lack of resistant germplasm, sterile nature, and long generation time. Transgenic technology can complement classical breeding for developing bacterial disease-resistant varieties. Some success has been achieved for developing host plant resistance in order to control banana *Xanthomonas* wilt (BXW) disease. Currently, the transgenic bananas expressing either sweet pepper *Pflp* or *Hrap* gene are under evaluation for resistance to *Xanthomonas* wilt disease in field trials in Uganda. Management of bacterial diseases through cultural practices like removal of male buds and use of pathogen-free seed material and disinfected cutting tools can contain outbreak of diseases although these are not absolute solutions for control of bacterial diseases. In this chapter, we have discussed various management practices as well as existing transgenic technologies to control bacterial diseases of banana.

Keywords

Banana • *Xanthomonas campestris* pv. *musacearum* • Transgenic technology

13.1 Introduction

Bacterial diseases are production constraint for many crops throughout the world. The outbreaks of some of the bacterial diseases have shattered

L. Tripathi (✉) • J.N. Tripathi
International Institute of Tropical Agriculture (IITA), 30709-00100 Nairobi, Kenya
e-mail: l.tripathi@cgiar.org

J. Kubiriba
National Agricultural Research Laboratories (NARL), 7084, Kampala, Uganda

S. Mohandas, K.V. Ravishankar (eds.), *Banana: Genomics and Transgenic Approaches for Genetic Improvement*, DOI 10.1007/978-981-10-1585-4_13

economies of several countries. The bacterial diseases may be systemic causing death of the entire plants, or it may be vascular causing yellowing and wilts or locally causing rot, spots, blights, and cankers or may be of a mixed nature. Bacterial diseases of bananas (Moko, blood, Bugtok, and *Xanthomonas* wilt) are very destructive, and they have devastated the banana production in various tropical and subtropical countries. Moko, Bugtok, and blood, caused by *Ralstonia solanacearum* species complex, are among the major biotic constraints for production of dessert and cooking bananas (Sequeira 1998). Moko and Bugtok diseases are caused by *Ralstonia solanacearum* phylotype II, whereas blood disease is caused by *Ralstonia solanacearum* phylotype IV (Fegan and Prior 2004). Moko is an important disease devastating banana and plantain production in Latin America, Caribbean, and the Philippines (Thwaites et al. 2000). However, Bugtok is an endemic disease affecting only cooking bananas in the Philippines (Thwaites et al. 2000). Blood disease has devastated banana production in Indonesian archipelago and there is risk of its spread to Australia and Papua New Guinea. Moko and blood diseases are systemic and their symptoms are similar: wilting of leaves, rotting of fruits, and vascular discoloration. However, symptoms of Bugtok disease differ from Moko and blood disease in that the symptoms are limited to the male buds and flowers of bananas; there is no wilting and vascular discoloration generally does not extend beyond peduncle (Thwaites et al. 2000). Banana *Xanthomonas* wilt (BXW) disease, caused by *Xanthomonas campestris* pv. *musacearum*, is only reported in Great Lakes region of East Africa (Tushemereirwe et al. 2004).

Generally, the plant diseases are managed through cultural practices, host plant resistance, chemicals, and regulations. The management practice varies from one disease to another depending upon the pathogen, the host, and the environmental factors. Once established, the bacterial diseases are difficult to control. Chemical control of bacterial disease is impossible, although several antibiotics have been tried. Antibiotics are not recommended for control of plant diseases because of the possibility of evolution of antibiotic-resistant strains of bacterial pathogens. Cultural practices including sanitation, crop rotation, host eradication, and improvement of crop environment can reduce pathogen levels or reduce the rate of disease development. The bacterial diseases can be managed cost-effectively through host plant resistance. Commercial growers and farmers favor resistant varieties if they are available. There is limited success in developing varieties resistant to bacterial diseases through conventional breeding due to lack of germplasm exhibiting resistance against most of the bacterial pathogens and also because of the appearance of new pathogen strains. Biotechnology has unraveled several new options to control plant bacterial diseases. Transgenic technologies to develop disease-resistant plants through improving host's regulatory mechanism or defense mechanism or expressing antimicrobial proteins are reported for several crops. This chapter reviews the attempts made for exploiting transgenic technologies for effective management of bacterial diseases and describes the progress on control measures of BXW disease.

13.2 Banana *Xanthomonas* Wilt (BXW) Disease

Among all the bacterial diseases damaging banana cultivation worldwide, BXW is considered as the major menace to banana production in Great Lakes region of East and Central Africa (Tripathi et al. 2009). The disease was first reported on *Ensete* (Yirgou and Bradbury 1968) and later on banana in Ethiopia (Yirgou and Bradbury 1974). Outside Ethiopia, BXW was first noticed in Uganda in 2001 (Tushemereirwe et al. 2004) and then in the Democratic Republic of the Congo (DR Congo) (Ndungo et al. 2006), Rwanda (Reeder et al. 2007), Kenya, Tanzania, and Burundi (Carter et al. 2009). The disease affects production of all types of banana including plantain, cooking, dessert, and juice-producing bananas (Ssekiwoko et al. 2006). BXW spreads very fast resulting in severe loses, unlike those of other diseases which gradually increase losses over years. The impact of BXW is due to absolute yield losses and death of the mother plant (Tripathi et al. 2009). The economic losses due to this disease were estimated at US$ 2–8 billion over a decade due to increase in price and production losses (Abele and Pillay 2007). The BXW disease causes yield losses through complete wilting of banana plant and rotting of fruits. Therefore, BXW disease jeopardizes the livelihood of millions of farmers growing bananas for food and cash.

13.2.1 Causal Agent

The causative organism of BXW disease is rod-shaped, gram-negative bacterium, named as *Xanthomonas campestris* (Yirgou et al. 1968, 1974). It has subsequently been retitled as *X. campestris* pv. *musacearum* (*Xcm*) (Young et al. 1991). The bacterium belongs to *Xanthomonadaceae* and produces yellow, circular, mucoid, slimy colonies (Tripathi et al. 2007). Previously, *X. campestris* pv. *musacearum* was considered to be a highly monomorphic pathogen with no genetic differences (Aritua et al. 2007). However, recent study on genome-wide sequencing reports the presence of two genetically distinct sub-lineages of the pathogen, one group including *X. campestris* pv. *musacearum* isolates from Burundi, Kenya, Tanzania, and Uganda and second group of isolates from DR Congo, Ethiopia, and Rwanda (Wasukira et al. 2012). This study indicates more than one introduction of BXW disease on banana in East Africa (Wasukira et al. 2012).

13.2.2 Symptoms of BXW

The main symptoms of BXW disease are yellowing and wilting of leaves, immature and uneven ripening, and rotting of fruits (Fig. 13.1a, b) with yellowish brown blotches and dark brown placental scars in the pulp (Tushemereirwe et al. 2004). Infection via insect transmission shows symptoms on different parts of flowers including shrinking and rotting of the male buds, wilting of bracts, and browning of peduncle first followed with further wilting of leaves occurring as the infection spreads to the complete plant. Cross sections of the pseudostem and rachis of infected plants show yellowish bacterial ooze. Ultimately, infected plants wilt completely and rot. Appearance of symptoms is very fast under favorable conditions and visible within 3–4 weeks in fields and 2–3 weeks after artificial inoculation of *X. campestris* pv. *musacearum* in glasshouse (Tripathi et al. 2008). Symptom development differs with cultivars, growth phase of plant, and mode of disease transmission (Tripathi et al. 2009).

13.2.3 Disease Transmission

The *X. campestris* pv. *musacearum* infects banana plant mainly through two ways. Farming tools (pangas, hoes, knives) and browsing animals that inflict damage to various parts of the plants (leaves, pseudostem, male bud, corm roots). When these tools inflict a wound to a diseased plant, they pick the bacteria and transmit them when they get to a clean plant. It is established that the bacteria survive for several days on the farming tools, therefore very effective in spreading the bacteria from infected plants to clean plants. In areas where there is excessive use of tools, this is a major means of disease transmission (e.g., Southeastern Uganda). In this case, the main agents spreading the disease are the farmers and the traders, but unfortunately they are unaware of this fact. The second main means of transmission is by insects that visit banana male buds to feed on sap that oozes out when the banana flowers fall off. As the insects feed, they mechanically bring in bacteria from the infected plants visited earlier. The insects and other agents like sapsucking birds and bats are able to spread the disease from field to field. This is the main means of disease transmission in all parts of Uganda where farmers do not intensively use farming tools in their plantations, e.g., in plantations of beer bananas (Pisang Awak commonly

Fig. 13.1 (**a**) BXW-infected plants showing wilting of leaves; (**b**) diseased plants showing rotting of fruits; (**c**) completely wilted control non-transgenic plant in glasshouse; (**d**) transgenic plants showing no BXW symptom after artificial inoculation with pathogen under glasshouse conditions. Pictures (**c** and **d**) were taken 8 weeks post artificial inoculation

known as Kayinja) in Central and Western Uganda.

The planting material from infected field can also spread disease. Sometimes, the infected plants do not show symptoms during incubation period, which can be up to 3 or more months.

13.3 Management of BXW Disease

The bacterial wilt diseases are systemic, and once pathogens have established in banana plantation, their control becomes very difficult (Eden-Green 2004). The infected banana plants should be removed immediately in order to limit spread of the disease. At present there are no chemicals, biocontrol tools, or resistant cultivars available to control BXW (Tripathi et al. 2009). Therefore, the disease management should involve combination of approaches like exclusion, eradication, cultivation of disease-tolerant banana varieties, and protection. Once BXW disease appears in a plantation, there is no option other than removal of all diseased plants and leaving the field unplanted or to follow crop rotation. Researchers at NARL, Kawanda, established that the bacteria survive no more than 45 days in soil rotting banana residues (Mwebaze et al. 2006). Conditions favoring rotting of plants also speed up natural destruction of the bacteria. There are only two known alternate hosts, *Canna* sp. and *Zebrina* sp., which are relatives of the banana plants (belong to the same plant family), of *X. campestris* pv. *musacearum* pathogen.

Currently, control of BXW is centered on protecting uninfected plants once the disease gets to an area as there is no effective curative measure for affected plants. The control measures target and block the bacteria from getting transmitted to uninfected plants. The male bud of banana plants presents the extreme risk for *X. campestris* pv. *musacearum* infection. Therefore, early removal of the male bud is an effective means of preventing the spread of bacterial pathogen through insects (Tinzaara et al. 2006). The removal of male buds is recommended for both affected and unaffected banana plants for limiting spread of the disease. The male buds should be removed by twisting the peduncle to break it with a forked wooden stick, instead of using cutting tools. BXW can be contained in banana fields if debudding is applied effectively. However, removal of male buds has not been adopted well by farmers (Kagezi et al. 2006), particularly in areas where farmers believe that the quality of the banana fruits especially for juice cultivars is affected by removal of male buds (Bagamba et al. 2006).

Previously, rouging the complete diseased banana mat was recommended to farmers (Tushemereirwe et al. 2004). Recently, it was demonstrated that it is feasible to only remove plants showing disease symptoms from the affected banana mat and the remaining uninfected plants on the same mat can be saved from BXW infection. This is feasible as insect-transmitted BXW infection starts from the flowers of the banana plant and takes some time before reaching to the corm of the plant, which is connected to the other plants on the mat (Sekiwoko et al. 2006). The technique of single-stem removal has been confirmed to be effective in control of BXW disease in Kenya, Democratic Republic of the Congo, and Uganda (Kubiriba and Tushemereirwe 2014). A 6-month fallow/unplanted period is recommended to minimize the risk of reinfection of newly planted banana plants in previously diseased fields (Tripathi et al. 2009). The mechanical spread of *X. campestris* pv. *musacearum* can be limited by disinfecting the cutting tools used for pruning or removing of diseased plants. Suspending the use of cutting tools in infected banana fields is crucial for control of BXW (Tushemereirwe et al. 2006). The *X. campestris* pv. *musacearum* bacterium can survive up to 22 days on cutting tools stored at room temperature (Buregyeya 2010).

Effective control of BXW disease is only possible if the cultural practices are advocated along with institutional approaches that effectively mobilize stakeholders using the technologies (Kubiriba et al. 2012). Other stakeholders from government and nongovernment organizations involved in rural development also contributed to the planning, generation, and promotion of these technologies. It also reported that sometimes farmers could change recommended practices in order to suit their conditions (Bagamba et al. 2006). The cultural practices are proved to be effective for managing the BXW disease in those areas where they have been deployed carefully and consistently (Tushemereirwe et al. 2006).

The farmers' knowledge and their decision to control plant diseases depend upon suitability of improved technologies (Sherwood 1997). More than 80 % of the farmers growing bananas in Uganda understand BXW disease and its diagnosis, spread, and control, due to massive awareness campaigns (Muhangi et al. 2006). Even though most of the banana farmers know control of BXW, only 30 % of them applied the knowledge to manage BXW (Tushemereirwe et al. 2006). Therefore, participatory approaches were used, which has resulted in control of BXW disease to below 5 % for over 3 years in some of the major banana-growing areas in Uganda (Kubiriba et al. 2012). Majority of the farmers participating in farmer's field school managed the BXW disease in their banana plantations.

The most striking strategy for controlling BXW is to use disease-resistant plants. Yet, no banana cultivars are known showing high resistance against *X. campestris* pv. *musacearum*. A few banana cultivars may escape insect-transmitted infection as they have male flowers with persistent bracts that protect the inflorescence against insects contaminated with *X. campestris* pv. *musacearum*. These include East African highland cooking banana cultivars such as "Nakitembe" and "Mbwazirume". The lack of natural resistance in banana against *X. campestris* pv. *musacearum*, along with the difficulties associated with conventional breeding of banana and labor demanding cultural practices, favors a transgenic technology to control BXW.

13.4 Transgenic Technologies for Resistance to Bacterial Diseases

Plants have well-developed defense mechanism to fight against pathogens. Transgenic technologies have potential to enhance host plant resistance against pathogens by either expressing resistance genes (Keen 1999) or overexpressing defense genes (Bent and Yu 1999; Rommens and Kishore 2000).

Resistance (R) genes have been identified and cloned from different plants for use against several pathogens (Bent 1996). These R-genes facilitate resistance against pathogens including bacteria, fungi, viruses, and pathogenic nematodes. Many R-gene products share structural motifs, which specify that resistance against various pathogens may be through similar molecular or chemical pathways. The R-gene *Pto* from tomato (*Lycopersicon esculentum* Miller) demonstrated resistance against several strains of *Pseudomonas syringae* pv. *tomato* (Martin et al. 1993; Kim et al. 2002). Transgenic tomato expressing *Pto* gene also showed resistance against bacterial pathogen *Xanthomonas campestris* pv. *vesicatoria* and the fungal pathogen *Cladosporium fulvum* (Mysore et al. 2003).

Several R-genes (*Bs1*, *Bs2*, and *Bs3*) conferring resistance in a "gene-for-gene" fashion have been cloned from pepper (Hibberd et al. 1987). Their corresponding avirulence genes (*avrBs1*, *avrBs2*, and *avrBs3*) have also been identified in *X. campestris* pv. *vesicatoria* and shown to be fundamental for resistance (Minsavage et al. 1990; Swanson et al. 1988). The pepper *Bs*2 gene confers resistance to bacterial spot disease of tomato caused by *X. campestris* pv. *vesicatoria*, which has the corresponding bacterial avirulence gene, *avrBs2* (Tai et al. 1999). The *RPS*4 gene from *Arabidopsis* provides resistance against *P. syringae* pv. *tomato* containing *avrRps4* (Gassmann et al. 1999).

Broad-spectrum durable disease resistance is very important for control of bacterial diseases. The maize *Rxo*1 gene provides resistance against *Xanthomonas oryzae* pv. *oryzicola* causing bacterial streak disease in rice and *Burkholderia andropogonis*, which causes bacterial disease in the sorghum and the maize (Zhao et al. 2005). The *Arabidopsis RRS1* gene provides broad-spectrum resistance to various strains of *R. solanacearum*, which also cause Moko disease in banana (Deslandes et al. 1998). Similarly, overexpression of *Arabidopsis NPR1* gene provides resistance to bacterial pathogens (Cao et al. 1998). Overexpression of *Arabidopsis NPR1* or the rice *NH1* gene enhanced resistance to the rice bacterial blight pathogen *X. oryzae* pv. *oryzae* and the blast pathogen *Magnaporthe grisea* Herbert (Chern et al. 2005; Yuan et al. 2007; Quilis et al. 2008; Feng et al. 2011). The rice *Xa21* gene also conferred resistance against *X. oryzae* pv. *oryzae*, causing bacterial leaf blight disease (Song et al. 1995; Wang et al. 1996; Yoshimura et al. 1998).

Once infected, plants employ different defense-related approaches in order to survive from pathogen attack. These approaches include induction of pathogenesis-related (PR) genes, strengthening of cell walls, accumulation of phytoalexins, and induction of salicylic acid (SA) pathway. A hypersensitive response (HR) is the initial defense response, which limits spread of pathogens from the infection site. HR leads to systemic acquired resistance (SAR), which induces expression of many PR genes, resulting in resistance against various pathogens (Durrant and Dong 2004).

Plant disease resistance can also be achieved by inserting antimicrobial peptides (AMPs) (Cary et al. 2000; Li et al. 2001) and enhancing the plant defense (Keller et al. 1999). AMPs have a broad-spectrum antimicrobial activity against bacterial and fungal pathogens and are safe to plants and mammals. Several AMPs like lysozymes (Düring et al. 1993), magainins (Zasloff 1987), cecropins (Boman and Hultmark 1987), attacins (Hultmark et al. 1983), thionins (Molina et al. 1993), and defensins (Broekaert et al. 1995) are well documented. Lysozyme, which can be obtained from bacteriophages, hen eggs, or cows, is a hydrolytic enzyme and part of the immune system of animals. Lysozyme attacks the murein layer of bacterial peptidoglycan, which results to death of both gram-negative and gram-positive bacteria by weakening their cell wall. Expression of the lysozyme gene showed resistance against plant pathogenic bacteria in transgenic potato (Düring et al. 1993), tobacco (Trudel et al. 1995), and apple (Ko 1999).

The magainins are antimicrobial proteins secreted from the skin of the African clawed frog (*Xenopus laevis*) (Zasloff 1987). Transgenic tobacco expressing magainin analog MSI-99 or Myp-30 showed resistance against bacterial and fungal pathogens (Li et al. 2001; De Gray et al. 2001; Chakrabarti et al. 2003). The cecropins isolated from hemolymph of *Hyalophora cecropia* are antibacterial lytic peptides, which kill gram-negative and gram-positive bacteria by creating transient ion channels in the outer phospholipid membranes (Durell et al. 1992). Different types of cecropins including native (cecropin B), synthetic (Shiva-1, D4E1), and mutant (SB-37, MB39) showed antimicrobial activity against bacterial pathogens like *Erwinia carotovora*, *E. amylovora*, *P. syringae*, *R. solanacearum*, *and X. campestris* (Nordeen et al. 1992; Kaduno-Okuda et al. 1995; Rajasekaran et al. 2001). Cecropin expressing transgenic tobacco plants conferred resistance against *P. syringae* pv. *tabaci* causing wildfire disease (Huang et al. 1997). Transgenic potato expressing Shiva-1 and SB-37 showed resistance against *E. carotovora* (Arce et al. 1999). The expression of SB-37 in transgenic apple also showed increased resistance to fire blight disease caused by *E. amylovora* (Norelli et al. 1998). Poplar plants expressing D4E1 conferred resistance to *Agrobacterium tumefaciens* and *X. populi* (Mentag et al. 2003).

The attacins are antibacterial proteins produced by *Hyalophora cecropia* pupae (Hultmark et al. 1983). Attacin expressed by transgenic pear and apple demonstrated enhanced resistance to fire blight disease caused by *E. amylovora*, under glasshouse and field conditions (Norelli et al. 1999; Reynoird et al. 1999). Transgenic apple expressing attacin in the intercellular space has conferred resistance to fire blight disease (Ko et al. 2000). The thionins is another group of plant antimicrobial proteins, which inhibit bacterial pathogens (Molina et al. 1993). Transgenic tobacco expressing alpha-thionin gene from barley conferred resistance against *P. syringae* (Carmona et al. 1993). Defensins isolated from mollusks, acari, arachnids, insects, mammals, and plants are also source of antimicrobial proteins. Plant defensins from *B. oleracea* and *B. campestris* demonstrated resistance to bacterial leaf blight disease of transgenic rice (Kawata et al. 2003).

Resistance genes enhance the hypersensitive response (HR) in nonhost plants through the release of the elicitor from bacterial pathogens. The HR is characterized by robust and localized cell death at the point of pathogen infection forming a physical barrier, which prevents spread of pathogen (Mehdy 1994). In addition, HR also induces plant defense responses in the surrounding and even distal parts of the plants leading to systemic acquired resistance (SAR) (Xie and Chen 2000). Elicitor-induced resistance can provide broad-spectrum resistance. Therefore, transgenic technologies using defense genes may be more ideal and it might provide resistance for bacterial as well as fungal pathogen.

The ferredoxin-like amphipathic protein (PFLP) and hypersensitive response-assisting protein (HRAP) isolated from the sweet pepper (*Capsicum annuum* L.) are plant proteins, which intensify HR (Lin et al. 1997; Chen et al. 2000). The expression of *Pflp* gene in transgenic tobacco, tomato, orchids, calla lily, and rice has demonstrated resistance against various bacterial pathogens such as *Erwinia*, *Pseudomonas*, *Ralstonia*, and *Xanthomonas* spp. (Huang et al. 2004, 2007; Liau et al. 2003; Tang et al. 2001; Yip et al. 2007). The transgenic plants overexpressing *Pflp* gene showed disease resistance through hypersensitive response and enhanced production of active oxygen species (AOS) (Dayakar et al. 2003; Haung et al. 2004).

The *Hrap* gene can protect plants from bacterial pathogen infection and belongs to hypersensitive cell death (HCD)-associated genes (Chen et al. 2000). The HRAP protein dissociates the multimeric forms of harpin$_{PSS}$ into dimers and monomers triggering hypersensitive cell death (Chen et al. 2000). The HRAP protein has a signal peptide, which leads its secretion into extracellular space and its transcription is activated early during incompatible interaction of sweet pepper with *P. syringae* (Chen et al. 1998, 2000). The expression of the *Hrap* gene in tobacco and *Arabidopsis* plants conferred resistance against virulent bacterial pathogens (Ger et al. 2002; Pandey et al. 2005).

13.5 Development of Transgenic Bananas Resistant to BXW Disease

Transgenic technologies facilitating transfer of useful genes across species have been shown to offer numerous advantages to circumvent the natural bottlenecks to breed banana for its improvement. It provides a cost-effective alternative to develop banana varieties resistant to BXW disease, due to lack of high resistance among banana genotypes. In order to stably integrate transgene into plant genome, availability of efficient and reproducible transformation and regeneration protocols is most important. Even though transforming a monocot species is difficult, genetic transformation of banana is routine in few labs (Becker et al. 2000; Sagi et al. 1995; Ganapathi et al. 2001; Khanna et al. 2004; Tripathi et al. 2012, 2015).

Researchers at IITA and NARL developed transgenic bananas expressing sweet pepper *Hrap* or *Pflp*, in order to determine if these genes can confer resistance to BXW disease. Both *Hrap* and *Pflp* genes have been demonstrated efficacy for control of several bacterial diseases. These genes have been acquired from the patent holder Academia Sinica through the African Agricultural Technology Foundation (AATF).

Embryogenic cell suspensions (ECS) of banana cultivars "Sukali Ndiizi" and "Nakinyika" were transformed through *Agrobacterium*-mediated procedure, and hundreds of independent transgenic lines expressing *Hrap* or *Pflp* gene were generated (Tripathi et al. 2010; Namakwaya et al. 2012). The generated transgenic events were validated for presence of gene by PCR analysis. The integration and low copy number of transgene was confirmed by Southern blot analysis. All the PCR-positive transgenic lines were evaluated for resistance against *X. campestris* pv. *musacearum* by artificial inoculation of in vitro plantlets in the laboratory as described by Tripathi et al. (2008). About 50–60 % of transgenic lines tested did not develop any disease symptoms, confirming that the *Hrap* or *Pflp* gene can enhance resistance against *X. campestris* pv. *musacearum*. The transgenic lines showing enhanced resistance were further evaluated in glasshouse using potted plants. The majority of transgenic lines showing enhanced resistance with in vitro assays also did not develop any symptoms of BXW disease after artificial inoculation of potted plants in the glasshouse, whereas control non-transgenic plants developed symptoms and wilted completely (Fig. 13.1c, d). The results from laboratory and glasshouse confirmed that transgenic bananas expressing *Hrap* or *Pflp* gene provide high resistance to BXW (Tripathi et al. 2010; Namukwaya et al. 2012). The transgenic lines exhibiting strong resistance to BXW were further characterized by molecular analysis to check the expression of transgenes.

The best 65 resistant lines showing high resistance against *X. campestris* pv. *musacearum* under screenhouse were further evaluated in a confined field trial at NARL, Kawanda, Uganda. The majority of transgenic lines showed significantly higher resistance in comparison to non-transgenic control plants (Tripathi et al. 2014b). Both mother and ratoon crops of 11 transgenic lines (7 *Hrap* lines and 4 *Pflp* lines) were found to be highly resistant showing 100 % disease resistance, whereas non-transgenic control plants wilted completely. The field trial results also confirmed the successful transfer of the disease resistance trait from mother to progeny in several lines. The flowering and yield (bunch weight and fruit size) of BXW-resistant transgenic banana lines was comparable to un-inoculated control non-transgenic plants of same varieties (Tripathi et al. 2014b). The best ten lines were further planted in second confined trial to test durability of disease resistance trait and agronomic performance. As bacterial pathogens evolve fast, there is risk of breaking down of resistance with single gene-based resistance. To minimize this, we are developing transgenic banana varieties using stacked genes (*Hrap-Pflp*) for durable high resistance to BXW disease.

We have also identified additional transgenes providing resistance to BXW. These genes can be stacked in order to develop enhanced and durable disease resistance. We tested the potential of rice pattern recognition receptor (PRR), XA21, for providing resistance against *X. campestris pv. musacearum* as it has been reported to confer resistance to the rice pathogen *X. oryzae* pv. *oryzae* (Ronald et al. 1992; Wang et al. 1996). Transgenic banana cultivar "Gonja manjaya" (AAB) expressing rice *Xa21* was generated and tested for resistance against *X. campestris* pv. *musacearum* through rapid bioassay and in glasshouse using potted plants. In both assays, about 50 % of the *Xa*21 transgenic lines tested showed complete resistance to BXW, whereas non-transgenic control plants develop severe symptoms and wilted completely. Our results confirmed that the expression of the rice *Xa21* gene in transgenic banana provides enhanced resistance to BXW disease (Tripathi et al. 2014a).

13.6 Conclusion

Banana *Xanthomonas* wilt (BXW) is one of the most significant diseases and is considered as the biggest menace to banana production in the East and Central Africa, which is the largest banana-producing region and where most of the bananas are grown by smallholder farmers for local consumption. The use of disease-resistant varieties is considered the timely, economically, and ecologically friendly approaches to control bacterial diseases. However, the development of banana varieties resistant to bacterial diseases through classical breeding is restricted, as sources of germplasm displaying resistance against bacterial pathogens has not been identified yet. As BXW is spreading rapidly and devastating banana production across East Africa, genetic engineering can provide an efficient, safe, and viable way to develop BXW-resistant banana varieties. Transgenic bananas expressing sweet pepper *Hrap* or *Pflp* gene demonstrated enhanced resistance to BXW under glasshouse and field conditions. These transgenic lines will be further tested in multi-location trials in order to test them in different environmental and climate conditions. These BXW-resistant varieties will enhance the resources available to control the disease and save livelihoods of millions of people who depend upon banana as staple food. The transgenic technology using *Hrap* and *Pflp* may also control other bacterial diseases of banana like Moko, Bugtok, or blood disease effecting banana production.

Acknowledgments The authors would like to thank US Agency for International Development (USAID) for financial support of BXW research. We acknowledge Academia Sinica, Taiwan, and the University of California, Davis, for providing gene constructs. We also acknowledge African Agricultural Technology Foundation (AATF) for negotiating license of *Pflp* and *Hrap* genes from Academia Sinica and for sublicensing the genes to International Institute of Tropical Agriculture (IITA).

References

Abele S, Pillay M (2007) Bacterial wilt and drought stresses in banana production and their impact on economic welfare in Uganda: implications for banana research in East African highlands. J Crop Improv 19:173–191. doi:10.1300/J411v19n01_09

Arce P, Moreno M, Gutierrez M, Gebauer M, Dell'Orto P, Torres H, Acuna I, Oliger P, Venegas A, Jordana X, Kalazich J, Holuigue L (1999) Enhanced resistance to bacterial infection by *Erwinia carotovora* subsp. *atroseptica* in transgenic potato plants expressing the attacin or the cecropin SB-37 genes. Am J Potato Res 76:169–177. doi:10.1007/BF02853582

Aritua V, Parkinson N, Thwaites R, Heeney JV, Jones DJ, Tushemereirwe W, Crozier J, Reeder R, Stead DE, Smith J (2007) Characterization of the *Xanthomonas* sp. causing wilt of enset and banana and its proposed reclassification as a strain of *X. vasicola*. Plant Pathol 57:170–177. doi:10.1111/j.1365-3059.2007.01687.x

Bagamba F, Kikulwe E, Tushemereirwe WK, Ngambeki D, Muhangi J, Kagezi GH, Ragama PE, Eden-Green S (2006) Awareness of banana bacterial wilt control in Uganda: farmers' perspective. Afr Crop Sci J 14:157–164. doi:10.4314/acsj.v14i2.27923

Becker DK, Dugdale B, Smith MK, Harding RM, Dale JL (2000) Genetic transformation of Cavendish banana (*Musa* sp. AAA group) cv. Grand Nain via microprojectile bombardment. Plant Cell Rep 19:229–234. doi:10.1007/s002990050004

Bent AF (1996) Plant disease resistance genes: function meets structure. Plant Cell 8:1757–1771. doi:10.1105/tpc.8.10.1757

Bent AF, Yu IC (1999) Applications of molecular biology to plant disease and insect resistance. Adv Agron 66:251–298. doi:10.1016/S0065-2113(08)60429-0

Boman HG, Hultmark D (1987) Cell free immunity in insects. Annu Rev Microbiol 41:103–126. doi:10.1146/annurev.mi.41.100187.000535

Broekaert WF, Terras FRG, Cammue BPA, Osborn RW (1995) Plant defensins: novel antimicrobial peptides as components of the host defence system. Plant Physiol 108:1353–1358. doi:10.1104/pp.108.4.1353

Buregyeya H (2010) Evaluation of the contribution of birds, bats and farm tools in the long distance transmission of banana bacterial wilt; Masters' thesis, Makerere University, Kampala Uganda

Cao H, Li X, Dong X (1998) Generation of broad-spectrum disease resistance by overexpression of an essential regulatory gene in systemic acquired resistance. Proc Natl Acad Sci U S A 95:6531–6536. doi:PNAS-1998-Cao-6531-6xccf

Carmona MJ, Molina A, Fernandez JA, Lopez-Fando JJ, Garcia-Olmedo F (1993) Expression of the alpha-*thionin* gene from barley in tobacco confers enhanced resistance to bacterial pathogens. Plant J 3:457–462. doi:10.1111/j.1365-313X.1993.tb00165.x

Carter BA, Reeder R, Mgenzi SR, Kinyua ZM, Mbaka JN, Doyle K, Nakato V, Mwangi M, Beed F, Aritua V, Lewis Ivey ML, Miller SA, Smith JJ (2009) Identification of *Xanthomonas vasicola* (formerly *X. campestris* pv. *musacearum*), causative organism of banana Xanthomonas wilt, in Tanzania, Kenya and Burundi. New Dis Rep 19:25. doi:10.1111/j.1365-3059.2009.02124.x

Cary JW, Rajasekaran K, Jaynes JM, Cleveland TE (2000) Transgenic expression of a gene encoding a synthetic antimicrobial peptide results in inhibition of fungal growth *in vitro* and in planta. Plant Sci 154:171–181. doi:10.1016/S0168-9452(00)00189-8

Chakrabarti A, Ganapathi TR, Mukherjee PK, Bapat VA (2003) MSI-99, a magainin analogue, imparts enhanced disease resistance in transgenic tobacco and banana. Planta 216:587–596. doi:10.1007/s00425-002-0918-y

Chen CH, Lin HJ, Feng TY (1998) An amphipathic protein from sweet pepper can dissociate harpinPSS multimeric forms and intensify the harpinPSS-mediated hypersensitive response. Physiol Mol Plant Pathol 52:139–149. doi:10.1006/pmpp.1997.0120

Chen CH, Lin HJ, Ger MJ, Chow D, Feng TY (2000) The cloning and characterization of a hypersensitive response assisting protein that may be associated with the harpin-mediated hypersensitive response. Plant Mol Biol 43:429–438. doi:10.1023/A:1006448611432

Chern M, Fitzgerald HA, Canlas PE, Navarre DA, Ronald PC (2005) Over-expression of a rice *NPR1* homolog leads to constitutive activation of defence response and hypersensitivity to light. MPMI 18:511–520. doi:10.1094/MPMI-18-0511

Dayakar BV, Lin HJ, Chen CH, Ger MJ, Lee BH, Pai CH, Chow D, Huang HE, Hwang SY, Chung MC, Feng TY (2003) Ferredoxin from sweet pepper (*Capsicum annuum* L.) intensifying harpin(pss)-mediated hypersensitive response shows an enhanced production of active oxygen species(AOS). Plant Mol Biol 51:913–924. doi:10.1023/A:1023061303755

De Gray G, Rajasekaran K, Smith F, Sanford J, Daniell H (2001) Expression of an antimicrobial peptide via the chloroplast genome to control phytopathogenic bacteria and fungi. Plant Physiol 127:852–862. doi:10.1104/pp.010233

Deslandes L, Pileur F, Liaubet L, Camut S, Can C, Williams K, Holub E, Beynon J, Arlat M, Marco Y (1998) Genetic characterization of RRS1, a recessive locus in *Arabidopsis thaliana* that confers resistance to the bacterial soilborne pathogen *Ralstonia solanacearum*. MPMI 11:659–667. doi:10.1094/MPMI.1998.11.7.659

Düring K, Porsch P, Fladung M, Lörz H (1993) Transgenic potato plants resistant to the phytopathogenic bacterium *Erwinia carotovora*. Plant J 3:587–598. doi:10.1046/j.1365-313X.1993.03040587.x

Durrant WE, Dong X (2004) Systemic acquired resistance. Annu Rev Phytopathol 42:185–209. doi:10.1146/annurev.phyto.42.040803.140421

Durell SR, Raghunathan G, Guy HR (1992) Modeling the ion channel structure of cecropin. Biophys J 63:1623–1631. doi:10.1016/S0006-3495(92)81730-7

Eden-Green S (2004) Focus on bacterial wilt. How can the advance of banana *Xanthomonas* wilt be halted? Infomusa 13:38–41

Fegan F, Prior P (2004) How complex is the *Ralstonia solanacearum* species complex. In: Allen C, Prior P, Hayward AC (eds) Bacterial wilt: the disease and *Ralstonia solanacearum* species complex. APS Press, St Paul

Feng JX, Cao L, Li J, Duan CJ, Xue-Mei Luo XM, Le N, Wei H, Liang S, Chu C, Pan Q, Tang JL (2011) Involvement of *OsNPR1/NH1* in rice basal resistance to blast fungus *Magnaporthe oryzae*. Eur J Plant Pathol 131:221–235. doi:10.1007/s10658-011-9801-7

Ganapathi TR, Higgs NS, Balint-Kurti PJ, Arntzen CJ, May GD, Van Eck JM (2001) Agrobacterium-mediated transformation of the embryogenic cell suspensions of the banana cultivars Rasthali (AAB). Plant Cell Rep 20:157–162. doi:10.1007/s002990000287

Gassmann W, Hinsch ME, Staskawicz BJ (1999) The *Arabidopsis RPS4* bacterial-resistance gene is a member of the TIR-NBSLRR family of disease-resistance genes. Plant J 20:265–277. doi:10.1046/j.1365-313X.1999.00600.x

Ger MJ, Chen CH, Hwang SY, Huang HE, Podile AR, Dayakar BV, Feng TY (2002) Constitutive expression of *Hrap* gene in transgenic tobacco plant enhances resistance against virulent bacterial pathogens by induction of a hypersensitive response. MPMI 15:764–773. doi:10.1094/MPMI.2002.15.8.764

Hibberd AM, Bassett MJ, Stall RE (1987) Allelism tests of three dominant genes for hypersensitive resistance to bacterial spot of pepper. Phytopathology 77:1304–1307. https://www.apsnet.org/publications/phytopathology/backissues/Documents/1987Articles/Phyto77n09_1304.PDF

Huang Y, Nordeen RO, Di M, Owens LD, McBeth JH (1997) Expression of an engineered cecropin gene cassette in transgenic tobacco plants confers disease resistance to *Pseudomonas syringae* pv. *tabaci*. Phytopathology 87:494–499. doi:10.1094/PHYTO.1997.87.5.494

Huang SN, Chen CH, Lin HJ, Ger MJ, Chen ZI, Feng TY (2004) Plant ferredoxin-like protein AP1 enhances Erwinia induced hypersensitive response of tobacco. Physiol Mol Plant Pathol 64:103–110. doi:10.1016/j.pmpp.2004.05.005

Huang HE, Ger MJ, Chen CY, Pandey AK, Yip MK, Chou HW, Feng TY (2007) Disease resistance to bacterial pathogens affected by the amount of ferredoxin-I protein in plants. Mol Plant Pathol 8:129–137. doi:10.1111/j.1364-3703.2006.00378.x

Hultmark D, Engström Å, Andersson K, Steiner H, Bennich H, Boman HG (1983) Insect immunity: Attacins, a family of antibacterial proteins from *Hyalophora cecropia*. EMBO J 2:571–576. http://www.ncbi.nlm.nih.gov/pmc/articles/PMC555063/pdf/emboj00257-0090.pdf, http://www.ncbi.nlm.nih.gov/pmc/articles/PMC555063/pdf/emboj00257-0090.pdf

Kaduno-Okuda K, Taniai K, Kato Y, Kotani E, Yamakaula M (1995) Effects of synthetic *Bombyx mor*i cecropin B on growth of plant pathogenic bacteria. J Invertebr Pathol 65:309–319. doi:10.1006/jipa.1995.1047

Kagezi GH, Kangire A, Tushemereirwe W, Bagamba F, Kikulwe E, Muhangi J, Gold CS, Ragama P (2006) Field assessment of banana bacterial wilt incidence in Uganda. Afr Crop Sci J 14:83–91. doi:10.4314/acsj.v14i2.27914

Kawata M, Nakajima T, Yamamoto T, Mori K, Oikawa T, Fukumoto F, Kuroda S (2003) Genetic engineering for disease resistance in rice (*Oryza sativa* L.) using antimicrobial peptides. Jpn Agric Res Q 37:71–76. https://www.jircas.affrc.go.jp/english/publication/jarq/37-2/37-02-01.pdf

Keen NT (1999) Plant disease resistance: progress in basic understanding and practical application. Adv Bot Res 30:292–328. doi:10.1016/S0065-2296(08)60230-X

Keller H, Pamboukdjian N, Ponchet M, Poupet A, Delon R, Verrier JL, Roby D, Ricei P (1999) Pathogen induced elicitin production in transgenic tobacco generates a hypersensitive response and non-specific disease resistance. Plant Cell 11:223–235. doi:10.1105/tpc.11.2.223

Khanna H, Becker D, Kleidon J, Dale J (2004) Centrifugation assisted *Agrobacterium tumefaciens* mediated transformation (CAA) of embryogenic cell suspensions of banana (*Musa* spp.) Cavendish AAA and Lady Finger AAB. Mol Breed 14:239–252. doi:10.1023/B:MOLB.0000047771.34186.e8

Kim YJ, Lin NC, Martin GB (2002) Two distinct *Pseudomonas* effector proteins interact with the Pto kinase and activate plant immunity. Cell 109:589–598. doi:10.1016/S0092-8674(02)00743-2

Ko K (1999) Attacin and T4 lysozyme transgenic 'Galaxy' apple: regulation of transgene expression and plant resistance to fire blight (*Erwinia amylovora*). Ph.D dissertation, Cornell University, New York

Ko K, Norelli JL, Reynoird JP, Boresjza-Wysocka E, Brown SK, Aldwinckle HS (2000) Effect of untranslated leader sequence of AMV RNA 4 and signal peptide of pathogenesis-related protein 1b on attacin gene expression, and resistance to fire blight in transgenic apple. Biotechnol Lett 22:373–381. doi:10.1023/A:1005672601625

Kubiriba J, Tushemereirwe WK (2014) Approaches for the control of banana Xanthomonas wilt in East and Central Africa. Afr J Plant Sci 8(8):398–404. doi:10.5897/AJPS2013.1106

Kubiriba J, Karamura EB, Jogo W, Tushemereirwe WK, Tinzaara W (2012) Community mobilization: a key to effective control of banana Xanthomonas wilt. J Dev Agric Econ 45(5):125–131. doi:10.5897/JDAE11.098

Li Q, Lawrence CB, Xing HY, Babbitt RA, Bass WT, Maiti IB, Everett NP (2001) Enhanced disease resistance conferred by expression of an antimicrobial magainin analogue in transgenic tobacco. Planta 212:635–639. doi:10.1007/s004250000480

Liau CH, Lu JC, Prasad V, Lee JT, Hsiao HH, You SJ, Yang NS, Huang HE, Feng TY, Chen WH, Chan MT (2003) The sweet pepper ferredoxin-like protein (pflp) conferred resistance against soft rot disease in Oncidium orchid. Transgenic Res 12:329–336. doi:10.1023/A:1023343620729

Lin HJ, Cheng HY, Chen CH, Huang HC, Feng TY (1997) Plant amphipathic proteins delay the hypersensitive response caused by harpinPss and *Pseudomonas syringae* pv. *syringae*. Physiol Mol Plant Pathol 51:367–376. doi:10.1006/pmpp.1997.0121

Martin GB, Brommonschenkel SH, Chunwongse J, Frary A, Ganal MW, Spivey R, Wu T, Earle ED, Tanksley SD (1993) Map-based cloning of a protein kinase gene conferring disease resistance in tomato. Science 262:1432–1436. doi:10.1126/science.7902614

Mehdy MC (1994) Active oxygen species in plant defense against pathogens. Plant Physiol 105:467–472. doi:10.1104/pp.105.2.467

Mentag R, Lukevich M, Morency MJ, Seguin A (2003) Bacterial disease resistance of transgenic hybrid poplar expressing the synthetic antimicrobial peptide

D4E1. Tree Physiol 23:405–411. doi:10.1093/treephys/23.6.405
Minsavage GV, Dahlbeck D, Whalen MC, Kearney B, Bonas U, Stasawicz BJ, Stall R (1990) Gene for gene relationships specifying disease resistance in *Xanthomonas campestris pv. vesicatoria*-pepper interactions. MPMI 3:41–47. doi:10.1094/MPMI-3-041
Molina A, Ahl Goy P, Fraile A, Sanchez-Monge R, Garcia-Olmedo F (1993) Inhibition of bacterial and fungal pathogens by thionins of types I and II. Plant Sci 92:169–177. doi:10.1046/j.1365-313X.1997.00669.x
Muhangi J, Nankinga C, Tushemereirwe WK, Rutherfold M, Ragama P, Nowakunda K, Abeyasekera S (2006) Impact of awareness campaigns for banana bacterial wilt in Uganda. Afr Crop Sci J 14(2):175–183. doi:10.4314/acsj.v14i2.27925
Mwebaze JM, Tusiime G, Tushmereirwe WK, Kubiriba J (2006) The survival of Xanthomonas campestris pv. musacearum in soil and plant debris. Afr Crop Sci J 14:121–127. doi:10.4314/acsj.v14i2.46166
Mysore KS, D'Ascenzo MD, He X, Martin GB (2003) Overexpression of the disease resistance gene *Pto* in tomato induces gene expression changes similar to immune responses in human and fruitfly. Plant Physiol 132:1901–1912. doi:10.1104/pp.103.022731
Namukwaya B, Tripathi L, Tripathi JN, Arinaitwe G, Mukasa SB, Tushemereirwe WK (2012) Transgenic banana expressing *Pflp* gene confers enhanced resistance to Xanthomonas wilt disease. Transgenic Res 4:855–865. doi:10.1007/s11248-011-9574-y
Ndungo V, Eden-Green S, Blomme G, Crozier J, Smith J (2006) Presence of banana Xanthomonas wilt (*Xanthomonas campestris* pv. *musacearum*) in the Democratic Republic of Congo (DRC). Plant Pathol 55:294. doi:10.1111/j.1365-3059.2005.01258.x
Nordeen RO, Sinden SL, Jaynes JM, Owens LD (1992) Activity of cecropin SB37 against protoplasts from several plant species and their bacterial pathogens. Plant Sci 82:101–107. doi:10.1016/0168-9452(92)90012-B
Norelli JL, Mills JZ, Momol MT, Aldwinkle HS (1998) Effect of cercropin-type transgenes on fire blight resistance of apple. Acta Hortic 489:273–278. doi:10.17660/ActaHortic.1999.489.47
Norelli JL, Borejsza-Wysocka E, Momol MT, Mills JZ, Grethel A, Alkwinckle HS, Ko K, Brown SK, Bauer DW, Beer SV, Abdul-Kader AM, Hanke V (1999) Genetic transformation for fire blight resistance in apple. Acta Hortic 489:295–296. doi:10.17660/ActaHortic.1999.489.50
Pandey AK, Ger MJ, Huang HE, Yip MK, Zeng J, Feng TY (2005) Expression of the hypersensitive response-assisting protein in *Arabidopsis* results in harpin-dependent hypersensitive cell death in response to *Erwinia carotovora*. Plant Mol Biol 59:771–780. doi:10.1007/s11103-005-1002-3
Quilis J, Penas G, Messeguer J, Brugidou C, Segundo BS (2008) The *Arabidopsis AtNPR1* inversely modulates defence responses against fungal, bacterial, or viral pathogens while conferring hypersensitivity to abiotic stress in transgenic rice. MPMI 21:1215–1231. doi:10.1094/MPMI-21-9-1215
Rajasekaran K, Stromberg KD, Cary JW, Cleveland TE (2001) Broad-spectrum antimicrobial activity *in vitro* of the synthetic peptide D4E1. J Agric Food Chem 49:2799–2803. doi:10.1021/jf010154d
Reeder R, Opolot O, Muhinyuza J, Aritua A, Crozier J, Smith J (2007) Presence of banana bacterial wilt (*Xanthomonas campestris* pv. *musacearum*) in Rwanda. New Dis Rep 14:52. doi:10.1111/j.1365-3059.2007.01640.x
Reynoird JP, Mourgues F, Norelli JL, Aldwinckle HS, Brisset MN, Chevreau E (1999) First evidence for improved resistance to fire blight in transgenic pear expressing the *attacin E* gene from *Hyalophora cecropia*. Plant Sci 149:23–31. doi:10.1016/S0168-9452(99)00139-9
Rommens CM, Kishore GM (2000) Exploiting the full potential of disease-resistance genes for agricultural use. Curr Opin Biotechnol 11:120–125. doi:10.1016/S0958-1669(00)00083-5
Ronald PC, Albano B, Tabien R, Abenes L, Wu K, McCouch S, Tanksley SD (1992) Genetic and physical analysis of the rice bacterial blight disease resistance locus, *Xa21*. Mol Gen Genet 236:113–120. doi:10.1007/BF00279649
Sagi L, Panis B, Remy S, Schoofs H, De Smet K, Swennen R, Cammue B (1995) Genetic transformation of banana (*Musa* spp.) via particle bombardment. Bio Technol 13:481–485. doi:10.1038/nbt0595-481
Sekiwoko F, Taligoola HK, Tushemereirwe WK (2006) *Xanthomonas campestris* pv *musacearum* host range in Uganda. Afr Crop Sci J 14:111–120. doi:10.4314/acsj.v14i2.46165
Sequeira L (1998) The missing element in international banana improvement programs. In: Prior P, Allen C, Elphinstone J (eds) Bacterial wilt disease: molecular and ecological aspect. Springer, Berlin
Sherwood SG (1997) Little things mean a lot: working with central American farmers to address the mystery of plant diseases. Agric Hum Values 14:181–189. doi:10.1023/A:1007383508503
Song WY, Wang GL, Chen LL, Kim HS, Pi LY, Holsten T, Gardner J, Holsten T, Wang B, Zhai WX, Zhu LH, Ronald P (1995) A receptor kinase-like protein encoded by the rice disease resistance gene, *Xa21*. Science 270:1804–1806. doi:10.1111/j.1467-7652.2007.00243.x
Swanson J, Kearney B, Dahlbeck D, Staskawicz BJ (1988) Cloned avirulence gene of Xanthomonas campestris pv. vesicatoria complements spontaneous race change mutant. MPMI 1:5–9. doi:10.1094/MPMI-1-005
Tai TH, Dahlbeck D, Clark ET, Gajiwala P, Pasion R, Whalen MC, Stall RE, Staskawicz BJ (1999) Expression of the *Bs2* pepper gene confers resistance to bacterial spot disease in tomato. Proc Natl Acad Sci U S A 23:14153–14158. doi:10.1073/pnas.96.24.14153

Tang K, Sun X, Hu Q, Wu A, Lin CH, Lin HJ, Twyman RM, Christou P, Feng TY (2001) Transgenic rice plants expressing the ferredoxin-like protein (AP1) from sweet pepper show enhanced resistance to *Xanthomonas oryzae* pv. *oryzae*. Plant Sci 160:1035–1042. doi:10.1016/S0168-9452(01)00351-X

Thwaites R, Eden-Green S, Black R (2000) Diseases caused by bacteria. In: Jones DR (ed) Diseases of banana, abaca and enset. CABI Publishing, Wallingford

Tinzaara W, Gold CS, Ssekiwoko F, Tushemereirwe W, Bandyopadhyay R, Abera A, Eden-Green S (2006) The possible role of insects in the transmission of banana Xanthomonas wilt. Afr Crop Sci J 14:105–110. doi:10.4314/acsj.v14i2.27916

Tripathi L, Tripathi JN, Tushemereirwe WK, Bandyopadhyay R (2007) Development of a semi-selective medium for isolation of *Xanthomonas campestris* pv. *musacearum* from banana plants. Eur J Plant Pathol 117:177–186. doi:10.1007/s10658-006-9083-7

Tripathi L, Odipio J, Tripathi JN, Tusiime G (2008) A rapid technique for screening banana cultivars for resistance to Xanthomonas wilt. Eur J Plant Pathol 121:9–19. doi:10.1007/s10658-007-9235-4

Tripathi L, Mwangi M, Abele S, Aritua V, Tushemereirwe WK, Bandyopadhyay R (2009) A threat to banana production in east and central Africa. Plant Dis 93:440–451. doi:10.1094/PDIS-93-5-0440

Tripathi L, Mwaka H, Tripathi JN, Tushemereirwe WK (2010) Expression of sweet pepper *Hrap* gene in banana enhances resistance to *Xanthomonas campestris* pv. *musacearum*. Mol Plant Pathol 11:721–731. doi:10.1111/J.1364-3703.2010.00639.X

Tripathi J, Muwonge A, Tripathi L (2012) Highly efficient regeneration and transformation protocol for plantain cv. 'Gonja Manjaya' (*Musa* spp. AAB) using embryogenic cell suspension. In Vitro Cell Dev Biol Anim 48:216–224. doi:10.1007/s11627-011-9422-z

Tripathi JN, Lorenzen J, Bahar O, Ronald P, Tripathi L (2014a) Transgenic expression of the rice Xa21 pattern-recognition receptor in banana (*Musa* sp.) confers resistance to *Xanthomonas campestris* pv. *musacearum*. Plant Biotech J 12:663–673. doi:10.1111/pbi.12170

Tripathi L, Tripathi JN, Kiggundu A, Korie S, Shotkoski F, Tushemereirwe WK (2014b) Field trial of Xanthomonas wilt disease-resistant bananas in East Africa. Nat Biotechnol 32(9):868–870. doi:10.1038/nbt.3007

Tripathi JN, Oduor RO, Tripathi L (2015) A high-throughput regeneration and transformation platform for production of genetically modified banana. Front Plant Sci 6:1025. doi:10.3389/fpls.2015.01025

Trudel J, Potvin C, Asselin A (1995) Secreted hen lysozyme in transgenic tobacco: recovery of bound enzyme and *in vitro* growth inhibition of plant pathogens. Plant Sci 106:55–62. doi:10.1016/0168-9452(95)04069-7

Tushemereirwe WK, Kangire A, Ssekiwoko F (2004) First report of *Xanthomonas campestris* pv. *musacearum* on banana in Uganda. Plant Pathol 53:802. doi:10.1111/j.1365-3059.2004.01090.x

Tushemereirwe WK, Okaasai O, Kubiriba J, Nankinga C, Muhangi J, Odoi N, Opio F (2006) Status of banana bacterial wilt in Uganda. Afr Crop Sci J 14(2):73–82. doi:10.4314/acsj.v14i2.27913

Wang GL, Song WY, Ruan DL, Sideris S, Ronald PC (1996) The cloned gene, *Xa21*, confers resistance to multiple *Xanthomonas oryzae* pv. Oryzae isolates in transgenic plants. MPMI 9:850–855. doi:10.1094/MPMI-9-0850

Wasukira A, Tayebwa J, Thwaites R, Paszkiewicz K, Aritua V, Kubiriba J, Smith J, Grant M, Studholme DJ (2012) Genome-wide sequencing reveals two major sub-lineages in the genetically monomorphic pathogen *Xanthomonas campestris* pathovar *musacearum*. Genes 3:361–377. doi:10.3390/genes3030361

Xie Z, Chen Z (2000) Harpin-induced hypersensitive cell death is associated with altered mitochondrial functions in tobacco cells. MPMI 13:183–190. doi:10.1094/MPMI.2000.13.2.183

Yip MK, Huang HE, Ger MJ, Chiu SH, Tsai YC, Lin CI, Feng TY (2007) Production of soft rot resistant calla lily by expressing a ferredoxin-like protein gene (*pflp*) in transgenic plants. Plant Cell Rep 26:449–457. doi:10.1007/s00299-006-0246-y

Yirgou D, Bradbury JF (1968) Bacterial wilt of enset (*Ensete ventricosum*) incited by *Xanthomonas musacearum* sp. nov. Phytopathology 58:111–112

Yirgou D, Bradbury JF (1974) A note on wilt of banana caused by the enset wilt organism *Xanthomonas musacearum*. East Afr Agric For J 40:111–114

Yoshimura S, Yamanouchi U, Katayose Y, Toki S, Wang ZX, Kono I, Kurata N, Yano M, Iwata N, Sasaki T (1998) Expression of *Xa1*, a bacterial blight-resistance gene in rice, is induced by bacterial inoculation. Proc Natl Acad Sci U S A 95:1663–1668. http://www.pnas.org/content/95/4/1663.full.pdf?sid=7aec7fc7-83c9-474c-8c99-d8920f7a97e8

Young JM, Bradbury JF, Davis RE, Dickey RS, Ercolani GL, Haywood AC, Vidaver AK (1991) Nomenclature revisions of plant pathogenic bacteria and list of names 1980–1988. Rev Plant Pathol 70:211–221. http://www.cabdirect.org/

Yuan Y, Zhong S, Li Q, Zhu Z, Lou Y, Wang L, Wang J, Wang M, Li Q, Yang D, He Z (2007) Functional analysis of rice *NPR1*-like genes reveals that *OsNPR1/NH1* is the rice orthologue conferring disease resistance with enhanced herbivore susceptibility. Plant Biotechnol J 5:313–324. doi:10.1111/j.1467-7652.2007.00243.x

Zasloff M (1987) Magainins, a class of antimicrobial peptides from Xenopus skin: isolation, characterization of two active forms and partial cDNA sequence of a precursor. Proc Natl Acad Sci USA 84:5449–5453. doi:pnas00330-0371

Zhao B, Lin X, Poland J, Trick H, Leach J, Hulbet S (2005) A maize resistance gene functions against bacterial streak disease in rice. Proc Natl Acad Sci U S A 43:15383–15388. doi:10.1073/pnas.0503023102

Engineering Resistance to *Fusarium* Wilt

14

T.R. Usharani, H.D. Sowmya, C. Sunisha, and Sukhada Mohandas

Abstract

Banana is an important crop grown worldwide. Panama disease caused by *Fusarium oxysporum* f.sp. *cubense* (Foc) is more devastating and causes 100 % yield loss in many cultivars of banana. Foc is known to exist as four important races (races 1, 2, 3 and 4) of which races 1 and 4 are of serious concern as it attacks the commercially acceptable banana cultivars across the globe. Foc has been continuously evolving and hence pathogen variability has posed a great challenge to banana industry. Genetic modification of banana has been a widely accepted tool due to the limited success of conventional breeding. Current review highlights on the application of genetic engineering for imparting resistance against *Fusarium* wilt which has been the main goal of researchers worldwide. We discuss the various strategies that have been employed involving PR-related genes (defensin gene), antimicrobial genes (Ace-AMP1 gene), antiapoptosis gene and RNAi-mediated approach – host-induced gene silencing (HIGS) that confers a certain level of tolerance towards pathogen infection. Further the cisgenic approach utilizing R genes and native cell-death genes from *Musa* spp. has also proven promising. However the field trials that help to assess the efficiency of genes in situ are lacking in many of the studies. The understanding of host-pathogen interaction in terms of defence and pathogen-related pathways and the study of pathogenicity mechanism help in identifying critical genes for targeting pathogen and to evolve resistant cultivars employing cisgenic and transgenic approaches.

Keywords

Fusarium • Banana • Transgenic • Cisgenic • Genetic engineering

T.R. Usharani (✉) • H.D. Sowmya • C. Sunisha
Indian Institute of Horticultural Research,
Hessaraghatta, Bangalore 560089, India
e-mail: usharanitr@gmail.com

S. Mohandas
Division of Biotechnology,
ICAR-Indian Institute of Horticultural Research,
Hessaraghatta Lake Post, Bangalore 560089, India

SM Plant Production Technologies Pvt Ltd.,
138 Kodi Halli, Madure Hobli, Doddaballapura
Taluk, Bangalore 562103, India

S. Mohandas, K.V. Ravishankar (eds.), *Banana: Genomics and Transgenic Approaches for Genetic Improvement*, DOI 10.1007/978-981-10-1585-4_14

14.1 Introduction

Bananas and plantains are perennial crops and vegetatively propagated and belong to Musaceae family of the order Zingiberales. The heart of the species is considered to be either from Malaysia (Simmonds 1962) or Indonesia (Horry et al. 1997). The majority of the cultivated or edible banana is derived by inter- and intraspecific hybridization of diploid species, namely, *Musa acuminata* (AA) and *Musa balbisiana* (BB) (Osuji et al. 1997). They are triploid and sterile and produce fruit by parthenocarpy (Simmonds 1962). The worldwide yield of banana is majorly affected by several abiotic and biotic factors. The abiotic factors responsible for low banana yield are poor soil fertility, drought, heat and salinity, and the biotic factors include fungal, bacterial, viral diseases and pests. The major fungal diseases are panama disease caused by *Fusarium oxysporum* f.sp. *cubense* (Foc), black Sigatoka caused by *Mycosphaerella fijiensis* and yellow Sigatoka caused by *Mycosphaerella musicola*. Among them, *Fusarium* wilt represents the top six important plant diseases in the world (Ploetz and Pegg 1997) and was first discovered in Brisbane, Queensland, Australia, during 1876 infecting banana plants var. Sugar (Silk AAB) (Bancroft 1876).

14.2 *Fusarium* Wilt Disease

14.2.1 National and International Scenario

Fusarium wilt is caused by a soilborne necrotrophic fungal pathogen *Fusarium oxysporum* f.sp. *cubense* (Snyder and Hansen 1940). By 1950, the pathogen had spread to all the banana-producing regions of the world with an exception of Mediterranean, Melanesia and Somalia islands in the South Pacific (Ploetz 2000) (Fig. 14.1). During 1890–1960 around 40,000 ha of banana cultivar Gros Michel was affected by Foc race 1, and estimated economic loss was around US$23

Fig. 14.1 Global widespread of *Fusarium oxysporum* f.sp. *cubense* (Source: http://panamadisease.org/map, Food and Agriculture Organization 2014)

billion. Banana industries depended on Cavendish cultivars, which was resistant to Foc race 1 instead of susceptible Gros Michel variety (Stover 1990). Later in 1990s, it was reported in Asia that an extremely virulent strain of Foc – TR4 – affected the Cavendish cultivars and other banana varieties which were susceptible to Foc race 1 and 2. The loss incurred by Foc – TR4 – was estimated to be about US$400 million. The disease has spread since then to Taiwan, Indonesia, Malaysia, Philippines, China and northern Australia. Recently, outbreak of this disease has been reported in Mozambique and Jordan (García Bastidas et al. 2014).

14.2.2 Pathogen and Disease Cycle

Fusarium oxysporum f.sp. *cubense* is a soilborne filamentous fungal phytopathogen causing wilt symptoms in banana plants. Based on their host specificity, Foc has been distinguished into four races referred to as 1–4 (Moore et al. 1993). There are three types of asexual spores produced by Foc, macroconidia, microconidia and chlamydospores (Fig. 14.2) (Nelson 1991). Macroconidia are fusiform, slender, foot-shaped basal and tapered apical cells, four to eight celled (Jones 2000). Chlamydospores are thick-walled, asexual, globose spores, usually formed singularly or in pairs, but may also be found in clusters or short chains.

Fig. 14.2 Schematic representation of different types of spores produced by Foc (Smith 2007)

14.2.2.1 Disease Cycle

The penetration of the host plant by *F. oxysporum* involves adhesion of the pathogen to the host surface. In response to primary and secondary root exudates, chlamydospores germinate and infect the root tips or epidermal cells near the root cap or within the zone of elongation (MacHardy and Beckman 1981; Beckman 1990). The pathogen moves inter- and intracellularly through the roots of parenchyma tissue until it reaches to protoxylem vessels (Mai and Abawi 1987). The fungus persists within the xylem by producing the microconidia and toxins and, thereafter, spreads to the neighbouring large reticulate vessels through the pits along with plant sap. The pathogen occludes the xylem vessel, causes wilt symptom and ultimately leads to the death of the plant. The disease cycle is repeated when the chlamydospores germinate and invade a new host plant (Stover 1962) (Fig. 14.3).

Studies on the mode of infection and colonization of Foc in banana roots and rhizome region have been carried out by Li et al. (2011a) with a green fluorescent protein (GFP)-tagged Foc TR4 isolate. It was found that chlamydospores get attached to banana roots and root hairs and then penetrate through root tips at the time of natural wound. A network of fungal hyphae is formed on root caps and elongation zone of banana roots at 11 and 15 days postinoculation (dpi) with Foc in cv. Brazil. The vascular space of rhizome filled by fungal spores and hyphae resulted in the disorganization of surrounding tissue. Xiao et al. (2013) reported maximum hyphal load in the psuedostem compared to roots and rhizome region of the infected plant.

14.3 Host-Pathogen Interaction

The availability of genome sequences of banana and different Foc races, as well as trancriptome and proteomic profiles of banana roots in response to Foc infection, is useful in understanding the molecular mechanisms underlying the Foc-banana interactions.

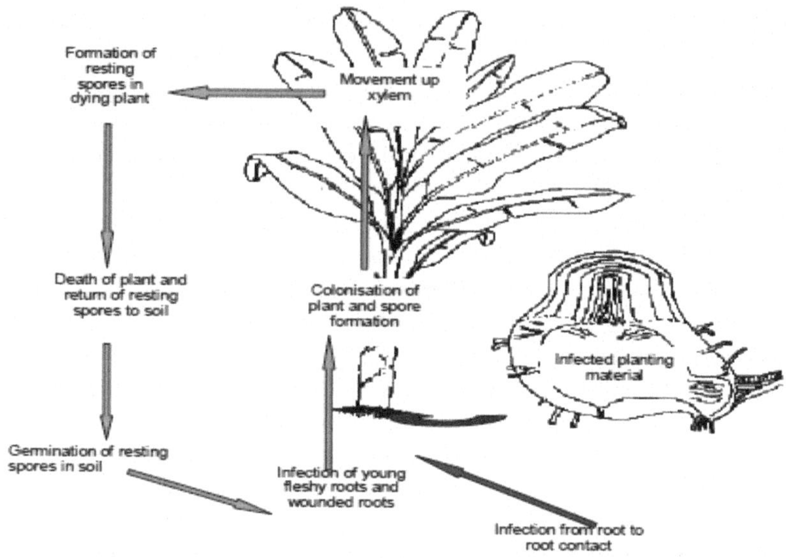

Fig. 14.3 Disease initiation and life cycle of Foc in a banana plant (http://www.nt.gov.au/d/Content/File/p/Plant_Pest/786)

14.3.1 Molecular and Biochemical Mechanism of the Host

The host response to pathogen infection is determined by the rate and extent of defence mechanism induced in the roots and vascular system of the banana. The pathogenicity mechanism of pathogen involved in banana vascular wilt disease development determines success or failure of resistance in the host.

14.3.1.1 Cell Wall Fortification

The banana genes involved in synthesis of phenyl propanoids (phenylalanine ammonia lyase) and cell wall strengthening (lignin-forming anionic peroxidase) were activated due to Foc infection as a first line of defence (Li et al. 2013a). The phenolic compounds produced in the phenyl propanoid pathway polymerize to form lignins, lignans, neolignans, flavonoids and anthocyanins that strengthen the cell wall to prevent the pathogen entry.

De Ascensao et al. (2000) showed that the increased flux through the phenyl propanoid pathway resulted in incorporation of esterified derivatives of cinnamic acid and benzoic acid into the cell wall fraction as part of the antimicrobial defences activated in the root tissue of banana cultivar 'Goldfinger' towards the cell wall-derived elicitor from the pathogen Foc, TR4. De Ascensao and Dubery (2003) reported the accumulation kinetics of polymerized phenolic monomers such as lignin and lignin-like polymers which reached maximum values after 24 h. Anterola et al. (2002) reported the upregulation of caffeoyl-CoA O-methyltransferase (CCOMT), a branch point enzyme of monolignol biosynthesis in moderately resistant banana 'Nongke No.1' in response to pathogen infection. Monolignols synthesized from PAL form lignins to fortify the cell walls implying that lignification played an important role in cell wall strengthening as an inducible defence mechanism of banana roots against Foc race 4 (Li et al. 2013b).

Melida et al. (2011) found alpha-1,4-glucan-protein synthase (UDP forming) associated with the formation of polysaccharoses, viz., hemicellulose and xylose, as cell wall components beneficial for the formation of physical barriers was upregulated in highly resistant 'Yueyoukang I' indicating complex cell wall formation as one of the effective defence strategies in response to fungal infection.

14.3.1.2 Reactive Oxygen Species

The accumulation of reactive oxygen species, namely, superoxide anion radical (O_2•–), hydroxyl radical (.OH) and hydrogen peroxide (H_2O_2) termed as 'oxidative burst', occurs first during Foc-banana interactions. Li et al. (2011b) reported 'oxidative burst' resulting in enhanced levels of H_2O_2 accumulation in the resistant isoline Williams-8818-1 as one of the key factors in banana resistance to Foc race 4 invasion. The genes encoding germin-like protein involved in the production of reactive oxygen species are among the strongest Foc-induced genes (Li et al. 2013a). The roles played by series of oxidative enzymes like peroxidase, superoxide dismutase, polyphenol oxidase and catalase that function in ROS scavenging were relatively higher in resistant hybrids than susceptible hybrids of banana (Kavino et al. 2007).

SOD transcripts were found upregulated until 72 h in the resistant isoline, but not in the susceptible isoline (Li et al. 2011a). However the protein IN2-1 that functions as glutathione peroxidase and helps in ROS scavenging in various stresses (Dixon et al. 2002) was upregulated in susceptible 'Brazil' and moderately resistant 'Nongke No.1' indicating that maintenance of elevated levels of ROS is a prerequisite for effective defence to combat pathogen infection. Li et al. (2012) reported the activity of membrane-localized NADPH oxidases (or respiratory burst oxidase) being upregulated in the resistant banana as a counter-effect of Foc TR4 infection corresponding to the previous observations in wheat, cotton and cucumber after infection by the *Fusarium* wilt fungal pathogen (Dowd et al. 2004).

ROS targets the Ca++ permeability in the cell membrane via activation of the elicitor-stimulated cyclic nucleotide-gated channels (CNGCs) – Ca^{2+} influx – viz. CNGC 1, CNGC 5 and CNGC 6, after PAMP perception in *Musa* that initiates all subsequent defence reactions (Li et al. 2012).

14.3.1.3 Salicylic Acid/Jasmonic Acid/ Ethylene Biosynthesis and Signalling

The plant immune system is regulated by three important hormone signalling molecules, such as salicylic acid (SA), jasmonic acid (JA) and ethylene (ET) in response to abiotic and biotic stresses (Chisholm et al. 2006). NPR1-like genes, the positive regulator of SA, namely, MNPR1A and MNPR1B, were reported from banana after Foc treatment. The reduced sensitivity to Foc in GCTCV-218 was partially attributed to the higher and an earlier expression of both MNPR1A and PR1 after Foc treatment.

JA/ET pathways are reported to be involved in defence response to necrotrophic pathogens, herbivore and injury in *Arabidopsis thaliana* (Zarate et al. 2007). Proteomic analysis revealed differential expression of proteins associated with hormone (JA, ABA and auxin) signalling pathways in banana. The lipoxygenase (LOX)-like gene expression in addition to allene oxide synthase (AOS)-like unigene was indicative of elevated levels of JA in the resistant mutant than in the susceptible wild type (Li et al. 2012). Interestingly, five genes encoding ethylene-responsive transcription factors (ERFs) involved in ethylene biosynthesis and signalling pathways were among the strongly induced genes by *Fusarium* infection, indicating the role of hormone in banana response to infection. Generally, the ethylene signalling pathway depicts major role for resisting necrotrophic pathogens such as *F. oxysporum* (van et al. 2006). Overexpression of ERF1 in *Arabidopsis*, a transcription factor that activates ethylene-responsive genes, enhances resistance to *F. oxysporum* f.sp. *conglutinans* and f.sp. *lycopersici* (Berrocal-Lobo and Molina 2004).

14.3.1.4 Signal Transduction

Host defence mechanisms were activated by signal pathways induced as a result of pathogen infection. R gene product activates plant defence mechanisms. To date, a number of R genes isolated from banana cultivars are known. Chen et al. (2007) reported wild cultivar *Musa acuminata* (AA) as source of six RCGs which showed resistance to *Fusarium* wilt caused by Foc race 4. The deduced amino acid sequence showed that two genes, WNB1 and WNB2, contained NR-ARC domain, whereas WST1 to WST4 contained serine/threonine kinase domain. The resistance showed by Cavendish banana may be attributed to the expression of NBS-type resistance protein RIN4/RPM1 complex. Also BAK1 (BRASSINOSTEROID INSENSITIVE 1-associated receptor kinase 1), a receptor kinase of brassinosteroid signalling pathway and the innate immune response, was highly induced in both Foc1 and Foc TR4-inoculated roots (Li et al. 2002; Chinchilla et al. 2007). The defence pathway in addition to PAMP-mediated innate immunity is negatively regulated by WRKY40-like gene (Pandey et al. 2010; Xu et al. 2006) which was suppressed by Foc, thus enhancing activation of the defence pathway. Wall-associated kinase (WAK)-like genes were found to be induced by both Foc races. The chitin elicitor-binding protein (CEBiP) and the chitin elicitor receptor kinase (CERK1) that recognizes chitin oligosaccharides, a representative general elicitor inducing defence responses, were upregulated in the resistant mutant 'Nongke No 1' (Li et al. 2012) (Fig. 14.4).

14.3.1.5 Pathogenesis-Associated Genes and Defence

Foc infection-induces expression of many PR genes such as thaumatin-like genes, antimicrobial phytoalexins and antifungal hydrolytic enzymes such as chitinases, 1,3-glucanases. A

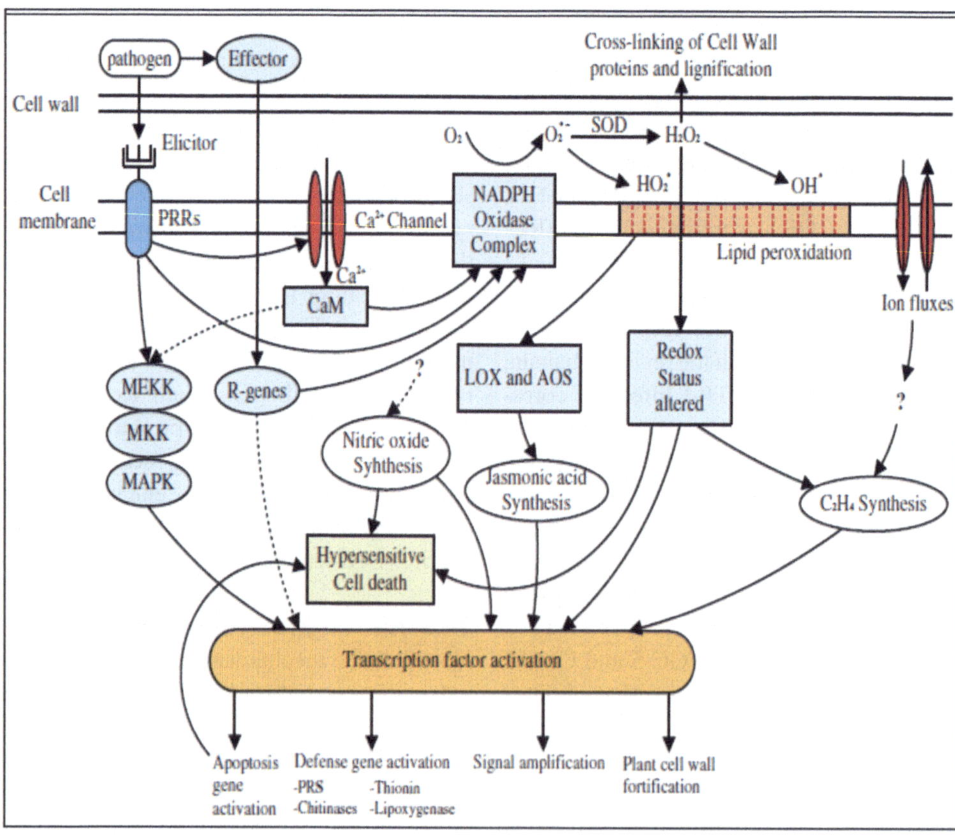

Fig. 14.4 Response of the resistant Cavendish banana mutant 'Nongke No 1' against Foc tropical race 4 (Foc TR4), (Hofius et al. 2007)

number of defence-related genes have been identified for their role in disease resistance for a number of fungal pathogens (Hoshikawa et al. 2012). During the initial stages of infection by fungus, the activity of defence gene PR1 and functions of the actin pathway play an important role in cell polarization defence by releasing and channelizing antimicrobial compounds and sediment barrier material to the site of infection for cell wall enforcement (Kobayashi and Hakuno 2003; Bae et al. 2005). This protein was found upregulated in moderately resistant 'Nongke No.1' suggesting protection offered by cell polarization hampers early-stage fungal infection in moderately resistant 'Nongke No.1' cultivar than in highly resistant 'Yueyoukang I' and susceptible 'Brazil'. Two PR5-like (thaumatin-like) and PR4-like (endochitinase) genes were found to be upregulated by both Foc 1 and Foc 4 strains at 1–2 days postinoculation.

14.3.2 Pathogenicity Mechanism of Foc

Genome analysis revealed a large number of putative virulence-associated genes in both Foc race 1 and 4 genomes, including genes putatively involved in root adhesion, cell wall degradation, detoxification of toxin, transport, secondary metabolite biosynthesis and signal transductions.

Guo et al. (2015) identified five and six adhesin genes in Foc1 and Foc4 genomes, respectively, required for adhering to host tissues. Three putative secreted proteins (SPs) encoded by Foc were found significantly homologous to INF2A, INF2B effectors of *P. infestans* (Huitema et al. 2005) and PosjNIPw of *Phytophthora sojae* (Qutob et al. 2002) which are able to elicit hypersensitive response or induce necrosis in host plants, suggesting the involvement of secreted proteins in Foc-banana interaction. The SIX coding genes secreted by Foc for pathogenicity involve three orthologs of SIX1, namely, Six1a, Six1b and Six1c in Foc4 genome while only one copy of SIX1 in Foc1 genome functional at vegetative and postinfection stage. Ma et al. (2010) reported chromosome 14 of Foc1 harbouring SIX effector genes, namely, *SIX5*, *SIX6* and *SIX7* as a 'pathogenicity' chromosome and demonstrated that the transfer of lineage-specific chromosomes between genetically isolated strains leads to the emergence of new pathogenic lineages in *F. oxysporum*. Fraser-Smith et al. (2014) used PCR and sequencing approach to identify variation in *Foc-SIX8* to differentiate race 4 from race 1 and 2 isolates and also subtropical and tropical races of Foc.

Fungi have evolved diversity of enzymes to overcome the chemical barrier synthesized in the host, viz., cytochrome P450s (CYP) of fungal origin is required for biosynthesis of secondary metabolites and detoxification of toxic compounds (Ichinose 2012). Both Foc genomes encode a great number of putative CYP genes similar to the fungal CYP sequences of monooxygenase gene BcBOT1 involved in phytotoxin biosynthesis in *Botrytis cinerea* and the demethylase genes PDAT9 and PDA6-1 that participate in detoxifying the phytoalexin pisatin from garden pea. Also few peroxidases were putatively implicated in the removal of toxins synthesized by the host. Foc1 encodes 25 peroxidases similar to the catalase peroxidases, VlcpeA from *Verticillium longisporum* (Singh et al. 2012) and CPXB from *M. oryzae* (Tanabe et al. 2011) that break down hydrogen peroxide (H_2O_2) generated by the host plant during the fungal infection. Li et al. (2013c) reported the presence of beauvericin and fusaric acid in all tissues of wilt-infected banana correlating with the virulence of FOC isolates. Importantly, relative to the Foc race 1 isolate (Foc1), the Foc race 4 isolate (Foc4) has evolved with few expanded gene families of transporters and transcription factors for transport of toxins and nutrients for better adoption to host environments and contribute to pathogenicity to banana.

In fungi, the two-component regulatory system serves significant roles in recognition and adaption of the environmental change. Accordingly it was found that Foc1 and Foc 4 could modulate the expression of different histidine kinase (HK) and response regulator (RR) genes resulting in activation of transcription or a

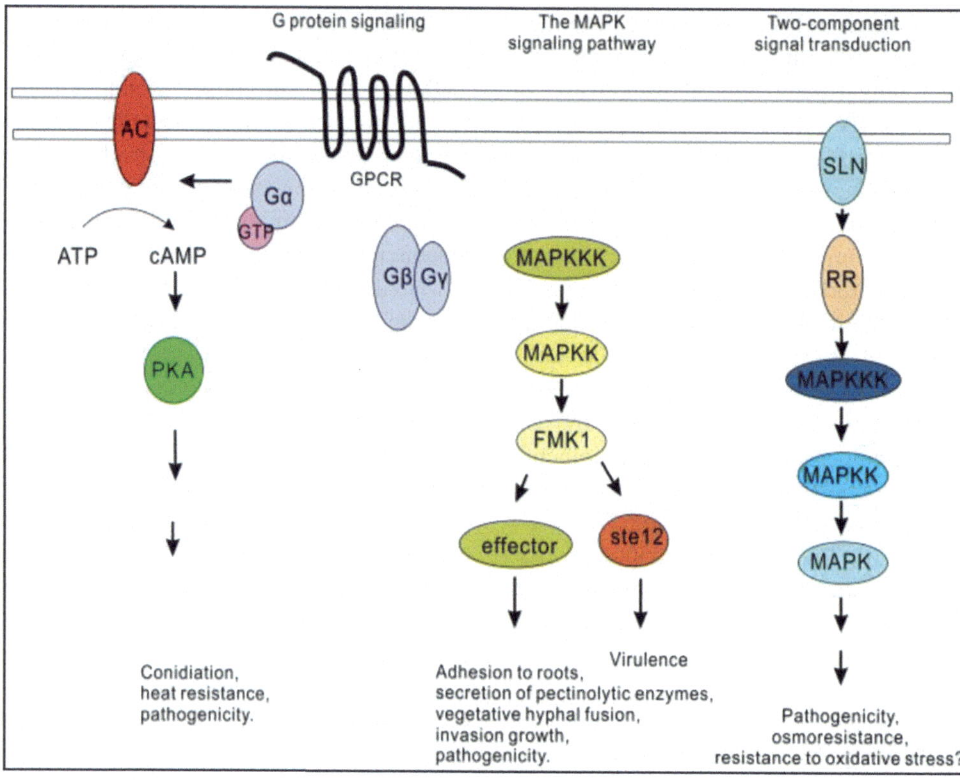

Fig. 14.5 Signalling pathway activated in Foc4 during infection (Guo et al. 2015)

mitogen-activated protein kinase cascade during infection (Catlett et al. 2003). Guo et al. (2015) revealed 115 genes encoding putative GPCRs indicating the conservation of G protein-mediated signalling pathway in ascomycete fungi (Fig. 14.5).

14.4 Biotechnological Approaches for Disease Resistance

Genetic modification of banana using biotechnological approaches is considered as a path towards increasing the value of this crop as conventional breeding is limited due to its long generation time (up to 2 years), narrow genetic base, seedless nature, male sterility, low female fertility and polyploidy nature. Genetic engineering has played a vital role in developing transgenic plant in banana as it is amenable to transformation due to ease of integration, potential low copy number and importantly integration into transcriptional active regions of chromosomes (Koncz et al. 1989). Isolation and stable expression of genes determine the success of transgenics. Table 14.1 summarizes *Agrobacterium*-mediated transformation of banana for *Fusarium* resistance. Different approaches enabled to achieve fungal resistance are discussed below.

14.4.1 Cisgenic Approach

Cisgenesis is the introduction of native genes and promoters isolated from related species or from the crop plant itself. Stacking of resistance genes as a strategy is more feasible as the existing varieties can be improved directly using genes from the gene pool of breeders. Marker-gene-free transformation protocols are main tools for cisgenesis in plants for resistance and quality breeding of vegetatively propagated heterozygous crops, such as potato and banana. The linkage-

Table 14.1 Summary of *Agrobacterium*-mediated transformation of banana for *Fusarium* resistance

Sl. no	Trait	Genotype and cultivar	Explant	Gene	*Agrobacterium* strain and OD	Vector type and promoter	Selection marker	Transformation efficiency %	References
1	*Fusarium* wilt	AAB cv. Rasthali	ECS	*Musa*DAD1	EHA105 (0.2)	pCAMBIA1301 (polyubiquitin promoter)	hpt	38	Ghag et al. (2014c)
				*Musa*BAG1				36	
				*Musa*BI1				33	
2		AAB cv. Rasthali	ECS	Defensin (Sm-AMP-D1)	EHA105 (0.2)	pCAMBIA1301 (polyubiquitin promoter)	hpt	–	Ghag et al. (2014b)
3		AAB cv. Rasthali	ECS	Petunia floral defensin (PhDef1/PhDef2)	EHA105 (0.2)	pCAMBIA1301 (polyubiquitin promoter)	hpt	–	Ghag et al. (2012)
4		AAB cv. Rasthali	ECS	velvet (vel)	EHA105 (0.2)	pCAMBIA1301 (polyubiquitin promoter)	hpt	11	Ghag et al. (2014a)
				Fusarium transcription factor 1 (Ftf1)				12	
5		AAA cv. Pei Chiao and cv. Gros Michel	Cauliflowerlike bud clumps	Plant ferredoxin-like protein (pflp)	EHA105 (0.8–1.0)	pBI (35SCaMV)	nptII	12.6	Yip et al. (2011)
								51.3	
				Arabidopsis root-type ferredoxin gene (Atfd3)	C58C1 (0.8–1.0)			13.1	
								4	
6		AAB cv. Lady Finger	ECS	Bcl-xL	AGL1 (0.5)	pCAMIA2301 (polyubiquitin promoter)	nptII	31	Paul et al. (2011)
7		ECS cv. Sukali Ndizi	ECS	Mced9 (modified form of ced9)	AGL1 (0.6)	pYC11 (polyubiquitin promoter)	nptII	42	Magambo (2012)
8		AAB Pisang Nangka	Cauliflower like buds	Thaumatin-like protein (tlps)	Particle bombardment	pCambia1304 (35SCaMV)	hpt	6.6	Mhadavi et al. (2012)
9		AA cv. Furenzhi		Endochitinase gene chit42	EHA105	pCAMBIA2301 (CaMV 35S)	nptII	186	Hu et al. (2013)
10		AAB	Multiple bud clumps	β-1-3-endoglucanase	LBA 4404	pROKla-Eg (CaMV 35S)	nptII	–	Mazhai et al. (2007)
11		AAA cv. Taijiao	Apical meristem	Human lysozyme	EHA105	pCAMBIA1301 (CaMV35S)	Hpt	0.95	Pie et al. (2005)
12		AAB cv. Rasthali	ECS	MSI-99 (magainin analogue synthetic peptide)	EHA105 (0.2)	pSAN164 and	nptII	35	Chakrbarti et al. (2003)
						pSAN168 (polyubiquitin promoter)		20	

drag-free cisgenic approach is the most attractive and embodies the possibility to stack resistance genes found in related wild species.

14.4.1.1 Resistance Gene (R Gene)

Resistance (R) genes in plants play an essential role in preventing disease. R genes encode receptors for Avr gene-dependent molecules of pathogen (Staskawicz et al. 1995). The pathogen-derived protein recognition leads R gene product to activate plant defence mechanisms. The majority of known plant resistance genes contain NBS and LLR domains. The conserved motifs within NBS/LLR domains have been exploited to identify new resistance gene analogue for effective genetic transformation and generation of disease-resistant plant using natural disease resistance gene pools which are prevailing in resistant germplasm. Peraza-Echeverria et al. (2008) characterized R gene candidate of NBS type (RGA 2) from *Musa acuminata* spp. *malaccensis*, a resistant cultivar for *Fusarium* wilt resistance. Sun et al. (2010) isolated 20 different types of RGAs from banana cv. Goldfinger resistant to Foc race 4. Phylogeny analysis of nucleotide sequence revealed that a homology percentage of 28–54 exists among RGAs known for *Fusarium* wilt resistance such as Fom2, 12c-1, 12c and 12 in other crops. *M. acuminata* ssp. *malaccensis*, which is a wild diploid subspecies of the *M. acuminata*, has been found to be highly resistant to Foc4, and the resistance is controlled by a single dominant gene (Smith and Hamill 1999). Isolation and utilization of such R gene in the genetic transformation of banana impart resistance to Foc4.

14.4.1.2 Native Cell Death Genes

Foc causes cell death in banana tissues probably by modulating genes that negatively regulate the plant cell-death (PCD) pathway (Lam et al. 2001; Thaler et al. 2004). The PCD-associated proteins in plants, namely, defender against death (DAD) domain protein, involved in glycosylation of proteins for correct folding and activity, have been widely reported in crop plants, viz., *Arabidopsis*, rice and pea (Dong et al. 1998). Other such proteins are the Bcl-2-associated athanogenes (BAG) which function as cochaperones in diverse physiological roles in cell survival, cell differentiation and stress responses (Doukhanina et al. 2006). The *Arabidopsis* BAG7 protein (AtBAG7) is an ER-resident protein that restores cellular homeostasis by facilitating efficient protein folding (Williams et al. 2010). BAX inhibitor (BI) is a membrane-bound protein that blocks the caspase-mediated cell-death process in animal cells. The rice suspension cells treated with the cell wall extract of rice blast fungus *Magnaporthe grisea showed* marked decrease in the BI-1 (OsBI-1) gene (Matsumura et al. 2003). Ghag et al. (2014a) postulated that transgenic banana constitutively overexpressing cell-death-related native genes can impart resistance to pathogenic fungi like Foc. In view of this, cell-death-related genes were isolated by treating the embryogenic cells with Foc culture and *Fusarium* toxin such as fusaric acid and beauvericin. The expressed cell-death-related genes like *MusaDAD1*, *MusaBAG1* and *MusaBI1* were overexpressed individually in transgenic banana cv. Rasthali plants. The resulted transgenic banana plants expressing *MusaBAG1* showed significantly enhanced resistance to *Fusarium* wilt disease compared to *MusaBI1*and *MusaDAD1* overexpressing transgenic plants.

14.4.2 Transgenic Approach (Heterologous Expression)

14.4.2.1 PR Protein Genes

Pathogenesis-related proteins are plant proteins produced in response to pathogen invasion. On the basis of their primary structure homology, 17 families of PR protein have been currently identified. Among the PR proteins, hydrolytic enzymes (chitinase and glucanase), thionins and defensins are especially important. Overexpression of genes encoding for PR proteins in banana for *Fusarium* resistance has met with some success.

Hu et al. (2013) developed banana plant resistant to wilt disease by introducing an endochitinase gene (*chit* 42) into cv. Furenzhi. The transgenic plants showed tolerance to wilt disease, even after 2 months of postinoculation.

Maziah et al. (2007) cloned gene encoding *β-1-3-endoglucanase* and expressed the same in banana cv. Rasthali to assess the level of tolerance against wilt caused by Foc race 1. The highest level of expression of the gene and enhanced disease resistance against Foc race 1 was obtained. Similarly, Sreeramanan et al. (2006) expressed both *chitinase* and *β-1,3-glucanase* gene in cv. Rasthali tolerance against Foc.

Ghag et al. (2012) conducted independent transformation of two floral defensin genes *phdef1* and *phdef2* into banana cv. Rasthali. The pot bioassay results demonstrated that transgenic plants expressing floral defensin gene showed less external and internal symptoms in contrast to control plants. Control plants succumbed to wilt disease after 6 weeks of postinoculation of Foc culture, whereas phenotypically normal transgenic plants showed mild symptoms and completely recovered within 3 weeks of postinoculation. These results indicated that the expression of defensin gene in the host plant enhanced resistance to *Fusarium* wilt, and it can be used with the other PR gene for gene stacking in order to develop pathogen-free trait. Recently, Ghag et al. (2014b) isolated a defensin gene *Sm-AMP-D1* from chickweed *Stellaria* media against *Fusarium oxysporum*. The *Sm-AMP-D1* defensin under in vitro condition showed inhibitory activity against Foc at IC50 value of 0.35, and the transgenic banana plants overexpressing *Sm-AMP-D1* showed enhanced tolerance to Foc under pot culture condition. Mahadavi et al. (2012) transformed thaumatin-like protein (*tlp*) gene isolated from the rice into the banana *Musa sapientum* cv. Nangka (AAB). The transgenic plant overexpressing *tlp* gene exhibited enhanced resistance to *Fusarium* wilt compared to control plants after 30 days of inoculation with Foc race 4 under pot culture condition.

14.4.2.2 Antimicrobial Peptide Gene

The antimicrobial peptides are small peptides, globular, cationic and amphipathic in nature, enriched with cysteine amino acids, encoded by multigenic families. AMPs have a broad range of activity and widely distributed among plant kingdom (Boman 1991; Hancock and Lehrer 1998). These peptides play a crucial role in the innate immunity and act as a first line of defence barrier in the plant system for various invading phytopathogens. Pei et al. (2005) transformed human *lysozyme* gene into banana plant cv. Taijiao (AAA). Transgenic plants overexpressing human *lysozyme* showed enhanced resistance against *Fusarium* wilt disease not only in pot condition but also in containment field condition. Overexpression of *pflp* gene in banana cv. Pei Chiao and cv. Gros Michel showed very less percentage of severity on exposure to Foc race 4 over a period of 9 weeks indicating their potentiality for resistance against Foc, apart from its role in controlling bacterial diseases (Yip et al. 2011).

Chakrabarti et al. (2003) cloned a magainin analogue (*MSI-99*), a small alpha helical peptide with 23 amino acids from the African clawed frog, that exerts broad spectrum activity to various bacteria, fungi and protozoa, besides having antitumorigenic activity. The authors utilized the potential of this peptide for imparting resistance against Foc in banana cv. Rasthali. They developed transgenic banana cv. Rasthali with *MSI-99* gene after determining the lytic dose of the peptide on the fungus where a concentration of 16 μg ml^{-1} and 128 μg ml^{-1} effectively inhibited spore formation and complete Foc inhibition, respectively. Most of the transgenic banana plants developed were tolerant against the Foc under primary pot culture evaluation.

The *Allium cepa* antimicrobial protein1 (Ace-AMP1) isolated from *Allium cepa* seeds shared a DNA sequence homology with nsLTPs. To date, successful utilization of *Ace-AMP1* in transgenic development has been reported for various bacterial and fungal diseases due to its strong antimicrobial activity against number of phytopathogens (Cammue et al. 1995). Mohandas et al. (2013) developed transgenic banana with Ace-AMP1 gene to assess the level of resistance against *Fusarium* wilt disease. ECS was transformed with the Ace-AMP1 gene by *Agrobacterium*-mediated transformation, and 148 individual transformants were developed. Gene integration and copy number were finally confirmed by Southern hybridization which indicated stable integration of Ace-AMP1 gene in seven transgenic plants derived from ECS. The expression level of Ace-AMP1 gene in three transgenic lines 19, 14 and 6 was around 4.13–4.83 μg mg^{-1}, and in other lines it was found less than 4 μg mg^{-1}. The transgenic lines showed significantly less percentage of disease index (PDI) compared to control plants over a period of 6 months under greenhouse condition. A contained field trial was conducted with the approval of Genetic Engineering Appraisal Committee (GEAC) and No Objection Certificate (NOC)from the state government of Karnataka with transgenic lines and non-transgenic plants challenged with the 150 g Foc inocula containing 2.8×10^6 spores and mycelia g^{-1}. The results revealed transgenic lines showing less incidence (PDI, 46.67–86.67 %, and VDI, 41–44 %) compared to control lines which showed 94 % of VDI. The transgenic lines challenged with the pathogens, however, did not flower under field conditions. Non-transgenic lines challenged or not challenged also got the infection and did not flower (unpublished data). The authors concluded that the role of dosage of the *Fusarium* spores to be used in such trials needs to be standardized and if expression level of Ace-AMP1 gene is increased in these plants higher level of resistance could be obtained.

14.4.2.3 Antiapoptosis-Related Proteins

Antiapoptosis genes have been shown as promising candidate to engineer resistance against necrotrophic fungal pathogens which prefer dead tissues to trigger nutrient leakage for survival. Apoptosis could be inhibited by overexpressing antiapoptotic members such as Bcl 2, Bcl-xl from mammalian and Ced9 from *Caenorhabditis elegans*. These proteins can regulate activation of Bax and Bak thus preventing cytochrome C release in the cytosol and subsequent apoptosis event. Paul et al. (2011) stably transformed banana cv. Lady Finger with apoptosis inhibition related, namely, *Bcl-xL*, *Ced-9* and *Bcl-2*. They investigated the effect of individual gene on disease resistance against Foc race 1. Transgenic lines examined for the challenge with Foc race 1 for a period of 12 weeks showed significantly less internal and external disease symptoms than the wild-type susceptible Lady Finger banana plants. Among them *Bcl2* 3′ UTR expressing transgenic line was highly resistant even after continuous exposure of transgenic plants for a period of 23 weeks.

Similarly, banana cv. Sukali Ndizi the most popular dessert banana in the East African region was transformed with a modified form of anti-apoptosis gene *ced9*. After 13 weeks of inoculation with Foc, transgenic plants showed significantly less disease severity compared to control plants (Magambo 2012).

14.4.2.4 RNAi-Mediated Host-Induced Silencing

The identification and silencing of important genes required for the invasion, growth and pathogenesis of fungi-infecting plants offer an ideal strategy for fungal pathogen control in crop plants. RNA interference (RNAi) can serve as an emerging strategy for control of pathogens, through silencing of a vital gene associated with pathogens. Mumbanza et al. (2013) developed synthetic dsRNAs for adenylate cyclase, the alpha and delta subunit of DNA polymerase, which showed varying levels of inhibition of spore germination of both the fungi Foc and *M. fijiensis* in vitro. Recently Ghag et al. (2014a) reported the application of host-induced gene silencing (HIGS) technique against *Fusarium* wilt resistance in banana cv. Rasthali. The transgenic plants with partial coding sequence of velvet family genes (*VeA*, *VelB* and *VosA*) involved in fungal morphogenesis and *Fusarium* transcription factor 1 (*ftf1*) required for fungal colonization and infection process were individually transformed into banana cv. Rasthali by *Agrobacterium*-mediated gene transformation. The resulted individual transgenic banana plants expressing fungal gene-targeted siRNAs showed significant tolerance towards *Fusarium* wilt disease even after 8 months post-Foc inoculation under greenhouse controlled condition. HIGS was also shown in *Arabidopsis* and in barley by expressing a dsRNA targeting the cytochrome P450 lanosterol C-14α-demethylase (*CYP51*) enzyme of fungal origin encoded by three genes, namely, *CYP51A*, *CYP51B and CYP51C*. The resulted transgenic plants expressing stacked three genes showed enhanced resistance to *Fusarium* head blight (FHB) and root rot disease caused by *Fusarium graminearum* (Koch et al. 2013).

14.5 Challenges and Opportunities for Enhancing Crop Disease Resistance via Transgenic Technology

Banana is an important monocot crop species, and molecular biotechnology offers great hope for improving banana for enhanced disease resistance and for other agronomically important traits by overcoming the constraints imposed by the sterility of cultivated banana. The unravelling of defence mechanism of banana in response to infection aids effective disease control by molecular breeding. Many defence-related genes and pathways in banana are different from existing pathways of rice and *Arabidopsis*, model plants suggesting that the mechanisms underlying host defence in plants may be variable. The genes that specifically expressed in the incompatible interaction could play an important role in the interaction of banana and Foc, and their spatial and temporal expressions require further study. A dominant resistance trait conferring gene in banana is the I-like R gene can offer great potential for imparting resistance to Foc.

Through transcriptomic profiling, the differentially expressed genes and putative signalling pathways identified will hopefully accelerate research on resistance in banana to Foc and contribute to a better understanding of the banana defence response to plant pathogens. Several defence-associated genes identified in resistant banana roots by suppression subtractive hybridization can be employed for better resistance to different Foc races by stacking genes. The proteomic profile analysis and whole genome sequence of banana are favourable to investigate the complex defence mechanism and the effect of pathogenesis in banana.

14.6 Conclusion

Genes that confer a broad spectrum resistance will be particularly helpful in genetic modification of *Musa* for developing disease-resistant plants. The transcriptome data and gene expression profiles of different Foc races and availability of the whole genome sequences of banana are very useful source of genes to be targeted for mediating resistance to *Fusarium* wilt disease.

Efficient transgenic approaches for resistance in banana against *Fusarium* wilt have been attempted by overexpressing petunia floral defensins, Ace-AMP1 gene, cell-death genes and HIGS in transgenic banana plants. Stacking of genes with different modes of action can also serve as an effective strategy to combat disease. Although the introduced genes are well defined, the field trials are lacking in many of the experiments.

Since contained field study will provide the opportunity to ascertain the effectiveness of gene in controlling the disease as well as enlighten on unexpected or undesirable consequences that have resulted from the transformation procedure, it is necessary to evaluate the effectiveness of genes in situ.

References

Anterola AM, Jeon JH, Davin LB, Lewis NG (2002) Transcriptional control of monolignol biosynthesis in pinus taeda: factors affecting monolignol ratios and carbon allocation in phenylpropanoid metabolism. J Biol Chem 277:18272–18280

Bae H, Herman E, Bailey B, Bae HJ, Sicher R (2005) Exogenous trehalose alters Arabidopsis transcripts involved in cell wall modification, abiotic stress, nitrogen metabolism, and plant defense. Physiol Plant 125:114–126

Bancroft J (1876) Report of the board appointed to enquire into the cause of disease affecting livestock and plants. In: Votes and proceedings 1877, Queensland, 3, pp 1011–1038

Beckman CH (1990) Host responses to the pathogen. In: Ploetz RC (ed) Fusarium wilt of banana. APS Press, St Paul, pp 93–105

Berrocal-Lobo M, Molina A (2004) Ethylene response factor 1 mediates Arabidopsis resistance to the soilborne fungus Fusarium oxysporum. Mol Plant Microbe Interact 17:763–770

Boman HG (1991) Antibacterial peptides: key components needed in immunity. Cell 65:205–207

Cammue BPA, Thevissen K, Hendriks M, Eggermont K, Goderis IJ, Proost P, VanDamme J, Osborn RW, Guerbette F, Kader JC, Broekaert W (1995) A potent antimicrobial protein from onion seeds showing sequence homology to plant lipid transfer proteins. Plant Physiol 109:445–455

Catlett NL, Yoder OC, Turgeon BG (2003) Whole-genome analysis of two component signal transduction genes in fungal pathogens. Eukaryot Cell 2:1151–1161

Chakrabarti A, Ganapathi TR, Mukherjee PK, Bapat VA (2003) MSI-99, a magainin analogue, imparts enhanced disease resistance in transgenic tobacco and banana. Planta 216:587–596

Chen YP, Chen YF, Zhao JT, Huang X, Huang XL (2007) Cloning and expression of resistance gene analogs (RGAs) from wild banana resistant to banana Fusarium wilt. J Plant Physiol Mol Biol 33:567–573

Chinchilla D, Zipfel C, Robatzek S, Kemmerling B, Nurnberger T, Jones JD, Felix G, Boller T (2007) A flagellin-induced complex of the receptor FLS2 and BAK1 initiates plant defence. Nature 448:497–500

Chisholm ST, Coaker G, Day B, Staskawicz BJ (2006) Host-microbe interactions: shaping the evolution of the plant immune response. Cell 124:803–814

De Ascensao AR, Dubery IA (2000) Panama disease: cell wall reinforcement in banana roots in response to elicitors from Fusarium oxysporum f.sp. cubense race four. Phytopathology 90:1173–1180

De Ascensao ARFDC, Dubery IA (2003) Soluble and wall-bound phenolics and phenolic polymers in *Musa acuminata* roots exposed to elicitors from *Fusarium oxysporum* f.sp. *cubense*. Phytochemistry 63:679–686

Dixon DP, Davis BG, Edwards R (2002) Functional divergence in the glutathione transferase superfamily in plants. J Biol Chem 277:30859–30869

Dong YH, Zhan XC, Kvarnheden A, Atkinson RG, Morris BA, Gardner RC (1998) Expression of a cDNA from apple encoding a homologue of DAD1, an inhibitor of programmed cell death. Plant Sci 139:165–174

Doukhanina EV, Chen S, Zalm E, Godzik A, Reed J, Dickman MB (2006) Identification and functional characterization of the BAG protein family in Arabidopsis thaliana. J Biol Chem 281:18793–18801

Dowd C, Wilson LW, McFadden H (2004) Gene expression profile changes in cotton root and hypocotyl tissues in response to infection with *Fusarium oxysporum f. sp vasinfectum*. Mol Plant Microbe Interact 17:654–667

Food and Agriculture organization (2014) (http://www.fao.org/news/archive/news-by-date/2014/en/)

Fraser-Smith S, Czislowski E, Meldrum RA, Zander M, O'Neill W, Balali GR, Aitken EAB (2014) Sequence variation in the putative effector gene *SIX8* facilitates molecular differentiation of *Fusarium oxysporum* f. sp. *cubense*. Plant Pathol 63:1044–1052

García-Bastidas F, Ordóñez N, Konkol J, Al-Qasim M, Naser Z, Abdelwali M, Waalwijk C, Ploetz RC, Kema GHJ (2014) First report of *Fusarium oxysporum* f. sp. *cubense* tropical race 4 associated with panama disease of banana outside Southeast Asia. Plant Dis 98:694

Ghag SB, Shekhawat UKS, Ganapathi TR (2012) Petunia floral defensins with unique prodomains as novel candidates for development of Fusarium wilt resistance in transgenic banana plants. PLoS One 7:e39557

Ghag SB, Shekhawat UKS, Ganapathi TR (2014a) Host induced posttranscriptional hairpin RNA-mediated gene silencing of vital fungal genes confers efficient resistance against Fusarium wilt in banana. Plant Biotechnol J 12:541–553

Ghag SB, Shekhawat UKS, Ganapathi TR (2014b) Transgenic banana plants expressing a Stellaria media defensing gene (Sm-AMP-D1) demonstrate improved resistance to *Fusarium oxysporum*. Plant Cell Tissue Organ 119:247–255

Ghag SB, Shekhawat UKS, Ganapathi TR (2014c) Native cell-death genes as candidates for developing wilt resistance in transgenic banana plants. AoB Plants 6:plu037. doi:10.1093/aobpla/plu037

Guo L, Han L, Yang L, Zeng H, Fan D, Zhu Y, Feng Y, Wang G, Peng C, Jiang X, Zhou D, Ni P, Liang C, Liu L, Wang J, Mao C, Fang X, Peng M, Huang J (2015) Genome and transcriptome analysis of the fungal pathogen fusarium oxysporum f. sp. cubense causing banana vascular wilt disease. PLoS ONE 10(1): e0117621. doi:10.1371/journal.pone.0117621

Hancock RE, Lehrer R (1998) Cationic peptides: a new source of antibiotics. Trends Biotechnol 16:82–88

Hofius D, Tsitsigiannis DI, Jones JD, Mundy J (2007) Inducible cell death in plant immunity. Semin Cancer Biol 17:166–187

Horry JP, Arnaud OE, Crouch JH, Ferris RSB, Jones DR, Mateo N, Picq C, Vuylsteke D (1997) Banana and plantain. In: Fuccillo D, Sears L, Stapleton P (eds) Biodiversity in trust. Conservation and use of plant genetic resources in CGIAR centres. Cambridge University Press, Cambridge, pp 67–81

Hoshikawa K, Ishihara G, Takahashi H, Nakamura I (2012) Enhanced resistance to gray mold (*Botrytis cinerea*) in transgenic potato plants expressing thionin genes isolated from *Brassicaceae* species. Plant Biotechnol 29:87–93

Hu CH, Wei YR, Huang YH, Yi GJ (2013) An efficient protocol for the production of chit42 transgenic Furenzhi banana (Musa spp. AA group) resistant to *Fusarium oxysporum*. In Vitro Cell Dev Biol Plant 49:584–592

Huitema E, Vleeshouwers VG, Cakir C, Kamoun S, Govers F (2005) Differences in intensity and specificity of hypersensitive response induction in Nicotiana spp. by INF1, INF2A, and INF2B of Phytophthora infestans. Mol Plant Microbe Interact 18:183–193

Ichinose H (2012) Molecular and functional diversity of fungal cytochrome P450s. Biol Pharm Bull 35:833–837

Jones DR (2000) History of banana breeding. In: Jones DR (ed) Disease of banana. Abaca and enset. CAB international, Wallingford

Kavino M, Kumar N, Damodaran T, Harish S, Saravanakumar D (2007) Biochemical markers as a useful tool for the early identification of *Fusarium oxysporum* f.sp. *cubense*, race1 resistance banana clones. Arch Phytopathol Plant Protect 42:1069–1078

Kobayashi I, Hakuno H (2003) Actin-related defense mechanism to reject penetration attempt by a non-pathogen is maintained in tobacco BY-2 cells. Soc Forces 217:340–345

Koch A, Kumar N, Weber L, Keller H, Imani J, Kogel KH (2013) Host induced gene silencing of cytochrome P450 lanosterol C14α-demethylase – encoding genes confers strong resistance to Fusarium species. Proc Natl Acad Sci U S A 110:19324–19329

Koncz C, Martini N, Mayerhofer R, Koncz-kalman Z, Korber H, Redei§ G, Schell J (1989) High-frequency T-DNA-mediated gene tagging in plants. Proc Natl Acad Sci U S A 86:8467–8471

Lam E, Kato N, Lawton M (2001) Programmed cell death, mitochondria and the plant hypersensitive response. Nature 411:848–853

Li J, Wen J, Lease KA, Doke JT, Tax FE, Walker JC (2002) BAK1, an Arabidopsis LRR receptor-like protein kinase, interacts with BRI1 and modulates brassinosteroid signaling. Cell 110(2):213–222

Li CY, Chen S, Zuo CW, Sun Q, Ye Q, Yi GJ, Huang BZ (2011a) The use of GFP-transformed isolates to study infection of banana with *Fusarium oxysporum* f. sp. *cubense* race 4. Eur J Plant Pathol 131:327–340

Li WM, Qian CM, Mo YW, Hu YL, Xie JH (2011b) Tolerance of banana for fusarium wilt is associated with early H_2O_2 accumulation in the roots. Afr J Biotechnol 10:11378–11387

Li CY, Deng GM, Yang J, Viljoen A, Jin Y, Kuang RB, Zuo CW, Lv ZC, Yang QS, Sheng O, Wei YR, Hu CH, Dong T, Yi GJ (2012) Transcriptome profiling of resistant and susceptible Cavendish banana roots following inoculation with *Fusarium oxysporum* f. sp. *cubense* tropical race 4. BMC Genomics 13:374

Li C, Zuo C, Deng G, Kuang R, Yang Q et al (2013a) Contamination of bananas with beauvericin and fusaric acid produced by *Fusarium oxysporum* f. sp. *cubense*. PLoS One 8(7):e70226

Li X, Bai T, Li Y, Ruan X, Li H (2013b) Proteomic analysis of *Fusarium oxysporum* f. sp. *cubense* tropical race 4-inoculated response to Fusarium wilts in the banana root cells. Proteome Sci 11:41

Li C, Shao J, Wang Y, Li W, Guo D, Yan B, Xia Y, Peng M (2013c) Analysis of banana transcriptome and global gene expression profiles in banana roots in response to infection by race 1 and tropical race 4 of *Fusarium oxysporum* f. sp. *cubense*. BMC Genomics 14:851

Ma LJ, van der Does HC, Borkovich KA, Coleman JJ, Daboussi MJ, Di Pietro A, Dufresne M, Freitag M et al (2010) Comparative genomics reveals mobile pathogenicity chromosomes in Fusarium. Nature 464:367–373

MacHardy WE, Beckman CH (1981) Vascular wilt Fusaria; infection and pathogenesis. In: Nelson PE, Toursoun TA, cook RT (eds) Fusarium; disease biology and taxonomy. The Pennsylvania State University Press, University Park/London, pp 365–390

Magambo B (2012) Generation of transgenic banana (cv. Sukali Ndizi) resistance to Fusarium wilt, B.Sc hn. Queensland University of Technology, Australia

Mahdavi F, Sariah M, Maziah M (2012) Expression of rice thaumatin-like protein gene in transgenic banana plants enhances resistance to Fusarium wilt. Appl Biochem Biotechnol 166:1008–1019

Mai WF, Abawi GS (1987) Interactions among root-knot nematodes and Fusarium wilt fungi on host plants. Annu Rev Phytopathol 25:317–338

Matsumura H, Nirasawa S, Kiba A, Urasaki N, Saitoh H, Ito M, Kawai-Yamada M, Uchimiya H, Terauchi R (2003) Overexpression of Bax inhibitor suppresses the fungal elicitor-induced cell death in rice (Oryza sativa L.) cells. Plant J 33:425–434

Maziah M, Sariah M, Sreeramanan S (2007) Transgenic banana Rastali (AAB) with β-1, 3-glucanase gene for tolerance to Fusarium wilt race 1 disease via *Agrobacterium*-mediated transformation system. Plant Pathol J 6:271–282

Melida H, Caparros-Ruiz D, Alvarez J, Acebes JL, Encina A (2011) Deepening into the proteome of maize cells habituated to the cellulose biosynthesis inhibitor dichlobenil. Plant Signal Behav 6:143–146

Mohandas S, Sowmya HD, Saxena AK, Meenakshi S, Rani RT, Mahmood R (2013) Transgenic banana cv. Rasthali (AAB, Silk gp) harboring Ace-AMP1gene imparts enhanced resistance to Fusarium oxysporum f.sp. cubense race. Sci Horticol 164:392–399

Moore NY, Pegg K, Allen AR, Irvin JAG (1993) Vegetative compatibility and distribution of *Fusarium oxysporum* f. sp. *cubense* in Australia. Aust J Exp Agric 33:797–802

Mumbanza FM, Kiggundu A, Tusiime G, Tushemereirwe WK, Nib-let C, Bailey A (2013) In vitro antifungal activity of synthetic dsRNA molecules against two pathogens of banana, *Fusarium oxysporum* f. sp. *cubense* and *Mycosphaerella fijiensis*. Pest Manag Sci 69:1155–1162

Nelson PE (1991) History of Fusarium systematics. Phytopathology 81:1045–1048

Osuji JO, Harrison G, Crouch J, Heslop-Harrison JS (1997) Identification of the genomic constitution of Musa L. lines (bananas, plantains and hybrids) using molecular cytogenetics. Ann Bot 80:787–793

Pandey SP, Roccaro M, Schon M, Logemann E, Somssich IE (2010) Transcriptional reprogramming regulated by WRKY18 and WRKY40 facilitates powdery mildew infection of Arabidopsis. Plant J 64:912–923

Paul JY, Becker DK, Dickman MB, Harding RM, Khanna HK, Dale JL (2011) Apoptosis-related genes confer resistance to Fusarium wilt in transgenic 'Lady Finger' bananas. Plant Biotechnol J 9:1141–1148

Pei XW, Chen SK, Wen RM, Ye S, Huang JQ, Zhang YQ, Wang BS, Wang ZA, Jia SR (2005) Creation of transgenic banana expressing human lysozyme gene for Panama wilt resistance. J Integr Plant Biol 47:971–977

Peraza-Echeverria S, Dale JM, Harding RM, Smith MK, Collet CC (2008) Characterization of disease resistance gene candidates of the nucleotide binding site (NBS) type from banana and correlation of a transcriptional polymorphism with resistance to *Fusarium oxysporum* f.sp. *cubense* race 4. Mol Breed 22:565–579

Ploetz RC (2000) Panama disease: a classic and destructive disease of banana. Plant Health Progress 10:1–7

Ploetz RC, Pegg KG (1997) Fusarium wilt of banana and Wallace's line: was the disease originally restricted to his Indo-Malayan region? Australas Plant Pathol 24:38–43

Qutob D, Kamoun S, Gijzen M (2002) Expression of a Phytophthora sojae necrosis-inducing protein occurs during transition from biotrophy to necrotrophy. Plant J 32:361–373

Simmonds NW (1962) Evolution of the bananas. Longmans, Green and Co, London

Singh S, Braus-Stromeyer SA, Timpner C, Valerius O, von Tiedemann A et al (2012) The plant host Brassica napus induces in the pathogen Verticillium longisporum the expression of functional catalase peroxidase which is required for the late phase of disease. Mol Plant Microbe Interact 25:569–581

Smith SN (2007) An overview of ecological and habitat aspects in the genus Fusarium with special emphasis on the soil-borne pathogenic forms. Plant Pathol Bull 16:97–120

Smith MK, Hamill SD (1999) Banana tissue culture for clean, sustainable production. In: Final report (FR96013) Horticultural Research and Development Corporation, Gordon

Snyder WC, Hansen HN (1940) The species concept in Fusarium. Am J Bot 27:64–67

Sreeramanan S, Maziah M, Abdullah MP, Sariah M, Xavier R (2006) Transient Expression of *gusA* and *gfp* gene in *Agrobacterium* mediated banana transformation using single tiny meristematic bud. Asian J Plant Sci 1:1–13

Staskawicz BJ, Ausubel FM, Baker BJ, Ellis JG, Jones JD (1995) Molecular genetics of plant disease resistance. Science 268:661–667

Stover RH (1962) Fusarial wilt (Panama disease) of bananas and other *Musa* species. In: Banana Plantain and Abaca diseases'. 9–26. Commonwealth Mycological Institute, Kew

Stover RH (1990) Fusarium wilt of banana: some history and current status of the disease. In: Ploetz RC (ed) Fusarium wilt of banana. APS Press, St Paul, pp 1–7

Sun YH, Zhang LM, Xie YJ (2010) Cloning and analysis of Fusarium wilt resistance gene analogs in 'Goldfinger' banana. Mol Plant Breed 7:1215–1222

Tanabe S, Ishii-Minami N, Saitoh K, Otake Y, Kaku H et al (2011) The role of catalase-peroxidase secreted by Magnaporthe oryzae during early infection of rice cells. Mol Plant Microbe Interact 24:163–171

Thaler JS, Owen B, Higgins VJ (2004) The role of the jasmonate response in plant susceptibility to diverse pathogens with a range of lifestyles. Plant Physiol 135:530–538

Van Loon LC, Geraats BP, Linthorst HJ (2006) Ethylene as a modulator of disease resistance in plants. Trends Plant Sci 11:184–191

Williams B, Kabbage M, Britt R, Dickman MB (2010) AtBAG7, an Arabidopsis Bcl-2-associated athanogene, resides in the endoplasmic reticulum and is involved in the unfolded protein response. Proc Natl Acad Sci U S A 107:6088–6093

Xiao RF, Zhu YJ, Li YD, Liu B (2013) Studies on vascular infection of *Fusarium oxysporum* f. sp. *cubense* race 4 in banana by field survey and green fluorescent protein reporter. ESci J Plant Pathol 2:44–51

Xu X, Chen C, Fan B, Chen Z (2006) Physical and functional interactions between pathogen-induced Arabidopsis WRKY18, WRKY40, and WRKY60 transcription factors. Plant Cell 18:1310–1326

Yip MK, Lee SW, Su KC, Lin YH, Chen TY, Feng TY (2011) An easy and efficient protocol in the production of pflp transgenic banana against Fusarium wilt. Plant Biotechnol Rep 5:245–254

Zarate SI, Kempema LA, Walling LL (2007) Silver leaf whitefly induces salicylic acid defenses and suppresses effectual jasmonic acid defenses. Plant Physiol 143:866–875

15 Engineering Resistance to Sigatoka

H.D. Sowmya, T.R. Usharani, C. Sunisha, and Sukhada Mohandas

Abstract

Bananas and plantains are the major staple food in the tropical and subtropical regions. Currently the global banana production is affected by both quality- and yield-limiting by foliar fungal disease, the black sigatoka caused by *Mycosphaerella fijiensis*. The management of the disease by chemical approach has been relatively successful; however indiscriminate usage of chemicals has negative impact on human life and environment. Heavy reliance on fungicides has increased the chance of development of pathogen resistance. The conventional breeding approach can be a relatively impressive strategy in the development of disease-resistant cultivars, but it has become increasingly difficult due to male sterility, narrow genetic bases, and polyploidy nature. The advances in molecular biological techniques have provided the deeper insights into pathogen and plant defense mechanism, which have opened new avenue for the improvement of banana through genetic engineering. In the current review, the potential strategies used to enhance the resistance to sigatoka through genetic engineering employing different approaches like overexpression of pathogenesis-related protein (PR) antifungal proteins and bacterial hydrolytic enzymes and utilization of RNAi-mediated gene silencing of pathogenesis-related protein encoded by pathogens are summarized.

Keywords

Antimicrobial peptides • *Musa* • *Mycosphaerella fijiensis* • RNAi-mediated gene silencing

H.D. Sowmya • T.R. Usharani • C. Sunisha
Indian Institute of Horticultural Research,
Hessaraghatta, Bangalore 560 089, India

S. Mohandas (✉)
Division of Biotechnology,
ICAR-Indian Institute of Horticultural Research,
Hessaraghatta Lake Post, Bangalore 560 089, India

SM Plant Production Technologies Pvt Ltd.,
138 Kodi Halli, Madure Hobli, Doddaballapura
Taluk, Bangalore 562103, India
e-mail: sukhada.mohandas@gmail.com

S. Mohandas, K.V. Ravishankar (eds.), *Banana: Genomics and Transgenic Approaches for Genetic Improvement*, DOI 10.1007/978-981-10-1585-4_15

15.1 Introduction

Globally, *Musa* spp. (bananas and plantains) are the fourth most economically important staple food crop after rice, wheat, and corn. Presently, it is cultivated in around 150 tropical and subtropical countries across the world in an area of more than 10 million hectares with a total production of around 144 million tonnes (FAOSTAT 2013). It is considered as one of the fundamental food sources for millions of people. About 80 % of banana produced is directed to domestic market and about 20 % enters into the export market. The production of banana is threatened by several abiotic and biotic factors. Foliar disease known as black sigatoka is one among the most destructive biotic factors causing huge economic loss in the world's banana production (Stover and Simmons 1987; Carlier et al. 2000; Jones 2003; Arzanlou et al. 2007). The black leaf streak disease (BLSD) or black sigatoka leaf spot disease caused by *Mycosphaerella fijiensis* (anamorph *Pseudocercospora fijiensis*) is the most destructive member of the "sigatoka disease complex," which includes *M. musicola* (anamorph *P. musae*), the causal agent of yellow sigatoka leaf spot and *M. eumusae* (anamorph *P. eumusae*), and the causal agent of *Septoria* leaf spot or *eumusae* leaf spot. The pathogen invades foliar system, significantly reduces the photosynthetic area, and ultimately causes the death of the leaves leading to premature ripening of the fruit and reduced bunch weight (Ploetz 2000; Carlier et al. 2003). The estimated yield loss by black sigatoka in global banana production is more than 50–100 % (Mourichon et al. 1997; Ploetz 2001; Burt et al. 2002; Marciel Cordeiro and Pires de Matos 2003). The disease was first reported in the Sigatoka Valley of Viti Levu in the Fiji Islands in 1963 (Leach 1964). The sigatoka disease severely affects the world's most popular banana Cavendish subgroups (AAA) as well as other genomic groups AA, AAB and ABB (Jones 2009). The control of disease by conventional breeding has been handicapped by the extreme low female fertility or male sterility, long generation, and polyploid nature of the plant (Ploetz 2006). However, the use of multiple fungicides at relatively high frequencies has been allowed to control the disease to some extent, but potentially damaging the environment and lives (Churchill 2011) and also increasing the cost of production by 25–30 % (Marı'n et al. 2003; Kema 2006). Meanwhile, genetic manipulation with diverse targeting approach (resistant genes) is the desirable option for obtaining resistance in susceptible plants.

15.2 The Pathogen

The genus *Mycosphaerella* that belongs to the Mycosphaerellaceae (Schoch et al. 2006) is represented as one of the largest genera of ascomycetes, containing nearly about 3,000 species and anamorph species of around 7,000 numbers (Aptroot 2006; Crous and Braun 2003; Crous et al. 2007). The pathogen is considered as the most destructive plant pathogens causing huge economic loss in important crops like acacia, banana, cereals, citrus, eucalypts, pines, sugar beet, strawberry, soybean, and many others (Farr et al. 1995; Crous and Braun 2003). The pathogen mainly infects foliar part of the plants, resulting in defoliation and thus reduced photosynthetic capacity. It also causes cankers on the stem and spots and specks/fruit lesions on fruits (Cortinas et al. 2006; Batzer et al. 2008). The infection of *Mycosphaerella* to the parts of the plants depends on the types of species under infection for, e.g., *M. citri* species, infects both foliar part and fruits. However, *M. nubilosa* infects only foliar part.

The taxonomy of *Mycosphaerella* is based on morphological characters of both anamorphs and teleomorphs (Crous 1998; Stewart et al. 1999; Crous et al. 2000). They belong to the class Dothideomycetes, order Capnodiales, and family Mycosphaerellaceae. The species have broad range of lifestyles ranging from saprobes and plant pathogens to fungal hyperparasites (Goodwin et al. 2001; Jackson et al. 2004) hemibiotrophic with a bipolar, heterothallic mating nature within the class. They are haploid for the major part of their life cycle, with a short dikaryotic and diploid phase during sexual reproduction. Besides sexual reproduction, some *Mycosphaerella* species also produce haploid conidia in closed fruiting bodies or on free conidiophores under asexual reproductive cycle. The recent molecular analysis revealed that the genus is poly- and paraphyletic in nature, involving at least two families, namely, Mycosphaerellaceae and Teratosphaeriaceae (Schoch et al. 2006; Crous et al. 2007).

The primary agents of *Mycosphaerella* species that are responsible for the black sigatoka disease complex in banana and plantains are *M. fijiensis*, *M. musicola*, and *M. eumusae*. *M. fijiensis* (M. Morelet) is a sexual, heterothallic fungus having *Pseudocercospora fijiensis* (M. Morelet) Deighton as the anamorph stage. Yellow sigatoka is caused by *Mycosphaerella musicola* Leach ex Mulder (anamorph *Pseudocercospora musae*) (Jones 2000). *M. eumusae* (anamorph *S. eumusae*) is the name proposed for the causal agent of *Septoria* leaf spot disease (Carlier et al. 2000).

15.2.1 Disease Symptoms

The three pathogens of the sigatoka disease complex cause unique symptoms on banana leaves, and it is difficult to distinguish (Henderson et al. 2006). Stages of development of black leaf streak disease are represented in Fig. 15.1. Initially, tiny chlorotic specks appear on the lower (abaxial) surface of the leaf laminar after 14–20 days of infection. Later small specks become streaks and are running parallel to the leaf veins which coalesce to form larger streaks that vary in color from rusty red to dark brown to black, depending on the pathogen. The time interval between the appearance of specks, development of streaks, and following enlargement of necrotic spots varies in length according to the type of cultivar and the severity of infestation (Mayorga 1990; Jacome and Schuh 1992; Fullerton 1994). These streaks eventually form fusiform or elliptical spots that aggregate, form a water-soaked border with variable degrees of a yellow halo, and, eventually, coalesce to cause immense leaf necrosis. This leads to considerable reduction in the photosynthetic capacity of the leaf and causes significant defoliation. As a consequence, it reduces the yield by slower filling of fingers, finger size, and premature ripening of fruits (Ramsey et al. 1990; Romero and Sutton 1997).

15.2.2 Life Cycle

Both *M. fijiensis* and *M. musicola* have unique nature of disease cycle with black sigatoka being much faster than yellow sigatoka in the earlier emergence of spots. Life cycle of the pathogens involves both asexual and sexual reproduction. Asexual reproduction (anamorph) occurs through the development of conidia which get disseminated for a short distance across the plant surface by water spill, whereas in sexual reproduction (teleomorph) ascospores are produced and disseminated over long distances by wind, and it is reported to be the most important stage in disease spread (Amil et al. 2007; Gauhl et al. 2000). Conidia are produced at a younger stage (stages 2–4) of the disease cycle, whereas ascospores are produced at later stage. Due to large amount of ascospores produced by *M. fijiensis* than *M. musicola*, ascospores are represented the main source of spread for *M. fijiensis*, while conidia are the main source of dispersal *M. musicola* (Stover and Dickson 1976).

Infections begin with spores germinating on cigar leaf under higher humidity condition and temperature ranging from 25 to 29 °C (Gauhl et al. 2000). The epiphytic growth stage begins and symptoms gradually evolve from stage 1 to stage 6 (Fouré 1982). The germ tube penetrates into the leaf via stomata pore by producing an appressorium through which it directs a hypha (Stahel 1937; Meredith 1970; Stover 1980) to penetrate the palisade region intercellularly. The vegetative hyphae colonized in the substomatal region produce conidiophores, which grow out through the stomatal pore on the lower surface or abaxial region (Stahel 1937). The production of conidiophores begins at initial specks (Fig. 15.2, stage 1) or at the first streak stage (Fig. 15.2, stage 2) up to the second spot stage (Fig. 15.2, stage 5). Further multiple conidia are produced from each conidiophore between the second streak (Fig. 15.2, stage 3) and second spot (Fig. 15.2, stage 5) stages (Meredith and Lawrence 1969). The production of spermatia (male gametes) occurs in spermagonia which start to develop from the substomatal chambers primarily on the lower leaf surface at the second streak

Fig. 15.1 Stages of development of black leaf streak disease (A) Stage 2 - first streak stage (B) Stage 3 - second streak stage (C) Stage 4 - first spot stage (D) Stage 5 - second spot stage (E) Stages 5 and 6 - third or mature spot stage (F) Coalescence of streaks and spots on a severely infected leaf. (G) All stages present on a single plant or leaf. (H & I) Rounded, nonstreak lesions and narrow leaf streak lesions present on young leaves of 'water' suckers. (J–L) Time course of symptom development on a 'Grande Naine' plant incubated in a growth chamber. (Source: Churchill 2011)

(Fig 15.2, stage 3) or first spot (Fig. 15.2, stage 4) (Liberato et al. 2009).

The first symptoms become apparent as small yellow specks or red-brown specks in *M. musicola* and *M. fijiensis* infection, respectively (Gauhl et al. 2000). The small specks further aggregate to produce streaks, and symptom progression depends on the environmental factors, susceptibility of the cultivar, and intensity of infestation (Meredith 1970). At the final disease stage, mature lesion starts to coalesce and dry the leaf tissue by producing spermagonia and perithecia. These spermagonia protrude through stomatal pore that produces hyaline and single-celled, rod-shaped spermatia through the ostiole and fertilize with adjacent female receptive hyphae known as trichogynes. After fertilization pseudothecia, perithecia-like structures are formed within the mature lesion which contains oblong to club-shaped bitunicate asci (two-layered cell wall) and

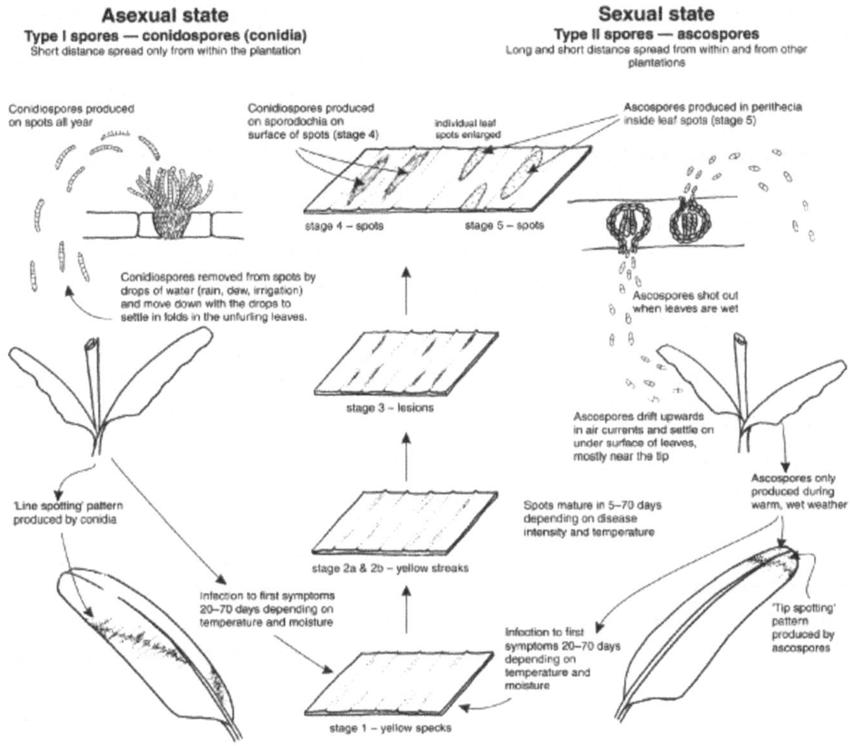

Fig. 15.2 Disease cycle of *M. musicola* on banana (Source: Henderson et al. 2006)

contains ascospores (colorless, contains one septum and fusiform) (Crous et al. 2009).

15.3 Management of Sigatoka Disease

A number of measures have been implemented to manage black sigatoka disease. Among them, conventional breeding and chemical control methods remain the two major strategies to control sigatoka disease (Nwauzoma et al. 2002; Marín et al. 2003). However, cultural practices are laborious, requiring expensive inputs and more labor, thus increasing the cost of production (Ploetz 2001).

The identification of resistant cultivar by conventional breeding is the most impressive and eco-friendly method, but is impeded due to its long generation time (up to 2 years), narrow genetic base, male sterility, low female fertility, and polyploidy nature variability (Rowe 1984; Swennen and Vuylsteke 1993; Pillay et al. 2002).

The control of sigatoka disease could be achieved by the frequent application of fungicides (Fouré 1983; Marín et al. 2003), such as benzimidazoles (benomyl, carbendazim), morpholines (tridemorph), triazoles (bitertanol, difenoconazole, cyproconazole, flusilazole, fenbuconazole, hexaconazole, propiconazole, triadimenol, and tebuconazole), and strobilurins (azoxystrobin, trifloxystrobin) along with the alternative application with protectant fungicide such as chlorothalonil and dithiocarbamates (mancozeb). But, the annual cost to control the disease was estimated to be around US$1,000/ha (Arias et al. 2003) which increased the production cost 25–30 % of the total (up to $1,800), which is not economical for small growers in developing countries (Marín et al. 2003; Kema 2006). Besides, the application of fungicides has an adverse impact not only on the terrestrial and aquatic ecosystems but on cultivation due the development of resistance by the fungus to these fungicides (Mukhopadhyay and Mukherjee 1996; Castillo et al. 2000; Wesseling et al. 2005;

Can˜as-Gutie'rrez et al. 2006; Vawdrey and Grice 2005; Amil et al. 2007).

15.4 Transgenic Approach

Development of resistance cultivars through non-conventional breeding strategy such as genetic engineering offers an alternative promising environmentally safe and sustainable approach for banana improvement (Swennen 2003; Passos et al. 2013; Kovács et al. 2013). The advances in the molecular biology techniques have enabled to apply the diverse targeting approaches to develop resistant plants against fungal pathogens.

15.4.1 Overexpression of Antimicrobial Peptides

The most commonly used approach to engineer fungal resistance in plants is through the overexpression of antimicrobial peptides. They are cysteine-rich small, globular cationic, amphipathic peptides encoded by multigenic families and widely distributed among the plant kingdom (Boman 1991; Hancock and Lehrer 1998). These peptides play an important role in the innate immunity and act as a first line of defense barrier in the plant system against various invading phytopathogens. AMP molecules have broad range of antimicrobial activity at minimal inhibitory concentration by damaging the cells either by targeting the cell membrane or targeting the intracellular region and arrest the nucleic acid or protein or enzyme activity (Powers and Hancock 2003; Brogden 2005). These proteins are evolutionarily conserved, expressed constitutively, or induced in response to various stress conditions. Due to their rapid mode of action at minimal inhibitory concentration, these peptides gained more importance for developing the crop resistance to phytopathogens. Antimicrobial peptides from various sources have been utilized to confer resistance against number of fungal pathogens in an array of genetically engineered plant species (Parkhi et al. 2010; Hoshikawa et al. 2012). Cammue et al. (1993) demonstrated the utilization of several antifungal peptide genes against *Mycosphaerella* spp. The resulted antifungal peptides showed inhibitory activity toward *Mycosphaerella* spp. under *in vitro* condition. Remy et al. (1998) developed the transgenic banana cv. Three Hand Planty expressing antifungal gene Dm-AMP1 isolated from *Dahlia merckii* under the control of different promoters, CaMV35S promoter, super promoter and maize ubiquitin promoter individually. The resulted transgenic banana plant expressing Dm-AMP1 gene driven by maize ubiquitin promoter showed 1.3 % Dm-AMP1 expression in the leaf extract and the rest two promoters showed 6 to 16 fold and 9 to 30 fold lowerexpressionsrespectively. Further these transgenic banana plants were tested with *M. fijiensis* under greenhouse condition to evaluate antifungal activity under *in vivo* condition. In another study, the potential of antimicrobial peptide MSI-99, a synthetic substitution analogue of magainin isolated from the skin secretions of the African clawed frog *Xenopus laevis*, was utilized to impart resistance against *M. musicola* in banana cv. Rasthali. During the experiment Chakrabarti et al. (2003) cloned MSI-99 gene into two different plant protein expression vectors (pMSI164 and pMSI168) to target the peptide into the cytoplasm and extracellular spaces of the cells. The resulted transgenic plants targeting antimicrobial peptide intracellular space showed 40 % reduction, whereas transformants targeting extracellular space showed 18–58 % reduction in lesion area development during in vitro detached leaf bioassay. Vishnevetsky et al. (2011) coexpressed the endochitinase gene (TnEn-42) from *Trichoderma harzianum* and stilbene synthase (StSy) gene from grapes under the control of CaMV35S promoter and inducible promoter PR-10, respectively, together with superoxide dismutase gene CuZn-SOD isolated from tomato, under control of the ubiquitin promoter in banana cv. Grand Nain. Out of 12 transgenic lines tested, six transgenic lines showed significantly reduced symptoms, and two transgenic lines exhibited significantly reduced symptoms throughout 4 years of field trial tested without any changes in the fruit quality and yield. The same transgenic lines also exhibited tolerance to *Botrytis cinereain vitro* bioassays.

Fungal cell wall-degrading enzymes like chitinases and glucanase are commonly used to produce different fungal resistance crops (Moravcˇı'kova' et al. 2004; Melander et al. 2006; Wally et al. 2009). Similar attempt was made in the control of sigatoka disease in banana. Kosky et al. (2010) introduced antifungal glucanase gene along with another agronomically important antifungal protein AP24 osmotin in banana cv. "Navolean" (*Musa* AAB). They obtained 25 primary transformants of which one transgenic line showed significantly enhanced tolerance to the foliar symptoms caused by sigatoka compared to susceptible control under field condition. Recently, Kovács et al. (2013) analyzed the potentiality of class I chitinase genes RCC2 (targeted intracellular) and RCG3 (extracellular) isolated from rice, in banana cv. Gros Michel against black sigatoka disease. The resulted transgenic banana plants were phenotypically normal. The plants expressing RCG3 gene showed considerable delay in lesion development and also exhibited ten times less lesion leaf area compared to susceptible control over a monitoring period of 108 days under *in vitro* detached leaf bioassay. Besides, transgenic banana plant expressing RCG3 was more tolerant than RCC2. It might be due to the extracellular expression nature of RCG3 gene.

15.4.2 RNAi-Mediated Host-Induced Gene Silencing

RNA-mediated resistance, that is, RNAi-induced gene silencing technique, emerged as a powerful tool to engineer pathogen-resistant plants by suppressing or knocking down the expression of targeted genes. It is a conserved regulatory mechanism that has been widely studied in eukaryotes, where small RNAs are utilized to silence the gene expression at post-transcriptional level. Using this strategy, it is possible to inhibit or modulate the expression of target genes in a phytopathogens which leads to reduction or complete inhibition of infection, growth development, or reproduction and eventually causes death of the pathogens. Proofs of concept of host-induced gene silencing (HIGS) using vital genes of fungal pathogens were recently reported on different fungal pathogens like *Fusarium oxysporum*, *Blumeria graminis*, and *Bremia lactucae* (Nowara et al. 2010; Ghag et al. 2014; Govindarajulu et al. 2015), suggesting trafficking of dsRNA or siRNA from host plants into fungal pathogens. Recently, Mumbanza et al. (2013) conducted an *in vitro* self- and cross-species RNA inhibitory bioassay to assess the effect of synthetic dsRNA on spore germination and colony formation by targeting different genes. During their experiment two target fungal pathogens *M. fijiensis* and *Fusarium oxysporum* were used. All the tested synthetic dsRNA molecules showed varying levels of reduction in spore germination. The dsRNA for adenyl cyclase, DNA polymerase alpha subunit, and DNA polymerase delta subunit showed maximum reduction in spore germination of both tested fungi during both self- and cross-species test. Inhibition ranges of 79.8–93.0 % and 19.0–57.8 % were observed when Foc dsRNA was applied to Foc spores and *M. fijiensis* spores, respectively. Similarly 34.4–72.3 % and 89.7–98.9 % of spore germination were observed when *M. fijiensis* dsRNA was treated to *M. fijiensis* and *F. oxysporum* f.sp. *cubense*, respectively.

15.5 Conclusion

Molecular biology has provided the new avenue for the development of transgenic plants. Various defense genes like chitinases, osmotin, and magainin from both plant and animal source have been effectively cloned and expressed in banana against sigatoka disease. Currently, the transgenic plants developed using single gene have provided more or less moderate level of resistance against pathogen. Development of the complete resistant plant has been not achieved so far. The gene pyrimidine approach of developing transgenic plants has shown an increased level of resistance which is promising. Hence, gene stacking approach in the future would provide a resistance source for banana. RNAi is another specific gene silencing tool that has been employed to silence the target gene that would reduce infectivity of the pathogen. Identification of the target gene for pathogen for RNAi is in progress and studies are limited to in vitro conditions. Presently, the use of such technologies against sigatoka disease has not been used in the development of transgenic plant. Far most importantly, understanding of the host-pathogen interaction, chemical moieties that involved in activating the plants, and innate defense system of the host plant is necessary. Recently it was reported that how pathogen modulates the innate immunity of host plant has been successfully demonstrated in rice against rice blast disease.

References

Amil AF, Heaney SP, Stanger C, Shaw MW (2007) Dynamics of QoI sensitivity in *Mycosphaerella fijiensis* in Costa Rica during 2000 to 2003. Phytopathology 97:1451–1457

Aptroot A (2006) *Mycosphaerella* and its anamorphs: 2. Conspectus of *Mycosphaerella*. CBS Biodivers Ser 5:1–231

Arias P, Dankers C, Liu P, Pilkauskas P (2003) The world banana economy 1895–2002. FAO Commodity studies. Food and Agriculture organization of the United Nations, Rome

Arzanlou M, Abeln ECA, Kema GHJ, Waalwijk C, Carlier J, de Vries I, Guzman M, Crous PW (2007) Molecular diagnostics for the Sigatoka disease complex of banana. Phytopathology 97:1112–1118

Batzer JC, Mercedes Diaz Arias M, Harrington TC, Gleason ML, Groenewald JZ, Crous PW (2008) Four species of *Zygophiala* (*Schizothyriaceae*, *Capnodiales*) are associated with the sooty blotch and flyspeck complex on apple. Mycologia 100:232–258

Boman HG (1991) Antibacterial peptides: key components needed in immunity. Cell 65:205–207

Brogden A (2005) Antimicrobial peptides: pore formers or metabolic inhibitors in bacteria? Nat Rev Microbiol 3:238–250

Burt PJA, Rosenberg LJ, Rutter GJ, Ramirez F, Gonzales H (2002) Forecasting the airborne spread of Mycosphaerella fijiensis, a cause of black Sigatoka disease on banana: estimations of numbers of perithecia and ascospores. Ann Appl Biol 135:369–377

Cammue BPA, De Bolle MFC, Terras FRG, Broekaert WF (1993) Fungal disease control in Musa: application of new antifungal proteins. In: Ganry J (ed) Breeding banana and plantain for resistance to diseases and pests. In: Proceedings of the international symposium on genetic improvement of bananas for resistance to diseases and pests, Organized by CIRAD-FLHOR, Montpellier, &-9 September 1992, pp 221–225

Can˜as-Gutie’rrez P, Patin˜o LF, Rodrı’guez-Arango E, Arango R (2006) Molecular characterization of benomyl-resistant isolates of Mycosphaerella fijiensis, collected in Colombia. J Phytopathol 154:403–409

Carlier J, Foure F, Gauhl F, Jones DR, Lepoivre P, Mourichon X, Pasberg-Gauhl C, Romero RA (2000) Black leaf streak. In: Jones RD (ed) Disease of Banana, Abaca and Enset. CABI Publishing, Wallingford, pp 37–47

Carlier J, Hayden H, Rivas G, Zapater M-F, Abadie C, Aitken E (2003) Genetic differentiation in Mycosphaerella leaf spot pathogens. In: Jacome L, Lepoivre P, Marin D, Ortiz, R, Romero R, Escalant JV (eds) Mycosphaerella leaf spot diseases of bananas: present status and outlook. In: Proceedings of the workshop on Mycosphaerella leaf spot diseases, San José, 20–23 May 2002. INIBAP, Montpellier. 123–129

Castillo LE, Ruepert C, Solis E (2000) Pesticide residues in the aquatic environment of banana plantation areas in the North Atlantic Zone of Costa Rica. Environ Toxicol Chem 19:9

Chakrabarti A, Ganapathi TR, Mukherjee PK, Bapat VA (2003) MSI-99, a magainin analogue, imparts enhanced disease resistance in transgenic tobacco and banana. Planta 216:587–596

Churchill (2011) Mycosphaerella fijiensis, the black leaf streak pathogen of banana: progress towards understanding pathogen biology and detection, disease development, and the challenges of control. Mol Plant Pathol 12:307–328

Cortinas MN, Crous PW, Wingfield BD, Wingfield MJ (2006) Multi-gene phylogenies and phenotypic characters distinguish two species within the *Colletogloeopsis zuluensis* complex associated with *Eucalyptus* stem cankers. Stud Mycol 55:133–146

Crous PW (1998) *Mycosphaerella* spp. and their anamorphs associated with leaf spot diseases of *Eucalyptus*. Mycol Mem 21:1–170

Crous PW, Braun U (2003) Mycosphaerella and its anamorphs. 1. Names published in Cercospora and Passalora. CBS Biodivers Ser 1:1–571

Crous PW, Aptroot A, Kang JC, Braun U, Wingfield MJ (2000) The genus *Mycosphaerella* and its anamorphs. Stud Mycol 45:107–121

Crous PW, Braun U, Groenewald JZ (2007) Mycosphaerella is polyphyletic. Stud Mycol 58: 1–32

Crous PW, Schoch CL, Hyde KD, Wood AR, Gueidan C, de Hoog GS, Groenewald JZ (2009) Phylogenetic lineages in the Capnodiales. Stud Mycol 64:17–47

FAOSTAT Agriculture Data (2013) http://faostat3.fao.org/download/Q/QC/E.

Farr DF, Bills GF, Chamuris GP, Rossman AY (1995) Fungi on plants and plant products in the United States. APS Press, St Paul

Fouré E (1982) Les cercosporioses du bananier et leurs traitements. Comportement des variétés. Etude de la sensibilité variétale des bananiers et plantains à Mycosphaerella fijiensis Morelet au Gabon (maladie des raies noires). II- Etude de quelques paramètres. Fruits 37:760–771

Fouré E (1983) Les cercosporioses du bananier et leurs traitements. Sélection de molecules fongicides nouvelles. Activités comparées de différentes molécules fongicides sur Mycosphaerella fijiensis Morelet, agent de la 'maladie des raies noires' des bananierset plantains au Gabon. Fruits 38:21–34

Fullerton RA (1994) Sigatoka leaf diseases. In: Ploetz RC, Zentmyer GA, Nishijinia WT, Rohrbach KG, Ohr HD (eds) Compendium of tropical fruit diseases. American Phytopathological Society, St. Paul, pp 12–14

Gauhl F, Pasberg-Gauhl C, Jones DR (2000) Disease cycle and epidemiology. In: Jones DR (ed) Diseases of Banana, Abacá and Enset. CAB International, Wallingford, pp 56–62

Ghag SB, Shekhawat UKS, Ganapathi TR (2014) Host induced post-transcriptional hairpin RNA-mediated gene silencing of vital fungal genes confers efficient resistance against Fusarium wilt in banana. Plant Biotechnol J 12:541–553

Goodwin SB, Dunkle DL, Zismann VL (2001) Phylogenetic analysis of *Cercospora* and *Mycosphaerella* based on the internal transcribed spacer region of ribosomal DNA. Phytopathology 91:648–658

Govindarajulu M, Epstein L, Wroblewski T, Michelmore RW (2015) Host-induced gene silencing inhibits the biotrophic pathogen causing downy mildew of lettuce. Plant Biotechnol J 13:875–883

Hancock RE, Lehrer R (1998) Cationic peptides: a new source of antibiotics. Trends Biotechnol 16:82–88

Henderson J, Pattemore JA, Porchun SC, Hayden HL, Van Brunschot S, Grice KRE, Peterson RA, Thomas-Hall SR, Aitken EAB (2006) Black Sigatoka disease: new technologies to strengthen eradication strategies in Australia. Australas Plant Pathol 35:181–193

Hoshikawa K, Ishihara G, Takahashi H, Nakamura I (2012) Enhanced resistance to gray mold (Botrytis cinerea) in transgenic potato plants expressing thionin genes isolated from Brassicaceae species. Plant Biotechnol 29:87–93

Jackson SL, Maxwell A, Neumeister-Kemp HG, Dell B, Hardy GESJ (2004) Infection, hyperparasitism and conidiogenesis of *Mycosphaerella lateralis* on *Eucalyptus globulus* in Western Australia. Australas Plant Pathol 33:49–53

Jacome LH, Schuh W (1992) Effects of leaf wetness duration and temperature on development of black Sigatoka disease on banana infected by Mycosphaerella fijiensis var. difformis. Phytopathology 82:515–520

Jones DR (2000) Diseases of banana, Abacá and Enset. CABI Publishing, New York

Jones DR (2003) The distribution and importance of the *Mycosphaerella* leaf spot diseases of banana. In: Jacome L, Lepoivre P, Marin D, Ortiz R, Romero R, Escalant JV (eds) Proceedings of 2nd international workshop on Mycosphaerella leaf spot disease of bananas, Costa Rica. INIBAP 20–23, San José, pp 25–42

Jones DR (2009) Disease and pest constraints to banana production. Acta Horticult 828:21–36

Kema GHJ (2006) The pesticide reduction plan for banana. In: Soprano E, Tcacenco FA, Lichtemberg LA, Silva MC (eds) Anais XVII Reunião Internacional ACORBAT. Bananicultura: um negócio sustentável ACORBAT, Joinville, pp 3–4

Kosky RG, Chong-Pérez B, López-Torres J, Reyes M et al (2010) Plantain (Musa spp. cv. 'Navolean' AAB) transgenic plants from Agrobacterium tumefaciens-mediated transformation of embryogenic cell suspensions. Biotecnol Veg 10:209–218

Kovács G, Sági L, Jacon G, Arinaitwe G, Busogoro JP, Thiry E, Strosse H, Swennen R, Remy S (2013) Expression of a rice chitinase gene in transgenic banana ('Gros Michel', AAA genome group) confers resistance to black leaf streak disease. Transgenic Res 22:117–130

Leach R (1964) A new form of banana leaf spot in Fiji, black leaf streak. World Crops 16(60):64

Liberato JR, Peterson RA, Gasparotto L, Ferrari JT, Grice K, Porchun SC, Shivas RG (2009) Black sigatoka of banana (Mycosphaerella fijiensis). Available at pest and diseases image library, species content page—http://www.padil.gov.au/viewPestDiagnosticImages.aspx?id=431; Plant Biosecurity Toolbox/Info Sheet—http://www.padilgov.au/pbt/index.php?q=node/46&pbtID=166

Maciel Cordeiro ZJ, Pires de Matos A (2003) Impact of Mycosphaerella spp. in Brazil. In: Jacome L, Lepoivre P, Marin D, Ortiz R, Romero R, Escalant JV (eds) Mycosphaerella leaf spot diseases of bananas: present status and outlook. In: Proceedings of the workshop on Mycosphaerella leaf spot diseases, San José, Costa Rica, 20–23 May 2002. INIBAP, Montpellier, pp 91–97

Marín DH, Romero RA, Guzmán M, Sutton TB (2003) Black Sigatoka: an increasing threat to banana cultivation. Plant Dis 87:208–222

Mayorga MH (1990) La Raya negra (Mycosphaerella fijiensis Morelet) del plátano y del banano. I. Ciclo de vida del patógeno bajo las condiciones del Urabá. Rev ICA 25:69–77

Melander M, Kamnert I, Happstadius I, Liljeroth E, Bryngelsson T (2006) Stability of transgene integration and expression in subsequent generations of doubled haploid oilseed rape transformed with chitinase and b-1,3-glucanase genes in a double-gene construct. Plant Cell Rep 25:942–952

Meredith DS (1970) Banana leaf spot disease (Sigatoka) caused by Mycosphaerella musicola Leach. Phytopathological papers, no. 11. Commonwealth Mycological Institute, Kew

Meredith DS, Lawrence JS (1969) Black leaf streak disease of bananas (Mycosphaerella fijiensis): symptoms of disease in Hawaii, and notes on the conidial state of the causal fungus. Trans Br Mycol Soc 52:459–476

Moravcˇı'kova' J, Matusˇı'kova' I, Libantova' J, Bauer M, Mlyna'rova' L (2004) Expression of cucumber class III chitinase and Nicotiana plumbaginifolia class I glucanase genes in transgenic potato plants. Plant Cell Tissue Organ Cult 79:161–168

Mourichon X, Carlier J, Fouré E (1997) Sigatoka leaf spot disease, Musa Disease. fact sheet, Inibap, Montpellier 8:4

Mukhopadhyay AN, Mukherjee PK (1996) Fungi as fungicides. Int J Trop Plant Dis 14:1–17

Mumbanza FM, Kiggundu A, Tusiime G, Tushemereirwe WK, Niblett C, Bailey A (2013) In vitro antifungal activity of synthetic dsRNA molecules against two pathogens of banana, Fusarium oxysporum f. sp. cubense and Mycosphaerella fijiensis. Pest Manag Sci 69:1155–1162

Nowara D, Gay A, Lacomme C, Shaw J, Ridout C, Douchkov D, Hensel G, Kumlehn J, Schweizer P (2010) HIGS: host-induced gene silencing in the obligate biotrophic fungal pathogen Blumeria graminis. Plant Cell 22:3130–3141

Nwauzoma AB, Tenkouano A, Crouch JH, Pillay M, Vuylsteke D, Daniel Kalio LA (2002) Yield and disease resistance of plantain (*Musa* spp., AAB group) somaclones in Nigeria. Euphytica 123:323–331

Parkhi V, Kumar V, LeAnne M, Campbell, Bell AA, Shah J, Rathore KS (2010) Resistance against various fungal pathogens and reniform nematode in transgenic cotton plants expressing Arabidopsis NPR1. Transgenic Res 19:959–975

Passos MAN, Cruz VO, Emediato FL, Teixeira CC et al (2013) Analysis of the leaf transcriptome of Musa acuminata during interaction with Mycosphaerella musicola: gene assembly, annotation and marker development. BMC Genomics 14:78

Pillay M, Hartman J, Tenkouano A (2002) Bananas and plantains: future challenges in Musa breeding. In: Kang MS (ed) Crop improvement, challenges in the twenty-first century. Food Products Press, New York, pp 223–252

Ploetz RC (2000) Black sigatoka. Pestic Outlook 11:19–23

Ploetz RC (2001) Black sigatoka of banana: the most important disease of a most important fruit. Plant Health Instructor. doi:10.1094/PHI-I-2001-0126-1001. Available at http://www.apsnet.org/publications/apsnetfeatures/Pages/blacksigatoka.aspx.

Ploetz RC (2006) Fusarium wilt of banana is caused by several pathogens referred to as *Fusarium oxysporum* f. sp. *cubense*. Phytopathology 96:653–656

Powers JP, Hancock RE (2003) The relationship between peptide structure and antibacterial activity. Peptides 24:1681–1691

Ramsey MD, Daniells JW, Anderson DJ (1990) Effects of Sigatoka leaf spot (*Mycosphaerella musicola* Leach) on fruit yields, field ripening and green life of bananas in North Queensland. Sci Hortic 41:305–313

Remy S, Buyens A, Cammue BPA, Swennen R, Sagi L (1998) Production of transgenic banana expressing antifungal proteins. Acta Horticult 490:433–436

Romero RA, Sutton TB (1997) Reaction of four Musa genotypes at three temperatures to isolates of *Mycosphaerella fijiensis* from different regions. Plant Dis 81:1139–1142

Rowe P (1984) Breeding bananas and plantains. In: Janick J (ed) Plant breeding reviews. Wiley, Hoboken

Schoch CL, Shoemaker RA, Seifert KA, Hambleton S, Spatafora JW, Crous PW (2006) A multigene phylogeny of the Dothideomycetes using four nuclear loci. Mycologia 98:1043–1054

Stahel G (1937) The banana leaf speckle in Surinam caused by Chloridium musae nov. spec and another related banana disease. Trop Agric 14:42–44

Stewart EL, Liu Z, Crous PW, Szabo L (1999) Phylogenetic relationships among some cercosporoid anamorphs of *Mycosphaerella* based on rDNA sequence analysis. Mycol Res 103:1491–1499

Stover RH (1980) Sigatoka leaf spots of bananas and plantains. Plant Dis 64:750–756

Stover RH, Dickson JD (1976) Banana leaf spot caused by *Mycosphaerella musicola* and *M. fijiensis* var. *difformis*: a comparison of the fírst Central American epidemics. FAO Plant Prot Bull 24:36–42

Stover RH, Simmons NW (1987) Bananas. Longman Group UK Ltd., Harlow

Swennen R (2003) Transgenic approaches for resistance to *Mycosphaerella* leaf spot diseases in *Musa* spp. In: Jacome L, Leproive P, Martin D, Ortiz R, Romero R, Escalant JV (eds) *Mycosphaerella* leaf spot diseases of bananas: present status and outlook. Proceeding of the workshop on *Mycosphaerella* leaf spot diseases held in San José, Costa Rica on 20–23 May 2002. INIBAP, Montpellier

Swennen R, Vuylsteke D (1993) Breeding Black Sigatoka resistant plantains with a wild banana. Trop Agric 70:74–77

Vawdrey LL, Grice K (2005) Field evaluation of strobilurins, triazoles and acibenzolar to control Sigatoka disease in Australia. Infomusa 14:11–15

Vishnevetsky J, Thomas L, White JR, Palmateer AJ, Flaishman M, Cohen Y, Elad Y, Velcheva M, Hanania U, Sahar N, Dgani O, Perl A (2011) Improved tolerance towards fungal diseases in transgenic Cavendish banana (Musa spp. AAA group) cv. Grand Nain. Transgenic Res 20:61–72

Wally O, Jayaraj J, Punja Z (2009) Comparative resistance to foliar fungal pathogens in transgenic carrot plants expressing genes encoding for chitinase, b-1,3-glucanase and peroxidise. Eur J Plant Pathol 123:331–342

Wesseling C, Corriols M, Bravo V (2005) Acute pesticide poisoning and pesticide registration in Central America. Toxicol Appl Pharmacol 207:S697–S705

Engineering Resistance to Viruses 16

James C. Green, Wayne Borth, and John S. Hu

Abstract

Banana viruses are limiting factors for banana production in many countries where bananas are grown. The most devastating viral-caused disease of bananas in Hawaii and many areas of Asia, Africa, and the Pacific is banana bunchy top caused by banana bunchy top virus (BBTV). Once BBTV is present in a plantation, it is very difficult to control. Control of the aphid vector using insecticides is inefficient and expensive and poses environmental risks. Removal of infected plants can effectively limit the spread of the virus but requires identification of early infections and involves increased labor and chemical costs. No naturally occurring resistance is known in banana. New control strategies for the efficient control of BBTV are needed. The recent development of transformation and regeneration systems for banana has made it possible to develop virus-resistant banana plants using genetic engineering. Several groups have investigated this possibility and utilized posttranscriptional gene silencing (PTGS) or RNA interference (RNAi) approaches to generate BBTV resistance in several cultivars. To date, no viable transgenic lines have been developed that display durable BBTV resistance. Recent efforts and approaches to develop BBTV-resistant transgenic banana are reviewed in this chapter, and novel strategies to produce virus-resistant transgenic banana are introduced. Future potentials of using transgenic banana as alternatives for management of banana virus diseases are discussed.

J.C. Green • W. Borth • J.S. Hu (✉)
Department of Plant and Environmental Protection Sciences, College of Tropical Agriculture and Human Resources, 3190 Maile Way, Honolulu, HI 96822, USA
e-mail: johnhu@hawaii.edu

S. Mohandas, K.V. Ravishankar (eds.), *Banana: Genomics and Transgenic Approaches for Genetic Improvement*, DOI 10.1007/978-981-10-1585-4_16

Keywords
Virus resistance • Banana bunch top virus (BBTV) • Post-transcriptional gene silencing (PTGS) • RNA interference (RNAi) • Virus activated cell death strategy

16.1 Introduction

Bananas (*Musa* spp.) are large monocotyledonous perennial herbs of the family Musaceae order Zingiberales found in the tropics and subtropics. Bananas and plantains are one of the world's most important food crops and represent a vital nutritional staple for millions of people in tropical and subtropical countries (Selvarajan and Balasubramanian 2014). Bananas are of significant value to resource-poor farmers because bananas produce fruit year-round (Kumar et al. 2014). Bananas are propagated primarily through asexual methods such as vegetative divisions or micropropagation through tissue culture. Four major diseases of viral etiology are recognized for their importance including banana bunchy top caused by banana bunchy top virus (BBTV), banana streak caused by banana streak virus (BSV), banana bract mosaic caused by banana bract mosaic virus (BBrMV), and banana mosaic caused by cucumber mosaic virus (CMV). Efforts have been made to develop resistant transgenic plants in banana to BBTV but not to other viruses, as far as we know. This review focuses on BBTV research activities.

16.2 Viruses of Banana

16.2.1 Banana Bunchy Top Virus

BBTV is the most serious of the viral diseases affecting banana (Dale 1987). In many countries BBTV is a significant limiting factor affecting banana production (Harding et al. 1991). BBTV is found in Southeast Asia, Taiwan, the Philippines, India, Africa, and most of the Pacific Islands including Hawaii, but has not yet been reported from the Americas (Burns et al. 1995). BBTV is transmitted by the banana aphid, *Pentalonia nigronervosa* (Beetham et al. 1999), in a semi-persistent manner with a minimum acquisition period of 4 h and a minimum inoculation period of 15 min (Hu et al. 1996). Insects can retain the virus for up to 20 days after feeding, and no evidence has been found for transovarial passage (Hu et al. 1996) or for multiplication of the virus in the vector (Hafner et al. 1995).

BBTV is in the genus *Babuvirus* in the family *Nanoviridae* (Randles et al. 2000). BBTV has a multipartite genome consisting of at least six, and as many as 11 separate circular single-stranded DNA (ssDNA) components, each of about 1 kb, which are packaged individually into small (18–20 nm) isometric virions (Burns et al. 1995). BBTV's genome components each contain one large open reading frame (ORF), except DNA-1 that encodes two proteins (Beetham 1999; Randles et al. 2000). The putative replication-associated protein (Rep) is encoded by DNA-1 and has been shown to both initiate and direct the replication of BBTV (Horser et al. 2001). No function for DNA-2 has yet been determined (Horser et al. 2001). DNA-3 encodes the capsid or coat protein (CP) (Wanitchakorn et al. 1997). DNA-3 is localized in the nucleus and cytoplasm together with DNA-6, although DNA-6 can be redirected to the cell periphery by DNA-4 (Wanitchakorn et al. 2000). DNA-4 encodes the putative movement protein (MP) and is localized in the cell periphery during infection (Wanitchakorn et al. 2000). DNA-5 encodes the cell cycle link protein (Clink) and contains an LXCXE motif involved in retinoblastoma-like plant protein (Rb) binding activity (Wanitchakorn et al. 2000). Clink may play an important role in the manipulation of the host plant's cell replication cycle during S-phase through binding with Rb (Aronson et al. 2000). DNA-6 encodes the nuclear shuttle protein (NSP) (Gronenborn 2004) that, when expressed without interaction with

DNA-4, is targeted to the nucleus (Wanitchakorn et al. 2000). The NSP is transported to the cell periphery before intercellular transport (Wanitchakorn et al. 2000). All of the six ssDNA components of the BBTV genome share several highly conserved regions, including a stem-loop, a TATA box, a highly conserved region, and a poly-A coding region, in addition to the major ORF (Randles et al. 2000).

16.2.2 Banana Streak Virus

BSV is in the *Badnavirus* genus of the *Caulimoviridae* family (James et al. 2011). BSV virions have a bacilliform shape, and the genome is a noncovalently closed double-stranded DNA of 7.4–8.0 kb and harbors three ORFs on the (+) sense strand (King 2012). ORF 1 encodes a 22 kDa protein of unknown function and ORF 2 encodes a 14 kDa protein, also of unknown function. ORF 3 encodes a large polyprotein of 210 kDa that is proteolytically cleaved into the capsid protein, an aspartic protease (AP), a reverse transcriptase with ribonuclease H (RH) activity, and an RNA binding (RB) protein (Harper et al. 1999; Harper and Hull 1998). There are two forms of BSV found in the genome of *Musa balbisiana* (B): episomal BSV and endogenous BSV (eBSV) (refs). eBSV is a type of endogenous retrovirus sequence (ERVs) found in the B genome of banana (Chabannes et al. 2013). The eBSV form of the virus can be activated by temperature stress, water stress, micropropagation manipulation, and genetic hybridization (Dahal et al. 2000; Dallot et al. 2001; Lheureux et al. 2003).

BSV has been confirmed to occur in Asia, Africa, Australia, the Americas, and Europe (Lockhart 1995). BSV can result in yield losses from 6 to 90 % although today's epidemics are limited primarily to Africa (Iskra-Caruana et al. 2010). Banana streak disease symptoms are characterized by abnormal bunch development, distortion of leaves and petioles, necrotic and chlorotic flecking on leaves, and stem cracking (Dahal et al. 2000). Three species of BSV, banana streak OL virus (BSOLV), banana streak MY virus (BSMYV), and banana streak GF virus (BSGFV), are recognized by the International Committee on the Taxonomy of Viruses (ICTV); however, additional species may exist (James et al. 2011). BSV is transmitted in a semi-persistent manner by the mealybugs (Meyer et al. 2008).

16.2.3 Banana Bract Mosaic Virus

BBrMV belongs to the *Potyvirus* genus of the *Potyviridae* family with virions that are flexuous filamentous rods (Marie-line et al. 2008; Fauquet et al. 2005). The genome is ssRNA in the sense (+) orientation and consists of 9,711 nucleotides. Symptoms of BBrMV include mosaic streaks on the bracts of the banana inflorescence, spindle-shaped streaks scattered along the petioles, and a mottled discoloration of the pseudostem. However, these BBrMV symptoms may be confused with the mosaic symptoms produced by CMV infections in banana (Rodoni et al. 1997). BBrMV is non-persistently transmitted by at least three species of aphids (Magnaye and Espino 1990; Munez 1992; Diekmann and Putter 1996). It has been reported from India and Sri Lanka (Diekmann and Putter 1996), from the Philippines (Bateson and Dale 1995), and from Thailand, Vietnam, and Western Samoa (Rodoni et al. 1999). The virus has been detected from infected banana plants by enzyme-linked immunosorbent assay (ELISA), immunomagnetic capture PCR (IC-PCR), and reverse transcription PCR (RT-PCR). Alternative hosts other than banana include cardamom (*Elettaria cardamomum*) in India (Siljo et al. 2012), abaca (*Musa textilis*) in the Philippines (Thomas et al. 1997), and flowering ginger (*Alpinia purpurata*) in Hawaii (Wang et al. 2010). The virus can be transmitted through the vegetative planting material and from banana to banana by sap inoculation, but attempts to mechanically inoculate herbaceous indicator plants have not been successful (Magnaye and Espino 1990; Munez 1992).

16.2.4 Cucumber Mosaic Virus

CMV is in the *Cucumovirus* genus of the *Bromoviridae* family and has virions that are isometric in shape and 28–30 nm in diameter (Lockhart and Jones 2000). The virions have the single-strand (+) sense RNA genome that is tripartite, consisting of three RNA molecules, designated RNA 1 (3,350 nucleotides), RNA 2 (3,050 nucleotides) and RNA 3 (2,200 nucleotides) that are encapsidated individually. In addition to the three genomic RNAs, a sub-genomic RNA (RNA 4) encoding the capsid protein is produced from RNA 3 (Roossinck 1999). RNA 1 contains a single ORF that encodes the replicase protein with methyltransferase and helicase domains. RNA 2 encodes another replicase component (2a), in addition to a suppressor protein (2b) that is expressed from sub-genomic RNA 4A. RNA 3 encodes the movement protein (3a) and coat protein (3b), but the CP is expressed from sub-genomic RNA 4. The host range of CMV includes more than 800 species of plants. CMV can be transmitted by many aphid species including *Aphis gossypii*, *Rhopalosiphum maidis*, *Myzus persicae*, *Macrosiphum pisi*, *Rhopalosiphum prunifolia*, and possibly *Pentalonia nigronervosa* (Magnaye and Valmayor 1995).

Different strains of CMV are known to occur throughout the world, where they can produce various symptoms depending on the host, the strain, and the temperature. The most common symptom is a mosaic on the leaves, but other symptoms may include leaf deformation or mosaic on the fruit. However, these symptoms may appear sporadically and be distributed unevenly throughout the plant. Other symptoms produced by severe strains include more severe leaf distortions and necrosis of emerging leaves or internal tissues that can lead to death of the plant (Lockhart and Jones 2000).

16.3 Transgenic Resistance to BBTV

16.3.1 Transgenic Approaches for Banana Bunchy Top Virus Resistance

The severity of BBTV epidemics and their resulting yield losses, together with the lack of natural resistance, warrants investigation into applying transgenic techniques to confer resistance in desirable cultivars. Several groups have investigated this possibility and utilized posttranscriptional gene silencing (PTGS) or RNA interference (RNAi) techniques in attempts to produce cultivars with durable resistance to BBTV (Elayabalan et al. 2013; Krishna et al. 2013; Shekhawat et al. 2012; Borth et al. 2011; Becker et al. 2000). Elayabalan et al. (2013) transformed the banana cultivar "Virupakshi" (AAB) using intron-hairpin RNA (ihpRNA) constructs with 440 bp of the 5′-end and 440 bp of the 3′-end, in sense and antisense orientation, respectively, of the BBTV Rep gene. The ubiquitin (Ubi) promoter was used to drive transcription, and a CRE intron was used as a spacer between the two partial fragments. These authors reported that the expression of ihpRNA led to enhanced resistance to BBTV in bananas in greenhouse. Krishna et al. (2013) transformed the Cavendish cultivar "Grand Nain" (AAA) using a similar approach utilizing hairpin structures. Their construct contained portions of DNA-1, DNA-3, DNA-4, and DNA-5 from BBTV that included a 642 bp intron as spacer region and the *CaMV* 35S promoter to drive transcription. They reported that viral symptoms were delayed by 3–6 months compared to wild-type untransformed plants. Shekhawat et al. (2012) transformed a "Rasthali" cultivar (AAB) using the ihpRNA approach with two constructs. One contained the full-length Rep coding sequence from DNA-1 and the other containing a partial Rep coding sequence together with its 5′-upstream regulatory region. These constructs were linked by the castor bean catalase gene as a spacer and driven by the *Zea mays* ubiquitin (Ubi) promoter. Partial BBTV resistance in these transgenic banana plants was indicated by the lack of symptom development in plants up to 6 months following virus challenge, and the detection of small interfering RNAs (siRNAs) derived from the ihpRNA sequence, establishing that RNAi was the mechanism of resistance in these plants.

Becker et al. (2000) transformed the Cavendish cultivars "Grand Nain" and "Williams" (AAA) using constructs containing components from BBTV's DNA-1, DNA-5, and DNA-6 that were co-transformed with constructs that contained the promoter from BBTV's DNA-6 and a construct containing an internal ORF frame from BBTV's DNA-1 or a construct containing an ORF from BBTVDNA-5 in an untranslatable form. These constructs all contained the Ubi intron first exon and intron from the maize polyubiquitin-1 gene and were driven by either the Ubi promoter from the maize polyubiquitin-1 gene or the Pro promoter from DNA-6 of BBTV. In this study, Becker et al. used microprojectile bombardment in lieu of *Agrobacterium*-mediated transformation, to eliminate the possibility of chimeric plants that may be produced by *Agrobacterium*-mediated transformation in banana. They reported transformation efficiencies of about 11 %. Although they were able to regenerate transgenic plants, no data was given on the resistance or susceptibility of the transgenic plants.

The majority of the work reporting the development of transgenic banana with BBTV resistance utilized *Agrobacterium*-mediated transformation. All protocols have utilized embryogenic cell suspensions (ECS) derived from immature male embryos as the starting tissue for transformation, with the exception of Krishna et al. (2013) who used "scalps" or shoots from multiple meristems of banana plants as their starting material. In addition to ECS, Elayabalan et al. (2013) utilized microcalli derived from ECS and noted that microcalli resulted in a high frequency of transformation. The use of ECS to establish a single cell origin of somatic embryos for regeneration of transgenic lines represents the most ideal explant due to the avoidance of regenerating chimeras (Huang et al. 2007). The regeneration of chimeras that may initially exhibit BBTV resistance but that later develop symptoms of BBTV infection after challenge is a problem that must be addressed in designing transformation protocols to develop BBTV-resistant germplasm in banana.

Shekhawat et al. (2012) reported that no DNAs from the CP, MP, or Rep genes could be detected by PCR in inoculated transgenic banana plants. Borth et al. (2011) reported that no CP gene sequences could be detected by PCR in inoculated transformed banana plants. This lack of expression of viral genes in inoculated transformed banana plants represents a significant accomplishment in the development of transgenic BBTV-resistant lines of banana.

Another strategy that utilizes a process termed "virus-activated cell death" has been developed by James Dale at the Queensland University of Technology (QUT) in Australia (Dale et al. 2001). In banana plants transformed with a vector that contains portions of the BBTV genome linked with a transgene that codes for barnase, an RNA endonuclease derived from *Bacillus amyloliquefaciens*, BBTV replication following infection, results in a hypersensitive response and subsequent death of the infected cells, thus preventing the spread of the virus within the plant tissues. The mechanism of this response is unique: when BBTV infects cells of banana plants transformed with this construct, the Rep protein produced by the replicating BBTV recognizes the stem-loop region (CR-SL) of the construct and initiates the replication of the entire transgenic sequence as a multicopy circular extrachromosomal DNA; barnase is then synthesized and degrades RNA in the infected cell resulting in death of the infected and potentially nearby cells. The CR-SL region is highly conserved among all known BBTV isolates and will be recognized by the Rep protein of all known BBTV strains. Because intron sequences are located adjacent to each CR-SL region, both the intron and the CR-SL regions will be spliced out of the transcript, leaving only the coding region of the barnase gene to be translated. The barnase transgene will only be activated during BBTV infection and will only be expressed in cells that are infected by BBTV. In the absence of BBTV infection, no functional transgene will be expressed. It is expected that the transgene protein will only be produced in inoculated and possibly adjacent cells and will not be present in other tissues such as fruit. Dale et al. (2001) then co-transformed embryogenic cell suspensions of banana with pRTBN6 and a selectable marker gene using microprojectile bombardment according to Becker et al. (2000). Plants surviving antibiotic selection were tested by PCR to confirm the complete integration of pRTBN6. Positive plants were multiplied and challenged with BBTV-viruliferous aphids. Several transgenic lines showed no typical symptoms of BBTV infection, but instead developed necrosis at the points of aphid feeding, presumably due to transgene-induced cell death. PCR and Southern hybridization failed to detect BBTV in these challenged transgenic lines, indicating that these plants are resistant to BBTV under greenhouse conditions (Dale et al. 2001).

16.3.2 Field Evaluations of Transgenic Banana in Hawaii

Borth et al. (2011) transformed a "Dwarf Brazilian" cultivar (AAB) using four different constructs from the Rep gene of BBTV's DNA-1: one construct contained the entire Rep ORF with a double point mutation; the second contained the entire Rep ORF in antisense orientation; the third contained 455 bp of the Rep ORF including the stem-loop conserved region (SL-CR), the TATA box, and 361 bp of the 5-end of the Rep coding region all in antisense orientation; and the fourth contained the third construct fused with an antisense construct of the Rep ORF. The *CaMV* 35S promoter and *AMV* enhancer sequence were used to drive transcription in all of the constructs. Embryogenic calli of banana cultivar "Dwarf Brazilian" that were initiated from immature male flowers were used to produce embryogenic cell suspensions (ECS). These cell suspensions were used as the source of explants for *Agrobacterium* transformations using the four constructs. Embryos formed from transformed ECS were germinated on media containing antibiotics to select transformed lines. These lines have been evaluated for BBTV resistance by challenge with viruliferous aphids in the greenhouse. Twenty transgenic lines displayed some degree of resistance to BBTV challenge. A 1-acre plot at the Waimanalo Field Station of University of Hawaii on the island of Oahu was used for field testing of the transgenic lines. All 20 putatively BBTV-resistant banana lines were planted in five separate field trials. At the conclusion of each of these field trials, there were still individual plants from a few transgenic lines that had not developed BBTV symptoms.

16.4 Conclusions

Banana viruses can cause devastating diseases of bananas in many parts of Asia, Africa, and the Pacific. Recently, BBTV has devastated the banana industry in several Africa countries. The development of BBTV-resistant banana plants through the use of genetic engineering offers the quickest way to develop banana plants with long-lasting, broad-spectrum resistance to the various strains of BBTV and other plant viruses afflicting banana. Transgenic banana plants that are resistant to virus challenge in the field and that also have good horticultural characteristics can be propagated and distributed to the public. The development of such cultivars will directly benefit banana growers in Asia, the Pacific, and Africa.

The work that has already been done by various workers to produce transgenic banana cultivars with virus resistance shares several things in common. The use of *Agrobacterium*-mediated transformation remains the method of choice in most of the studies with the exception of the work of Becker et al. (2000) who utilized microprojectile bombardment. Embryogenic cell suspensions derived from immature male flowers are the most common tissue transformed because it minimizes the occurrence of chimeras in the resulting transgenic lines. Several strategies have been used to generate virus resistance in banana. Constructs utilizing the Rep gene from BBTV DNA-1, in both sense and antisense orientation, and constructs designed to generate ihpRNA structures in transgenic tissues appear to be the most promising approaches to achieve virus resistance by exploiting RNA silencing. The constitutive promoter *CaMV* 35S has often been used to drive transcription of introduced transgenes when high levels of transcription are needed.

New techniques for genetic engineering plants have become available in recent years. One of these new technologies involves the use of clustered regularly interspaced short palindromic repeats (CRISPR) and CRISPR-associated proteins (Cas proteins) (Richter et al. 2012). CRISPR-Cas molecules act as programmable endonucleases that make site-specific double-strand breaks (DSBs) in DNA (Jinek et al. 2012). Short CRISPR RNAs (crRNAs) are processed from CRISPR transcripts (pre-crRNAs) and serve as homing oligonucleotides impeding the replication of invading viruses (Doudna and Charpentier 2014). Site-specific DSBs made by CRISPR-Cas cause single point mutations generated through nonhomologous end joining (NHEJ) ultimately resulting in the activation of gene silencing directed against the target sequences. In transgenic plants the CRISPR-Cas system has been shown to reduce viral accumulation to confer resistance to ssDNA viruses: bean yellow dwarf virus (BeYDV), beet severe curly top virus (BSCTV), and tomato yellow leaf curl virus (TYLCV) (Baltes et al. 2015; Ji et al. 2015; Ali et al. 2015). Unique CRISPR-Cas constructs that target conserved BBTV sequences may be designed and engineered into banana plants to produce transgenic banana plants that are resistant to BBTV. In addition to NHEJ, site-specific homologous recombination (HR) is possible using CRISPR-Cas. The advantages of site-specific insertions are that they effectively eliminate position effects that may act upon the inserted sequences and prevent the incorporation of transgenes into undesirable chromosomal locations (Matzke and Matzke 1998). Successful strategies that are based upon PTGS or RNAi mechanisms suggest that site-specific insertion directed by CRISPR-Cas of BBTV constructs could produce non-chimeric lines that have high transgene expression levels. CRISPR-Cas offers the potential to transform plants more precisely, allowing the introduction of transgenes at specific sites (Wilson et al. 1990; Lee et al. 2015).

Additionally, a Cry-related protein may be useful for conferring resistance to *P. nigronervosa* in banana, potentially limiting the ability of the vector to spread BBTV between plants. The use of "stacked" or "pyramided" traits in transgenic plants to induce resistance to viruses and their vectors is another approach to create germplasm potentially resistant to BBTV. Palma et al. (2014) showed that the LC50 value of a Cry-related protein derived from *Bacillus thuringiensis* (*Bt*) against *Myzus persicae* was 32.7 μg/mL, the lowest value yet reported for Cry proteins against any member of the *Aphididae*. Cry toxins from *Bt* consist of the N-terminal three-domain "toxic core" and the proteolytically labile C-terminal segment (Evdokimov et al. 2014). The high pH found in the gut of insects cleaves the two segments and activates the core toxin that binds with receptors in the gut membrane ultimately, forming transmembrane pores that lead to cell death through osmotic shock (Gomez et al. 2007). Cry-type proteins expressed in the phloem of the banana at high levels may provide resistance to the vector of BBTV.

Another newly developed technique with great potential benefit is the use of an internal ribosome entry site (IRES) that may facilitate the stable long-term expression of transgenes in plants, but has yet to be tried in banana (Renaud-Gabardos et al. 2015). These novel approaches to generate durable BBTV and vector resistance hold great potential for the successful development of transgenic BBTV-resistant germplasm in the future.

Acknowledgements This work was supported in part by grants from the National Institute of Food and Agriculture, U.S. Department of Agriculture, Hatch HAW09025-H under Accession no. 1001478 and from the U.S. Department of Agriculture, Agricultural Research Service under award number 58-5320-4-012.

References

Ali Z, Abulfaraj A, Idris A, Ali S, Tashkandi M, Mahfouz MM (2015) CRISPR/Cas9-mediated viral interference in plants. Genome Biol 16:238 doi:10.1186/s13059-015-0799-6

Aronson MN, Meyer AD, Györgyey J, Katul L, Vetten HJ, Gronenborn B, Timchenko T (2000) Clink, a nanovirus-encoded protein, binds both pRB and SKP1. J Virol 74:2967–2972

Baltes NJ, Hummel AW, Konecna E, Cegan R, Bruns AN, Bisaro DM, Voytas DF (2015) Conferring resistance to geminiviruses with the CRISPR–Cas prokaryotic immune system. Nat Plants 1:15145. doi:10.1038/nplants.2015.145

Bateson MF, Dale JL (1995) Banana bract mosaic virus: characterisation using potyvirus specific degenerate PCR primers. Arch Virol 140:515–527

Becker DK, Dugdale B, Smith MK, Harding RM, Dale JL (2000) Genetic transformation of Cavendish banana (Musa spp. AAA group) cv'Grand Nain' via microprojectile bombardment. Plant Cell Rep 19:229–234

Beetham PR, Harding RM, Dale JL (1999) Banana bunchy top virus DNA-2 to 6 are monocistronic. Arch Virol 144:89–105

Borth W, Perez E, Cheah K, Chen Y, Xie WS, Gaskill D, Khalil S, Sether D, Melzer M, Wang M et al (2011) Transgenic banana plants resistant to banana bunchy top virus infection. Acta Horticult 897:449–457

Burns TM, Harding RM, Dale JL (1995) The genome organization of banana bunchy top virus: analysis of six ssDNA components. J Gen Virol 76:1471–1482

Chabannes M, Baurens F-C, Duroy P-O, Bocs S, Vernerey M-S, Rodier-Goud M, Barbe V, Gayral P, Iskra-Caruana M-L (2013) Three infectious viral species lying in wait in the banana genome. J Virol 87:8624–8637. doi:10.1128/JVI.00899-13

Dahal G, Ortiz R, Tenkouano A, Hughes J d'A, Thottappilly G, Vuylsteke D, Lockhart BEL (2000) Relationship between natural occurrence of banana streak badnavirus and symptom expression, relative concentration of viral antigen, and yield characteristics of some micropropagated Musa spp. Plant Pathol 49:68–79

Dale JL (1987) Banana bunchy top: an economically important tropical plant virus disease. Adv Virus Res 33:301–326

Dale JL, Dugdale B, Hafner GJ, Harding RM, Becker DK, Hermann SR, Chowpongpang S (2001) A construct capable of release in closed circular form from a larger nucleotide sequence permitting site specific expression and/or developmentally regulated expression of selected genetic sequences. Patent Application Number WO0172996

Dallot S, Acuna P, Rivera C, Ramirez P, Côte F, Lockhart BEL, Caruana M-L (2001) Evidence that the proliferation stage of micropropagation procedure is determinant in the expression of Banana streak virus integrated into the genome of the FHIA 21 hybrid (Musa AAAB). Arch Virol 146:2179–2190

Diekmann DR, Putter CAJ (1996) Musa, 2nd edn. FAO/IPGRI technical guidelines for the safe movement of germplasm, no. 15, Food and Agriculture Organization of the United Nations, Rome/International Plant Genetic Resources Institute, Rome, 28 pp

Doudna JA, Chapentier E (2014) The new frontier of genome engineering with CRISPR-Cas9. Science 346(6213):1077–1085. doi:10.1126/science.1258096

Elayabalan S, Kalaiponmani K, Subramaniam S, Selvarajan R, Panchanathan R, Muthuvelayoutham R, Kumar KK, Balasubramanian P (2013) Development of Agrobacterium-mediated transformation of highly valued hill banana cultivar Virupakshi (AAB) for resistance to BBTV disease. World J Microbiol Biotechnol 29:589–596. doi:10.1007/s11274-012-1214-z

Evdokimov AG, Moshiri F, Sturman EJ, Rydel TJ, Zheng M, Seale JW, Franklin S (2014) Structure of the full-length insecticidal protein Cry1Ac reveals intriguing details of toxin packaging into in vivo formed crystals: full-length Cry1Ac structure. Protein Sci 23:1491–1497. doi:10.1002/pro.2536

Fauquet CM, Mayo MA, Maniloff J, Desselberger U, Ball LA (2005) Virus taxonomy, VIIIth report of the ICTV. Elsevier Academic Press, London, pp 751–756

Gomez I, Pardo-Lopez L, Munoz-Garay C, Fernandez LE, Perez C, Sanchez J, … Bravo A (2007) Role of receptor interaction in the mode of action of insecticidal Cry and Cyt toxins produced by Bacillus thuringiensis. Peptides 28(1):169–173. doi:10.1016/j.peptides.2006.06.013

Gronenborn B (2004) Nanoviruses: genome organisation and protein function. Vet Microbiol 98:103–109. doi:10.1016/j.vetmic.2003.10.015

Hafner GJ, Harding RM, Dale JL (1995) Movement and transmission of banana bunchy top virus DNA component one in bananas. J Gen Virol 76:2279–2285

Harding RM, Burns TM, Dale JL (1991) Virus-like particles associated with banana bunchy top disease contain small single-stranded DNA. J Gen Virol 72:225–230

Harper G, Hull R (1998) Cloning and sequence analysis of banana streak virus DNA. Virus Genes 17:271–278

Harper G, Dahal G, Thottappilly G, Hull R (1999) Detection of episomal banana streak badnavirus by IC-PCR. J Virol Methods 79:1–8

Horser CL, Harding RM, Dale JL (2001) Banana bunchy top nanovirus DNA-1 encodes the "master" replication initiation protein. J Gen Virol 82:459–464

Hu JS, Wang M, Sether D, Xie W, Leonhardt KW (1996) Use of polymerase chain reaction (PCR) to study transmission of banana bunchy top virus by the banana aphid (Pentalonia nigronervosa). Ann Appl Biol 128:55–64

Huang X, Huang X-L, Xiao W, Zhao J-T, Dai X-M, Chen Y-F, Li X-J (2007) Highly efficient Agrobacterium-mediated transformation of embryogenic cell suspensions of Musa acuminata cv. Mas (AA) via a liquid co-cultivation system. Plant Cell Rep 26:1755–1762. doi:10.1007/s00299-007-0376-x

Iskra-Caruana M-L, Baurens F-C, Gayral P, Chabannes M (2010) A four-partner plant-virus interaction: enemies can also come from within. Mol Plant Microbe Interact 23:1394–1402

James AP, Geijskes RJ, Dale JL, Harding RM (2011) Molecular characterisation of six badnavirus species associated with leaf streak disease of banana in East Africa. Ann Appl Biol 158:346–353. doi:10.1111/j.1744-7348.2011.00466.x

Ji X, Zhang H, Zhang Y, Wang Y, Gao C (2015) Establishing a CRISPR–Cas-like immune system conferring DNA virus resistance in plants. Nat Plants doi:10.1038/nplants.2015.144

Jinek M, Chylinski K, Fonfara I, Hauer M, Doudna JA, Charpentier E (2012) A programmable dual-RNA-guided DNA endonuclease in adaptive bacterial immunity. Science 337:816–821. doi:10.1126/science.1225829

King A (ed) (2012) Virus taxonomy: classification and nomenclature of viruses; ninth report of the International Committee on Taxonomy of Viruses, report of the International Committee on Taxonomy of Viruses. Elsevier, Amsterdam

Krishna B, Kadu AA, Vyavhare SN, Chaudhary RS, Joshi SS, Patil AB, Subramaniam VR, Sane PV (2013) RNAi-mediated resistance against banana bunchy top virus (BBTV) in'GrandNain'banana. In: II genetically modified organisms in horticulture symposium, vol 974, pp 157–164

Kumar PL, Selvarajan R, Iskra-Caruana M-L, Chabannes M, Hanna R (2014) Biology, etiology, and control of virus diseases of banana and plantain. In: Advances in virus research. Elsevier, Burlington, pp 229–269

Lee JS, Kallehauge TB, Pedersen LE, Kildegaard HF (2015) Site-specific integration in CHO cells mediated by CRISPR/Cas9 and homology-directed DNA repair pathway. Sci Rep 5:8572. doi:10.1038/srep08572

Lheureux F, Carreel F, Jenny C, Lockhart B, Iskra-Caruana M (2003) Identification of genetic markers linked to banana streak disease expression in inter-specific Musa hybrids. Theor Appl Genet 106:594–598

Lockhart BE (1995) Banana streak badnavirus infection in Musa: epidemiology, diagnosis and control. ASPAC Food & Fertilizer Technology Center, Taipei City

Lockhart BEL, Jones DR (2000) Bract mosaic. In: Jones DR (ed) Diseases of banana abaca and ensete. CABI Publishing, New York, pp 256–263

Magnaye LV, Espino RCC (1990) Note: banana bract mosaic, a new disease of banana. I. Symptomatology. Philipp Agric 73:55–59

Magnaye LV, Valmayor RV (1995) BBTV, CMV and other viruses affecting banana in Asia and the Pacific. ASPAC Food & Fertilizer Technology Center, Taipei City

Marie-line IC, Serge G, Nathalie L (2008) A reliable IC one-step RT-PCR method for the detection of BBrMV to ensure safe exchange of *Musa* germplasm. J Virol Methods 153:223–231

Matzke AJ, Matzke MA (1998) Position effects and epigenetic silencing of plant transgenes. Curr Opin Biotechnol 1:142–148

Meyer JB, Kasdorf GGF, Nel LH, Pietersen G (2008) Transmission of activated-episomal banana streak ol (badna) virus (bsolv) to cv. Williams banana (musa sp.) by three mealybug species. Plant Dis 92:1158–1163

Munez AR (1992) Symptomatology, transmission and purification of banana bract mosaic virus (BBMV) in 'Giant Cavendish' banana. MSc thesis, University of the Philippines, Los Banos, PhilippinesPalma, L., Munoz, D., Berry, C., Murillo, J., de Escudero, I. R., & Caballero, P. (2014). Molecular and insecticidal characterization of a novel Cry-related protein from Bacillus thuringiensis toxic against Myzus persicae. Toxins (Basel), 6(11), 3144-3156. doi:10.3390/toxins6113144

Palma L, Munoz D, Berry C, Murillo J, de Escudero IR, Caballero P (2014) Molecular and insecticidal characterization of a novel Cry-related protein from Bacillus thuringiensis toxic against Myzus persicae. Toxins (Basel) 6(11):3144–3156. doi:10.3390/toxins6113144

Randles JW et al (2000) Genus *Nanovirus*. In: van Regenmortel MHV, Fauquet CM, Bishop DHL, Carstens EB, Estes MK, Estes SM, Lemon SM, Maniloff J, Mayo MA, McGeoch DJ, Pringle CR, Wickner RB (eds) Virus taxonomy. Seventh report of the International Committee on Taxonomy of Viruses. Academic, San Diego, pp 303–309

Renaud-Gabardos E, Hantelys F, Morfoisse F, Chaufour X, Garmy-Susini B, Prats AC (2015) Internal ribosome entry site-based vectors for combined gene therapy. World J Exp Med 5:11. doi:10.5493/wjem.v5.i1.11

Richter C, Chang JT, Fineran PC (2012) Function and regulation of Clustered Regularly Interspaced Short Palindromic Repeats (CRISPR)/CRISPR Associated (Cas) Systems. Viruses 4:2291–2311. doi:10.3390/v4102291

Rodoni BC, Ahlawat YS, Varma A, Dale JL, Harding RM (1997) Identification and characterization of banana bract mosaic virus in India. Plant Dis 81:669–672

Rodoni BC, Dale JL, Harding RM (1999) Characterization and expression of the coat protein-coding region of banana bract mosaic potyvirus, development of diagnostic assays and detection of the virus in banana plants from five countries in southeast Asia. Arch Virol 144:1725–1737

Roossinck MJ (1999) Cucumoviruses (*Bromoviridae*)-molecular biology. In: Granoff A, Webster RG (eds) Encyclopedia of virology, 2nd edn. Academic, San Diego, pp 320–324

Selvarajan R, Balasubramanian V (2014) Host–virus interactions in banana-infecting viruses. In: Plant virus–host interaction. Elsevier, San Diego, pp 57–78

Shekhawat UKS, Ganapathi TR, Hadapad AB (2012) Transgenic banana plants expressing small interfering RNAs targeted against viral replication initiation gene display high-level resistance to banana bunchy top virus infection. J Gen Virol 93:1804–1813. doi:10.1099/vir.0.041871-0

Siljo A, Bhat AI, Biju CN, Venugopal MN (2012) Occurrence of banana bract mosaic virus on cardamom. Phytoparasitica 40:77–85. doi:10.1007/s12600-011-0193-1

Thomas JE, Geering ADW, Gambley CF, Kessling AF, White M (1997) Purification, properties, and diagnosis of banana bract mosaic potyvirus and its distinction from abaca mosaic potyvirus. Phytopathology 87:698–705

Wang IC, Sether DM, Melzer MJ, Borth WB, Hu JS (2010) First report of banana bract mosaic virus in flowering ginger in Hawaii. Plant Dis 94:921

Wanitchakorn R, Harding RM, Dale JL (1997) Banana bunchy top virus DNA-3 encodes the viral coat protein. Arch Virol 142:1673–1680

Wanitchakorn R, Hafner GJ, Harding RM, Dale JL (2000) Functional analysis of proteins encoded by banana bunchy top virus DNA-4 to-6. J Gen Virol 81:299–306

Wilson C, Bellen HJ, Gehring WJ (1990) Position effects on eukaryotic gene expression. Ann Rev Cell Biol 6:679–714

Transgenic Approaches to Improve Resistance to Nematodes and Weevils

17

Hugh Roderick, Leena Tripathi, and S. Poovarasan

Abstract

Banana and plantain are important staple crops for Africa and important fruit crops for Asia, Latin America and Caribbean islands. Several nematode species and rhizome weevil (*Cosmopolites sordidus*) are major pests in banana that cause heavy damage and revenue loss. Pesticides and bio-control agents control the pests, but pesticide residues pose severe environmental problems. Conventional breeding is a difficult and slow process due to the limited sources of resistance, sterility of cultivated banana varieties, polyploidy levels, long cropping cycle and the lack of rapid screening methods. Genetic engineering is considered as one of the eco-friendly and safer methods to control these pests. This review discusses the seriousness of the problem, the status and source of pest resistance and the mechanisms involved. The availability of various genes with potential to control nematodes and weevils is discussed. Further, current efforts and future prospects for identifying natural resistance genes and RNAi-based defences with potential to control nematode and banana weevil in a transgenic approach are outlined and discussed. Nematode-resistant transgenic banana cultivars expressing rice or maize cystatin genes and peptides evaluated under field conditions and those weevil-resistant cultivars

H. Roderick
Centre for Plant Sciences, University of Leeds, Leeds, UK

L. Tripathi (✉)
International Institute of Tropical Agriculture (IITA), 30709-00100 Nairobi, Kenya
e-mail: l.tripathi@cgiar.org

S. Poovarasan
Indian Institute of Horticultural Research, Bangalore 560089, India

S. Mohandas, K.V. Ravishankar (eds.), *Banana: Genomics and Transgenic Approaches for Genetic Improvement*, DOI 10.1007/978-981-10-1585-4_17

developed using papaya cystatin gene with enhanced inhibitory potential are discussed in the light of biosafety concerns.

Keywords
Nematode and weevil resistance • Protein and peptide based transgenic defences • Cystatin genes and peptides • Plant lectins • Insecticidal proteins • Alpha-amylase inhibitors and chitinase enzymes • RNAi-based defences

17.1 Introduction

Banana and plantain (*Musa* spp.) are cultivated in over 130 countries worldwide covering approximately 10 million hectares with an annual production of 139 million tons (FAOSTAT 2014). It forms an important staple diet of Latin American and Caribbean islands, grown mainly for subsistence and for local sale. In recent years, the income through banana cultivation has reduced drastically due to severe problems affecting banana cultivation. Several nematode species and rhizome weevil (*Cosmopolites sordidus*) are major pests that cause huge production losses.

Nematodes are key pests in many commercial dessert banana plantations, and they also damages both cooking banana and plantains. Root systems damaged by nematodes are less able to utilize nutrients and water, become susceptible to secondary infection and provide weakened anchorage to the plant. Plants with weakened root systems are prone to toppling, especially in strong winds and when bearing fruit leading to loss of the fruit. Banana suffers estimated losses to nematodes of 6 M tonnes/year, representing the consumption need of 60 M people in banana-dependent countries. Weevil is most severe in plantains and East African highland bananas. Weevils damage the corm of plants by making tunnels and rootstock. Damaged corm interferes with root initiation and sap flow in the plant, resulting in yellowing of leaves and wilting of plants particularly the young suckers. The suckers finally die, whereas the older plants are retarded in their growth and produce small bunches. The weevil-damaged plants can easily be blown over by the wind. Pesticides are usually used to control nematodes and weevils, but pesticide residue is a main concern considering the environmental and health problems. Genetic transformation is considered as one of the eco-friendly approaches for controlling weevils and nematodes.

17.2 Nematodes

Nematodes cause heavy damage to banana and plantain crops with approximately a 20 % reduction in productivity globally (Sasser and Freckman 1987). In areas prone to tropical storms and particularly in Africa, losses of 40 % or greater can frequently occur. Nematicide application experiments in West Africa have shown that after three crop-cycles, the potential reduction in yields is 71 ± 16 % (Atkinson 2003), a region of the world where bananas and plantains provide >25 % of the carbohydrate intake of approximately 70 million people, corresponding to 10 % of their food energy (Ortiz and Vuylsteke 1996; Robinson 1996). Several nematodes are major pests of banana and plantain crops (Atkinson 2003; Brentu et al. 2004; Gowen and Quénéhervé 1990). *Radopholus similis* is considered the most damaging species. It has a life cycle of 20–25 days which makes its population multiply rapidly and cause severe crop loss (Bridge et al. 1995; Haegeman et al. 2010; Price 2006). In severe infections with *R. similis*, yield losses due to stunted growth are compounded by increased plant toppling in strong winds as a result of reduced root system anchorage (Gowen and Quénéhervé 1990). In the absence of *R. similis, Pratylenchus coffeae*, *Helicotylenchus multicinc-*

tus and *Meloidogyne* spp. can also become severe constraints on banana plantations (Bridge et al. 1995; McSorley and Parrado 1983; Brentu et al. 2004; Price 2006). Secondary fungal and bacterial infections frequently compound the direct damage nematodes cause (Duncan and Moens 2006).

Nematicides are widely applied to soils in intensive banana plantations, but they are environmentally toxic and a risk to human health (Atkinson et al. 2003). The prevalence of banana and plantain cultivation on small plots across the tropics also means that nematicides are inappropriate on the basis of both cost and grower safety. Nematode-tolerant and nematode-resistant banana cultivars have been identified, though the cultivars are only effective against single species of nematode (Lorenzen et al. 2010; Pinochet 1988). Concomitant nematode species infection is common in banana and plantain growing areas. In Uganda, the largest producer of bananas and plantains in Africa (FAOSTAT 2014), *R. similis*, *H. multicinctus* and *Meloidogyne* spp. are present in all banana-growing regions (Kashaija et al. 1994). Work in Costa Rica on the dessert banana cultivar 'Grand Naine' identified *R. similis, P. coffeae, M. incognita* and *H. multicinctus* causing significant reductions in yield, with the multispecies infections causing greater damage than single species (Moens et al. 2006). Similar work with the plantain cultivar 'Apantu-pa' in Ghana found that while *P. coffeae* caused the largest losses, the greatest necrosis and toppling, co-infection with *M. javanica* and *H. multicinctus* caused an increase in damage to the plant over any single species infection (Brentu et al. 2004). Conventional breeding has struggled to introduce the single species resistance into economically important cultivars or develop hybrids with broad nematode resistance (Lorenzen et al. 2010). A transgenic approach for nematode resistance is strongly favoured for bananas and plantains, both because of the difficulty of developing an effective resistance through breeding and the availability of several proven anti-nematode genes.

17.3 Banana Weevil

Banana and plantain production in Africa is effected significantly by rhizome weevil which is a serious pest (Ostmark 1974; Gold 1998; Gold and Messiaen 2000; Swennen and Vulysteke 2001; Fogein et al. 2002). The banana plantation decline (Gold et al. 1999) called 'yield decline syndrome' in Africa is associated with this weevil. By the time the crop reaches the 4th cycle, the yield loss reaches 44 % (Rukazambuga et al. 1998). Crop losses up to 100 % are also reported in cases of severe infestation. Establishment of new plantations is difficult due to the persistence of this infection in soil (Sengooba 1986; Price 1994). The causal organism of this devastating disease is the weevil *Cosmopolites sordidus* (Germar 1824) (Coleoptera: Curculionidae). The weevil in its adult form is free living and black in colour and measures 10–15 mm. It is also associated with crop debris and is nocturnally active as it becomes desiccated easily. The adults inhabit a particular location and remain there for long period of time. Weevils generally do not fly. Infected planting materials disseminate the weevil. Banana weevils have long life span and low fecundity. Some may live up to 4 years, though generally for 1 year. Adults generally feed on dead or dying banana plants. They are found living under newly cut or rotting pseudostems. They can survive without feeding for several months with little moisture. They lay more than one egg per week on flowering plants and crop residues. In the holes made by the rostrum, the females place their white, oval eggs singly. On the leaf sheaths and rhizome surfaces also oviposition is noticed. The emerging larvae feed on the rhizome, stem and the pseudostem. The larvae pass through five to eight instars. The adults emerge within 5–7 weeks out under tropical conditions. Eggs do not develop below 12 °C (Gold and Messiaen 2000).

The volatiles released by the host plants attract the adult weevils which enter banana plants through cut rhizomes. The weevil attack effects roots completely resulting in reduced nutrient uptake which results in reduced plant vigour and delayed flowering increasing plant's susceptibility to other pests and diseases. Yield is reduced due to rhizomes weakening resulting in toppling and plant death.

Several cultural practices such as keeping the plantation clean by trapping the weevils are followed, but are cost intensive. Crop sanitization will remove weevil refuges and breeding sites. Application of neem (20 %) is found to be beneficial in reducing the population and oviposition. Chemical fertilizers are not affordable by farmers, besides developing insect resistance. Development of resistant plants has been suggested as a potential long-term solution for controlling the weevil.

There are not many biocontrol agents known which can control these weevils. *Tetramorium guineense* and *Pheidole megacephala*, the myrmicine ants, have been reported to have successfully controlled the weevil in Cuba. Antifeedants play a significant role in weevil resistance (Ortiz et al. 1995). Kiggundu (2000) observed that corm size, hardiness, resin/sap production and their suckering ability are significant parameters in resistance response of clones introduced in Africa. Large corm size provides greater resistance (Balachowsky 1963). Certain toxic compounds present in BB genomes imparted weevil resistance which was absent in AA genome. Corm extracts from weevil-resistant AB (Kisubi) and ABB (Pisang awak) genomes showed HPLC peaks which were absent in susceptible and resistant AA clones. African cultivars have been found to be highly susceptible to weevil infection and few Indian cultivars (Karumpoovan and Poozhachendu) highly resistant (Padmanaban et al. 2001). Certain resistant cultivars are reported in Cameroon (Fogain and Price 1994). Kiggundu et al. (2003a, b) reported that some wild diploid banana (Calcutta-4), three diploid banana hybrids (TMB2×6142-1, TMB2×8075-7 and TMB2×7197-2) and cultivars like Yangambi-Km5 and Cavendish which possess high level of resistance may be exploited as resistance source. Some cultivars like Tereza, Nalukira and Nsowe possess intermediary resistance. *Musa accuminata*, AA genome progenitor, was found to be more susceptible to weevils than *M. balbisiana*, the BB progenitor (Mesquita et al. 1984).

Laboratory studies conducted by Kiggundu et al. (2006) on the modalities of resistance to banana weevil revealed that all cultivars were attractive to the weevil and females oviposited on all cultivars. The resistant cultivars showed lower survivorship compared to susceptible ones. Antibiosis mechanisms existed in insect resistance and not antixenosis. Larval development on corms of susceptible cultivars was inhibited by methanol extracts from resistant cultivars in the laboratory.

17.4 Genes Available for Nematodes and Weevils Resistance

17.4.1 Genes for Nematode Resistance

17.4.1.1 Protein- and Peptide-Based Transgenic Defences

Proteinase inhibitors (PI) of plant origin like trypsin, serine and cysteine, which inhibit nematode development and reduce fecundity of females, have been used to control nematodes (Kiggundu et al. 2003a, b; Urwin et al. 1997). Cysteine proteinase inhibitors (cystatins), which prevent proper intestinal digestion of dietary protein in nematodes, are well developed as anti-nematode proteins (Atkinson 2000; Urwin et al. 1995). Transgenic plants expressing cystatins can provide effective control of both cyst and root-knot nematodes. The cystatins have demonstrated effectiveness against a range of major nematode pests including in field trials with *Globodera* spp. on potato expressing an engineered rice grain cystatin (Urwin et al. 2001, 2003) and *R. similis* and *H. multicinctus* on plantain expressing a maize kernel cystatin (Tripathi et al. 2015). In glasshouse trials, the engineered rice cystatin has also provided 75 ± 5 % resistance in lily against *P. penetrans* (Vieira et al. 2014), and the maize cystatin has provided 84 ± 8 % resistance in plantain to a mixed population of nematodes *R. similis*, *H. multicinctus* and *Meloidogyne* spp. (Roderick et al. 2012). Vain et al. (1998) demonstrated 55 % reduction in egg production by *Meloidogyne incognita* in transgenic rice plants expressing rice cystatin. Transgenic tomato expressing taro cystatin showed resistance against *Meloidogyne* spp. (Chan et al. 2010). The transgenic plant expressing dual proteinase inhibitor transgenes demonstrated enhanced resistance to nematodes (Urwin et al. 1998). On current evidence, it is likely that a cystatin-based defence should be effective against all economically important parasitic nematode pest of banana and plantain.

A potential drawback to a cystatin defence is that the nematode is not exposed to it until after invasion of the root; consequently, young banana plants may suffer stunting given a large enough initial inoculation of nematodes. Peptides that overcome this problem by disrupting localization and invasion of host roots by plant parasitic nematodes have also been developed. Nematodes utilize a range of chemical signals produced by plant roots to achieve a successful parasitic interaction (Reynolds et al. 2011). Two synthetic peptides with distinct modes of action have been identified that interfere with the nematode cholinergic nervous system by binding to either acetylcholinesterase or nicotinic acetylcholine receptors (nAChRs) to disrupt chemoreception (Winter et al. 2002). The acetylcholinesterase-inhibiting peptide reduces the number of female *Heterodera schachtii* on *Arabidopsis thaliana* by more than 80 %. In the same set of experiments, expression of this peptide in the root tips of potato plants resulted in almost 95 % resistance to *Globodera pallida* (Lilley et al. 2011a). The mode of uptake of acetylcholinesterase-inhibiting peptide is well documented for *H. schachtii* and *R. similis* (Wang et al. 2011; Roderick et al. 2012) and likely a highly conserved process across plant parasitic nematodes. The nAChR-binding peptide is taken up by the open-ended chemosensory sensilla within the anterior amphidial pouches and is then transported along chemoreceptive neurons to their cell bodies where nAChRs are located (Wang et al. 2011; Roderick et al. 2012). Chemoreception is only impaired when that transport had been completed (Wang et al. 2011). This peptide, when expressed in root tips of potatoes with a cellular export signal peptide, achieved up to 77 % resistance against *G. pallida* in glasshouse and field trials (Green et al. 2012). Plantains expressing the nAChR-binding peptide achieved 69 ± 6 % resistance to a mixed population of *R. similis*, *H. multicinctus* and *Meloidogyne* spp. in screen house challenges (Roderick et al. 2012) and 99 ± 1 % resistance to a concomitant infection with *R. similis* and *H. multicinctus* in the field (Tripathi et al. 2015). Migratory plant parasitic nematode species like *R. similis*, *H. multicinctus* and *Pratylenchus* spp. that remain motile and infective during all developmental stages may be affected throughout their lifecycle, while sedentary endoparasitic nematodes, such as *Meloidogyne* spp., are vulnerable to sensory intervention during their infective stages prior to feeding cell initiation. Both the cystatin and chemoreception disrupting peptide defences are being deployed as a stacked defence in plantain (Tripathi et al. 2015) and East African Highland banana to ensure a broad and durable defence. However, there does not appear to be a cumulative level of resistance from having two defences present (Roderick et al. 2012; Tripathi et al. 2015).

Bt proteins have effects on free-living bacterial feeding nematodes (Marroquin et al. 2000). The Cry5B protein is toxic to wild-type *Caenorhabditis elegans*, whereas some mutants of *C. elegans* are resistant to it but susceptible to Cry6A toxin (Marroquin et al. 2000). Cry55Aa, Cry6Aa and Cry5Ba showed toxicity to *M. hapla* in an induced uptake study (Zhang et al. 2012), and Cry6Aa2 reduced *M. hapla* numbers when applied as a soak (Yu et al. 2015). Transgenic Cry5B expressed in tomato hairy roots reduced *M. incognita* numbers by 75 % compared to controls (Li et al. 2008). Plant parasitic nematode control using Bt Cry proteins has potential, but the evidence base for broad nematode species control or for efficacy in the field has not yet been developed (Wei et al. 2003). The lectin concanavalin A has been shown to suppress *M. incognita* multiplication. and others, such as snowdrop lectins, have biological activity against nematodes. Many lectins, however, have toxic effects on insects and mammals (Burrows and de Waele

1997). Toxicological safety of lectins is a major concern for commercial development and needs to be studied in depth. Transgenic expression of lectins have not yet shown enough promise to make it into crop field trials (Fuller et al. 2008; Atkinson et al. 2009; Lilley et al. 2011b).

Natural resistance genes can also offer a strategy for combating plant parasitic nematodes, and several R-genes against nematodes have been identified. The sugar beet gene *Hs1pro-1* confers resistance to the cyst nematode *Heterodera schachtii* (Cai et al. 1997). The tomato *Mi-1.2* gene confers resistance against *Meloidogyne* species and has been introduced into cultivated tomato, *Lycopersicon esculentum*, by an interspecies cross from the wild species *L. peruvianum* (Milligan et al. 1998). The *Gpa2* gene also conferred resistance against potato cyst nematode *Globodera pallida* (van der Vossen et al. 2000). The main drawbacks of R-genes are a lack of genes for resistance to banana nematodes and a tendency for a highly species-specific effect.

17.4.1.2 RNAi-Based Transgenic Defences

RNA interference (RNAi) results when double-stranded RNA (dsRNA) triggers the degradation of messenger RNA (mRNA) resulting in the silencing of specific target genes. It has proven a useful tool for functional analysis of nematode genes, including for plant parasitic nematodes (Rosso et al. 2009). Triggering of RNAi silencing in nematodes that feed on plants expressing dsRNA targeting nematode genes is currently being developed as a nematode control strategy (Lilley et al. 2007). Experiments with transgenic *Arabidopsis* expressing dsRNA from inverted repeat hairpin constructs have identified six *Heterodera schachtii* genes that when suppressed result in significant reductions in female numbers up to 64 % (Patel et al. 2008, 2010; Sindhu et al. 2009). Suppression of *H. glycines* by 81–93 % has been achieved in soybean plants expressing dsRNA targeting one of two ribosomal proteins, a spliceosomal protein or synaptobrevin (Klink et al. 2009), while a similarly high reduction in egg production was achieved by targeting mRNA splicing factor *prp-17* or an uncharacterized gene *cpn-1* (Li et al. 2010).

Similar positive results have been seen for banana parasitic nematodes, particularly for *Meloidogyne* spp. A high level of resistance resulted from targeting the 16D10 gene expressed in the subventral gland cells and required for parasitism in *Meloidogyne incognita. Arabidopsis thaliana* plants expressing dsRNA targeting the 16D10 gene achieved 63–90 % reduction in gall number and size with a corresponding reduction in egg production in *M. incognita*, *M. javanica*, *M. arenaria* and *M. hapla* (Huang et al. 2006). Tobacco plants expressing dsRNA targeting either a splicing factor or an integrase gene of *Meloidogyne incognita* achieved a high level of resistance to that nematode (Yadav et al. 2006). Reduction in gall numbers by >90 % for *M. incognita* on transgenic soybean roots has also been shown (Ibrahim et al. 2011). However, not all plants delivered dsRNA targeting *Meloidogyne* genes have resulted in a resistance phenotype. Silencing of the putative transcription factor MjTis11 of *M. javanica* did not significantly affect the nematodes (Fairbairn et al. 2007). Only partial resistance was achieved when *A. thaliana* plants targeted either a dual oxidase gene with a probable role in cuticle formation or a subunit of a signal peptidase, a protein complex required for the processing of secreted proteins targeted individually in *M. incognita.* Higher levels of resistance were achieved by crossing transgenic lines expressing these two defences (Charlton et al. 2010). However, a combinatorial RNAi targeted at *H. glycines* did not deliver that benefit (Bakhetia et al. 2008). Possibly transgenic silencing RNAs may saturate the RNA silencing complex reducing efficacy.

Radopholus similis is also susceptible to RNAi although the extent of silencing can vary by the region of the nematode gene targeted and from experiment to experiment (Haegeman et al. 2009). Reductions by 60 % in infection to *Medicago truncatula* were achieved after soaking *R. similis* in dsRNA targeting a gland cell xylanase gene (Haegeman et al. 2009). Tobacco plants expressing an inverted repeat hairpin construct targeting an *R. similis* cathepsin B cysteine proteinase transcript achieved 80 % reduction in the number of nematodes recovered (Li et al. 2015a). Targeting an *R. similis* calreticulin resulted in a 75 % reduction in *R. similis* numbers recovered on transgenic tomato plants (Li et al. 2015b). There has been no report of control of *Pratylenchus* spp. by RNAi on transgenic plants. Experiments that induce in vitro uptake of dsRNA have demonstrated transcript knockdown for troponin C (*pat-10*) and calponin (*unc-87*) genes, required for muscle structure and contraction, in *P. coffeae, P. thornei* and *P. zeae*. Following treatment with dsRNA nematode movement was aberrant, and multiplication on carrot discs was significantly reduced (Joseph et al. 2012; Tan et al. 2013).

The susceptibility of *H. multicinctus* to RNAi has yet to be investigated and is hampered by the scarcity of genetic sequences available for the nematode. In contrast to the availability of complete genome sequences for the *P. coffeae* (Burke et al. 2015a, b) and *Meloidogyne* spp. (Abad et al. 2008; Opperman et al. 2008) and the ongoing sequencing of *R similis* genome (Bird et al. 2015), a complete mitochondrial genome for *R. similis* is available (Jacob et al. 2009), which greatly increases the likelihood of identifying suitable targets.

17.4.2 Genes for Weevil Resistance

Studies on differentially expressed genes following weevil infestation was initiated in a joint project between UBBP (Uganda Banana Biotechnology Project) and FABI (Forestry and Agricultural Biotechnology Institute) of the University of Pretoria. Genes expressed during weevil infestation were compared in resistant and susceptible Musa varieties using techniques called cDNARDA (Representational Difference Analysis of cDNA) (Hubank and Schatz 1994).

17.4.2.1 Proteinase Inhibitors

Protein inhibitors are used for obtaining defence against weevils also. Expression of proteinase inhibitors naturally in plants when insect attacks was studied during insect wounding and herbivory by Ryan (1990), Pernas et al. (2000) and Ashouri et al. (2001). As mentioned under nematodes, cysteine proteases are enzymes in the mid gut of coleopteran insects such as the banana weevil, important in the breakdown of dietary proteins. Two major proteinase classes, serine and cysteine, are present in the digestive system of insects. Lepidoptera, Dictyoptera and Hymenoptera belong to Serine proteinase, while Odoptera and Hemiptera possess cysteine proteinase activity. Cysteine proteinases are used by Coleopteran insects (Gatehouse et al. 1985; Murdock et al. 1987). A combination of both serine and cysteine proteinases is also used (Gerald et al. 1997) by pyramiding, to harvest the double advantage of both the proteins, to combat the weevil problem (Gerald et al. 1997).

The potential of phytocystatins (OC-I and papaya cystatin) in controlling the banana weevil was studied by Kiggundu et al. (2002) who analyzed, in the gut of banana weevil, protease activity. Hydrolysis of casein at an acidic pH optimum (pH) was observed in extracts from complete weevil larval guts. Alkaline pH (pH 8.0) showed lesser activity. The presence of cathepsin L and B and cysteine protease in the larval gut was evident by the hydrolysis of the specific substrates Z-Phe-Arg-MCA and Z-Arg-Arg-MCA. In addition, by using specific Bz-Arg-MCA and N-uc-Ala-Ala-Pro-Phe-MCA substrates, trypsin- and chymotrypsin-like protease activity was observed. OC-I and cystatin were produced as fusion protein with histidine tag in *E. coli* and purified. These purified proteins at 1×10^{-5} ngml^{-1} and 2.1×10^{-5}ng ml^{-1} showed 66.2 % and 81.6 % with LD50 inhibition of cysteine protease activity in the banana weevil gut homogenate. Purified OC-I at 0.6 mg cystatin g fresh weight^{-1} inhibited larval weight gain per day when fed on banana stem disc vacuum infiltrated with the inhibitory protein. This study demonstrated that cysteine proteases are used instead of cathepsin L and B by the banana weevil for protein digestion and metabolism in the gut, while phytocystatins are potential control agents for banana weevil growth. The importance of papain-like-cysteine proteases, trans-epoxysuccinyl-L-leucylamido-

(e-guanidino) butane (E-64) on the growth and development of several coleopteran insects was reported by Fabrick et al. (2002). In GM plants, for insect control, exogenous cysteine proteinase inhibitors were used (Leple et al. 1995). The plant cystatins OsCys I and OpCys II showed extensive growth delay on *Cosmopolites sordidus* on cystatin extract media (Kiggundu et al. 2010).

17.4.2.2 Chitinase Enzymes and Alpha-Amylase Inhibitors (AI)

Chitinase enzymes and alpha-amylase inhibitors (AI) also act potentially against the weevil infestation. The chitinolytic activity on the insect cell wall protects the plants from the further damage; genetic transformation of these anti-insecticidal genes may help to develop GM banana with enhanced resistance to weevil (Morton et al. 2000). Enhanced resistance to Lepidopteran insects was seen in transgenic plants expressing enhanced chitinase activity (Ding et al. 1998). Alpha-amylase inhibitors inhibit AL-1 and AL-2, two types of amylases isolated from wild beans (*Phaseolus vulgaris*) (Le Berre-Anton et al. 1997; Morton et al. 2000). Enhanced resistance to coleopteran insects was observed in transgenic azuki beans expressing seed alpha-amylase (Ishimoto et al. 1996).

17.4.2.3 Plant Lectins and Insecticidal Proteins

Plant lectins are inhibitory to a number of organisms (Sharma et al. 2000). Lectins isolated from pea, wheat, rice and soybean are toxic to insects due to their carbohydrate binding capabilities. A lectin from snowdrop, *Galanthus nivalis* agglutinin (GNA), is toxic to several Homoptera, Coleoptera and Lepidopteran insect pests (Tinjuangjun 2002). GNA has been found to be useful in developing transgenic potato and sugarcane resistant to Peach potato aphid and sugarcane grub (*Antitrogus consanguineus*), respectively (Gatehouse et al. 1997; Nutt et al. 1999). Some of the lectins are toxic to mammals (Jouanin et al.1998) which is a major concern (Boulter 1993), while garlic lectins are toxic only to insects, a major concern about the use of lectins, and are potential candidates for weevil control (Kiggundu 2003).

For Lepidopteran control in GM crops, Bt gene technology is the most widely used (Krettiger 1997). There are more than 50 insecticidal crystal proteins among Bt genes. The proteins are solubilized in the alkaline environment of the insect's midgut when an insect feeds on Bt endotoxin protein in GM plant and become toxic to the insect causing its death. Transgenic rice developed with cry1Ab gene was found resistant to rice leaf folder (*Cnaphalocrocis medinalis*) (Ye et al. 2003). Transgenic potato and cotton-carrying Cry03Aa offered resistance to Colorado beetle and boll weevil respectively (Wilson et al. 1992). Against the banana weevil, Bt gene with high toxic effects has not been identified so far.

Bakaze (2010) developed a diet to evaluate resistance in banana germplasm and in vitro efficacy of a *Bacillus thuringiensis* endotoxin Cry6A as well as *Carica papaya* cystatin (CpCYS) against *C. sordidus*. The artificial diet enabled banana weevil larvae to develop to adults in 48 days compared to 36 days on natural banana stem diet. The survival rate and the life cycle completion of the neonate were found more in media mixed with susceptible banana (Mbwazirume) corm powder than the resistant variety (Cavendish). The individual expression of these proteins was done in M15 cells by using pQE9/pQE30x9 expression system. The expressed proteins were purified and mixed at different concentration (1 ppm, 2 ppm) with diet. More than 65 % of mortality was observed at 1 ppm concentration of Cry6A followed by *CpCYS*. A concentration of 2 ppm showed 83 % mortality with *CpCYS* and 75 % with Cry6A. Neonate mortality did not increase significantly when these proteins were mixed and used. Fifty percent mortality (LD50) of neonate larvae was observed at 0.24 ppm and 0.15 ppm for Cry6A and CpCYS, respectively. The outcome of this study showed that two genes together were more effective in combating *C. sordidus*.

Vegetative insecticidal proteins (VIPs) are another class of proteins which cause gut paralysis and lysis of the gut epithelium cells, thus arresting gut function fully leading to the death of the insect (Duck and Evola 1997).

17.4.2.4 RNAi-Based Approaches

Ocimati et al. (2004) reported the 100 % mortality on banana weevil growth by using dsRNA synthesized from E2 ubiquitin gene which plays major role in protein catabolism of banana weevil. Significant growth retardation was observed with 50 and 100 ng/µl of dsRNA concentration in in vitro bioassay. Through this mechanism, essential genes can be silenced across the species, thus providing a molecular approach for great promise for the control of plant disease and pest (Whyard et al. 2009).

17.5 Genetic Transformation of Banana and Plantain for Nematode and Weevil Resistance

Transgenic banana of cultivar 'Cavendish Williams' expressing rice cystatin (OcIΔD86) was developed and tested in glasshouse for nematode resistance. These transgenic lines showed about 69–70 % resistance against *R. similis* (Atkinson et al. 2004b). Transgenic plantain of cultivar 'Gonja manjaya' was generated using a maize cystatin or the nAChR inhibiting peptide or both these traits stacked together (Roderick et al. 2012). Evaluation in the screen house for resistance against mixed population of the banana nematodes *R. similis*, *H. multicinctus* and *Meloidogyne* spp. identified several transgenic lines that provided 70–84 % resistance to *R. similis*. Numbers of *H. multicinctus* and *Meloidogyne* spp. were also suppressed, though the population on controls was too small to show statistical significance (Roderick et al. 2012). Further evaluation of the 12 lines with high levels of resistance in a confined field trial in Uganda identified a number of lines that matched resistance seen in the screen house. The best line, an nAChR inhibiting peptide line, reduced *R. similis* numbers by 99 % and provided a 186 % increase in yield compared to control plants (Tripathi et al. 2015). This is the first field-based evidence of transgenic banana for resistance against nematodes.

Namuddu et al. (2013) developed transgenic banana cultivar 'Sukali Ndiizi' (ABB) using papaya cystatin (CpCYS-Mut89). This gene has been previously modified to improve its inhibitory potential against banana pests (Kiggundu et al. 2010). A total of 57 transgenic lines were generated. Putatively transgenic plants were validated by PCR, and gene integration was further confirmed by Southern blot hybridization. These transgenic lines are yet to be evaluated for resistance to weevils and nematodes.

17.6 Biosafety of Transgenic Nematode Resistance and Conclusion

Biosafety of transgenic approaches is an important consideration, particularly when intended for commercial release in a food crop. The safety of the engineered rice cystatin used in potato and banana field trials is well established. Cystatins are not toxins (Atkinson et al. 2004a), and plants expressing transgenic cystatins do not harm a range of nontarget organisms (Atkinson et al. 2009). Studies of impacts on free-living soil nematodes by transgenic anti-nematode plants have also been developed due to the high sensitivity of this group of organisms to changes in the soil microenvironment and them being the most likely group to be affected by anti-nematode defences (Ingham 2000). Nematode faunal analysis is used to quantify the shifts in free-living nematode genera across the different trophic groups present in the soil. Genera are split into those that respond rapidly to environmental change, which are used to calculate an enrichment index, and those that prefer undisturbed habitats, which are used to calculate a structural index. The ratio of enrichment to structural values indicates the state of the soil, and a shift in the ratio indicates disturbance (Ferris et al. 2001). When applied to a transgenic potato field trial with an nAChR-binding peptide and an engineered rice cystatin, the transgenic lines did not significantly shift the structural or enrichment indices (Green et al. 2012).

Biosafety can be increased by reduction of transgenic defence expression outside of the roots through the use of tissue-specific promoters. Promoters have been used to deliver cystatins to the feeding sites of root-knot nematodes (Lilley et al. 2004). The tobacco cellulase promoter is expressed the syncytia of *H. schachtii* and has been used to develop an RNAi defence (Patel et al. 2008, 2010). The MDK4-20 promoter of *A. thaliana* has been used to target expression of an nAChR-binding peptide to root tips and also the root border cells that detach from the roots of many crops (Lilley et al. 2011a, b); when used to drive expression in potato, this promoter also provided greater levels of resistance compared to lines utilizing a constitutive promoter (Green et al. 2012). Such promoters can lower the burden of the transgene expression in transgenic plants, reduce nontarget organism exposure and increase food safety by preventing or reducing the presence of the transprotein in edible tissues.

RNAi-based defences lack the potential of allergenicity inherent in protein-based defences. It is, however, at risk of off-target effects both within the target organism and on nontarget organisms. Each double-stranded RNA (dsRNA) molecule needs to be carefully designed to reduce sequence identity between the target gene of the nematode and that any other sequence likely to be exposed to the dsRNA. Both of these considerations could be particularly important if small RNA molecules are shown to have prolonged environmental persistence (Auer and Frederick 2009).

Acknowledgement The author, Poovarasan, S., acknowledges Dr. A. Kiggundu for providing information on weevils which facilitated this writing.

References

Abad P, Gouzy J, Aury JM, Castagnone-Sereno P, Danchin EG, Deleury E, Perfus-Barbeoch L, Anthouard V, Artiguenave F, Blok VC (2008) Genome sequence of the metazoan plant-parasitic nematode *Meloidogyne incognita*. Nat Biotechnol 26:909–915

Ashouri A, Michaud D, Cloutier C (2001) Recombinant and classically selected factors of potato plant resistance to the Colorado potato beetle, *Leptinotarsa decemlineata*, variously affect the potato aphid parasitoid *Aphidius nigripes*. Biocontrol 46:401–418

Atkinson HJ (2003) Strategies for resistance to nematodes in *Musa* spp. In: Atkinson HJ, Dale HJ, Harding R, Kiggundu A, Kunert K, Muchwezi JM, Sagi L, Viljoen A (eds) Genetic transformation strategies to address the major constraints to banana and plantain production in Africa. INIBAP, Montpellier, pp 74–107

Atkinson HJ, Johnston KA, Robbins M (2004a) *Prima facie* evidence that a phytocystatin for transgenic plant resistance to nematodes is not a toxic risk in the human diet. J Nutr 134:431–434

Atkinson HJ, Grimwood S, Johnston K, Green J (2004b) Prototype demonstration of transgenic resistance to the nematode *Radopholus similis* conferred on banana by a cystatin. Transgenic Res 13:135–142

Atkinson HJ, Urwin PE, Hussey RS (2009) Plant biotechnology and control. In: Perry RN, Moens M, Starr JL (eds) Root knot nematodes. CABI, Wallingford, pp 338–362

Auer C, Frederick R (2009) Crop improvement using small RNAs: applications and predictive ecological risk assessments. Trends Biotechnol 27:644–651

Bakaze E (2010) Evaluation of *Bacillus thuringiensis* crystal (CRY6A) and carica papaya cystatin toxins against the banana weevil (*Cosmopolites sordidus*) using a novel diet and construction of a plant transformation vector with two stacked toxin genes. http://hdl.handle.net/10570/3967

Bakhetia M, Urwin PE, Atkinson HJ (2008) Characterisation by RNAi of pioneer genes expressed in the dorsal pharyngeal gland cell of *Heterodera glycines* and the effects of combinatorial RNAi. Int J Parasitol 38:1589–1597

Balachowsky AS (1963) Entomologie appliquée à l'agriculture. Traité I 2:1099–1114

Bird DM, Williamson VM, Opperman CH (2015) Exploiting solved genomes of plant-parasitic nematodes to understand parasitism. Adv Bot Res 73:241–258

Boulter D (1993) Insect pest control by copying nature using genetically engineered crops. Phytochemistry 34:1453–1466

Brentu CF, Speijer PR, Green K, Hemeng B, De Waele D, Coyne DL (2004) Micro-plot evaluation of the yield reduction potential of *Pratylenchus coffeae*, *Helicotylenchus multicinctus* and *Meloidogyne javanica* on plantain cv. Apantu-pa (*Musa* spp., AAB-group) in Ghana. Nematology 6:455–462

Bridge J, Price NS, Kofi P (1995) Plant parasitic nematodes of plantain and other crops in Cameroon, West Africa. Fundam Appl Nematol 18:251–260

Burke M, Scholl EH, Bird DM, Schaff JE, Coleman S, Crowell R, Diener S, Gordon O, Graham S, Wang X,

Windham E (2015a) The plant parasite *Pratylenchus coffeae* carries a minimal nematode genome. Nematology 17:621–637

Burke M, Scholl EH, Bird DM, Schaff JE, Coleman S, Crowell R, Diener S, Gordon O, Graham S, Wang X, Windham E, Wright GM, Opperman C (2015b) The plant parasite *Pratylenchus coffeae* carries a minimal nematode genome. Nematology 00:1–17

Burrows PR, de Waele D (1997) Engineering resistance against plant parasitic nematodes using anti-nematode genes. In: Fenoll C, Grundler FMW, Ohl SA (eds) Cellular and molecular aspects of plant-nematode interactions. Kluwer Academic Press, Dordrecht, pp 217–236

Cai DG, Kleine M, Kifle S, Harloff HJ, Sandal NN (1997) Positional cloning of a gene for nematode resistance in sugar beet. Science 275:832–834

Chan YL, Yang AH, Chen JT, Yeh KW, Chan MT (2010) Heterologous expression of taro cystatin protects transgenic tomato against *Meloidogyne incognita* infection by means of interfering sex determination and suppressing gall formation. Plant Cell Rep 29:231–238

Charlton WC, Harel HYM, Bakhetia M, Hibbard JK, Atkinson HJ, McPherson MJ (2010) Additive effects of plant expressed double-stranded RNAs on root-knot nematode development. Int J Parasitol 40:855–864

Ding X, Gopalakrishnan B, Johnson LB, White FF, Wang X, Morgan TD, Kramer KJ, Muthukrishnan S (1998) Insect resistance of transgenic tobacco expressing an insect chitinase gene. Transgenic Res 7:77–84

Duck N, Evola S (1997) Use of transgenes to increase host plant resistance to insects opportunities and challenges. In: Carozi N, Koziel M (eds) Advances in insect control the role of transgenic plants. Taylor and Francis, London, pp 45–56

Duncan L, Moens M (2006) Migratory endoparasitic nematodes. In: Perry RN, Moens M (eds) Plant nematology. CABI Publishing, Wallingford, pp 123–152

Fabrick J, Behnke C, Czapla T, Bala K, Rao AG, Kramer KJ, Reeck GR (2002) Effects of a potato cysteine proteinase inhibitor on midgut proteolytic enzyme activity and growth of the southern corn rootworm, *Diabrotica undecimpunctata howardi* (Coleoptera: Chrysomelidae). Insect Biochem Mol Biol 32:405–415

Fairbairn DJ, Cavallaro AS, Bernard M, Mahalinga-Iyer J, Graham MW, Botella JR (2007) Host-delivered RNAi: an effective strategy to silence genes in plant parasitic nematodes. Planta 226:1525–1533

FAOSTAT (2014) faostat.fao.org/default.aspx

Ferris H, Bongers T, de Goede RGM (2001) A framework for soil food web diagnostics: extension of the nematode faunal analysis concept. Appl Soil Ecol 18:13–29

Fogain R, Price NS (1994) Varietal screening of some *Musa* cultivars for susceptibility to the banana weevil, *Cosmopolites sordidus* (Coleoptera: Curculionidae). Fruits 49:247–251

Fogain R, Messiaen S, Foure E (2002) Studies on the banana borer weevil in Cameroon. InfoMusa 11(1):XII

Fuller VL, Lilley CJ, Urwin PE (2008) Nematode resistance. New Phytol 180:27–44

Gatehouse AMRK, Butler KJ, Fenton KA, Gatehouse JA (1985) Presence and partial characterization of a major proteolytic enzyme in the gut of larval *Callosobruchus maculatus*. Entomol Exp Appl 39:279–286

Gatehouse AMR, Daavison GM, Newell CA, Merryweather A, Hamilton WDO, Burgues EPJ, Gilbert RJC, Gatehouse JA (1997) Transgenic potato plants with enhanced resistance to tomato moth, *Lacanobia oleracea*, growth room trials. Mol Breed 3:49–63

Gerald RR, Kramer KJ, Baker JE, Kanost JF, Behke CA (1997) Proteinase inhibitors and resistance of transgenic plants to insects. In: Carozi N, Koziel M (eds) Advances in insect control the role of transgenic plants. Taylor and Francis, London, pp 157–183

Gold CS (1998) Banana weevil: ecology pest status and prospects for integrated control with emphasis on East Africa. In: Saini SK (ed) Proceedings of a symposium on biological control in tropical habitats: third international conference on tropical entomology. ICIPE, Nairobi, pp 49–74

Gold CS, Messiaen S (2000) The banana weevil *Cosmopolites sordidus*. MusaPest INIBAP Fact Sheet No 4

Gold CS, Karamura EB, Kiggundu A, Bagamba F, Abera AMK (1999) Geographic shifts in highland cooking banana (*Musa* spp. group AAA-EA) production in Uganda. Int J Sustain Dev World Ecol 6:45–59

Gowen S, Quénéhervé P (1990) Nematode parasites of bananas, plantains and abaca. In: Luc M, Sikora RA, Bridge J (eds) Plant parasitic nematodes in subtropical and tropical agriculture. CAB International, Wallingford, pp 431–460

Green J, Wang D, Lilley CJ, Urwin PE, Atkinson HJ (2012) Transgenic potatoes for potato cyst nematode control can replace pesticide use without impact on soil quality. PLoS One 7:e30973

Haegeman A, Vanholme B, Gheysen G (2009) Characterization of a putative endoxylanase in the migratory plant-parasitic nematode *Radopholus similis*. Mol Plant Pathol 10:389–401

Haegeman A, Elsen A, De Waele D, Gheysen G (2010) Emerging molecular knowledge on *Radopholus similis*, an important nematode pest of banana. Mol Plant Pathol 11:315–323

Huang GZ, Allen R, Davis EL, Baum TJ, Hussey RS (2006) Engineering broad root-knot resistance in transgenic plants by RNAi silencing of a conserved and essential root-knot nematode parasitism gene. Proc Natl Acad Sci U S A 103:14302–14306

Hubank M, Schatz DG (1994) Identifying differences in mRNA expression by representational difference analysis of cDNA. Nucleic Acids Res 22:5640–5648

Ibrahim HMM, Alkharouf NW, Meyer SLF, Aly MAM, Gamal El-Din AY, Hussein EHA, Matthews BF (2011)

Post-transcriptional gene silencing of root-knot nematode in transformed soybean roots. Exp Parasitol 127:90–99

Ingham ER (2000) Nematodes. In: Tugel A, Lewandowski A, Happe-von Arb D (eds) Soil biology primer (rev ed.). Soil and Water Conservation Society, Ankeny

Ishimoto M, Sato T, Chrispeels MJ, Kitamura K (1996) Bruchid resistance of transgenic azuki bean expressing seed -amylase inhibitor of common bean. Entomol Exp Appl 79:309–315

Jacob JEM, Vanholme B, Van Leeuwen T, Gheysen G (2009) A unique genetic code change in the mitochondrial genome of the parasitic nematode *Radopholus Similis*. BMC Res Notes 2:192

Joseph S, Gheysen G, Subramaniam K (2012) RNA interference in *Pratylenchus coffeae*: knock down of Pc-pat-10 and Pc-unc-87 impedes migration. Mol Biochem Parasitol 186:51–59

Jouanin L, Bonade-Bottino M, Girard C, Morrot G, Giband M (1998) Transgenic plants for insect resistance. Plant Sci 131:1–11

Kashaija IN, Speijer PR, Gold CS, Gowen SC (1994) Occurrence, distribution and abundance of plant parasitic nematodes of bananas in Uganda. Afr Crop Sci J 2:99–104

Kiggundu A (2000) Host-plant interactions and resistance mechanisms to banana weevil *Cosmopolites sordidus* (Germar) in Ugandan *Musa* germplasm. M.Sc. Thesis. University of the Orange Free State, Bloemfontain, South Africa

Kiggundu A, Kunert K Viljoen A, Pillay M, Gold CS (2002) Designing inhibitors for banana weevil control p21. Abstracts 3rd International symposium on molecular and cellular biology of bananas

Kiggundu A, Gold CS, Labuschagne M, Vuylsteke D, Louw SVDM (2003a) Levels of host plant resistance to banana weevil *Cosmopolites sordidus* (Germar) (Coleoptera: Curculionidae) in African *Musa* germplasm. Euphytica 133:267–277

Kiggundu A, Pillay M, Viljoen A, Gold C, Tushemereirwe W, Kunert K (2003b) Enhancing banana weevil (*Cosmopolites sordidus*) resistance by genetic modification: a perspective. Afr J Biotechnol 2:563–569

Kiggundu A, Goulet MC, Goulet C, Dubuc J-F, Rivard D, Benchabane M, Pe'pin G, Van der Vyver C, Kunert K, Michaud D (2006) Modulating the proteinase inhibitory profile of a plant cystatin by single mutations at positively selected amino acid sites. Plant J 48:403–413

Kiggundu A, Muchwezi J, Van Der Vyvr C, Viljoen A, Vorster J, Schluter U, Kunert K, Michaud D (2010) Deleterious effects of plant cystatins against the banana weevil *Cosmopolites sordidus*. Arch Insect Biochem Physiol 73(2):87–105

Klink VP, Kim KH, Martins V, MacDonald MH, Beard HS, Alkharouf NW, Lee SK, Park SC, Matthews BF (2009) A correlation between host-mediated expression of parasite genes as tandem inverted repeats and abrogation of development of female *Heterodera glycines* cyst formation during infection of *Glycine max*. Planta 230:53–71

Krettiger AF (1997) Insect resistance in crops: a case study of Bacillus (Bt.) and its transfer to developing countries. International Service for the Acquisition of Agri-biotech Applications, Ithaca

LeBerre-Anton V, Bompard-Gilles C, Payan F, Rouge P (1997) Characterization and functional properties of the alpha-amylase inhibitor (alpha A-1) from kidney bean (*Phaseolus vulgaris*) seeds. Biochim Biophys Acta 1343:31–40

Leple JC, Bonadebottino M, Augustin S, Pilate G, Letan VD, Delplanque A, Cornu D, Jouanin L (1995) Toxicity to *Chrysomela tremulae* (Coleoptera, Chrysomelidae) of transgenic poplars expressing a cysteine proteinase inhibitor. Mol Breed 1:319–328

Li X-Q, Tan A, Voegtline M, Bekele S, Chen C-S, Aroian (2008) Expression of Cry5B protein from *Bacillus thuringiensis* in plant roots confers resistance to root-knot nematode. Biol Control 47:97–102

Li J, Todd TC, Oakley TR, Lee J, Trick HN (2010) Host-derived suppression of nematode reproductive and fitness genes decreases fecundity of *Heterodera glycines* Ichinohe. Planta 232:775–785

Li Y, Wang K, Xie H (2015a) Cathepsin B cysteine proteinase is essential for the development and pathogenesis of the plant parasitic nematode *Radopholus similis*. Int J Biol Sci 11:1073–1087

Li Y, Wang K, Xie H, Wang YT, Wang DW (2015b) A nematode calreticulin, Rs-CRT, is a key effector in reproduction and pathogenicity of *Radopholus similis*. PLoS One 10:e0129351

Lilley CJ, Urwin PE, Johnston KA, Atkinson HJ (2004) Preferential expression of a plant cystatin at nematode feeding sites confers resistance to *Meloidogyne incognita* and *Globodera pallida*. Plant Biotechnol J 2:3–12

Lilley CJ, Bakhetia M, Charlton WL, Urwin PE (2007) Recent progress in the development of RNA interference for plant parasitic nematodes. Mol Plant Pathol 8:701–711

Lilley CJ, Wang D, Atkinson HJ, Urwin PE (2011a) Effective delivery of a nematode-repellent peptide using a root cap-specific promoter. Plant Biotechnol J 9:151–161

Lilley CJ, Kyndt T, Gheysen G (2011b) Nematode resistant GM crops in industrialised and developing countries. In: Jones J, Gheysen G, Fenoll C (eds) Genomics and molecular genetics of plant-nematode interactions. Springer, Heidelberg, pp 515–539

Lorenzen J, Tenkouano A, Bandyopadhyay R, Vroh B, Coyne D, Tripathi L (2010) Overview of banana and plantain (*Musa* spp.) improvement in Africa: past and future. Acta Hortic 879:595–603

Marroquin LD, Elyassnia D, Griffitts JS, Feitelson JS, Aroian RV (2000) *Bacillus thuringiensis* (Bt) toxin susceptibility and isolation of resistance mutants in the nematode Caenorhabditis elegans. Genetics 155:1693–1699

McSorley R, Parrado JL (1983) The spiral nematode, *Helicotylenchus multicinctus*, on bananas in Florida and its control. Proc Fla State Hortic Soc 96:201–207

Mesquita ALM, Alvers EJ, Caldas RC (1984) Resistance of banana cultivars to *Cosmopolites sordidus* (GERMAR, 1824). Fruits 39:254–257

Milligan SB, Bodeau J, Yaghoobi J, Kaloshian I, Zabel P, Williamson VM (1998) The root knot nematode resistance gene Mi from tomato is a member of the leucine zipper, nucleotide binding, leucine-rich repeat family of plant genes. Plant Cell 10:1307–1319

Moens T, Araya M, Swennen R, De Waele D (2006) Reproduction and pathogenicity of *Helicotylenchus multicinctus*, *Meloidogyne incognita* and *Pratylenchus coffeae*, and their interaction with *Radopholus similis* on *Musa*. Nematology 8:45–58

Morton RL, Schoroeder HE, Bateman KS, Chrispeels MJ, Armstrong E, Higgins TJV (2000) Bean-amylase inhibitor-I in transgenic peas (*Pisum sativum*) provided complete protection from pea weevil (*Bruchus pisorum*) under field conditions. Proc Natl Acad Sci 97:3820–3825

Murdock LL, Brookhart G, Dunn PE, Foard DE, Kelley S, Kitch L, Shade RE, Shuckle RH, Wolfson JL (1987) Cysteine digestive proteinases in Coleoptera. Comp Biochem Physiol 87B:783–787

Namuddu A, Kiggundu A, Mukasa SB, Kurnet K, Karamura E, Tushemereirwe W (2013) *Agrobacterium* mediated transformation of banana (*Musa* sp.) cv. Sukali Ndiizi (ABB) with a modified Carica papaya cystatin (CpCYS) gene. Afr J Biotechnol 12(15):1811–1819

Nutt KA, Allsopp PG, McGhie TK, Shepaerd KM, Joyce PA, Taylo GO, McQualter RB, Smith GR, Ogarth DM (1999) Transgenic sugarcane with increased resistance to canegrubs. Proceedings of the 1999 Conference of Australian Society of Sugarcane Technologies, Brisbane, Australia, pp 171–176

Ocimati W, Kiggundu A, Bailey A, Niblett CL, Pedun H, Tazuba AF, Tushemereirwe W, Karamura E (2004) Suppression of the ubiquitin E2 gene through RNA interference causes mortality in the banana weevil, *Cosmopolites sordidus* (Germar)

Opperman CH, Bird DM, Williamson VM, Rokhsar DS, Burke M, Cohn J, Cromer J, Diener S, Gajan J, Graham S (2008) Sequence and genetic map of *Meloidogyne hapla*: a compact nematode genome for plant parasitism. Proc Natl Acad Sci U S A 105:14802–14807

Ortiz R, Vuylsteke D (1996) Improving plantain and banana-based systems. In: Ortiz R, Akoroda MO (eds) Plantain and banana production and research in West and Central Africa. IITA, Ibadan, pp 2–7

Ortiz R, Vuylsteke D, Dumpe B, Ferris RSB (1995) Banana weevil resistance and corm hardness in *Musa* germplasm. Euphytica 86:95–102

Ostmark HE (1974) Economic insect pests of bananas. Ann Rev Entomol 19:161–176

Padmanaban B, Sundararaju P, Velayudhan KC, Sathiamoorthy S (2001) Evaluation of Musa germplasm against banana weevil borers. InfoMusa 10:26–28

Patel N, Hamamouch N, Li CY, Hussey RS, Mitchum M, Baum T, Wang X, Davis EL (2008) Similarity and functional analyses of expressed parasitism genes in *Heterodera schachtii* and *Heterodera glycines*. J Nematol 40:299–310

Patel N, Hamamouch N, Li CY, Hewezi T, Hussey RS, Baum T, Mitchum M, Davis EL (2010) A nematode effector protein similar to annexins in host plants. J Exp Bot 61:235–248

Pernas M, Sanchez-Mong R, Salcedo G (2000) Biotic and abiotic stress can induce cystatin expression in chestnut. FEBS Lett 467:206–210

Pinochet J (1988) Comments on the difficulty in breeding bananas and plantains for resistance to nematodes. Rev Nematol 11:3–5

Price NS (1994) Alternate cropping in the management of Radopholus similis and Cosmopolites sordidus two important pests of banana and plantain. Int J Pest Manage 40:237–244

Price NS (2006) The banana burrowing nematode, *Radopholus similis* (Cobb) Thorne, in the Lake Victoria region of East Africa: its introduction, spread and impact. Nematology 8:801–817

Reynolds AM, Dutta TK, Curtis RHC, Powers SJ, Gaur HS, Kerry BR (2011) Chemotaxis can take plant-parasitic nematodes to the source of a chemo-attractant via the shortest possible routes. J R Soc Interface 8:568–577

Robinson JC (1996) Bananas and plantain. CABI, Wallingford

Roderick H, Tripathi L, Babirye A, Wang D, Tripathi JN, Urwin PE, Atkinson HJ (2012) Generation of transgenic plantain (*Musa* spp.) with resistance to plant pathogenic nematodes. Mol Plant Pathol 13:842–851

Rosso MN, Jones JT, Abad P (2009) RNAi and functional genomics in plant parasitic nematodes. Annu Rev Phytopathol 47:207–232

Rukazambuga NDTM, Gold CS, Gowen SR (1998) Yield loss in East African Highland Banana (Musa spp., AAA-EA group) caused by the banana weevil, *Cosmopolites sordidus* Germar. Crop Protect 17:1–9

Ryan CA (1990) Protease inhibitors in plants: genes for improving defenses against insects and pathogens. Annu Rev Plant Physiol Plant Mol Biol 28:425–449

Sasser JN, Freckman DW (1987) A world perspective on nematology: the role of the society. In: Veech JA, Dickerson DW (eds) Vistas on nematology. USA Society of Nematologists, Hyatsville, pp 7–14

Sengooba T (1986) Survey of banana pest problem complex in rakai and masaka districts, Preliminary trip report, Uganda ministry of agriculture, Kawanda Agricultural Research Station, 10 pp

Sharma HC, Sharma KK, Seetharama N, Ortiz R (2000) Prospects for using transgenic resistance to insects in

crop improvement. EJB Electron J Biotechnol 3(2) http://www.ejb.org/content/vol3/issue2/full/3
Sindhu AS, Maier TR, Mitchum MG, Hussey RS, Davis EL, Baum TJ (2009) Effective and specific in planta RNAi in cyst nematodes: expression interference of four parasitism genes reduces parasitic success. J Exp Bot 60:315–324
Swennen R, Vuylsteke D (2001) Banana *Musa* L. In: Raemaekers RH (ed) Crop production in tropical Africa. DGIC Brussels, Belgium, pp 530–552
Tan JACH, Jones MGK, Fosu-Nyarko J (2013) Gene silencing in root lesion nematodes (*Pratylenchus* spp.) significantly reduces reproduction in a plant host. Exp Parasitol 133:166–178
Tinjuangjun P (2002) Snowdrop lectin gene in transgenic plants: its potential for Asian agriculture Agbiotechnet.com (ABN091)
Tripathi L, Babirye A, Roderick H, Tripathi JN, Changa C, Urwin PE, Tushemereirwe WK, Coyne D, Atkinson HJ (2015) Field resistance of transgenic plantain to nematodes has potential for future African food security. Sci Rep 5:8127
Urwin PE, Atkinson HJ, Waller DA, McPherson MJ (1995) Engineered oryzacystatin I expressed in transgenic hairy roots confers resistance to Globodera pallida. Plant J 8:121–131
Urwin PE, Lilley CJ, McPherson MJ, Atkinson HJ (1997) Resistance to both cyst and root-knot nematodes conferred by transgenic Arabidopsis expressing a modified plant cystatin. Plant J 12:455–461
Urwin PE, McPherson MJ, Atkinson HJ (1998) Enhanced transgenic plant resistance to nematodes by dual protease inhibitor constructs. Planta 204:472–479
Urwin PE, Troth KM, Zubko EI, Atkinson HJ (2001) Effective transgenic resistance to *Globodera pallida* in potato field trials. Mol Breed 8:95–101
Urwin PE, Green J, Atkinson HJ (2003) Expression of a plant cystatin confers partial resistance to *Globodera*, full resistance is achieved by pyramiding a cystatin with natural resistance. Mol Breed 12:263–269
Vain P, Worland B, Clarke MC, Richard G, Beavis M, Liu H, Kohli A, Leech M, Snape J (1998) Expression of an engineered cysteine Proteinase inhibitor (Oryzacystatin-IΔD86) for nematode resistance in transgenic rice plants. Theoretical Appl Genet 96(2):266–271
van der Vossen EAG, van der Voort J, Kanyuka K, Bendahmane A, Sandbrink H (2000) Homologues of a single resistance-gene cluster in potato confer resistance to distinct pathogens: a virus and a nematode. Plant J 23:567–576
Vieira P, Wantosh S, Lilley CJ, Chitwood DJ, Atkinson HJ, Kamo K (2014) Expression of a cystatin transgene can confer resistance to root lesion nematodes in *Lilium longiflorum* cv. Nellie White. Transgenic Res 24(3):421–432
Wang D, Jones LM, Urwin PE, Atkinson HJ (2011) A synthetic peptide shows retro- and anterograde neuronal transport before disrupting thermo chemosensation of plant-pathogenic nematodes. PLoS One 6:e17475
Wei JZ, Hale K, Cara L, Platzer E, Wong C, Fang SC, Raffi V (2003) *Bacillus thuringiensis* crystal proteins that target nematodes. Proc Natl Acad Sci U S A 10:2760–2765
Whyard S, Singh AD, Wong S (2009) Ingested double-stranded RNAs can act as species-specific insecticides. Insect Biochem Mol Biol 39:824–832
Willson WD, Flint HM, Deaton RW, Fischhoff DA, Perlak FJ, Armstrong TA, Fusch RL, Berberich SA, Parks NJ, Stapp BR (1992) Resistance of cotton lines containing Bacillus thuringinensis toxin to pink bollworm (Lepidoptera: Gelechiidae) and other insects. J Econ Entomol 85:1516–1521
Winter MD, McPherson MJ, Atkinson HJ (2002) Neuronal uptake of pesticides disrupts chemosensory cells of nematodes. Parasitology 125:561–565
Yadav BC, Veluthambi K, Subramaniam K (2006) Host-generated double stranded RNA induces RNAi in plant-parasitic nematodes and protects the host from infection. Mol Biochem Parasitol 148:219–222
Ye GY, Yao HW, Shu QY, Cheng X, Hu C, Xia YW, Gao MW, Altosaar I (2003) High levels of stable resistance in transgenic rice with cry1Ab gene from *Bacillus thuringiensis* Berliner to rice leaffolder, *Cnaphalocrocis medilinalis* (guenee) under field conditions. Crop Protect 22:171–178
Yu Z, Xiong J, Zhou Q, Luo H, Hu S, Xia L, Sun M, Li L, Yu Z (2015) The diverse nematicidal properties and biocontrol efficacy of *Bacillus thuringiensis* Cry6A against the root-knot nematode *Meloidogyne hapla*. J Invertebr Pathol 125:73–80
Zhang F, Peng D, Ye X, Yu Z, Hu Z, Ruan L, Sun M (2012) In vitro uptake of 140 kDa *Bacillus thuringiensis* nematicidal crystal proteins by the second stage juvenile of *Meloidogyne hapla*. PLOS One. http://dx.doi.org/10.1371/journal.pone.0038534

18 Molecular Farming: Prospects and Limitation

Himanshu Tak, Sanjana Negi, T.R. Ganapathi, and V.A. Bapat

Abstract

Plant molecular farming is the production of recombinant pharmaceutical and nonpharmaceutical proteins of commercial importance utilizing plants as bioreactors. Research and development on plant-derived recombinant proteins have gained momentum in recent years. Advantages of employing plants as bioreactors for recombinant protein generation are many including low cost of production, easier scale-up, cost-effective storage, and absence of animal pathogens in protein preparations. This article reviews the various technologies developed for employing plants as bioreactors, different plant systems being used as expression host, and limitations and research advances to overcome these limitations. An overview of different plant-derived products whether currently in market or are in different stages of development, including phases of clinical trials, is described. Special emphasis has been given on banana being used as an expression host, advantages and limitations of using banana in plant molecular farming, and different approaches which can be utilized to overcome those limitations have been described.

Keywords

Molecular farming • Recombinant pharmaceuticals • Magnification • Glycosylation

18.1 Introduction

Molecular farming is the production of commercially important proteins using plants or plant cell cultures and utilized for industrial and medicinal purposes (Daniell et al. 2001a; Horn et al. 2004). The field of molecular farming has provided immense opportunities for production

H. Tak • S. Negi • T.R. Ganapathi (✉)
Plant Cell Culture Technology Section,
Nuclear Agriculture and Biotechnology Division,
Bhabha Atomic Research Centre,
Trombay, Mumbai 400 085, India
e-mail: trgana@barc.gov.in

V.A. Bapat
Department of Biotechnology, Shivaji University,
Kolhapur 416 004, India

S. Mohandas, K.V. Ravishankar (eds.), *Banana: Genomics and Transgenic Approaches for Genetic Improvement*, DOI 10.1007/978-981-10-1585-4_18

of affordable pharmaceuticals worldwide. Pharmaceutically important proteins are largely produced in mammalian cell cultures to match the glycosylation pattern and other biochemical properties of derived protein with that of protein from human origin. Limitations of scaling up and higher production cost have prevented the extensive utilization of mammalian cells for production of therapeutic proteins preventing affordability by poor people worldwide. Plant-based therapeutic protein production can be superior to mammalian or bacterial expression due to reasons like proper posttranslational modifications and folding, ease of scale-up, lower production and processing cost, higher acceptance due to less ethical issues than animal-based protein production, and less chances of biotic contaminants like viruses and bacterial toxins (Doran 2000). The technology involving systematic steps like plant transformation, growing, harvesting, and downstream processing (Wilde et al. 2002) was first demonstrated for production of human growth hormone in transformed tobacco and sunflower callus tissue (Barta et al. 1986). The first plant-based recombinant protein production for commercial purpose was carried out in 1997, when egg protein, avidin, was expressed in transgenic maize (Hood et al. 1997). These scientific reports indicated the possibility of using plants for large-scale production of recombinant proteins with the advantage of production of complex and multimeric proteins due to the fact that plants can carry out proper folding and necessary posttranslational modifications needed for functioning of such proteins (Lienard et al. 2007). Depending on the protein, the expression system, and the plant system, the cost of the recombinant protein production could be less than US\$50 g^{-1} (Daniell et al. 2001a). Similar protein production using animal cell culture technology could cost approximately US\$5,000 g^{-1}, and such higher production cost coupled with inefficient production has resulted in shortage in supply of such therapeutic proteins. This has lead to increased active research and development in the field of molecular farming aimed at utilizing plants as bioreactors for production of therapeutic proteins. Around the world, approximately 120 universities, institutes, and private companies are actively working in the area of molecular farming (Basaran and Rodrı́guez-Cerezo 2008). This review discusses about the important plant expression systems, different plant-derived proteins, and different technologies being employed in the production of plant-based recombinant proteins.

18.2 Gene Transfer Methods Used for Molecular Farming

18.2.1 Nuclear Transformation

Stable nuclear transformation incorporates the foreign gene into the nuclear genome of the plant allowing continuous production of the recombinant protein, thereby reducing the production cost (Tremblay et al. 2010). Stable nuclear integration has been the method of choice for molecular farming till date because of factor like ease of scale-up as many crops can be grown year round at different locations. Factors like potential out-crossing with native species (Obembe et al. 2011) and long production cycle of some crops are the limiting factors. The method of plant transformation can be either *Agrobacterium tumefaciens* based or using biolistic method with the help of a particle gun.

18.2.2 Chloroplast Transformation

This method is a highly efficient expression system because of unique feature of site-specific integration of foreign gene in chloroplast genome by homologous recombination. This site-specific recombination eliminates the problem of gene silencing and positional effect which can be the source of variation in expression among different transgenic lines (Ruhlman et al. 2010). Other advantages are transgene containment because of maternal inheritance observed in case of plastids and possibilities of multigene engineering because of prokaryotic nature of the chloroplast (Chan and Daniell 2015; Kumar et al. 2012). Very high level of recombinant protein expression can be observed with this system because of the fact that each cell contains multiple chloroplasts and each chloroplast contains numerous copies of genome (Bendich 1987). Tobacco has been used with ease for plastid transformation; however, a major limitation of molecular farming in tobacco is non-suitability for oral delivery of recombinant protein (Daniell et al. 2002). Second limitation is that the edible region of many crops is not green and not photosynthetic and nongreen plastids like chromoplasts have generally very low level of gene expression resulting from suppression of plastid gene expression (Caroca et al. 2013; Chan and Daniell 2015). This generally results in very low expression of recombinant protein in nongreen edible portions, and hence green edible leafy vegetables are best candidates for molecular farming utilizing plastid transformation (Chan and Daniell 2015). The plastid transformation method requires a biolistic method to transfer the foreign gene into plastid genome.

18.2.3 Agroinfiltration Method

This method involves vacuum infiltration of *Agrobacterium* into the leaf tissue resulting in transfer of T-DNA to a large number of cells. The T-DNA does not get integrated into chromosomes and transiently expresses at very high level resulting in higher expression of recombinant protein (Kapila et al. 1997). This method is rapid and high yielding and also has the advantage that with the expression of multiple genes in one tissue, the assembly of multimeric proteins can be analyzed in planta (Rybicki 2010; Vézina et al. 2009; Vaquero et al. 1999).

18.2.4 Transient Expression Using Virus Infection

RNA viruses can deliver foreign genes fused with the coat protein coding gene into plant causing rapid production of recombinant protein as viruses can replicate to very high copy number in the host. The expression system for this approach has been developed using viruses like tobacco mosaic virus, cowpea mosaic virus, and tomato bushy stunt virus (Canizares et al. 2005) resulting in high expression without integration of foreign gene (Porta and Lomonossoff 2002). The limitations of the approach are instability of recombinant protein and immediate tissue processing to prevent its degradation. Research for development of idiotype vaccines for treatment of B-cell non-Hodgkin's lymphoma was carried out at a pharmaceutical company using this approach (McCormick et al. 2008).

18.2.5 Magnifection Technology

The inability of virus and *Agrobacterium*-mediated recombinant protein expression in plant to yield high-level expression of multiple peptides required for assembly of multimeric proteins (Giritch et al. 2006) has prompted the development of a recent and robust technique known as magnifection. This method involve employing *Agrobacterium* infection as a source for delivery of viral vectors to the entire host plant (Gleba et al. 2005). Magnifection resulted in enhanced infectivity and greater amplification and thus generate higher expression of multiple proteins simultaneously, which can assemble into functional multimeric protein. This method has shown success with a number of proteins like IgG (Giritch et al. 2006), *Yersinia pestis* antigen fusion protein F1–V (Rosales-Mendoza et al. 2010; Santi et al. 2006), and hepatitis B core antigen (Huang et al. 2006).

18.3 Plant Cell Suspension Cultures

Plant cell suspension-based molecular farming has advantages like sterile conditions, simple and cost-effective production of recombinant protein (Kim et al. 2008), and greater transgene containment (Franconi et al. 2010). Also this method is faster owing to the fact that time invested in regeneration and characterization of transgenic lines is not required (Shaaltiel et al. 2007). In banana embryogenic cell suspension, cultures have been very well established in a few varieties, which can be exploited for the production of recombinant proteins. However, to the best of our knowledge, there is no published information on this aspect. Availability of a very few well-characterized plant cell lines for their potential application in the field of molecular farming (Breyer et al. 2009) and reduced yield of recombinant protein due to elevated photolytic activity during stationary phase are major challenges in application of plant cell lines for molecular farming (Corrado and Karali 2009).

18.4 Host Systems Explored for Molecular Farming

The efficiency of recombinant protein production depends much on the host selected for molecular farming. Desirable characteristics in a host are the ease of transformation and regeneration, stability of foreign gene, greater yield of desired protein, potential to scale-up, low processing and purification cost, and absence of antinutritional factors. Other factors of consideration are total biomass production, facilities of storage, value of the recombinant protein, processing and transportation, cost of scale-up, containment, labor, land, and maintenance (Schillberg et al. 2005). Till date there is not a single perfect host; the most suitable host has to be selected based on the above mentioned factors along with the value of the recombinant protein (Delaney 2002; Schillberg et al. 2005).

Tobacco is a choice of host for molecular farming because of factors like established culture and transformation protocols along with the advantages like high biomass yield, established protein extraction protocols, ease to scale-up, availability of a range of germplasm for genetic manipulation, nonfood crop status preventing entry into food chain, and year-round growth (Biemelt and Sonnewald 2005; Stoger et al. 2005). Research has indicated that soluble protein content in tobacco can be in the range of 2.3–2.8 % and yield of extracted protein can be from 155 to 228 kg/ha (Woodleif et al. 1981). A lot of research and development has happened in the field of molecular farming using tobacco as host for the production of therapeutic proteins like cytokines, antibodies, and vaccines. Important therapeutic proteins like human serum albumin (Sijmons et al. 1990), human somatotropin (Staub et al. 2000), human protein C (Cramer et al. 1996), human erythropoietin (Matsumoto et al. 1995), epidermal growth factor (Higo et al. 1993), IL-2 and IL-4 (Magnuson et al. 1998), human lactoferrin (Salmon et al. 1998), and angiotensin converting enzyme-1 (Hamamoto et al. 1993) are few among others which are expressed in tobacco for molecular farming. Limitations in using tobacco as the host for molecular farming is the necessity to remove toxic compounds like nicotine during purification of recombinant protein and the negative image of tobacco in community.

Another host which is being explored for molecular farming is alfalfa, which is a leafy crop capable of symbiotic nitrogen fixation. The advantage with alfalfa is easy clonal propagation by stem cutting and lower secondary metabolite content. However, the recombinant proteins are less stable in leafy portions, and hence the purification process has to be carried out soon after harvesting or else the harvest needs to be frozen till the time of processing (Fischer et al. 2004). There is a concern about entry of these therapeutic proteins into the food chain by potential consumption of leafy host expressing recombinant protein by herbivores.

Many other hosts like tomatoes, banana, and carrot are being explored where the recombinant proteins like vaccines can be delivered directly by consumption of fruits, eliminating the need for extraction and purification (Mason et al. 2002). Banana is being considered as a potential fruit crop for production of edible vaccine. The features like easy digestibility and palatability by infants, year-round availability, and high level of foreign gene containment because of vegetative mode of propagation in many of the edible ones (Kumar et al. 2004, 2005) make it an ideal host for the production of edible vaccines. Edible vaccine from banana can be particularly useful in developing countries where it is widely grown, thus eliminating the need of transportation and low temperature storage generally required in conventional vaccines (Biemelt and Sonnewald 2005). One of the limitations in using banana for molecular farming is low expression of recombinant protein in banana fruit which might be effectively eliminated by employing promoters of genes coding for fruit-specific proteins (Clendennen et al. 1998; Peumans et al. 2002). Potato can be a useful candidate for edible vaccine production. The advantages of potato as host for molecular farming are ease of genetic transformation and clonal propagation, short regeneration time, ease of storage, availability of information about tissue-specific promoters, and product stability inside the potato tuber (Sparrow et al. 2007). Potato-derived therapeutic proteins have entered into clinical trials indicating that potato can be the choice of host for molecular farming in the future (Streatfield and Howard 2003; Walmsley and Arntzen 2003). Another crop which is being explored for molecular farming is tomato. Tomato has the potential to reduce the processing cost as it is cheaper to process watery tissue than dried tissue (Schillberg et al. 2005; Yano et al. 2010). The other advantages of tomato as host for molecular farming are high biomass yield, short life cycle, and high-level containment because of easy growth in the greenhouse, and different food formulations can be prepared using tomato.

Another strategy which is being explored for production of plant-based recombinant protein is the seed-based expression. Advantages for expression of protein in seeds include storage of protein in small volume allowing easier and rapid purification, ease of storage as freeze storage or drying is not required, ability of long-term storage because of protection against proteases, proper folding of protein due to the presence of chaperons and disulfide isomerases, and lower levels of phenolics and other secondary metabolites which can interfere in purification process (Stoger et al. 2002; Nochi et al. 2007; Muntz 1998). Another advantage which can be observed with the seed of certain plants like sunflower is the presence of oil bodies in the seeds of these plants which can allow easier purification of recombinant protein by oleosin technology (Stoger et al. 2005). Moreover, seed-based expression of vaccine antigens and other therapeutic proteins has the ability of oral delivery of these recombinant proteins for immunization or disease treatment (Lamphear et al. 2002). Cereal seeds like rice, maize, barley, and wheat have been analyzed to test their potential for molecular farming. Rice is an ideal crop for molecular farming owing to many advantages like high biomass yield, ease of genetic transformation, self-pollinating crop preventing undesirable gene transfer, ease of scale-up, and availability of information about seed-specific promoters. Maize is another good crop for molecular farming because of high biomass yield, easy transformation protocols, and scaling up (Ramessar et al. 2008). Barley and wheat are other attractive crops for molecular farming as the production cost is low and also barley has high seed protein content. Legume crops like soybean and pea are attractive candidates for molecular farming owing to very high seed protein content. However, with soybean, high oil content may limit its use in molecular farming. Oil seed-based recombinant protein production may have the advantage of easier purification of expressed protein if molecular farming is coupled with oleosin fusion technology. The oleosin technique utilizes the fusion of desired protein to the oleosins (unique membrane proteins) and their targeting to oil bodies which can be further purified easily (Bhatla et al. 2010).

Hairy root-based recombinant production is another attractive option which offers advantages like high degree of gene containment, controlled environmental conditions, and homogeneity in production of recombinant proteins (Schillberg et al. 2013). Other advantages with hairy roots are the ability to grow in relatively simpler medium, high genetic stability, and easier scale-up (Guillon et al. 2006). In vitro secretion-based recombinant protein production using hairy roots can reduce the cost and labor of tedious purification from plant tissue. The secretion of protein into the culture medium continuously by hairy roots is known as rhizosecretion and depends on the biosynthetic potential of the roots (Gaume et al. 2003; Drake et al. 2009).

Another host which is being explored for molecular farming is the moss *Physcomitrella patens* (a haploid bryophyte). This moss can be grown in bioreactor under controlled environmental condition providing high gene containment along with the advantage of secretion of recombinant protein in medium which can reduce the cost of downstream processing (Decker and Reski 2004, 2008; Hohe et al. 2002). Other advantages of using moss as a host are easy transformation protocols, shorter production cycle, availability of information related to regulatory sequences like promoters and signal peptides (Jost et al. 2005), and amenability for homologous recombination allowing stable transformation of foreign genes and easier disruption of endogenous genes by gene knockout (Schaefer 2002). The moss *Physcomitrella* is being used by Greenovation Biotech (German biotech company) as a host for molecular farming because of availability of a genetically engineered moss strain with modified glycosylation pathway (preventing the moss to add immunogenic glycan chains) (Koprivova et al. 2004; Faye et al. 2005). The company has branded the moss-based biopharmaceutical production as bryotechnology (http://www.greenovation.com/technology.html).

Another plant as a host for recombinant protein production is duckweed (*Lemna minor*) with advantages like fast rate of biomass generation due to short doubling time (around 36 h) (Lienard et al. 2007). Other advantages include growth in bioreactors under controlled conditions, availability of transformation and genetic engineering protocols (Cox et al. 2006), growth in simpler medium, and consistency in production due to defined growth rate of the plant. The *Lemna*-based recombinant protein production is developed by the US biotechnology company Biolex Inc. (Gasdaska et al. 2003).

A novel expression system based on lower eukaryote *Chlamydomonas reinhardtii* has been developed recently. Such microalgae can carry out the required posttranslational modifications of the therapeutic proteins. One of published report demonstrated the production of monoclonal antibodies in chloroplast of *Chlamydomonas reinhardtii* (Mayfield et al. 2003). Advantages of using algae are easy and quick genetic transformation, easy scale-up, possibility of scaling up to even 5,00,000 l in cost-effective manner as algae growth can be sustained in simpler medium, possibility of different protein productions simultaneously after transformation of chloroplast and nuclear genome, availably of characterized genetic factors like inducible promoters, and the inclusion of green algae in GRAS (generally recognized as safe) category (Franklin and Mayfield 2005).

18.5 Optimization of Plant Production Systems

There are limitations of plant-based protein production making their social acceptance lower. Problems like lower protein production, nonmammalian glycosylation pattern, and choice of most appropriate expression host for a particular protein are a few among the other.

18.5.1 Overcoming the Problem of Low Yield and Optimizing Transgene Expression

The low yield can be increased to a certain extent by improving the transcript level and recombinant protein stability in the host system. To increase the yield of recombinant protein from the expression host, optimization of transgene expression is a paramount factor. Promoters are the most important factor in the expression cassette as they determine the amount of transgene transcript being made in the host. Strong and constitutive promoters like cauliflower mosaic virus *35S* promoter (*CaMV 35S*) for dicots (Twyman et al. 2005) and maize *ubiquitin-1* promoter (*ubi1*) for monocots (Christensen and Quail 1996; Fischer et al. 2004) are generally used. Vectors of the *pPLEX* series have been modified to work both in dicots and monocots by addition of certain regulatory factors like *Ubi1* or *Act1* introns and GC-rich enhancer sequences from banana bunchy top virus (BBTV) (Schunmann et al. 2003).

Regulatory promoters like organ- and tissue-specific promoters prevent the expression of transgene in undesired tissue or organ. Most common organ-specific promoters used for molecular farming are seed-specific (De Jaeger et al. 2002) and fruit-specific promoters (He et al. 2008). Seed-specific promoter allows expression of recombinant protein in seeds increasing the biosafety toward non-target organisms and additionally preventing unwanted growth inhibition of expression host (Commandeur et al. 2003). Inducible promoter is also an attractive option for molecular farming because the expression of recombinant protein can be regulated by chemical or external stimuli (Padidam 2003). The other advantage of the inducible promoter is the possibility to regulate expression in cell suspensions (Nara et al. 2000) if there is any lethality problem (Corrado and Karali 2009) or when the expression is desirable during harvest.

Use of other strategies for increasing the transgene expression include using DNA regulatory sequences like terminators of heats shock gene (Nagaya et al. 2010) and use of transcription factors (Yang et al. 2001). Endogenous inducible promoters may be very strong and further may be prone to some degree of leakiness leading to unwanted background expression. In this regard use of promoters of bacterial operons may be useful. Transcript stability is also one of the important factors in ensuing elevated yield of recombinant protein. Inclusion of 5′ untranslated leader sequence of rice polyubiquitin gene *RUB13* (Lu et al. 2008) and alfalfa mosaic virus (Sharma et al. 2008) has been studied. Recombinant proteins expressed as fusion of glutathione-S-transferase, maltose-binding protein (LaVallie and McCoy 1995), and ubiquitin (Mishra et al. 2006) proteins resulted in more yield of recombinant protein perhaps because of enhanced solubility imparted by fusion partner. Suppression of gene silencing by p19 protein of tomato bushy stunt virus resulted in enhanced transcript stability (Voinnet et al. 2003). Also adapting to the codon usage of the expression host increase the translational efficiency resulting in enhanced recombinant protein yield (Sharma and Sharma 2009). The sequences around the initiation codon of mRNA influence the translation efficiency. Additionally the consensus sequences for binding of the 40S pre-initiation complexes in animals and plants differ (Lutcke et al. 1987). Hence, modification of this consensus sequence for binding of 40S pre-initiation complexes to match that of the plant genes may increase the yield. An ideal sequence surrounding the initiation codon in plant genes for efficient translation has been reported as AACAAUGGC (Lutcke et al. 1987). The expression of transgene in different transgenic plants derived from same construct may vary because of the factors like position effect, copy number of foreign gene, and gene silencing (Bhat and Srinivasan 2002). The uniformity in expression can be obtained by inclusion of the nuclear matrix attachment regions (MAR) in the T-DNA region. Nuclear matrix attachment regions (MAR) allow placing of the T-DNA region in the transcriptionally active region of the nucleus, thus enhancing the transgene expression (Spiker and Thompson 1996). Transgenic lines with single-copy insertion of T-DNA can be generated by the use of elements of *CRE/loxP* recombinase (De Paepe et al. 2009). Position effect and gene silencing leading to reduced transgene expression can also be eliminated by use of artificial minichromosomes (Ananiev et al. 2009) and targeting of transgene into plastids as gene silencing has not been observed in plastids (Daniell et al. 2001b).

18.5.2 Optimizing Recombinant Protein Stability

Stability of foreign protein can be increased by targeting the protein to subcellular compartments like endoplasmic reticulum (ER). The protein can be fused to ER retention signal which is a C-terminal tetrapeptide, usually KDEL (Pelham 1990). Another ER retention signal is recently isolated from γ-zein which is a prolamin protein from maize and was shown to be more superior than KDEL (Mainieri et al. 2004). The foreign protein can be targeted to oil bodies for easier purification after fusing the foreign protein with oleosin (Murphy 1993). The fusion of the foreign protein with oleosin prevents posttranslational modifications as well as cytosolic degradation. Plant seed vacuoles for protein storage are also attractive options for accumulation of foreign protein. Certain signal sequences have been identified which are useful for targeting the protein to the protein storage vacuoles (Frigerio et al. 1998; Koide et al. 1997).

18.6 Glycosylation of Recombinant Proteins

Many therapeutic proteins require addition of glycan chains for their biological activity, solubility, and increased stability (Elliott et al. 2003). Glycosylation occurs in the endoplasmic reticulum and Golgi apparatus. The N-glycan composition differs between animals and plants as plant's N-glycan lacks α (1,6) fucose, glucose, and sialic acid residues (present in animal-derived N-glycan) and contains α(1,3) fucose and β(1,2) xylose residues (Gomord et al. 2004) which are added in the Golgi complex. Such difference in N-glycan can give rise to immunogenic-related problems in patients (Krapp et al. 2003), and hence the glycosylation machinery of plant must be modified to match N-glycan composition of therapeutic protein to the animal counterpart. Strategies to modify this N-glycan pattern of plants are retention of recombinant protein in ER by ER retention signal (Floss et al. 2009), use of knockout mutant of expression host lacking the ability to add xylose and fructose residues to N-glycan (Kang et al. 2008), RNA interference to inhibit addition of plant-specific N-glycan epitopes (Sourrouille et al. 2008), inactivation of N-acetylglucosaminyltransferase by antisense expression (Wenderoth and von Schaewen 2000), and expression of human 1,4-galactosyltransferase (Palacpac et al. 1999).

18.7 Banana as an Expression Host for Production of Recombinant Proteins

Banana is one of the ideal hosts for production of recombinant proteins because of factors like year-round availability, easy digestibility and palatability by the infants, large-scale cultivation in developing countries like Africa, high gene containment because of generally vegetative propagation, and well-developed and well-established genetic transformation protocol (Kumar et al. 2004; Ganapathi et al. 2001). Banana has been tested in recent years as an expression host for generation of hepatitis B surface antigen (HBsAg) as production of vaccine in edible part of plant has advantages like easier delivery of recombinant protein, generation of mucosal immunity, etc. (Kumar et al. 2004). Transgenic banana expressing hepatitis B surface antigen (HBsAg) was generated and analyzed for expression of the transgene (Kumar et al. 2005). The expression of *HBsAg* with or without endoplasmic reticulum (ER) retention signal (*HER*) was derived by either *ubq3* promoter of *Arabidopsis* or *ethylene-forming enzyme* gene (*EFE*) promoter (May and Kipp 1997) of banana (Kumar et al. 2005). Expression of *HBsAg* was higher under in vitro conditions with *EFE* promoter (38 ng/g F.W.) and under greenhouse conditions (19.92 ng/g F.W.) with *ubq3* promoter (Kumar et al. 2005). Conformational integrity of immunologically important epitopes of HBsAg is dependent on formation of disulfide bonds (Kumar et al. 2005; Smith et al. 2002). Retention of HBsAg in endoplasmic reticulum (ER) by C-terminal ER retention signal resulted in facilitation of disulfide bond formation due to the presence of disulfide isomerase (Kumar et al. 2005) which further resulted in greater yield of immunologically reactive HBsAg (67.87 %) than previously reported yield of 37 % (Smith et al. 2003). The study also showed that banana leaf-derived

HBsAg was similar to human serum-derived HBsAg as buoyant densities (1.146 g/ml for banana-derived HBsAg and 1.2 g/ml for human serum derived) of both were almost similar (Kumar et al. 2005; Valenzuela et al. 1982). The expression of *HBsAg* was low in banana fruit indicating that expression of recombinant protein under *Arabidopsis ubq3* promoter or banana *ethylene-forming enzyme* (*EFE*) promoter may not be suitable for expression in fruit (Kumar et al. 2005). One of the early report has demonstrated strong GUS staining in fruit of transgenic banana wherein the *GUS* expression was under the control of *Gelvin* promoter (Ni et al. 1995; Ganapathi et al. 2001).

Higher level of recombinant protein in fruit tissue can be achieved by employing promoter of abundant fruit-specific proteins. Promoter of a 31 kDa protein (P31) present as a major banana pulp protein, identified as a homologous to class III acidic chitinase, can be a potential candidate for banana fruit (Clendennen et al. 1998). One of the studies has identified genes like *chitinase-related protein* (*CRP*), *β-1,3-glucanase*, a class I *chitinase*, and a *mannose-binding lectin* expressing at high level in banana fruit suggesting that promoters of such genes can drive high expression of recombinant protein in banana fruit at various stages of fruit development (Peumans et al. 2002). Recently promoters of three genes (*chitinase*, *glucanase*, or *expansin*) coding for banana fruit pulp abundant proteins were analyzed for their ability to derive *GUS* expression in banana embryogenic cells (Ghosh et al. 2012). The above mentioned study also provided a rapid method to screen the activity of different banana promoters including the promoters of fruit-specific promoters. Approaches like codon optimization of the transgene, employing untranslated regions (*UTRs*) of banana genes, overcoming gene silencing and position effect related issues, etc. for optimization of recombinant protein expression have been already discussed in the context of optimization of recombinant protein expression in expression host. These strategies need to be tested in banana to make it a promising host for the production of edible/oral vaccines or any other important pharmaceutical proteins.

The major limitation in using banana as a production platform is the low protein content in the fruits (1.1 %). For increasing the recombinant protein content in fruit, the major challenge would be to increase the total protein content, and then it may be possible to achieve the reasonably higher amount of recombinant protein in fruits, which can be delivered orally. Alternatively, using the strong constitutive promoters for expression throughout the plant and further purification can be a viable option. As banana has a lot of biomass compared to other crops, the recombinant protein can be extracted and used for therapeutic use.

18.8 Conclusion

Recent increase of studies and development of multiple plant-based therapeutic proteins indicated that plant-based molecular farming has raised much interest of multiple research institutes and biotechnology companies. Many pharmaceutical companies are now engaged in developing plant-based recombinant proteins (Kaiser 2008). The company Bayer which has taken over Icon Genetics, the developer of the MagnICON technology is collaborating with different biotechnology companions (Kaiser 2008; Obembe et al. 2011). Advantages of molecular farming in plants include ease of scale-up, easier storage of recombinant protein as seeds, absence of animal pathogens in protein preparation, etc. (Ma et al. 2003; Chen et al. 2005; Sharma and Sharma 2009). Though a lot of progress has been made in the field of plant-derived recombinant proteins, certain challenges like unsuitability of glycosylation patterns of plant-derived proteins, low yield of recombinant protein, gene containment issues, consumer perception, and cost of downstream processing still need to be addressed properly. The success in the field of plant-derived recombinant proteins in the future will lie on the successful clinical trials and approvals for human use. Some of the plant-derived proteins are already in the market (Table 18.1), and commercial success of such proteins due to lower cost than animal-derived proteins indicated toward bright scope of other products being developed at present (Table 18.2). As described in this chapter,

Table 18.1 Recombinant proteins derived from plants and commercially available in the market

Product	Expression host	Commercial name	Company
Recombinant human intrinsic factor	*Arabidopsis*	Coban	Cobento Biotech AS
Recombinant lipase	Corn	Merispase®	Meristem Therapeutics
Aprotinin	Corn, tobacco	AproliZean	Prodigene
Recombinant human lysozyme	Rice	Lysobac™	Ventria Bioscience
Recombinant human lactoferrin	Corn, rice	Lacromin™	Meristem Therapeutics, Ventria Bioscience
Avidin	Corn	Avidin	Prodigene
Trypsin	Corn	TrypZean™	Prodigene
β-Glucuronidase	Corn	GUS	Prodigene
Antibody against hepatitis B	Tobacco		CIGB, Cuba
Human growth factor	Barley	ISOkine™, DERMOkine™	ORF Genetics

Information obtained from Sharma and Sharma (2009), Obembe et al. (2011), Yao et al (2015), Faye and Gomord (2010), and Key et al. (2008)

Table 18.2 Plant-derived pharmaceuticals in clinical trials

Product	Expression host	Company	Phase
Taliglucerase alfa	Carrot cell suspension	Protalix	Phase 3 completed
ZMApp	Tobacco	National Institute of Allergy and Infectious Diseases (NIAID), USA	Phase 1 and 2
HAI-05	Tobacco	Center for Molecular Biotechnology, Plymouth, USA	Phase 1
Hepatitis B antigen (HBsAg)	Lettuce	Thomas Jefferson University, USA	Phase 1
Cancer vaccine	Tobacco	Large Scale Biology, USA	Phase 2
Heat-labile toxin B subunit of *Escherichia coli*	Maize	ProdiGene, USA	Phase 1
HN protein of Newcastle disease virus	Tobacco suspension	Dow Agro Sciences, USA	USDA Approved
Poultry vaccine for *coccidiosis* infection	Canola	Guardian Biosciences, Canada	Phase 2
H5N1 vaccine	Tobacco	Medicago, USA	Phase 1
DoxoRX	Tobacco	Planet Biotechnology, USA	Phase 1
RhinoRX	Tobacco	Planet Biotechnology, USA	Phase 1
IgG	Tobacco	Planet Biotechnology, USA	Phase 1
α-Galactosidase	Tobacco	Planet Biotechnology, USA	Phase 1
Lactoferon™	Duckweed	Biolex, USA	Phase 2
Insulin	Safflower	SemBioSys, Canada	Phase 3
Apolipoprotein	Safflower	SemBioSys, Canada	Phase 1

Source: Sharma and Sharma (2009), Obembe et al. (2011), Yao et al. (2015), Faye and Gomord (2010), and Key et al. (2008)

banana has many advantages as a host for molecular farming; however, the limitations need to be addressed to make it one of the attractive platforms for the production of pharmaceutical proteins.

Acknowledgment Authors thank Dr. SP Kale, Head, Nuclear Agriculture and Biotechnology Division, BARC, for his constant encouragement. VAB thanks Indian National Science Academy, New Delhi for Senior Scientist Fellowship.

References

Ananiev EV, Wu C, Chamberlin MA, Svitashev S, Schwartz C, Gordon-Kamm W et al (2009) Artificial chromosome formation in maize (*Zea mays* L.). Chromosoma 118:157–177

Barta A, Sommengruber K, Thompson D, Hartmuth K, Matzke MA, Matzke AJM (1986) The expression of a napoline synthase human growth hormone chimeric gene in transformed tobacco and sunflower callus tissue. Plant Mol Biol 6:347–357

Basaran P, Rodríguez-Cerezo E (2008) Plant molecular farming: opportunities and challenges. Crit Rev Biotechnol 28:153–172

Bendich AJ (1987) Why do chloroplast and mitochondria contain so many copies of their genome? Bioessays 6:279–282

Bhat SR, Srinivasan S (2002) Molecular and genetic analyses of transgenic plants: considerations and approaches. Plant Sci 163:673–681

Bhatla SC, Kaushik V, Yadav MK (2010) Use of oil bodies and oleosins in recombinant protein production and other biotechnological applications. Biotechnol Adv 28:293–300

Biemelt S, Sonnewald U (2005) Molecular farming in plants. Nature encyclopedia of life sciences. Nature Publishing Group, London. doi:10.1038/npg.els.0003365. http://www.els.net/

Breyer D, Goossens M, Herman P, Sneyers M (2009) Biosafety considerations associated with molecular farming in genetically modified plants. J Med Plants Res 3:825–838

Canizares MC, Nicholson L, Lomonossoff GP (2005) Use of viral vectors for vaccine production in plants. Immunol Cell Biol 83:263–270

Caroca R, Howell KA, Hasse C, Ruf S, Bock R (2013) Design of chimeric expression elements that confer high-level gene activity in chromoplasts. Plant J 73:368–379

Chan HT, Daniell H (2015) Plant-made oral vaccines against human infectious diseases-are we there yet? Plant Biotechnol J 13:70

Chen M, Liu M, Wang Z, Song J, Qi Q, Wang PG (2005) Modification of plant N-glycans processing: the future of producing therapeutic proteins in transgenic plants. Med Res Rev 25:343–360

Christensen AH, Quail PH (1996) *Ubiquitin* promoter-based vectors for high-level expression of selectable and/or screenable marker genes in monocotyledonous plants. Transgenic Res 5:213–218

Clendennen SK, Lopez-Gomez R, Gomez-Lim M et al (1998) The abundant 31-kilodalton banana pulp protein is homologous to class III acidic chitinases. Phytochemistry 47:613–619

Commandeur U, Twyman RM, Fischer R (2003) The biosafety of molecular farming in plants. AgBiotechNet 5:ABN 110

Corrado G, Karali M (2009) Inducible gene expression systems and plant biotechnology. Biotechnol Adv 27:733–743

Cox KM, Sterling JD, Regan JT, Gasdaska JR, Frantz KK, Peele CG et al (2006) Glycan optimization of a human monoclonal antibody in the aquatic plant Lemna minor. Nat Biotechnol 24:1591–1597

Cramer CL, Weissenborn DL, Oishi KK et al (1996) High level production of enzymatically active human lysosomal proteins in transgenic plants. In: Owen MRL, Pen J (eds) Transgenic plants: a production system for industrial and pharmaceutical protein. Wiley, Chichester, pp 299–310

Daniell H, Streatfield SJ, Wycoff K (2001a) Medical molecular farming: production of antibodies, biopharmaceuticals and edible vaccines in plants. Trends Plant Sci 6:219–226

Daniell H, Lee SB, Panchal T, Wiebe PO (2001b) Expression of the native cholera toxin B subunit gene and assembly as functional oligomers in transgenic tobacco chloroplasts. J Mol Biol 311:1001–1009

Daniell H, Khan MS, Allison L (2002) Milestones in chloroplast genetic engineering: an environmentally friendly era in biotechnology. Trends Plant Sci 7:84–91

De Jaeger G, Scheffer S, Jacobs A, Zambre M, Zobell O, Goossens A et al (2002) Boosting heterologous protein production in transgenic dicotyledonous seeds using *Phaseolus vulgaris* regulatory sequences. Nat Biotechnol 20:1265–1268

De Paepe A, De Buck S, Hoorelbeke K, Nolf J, Peck I, Depicker (2009) A high frequency of single copy T-DNA transformants by floral dip in *CRE*-expressing *Arabidopsis* plants. Plant J 59:517–527

Decker EL, Reski R (2004) The moss bioreactor. Curr Opin Plant Biol 7:166–170

Decker EL, Reski R (2008) Current achievements in the production of complex biopharmaceuticals with moss bioreactors. Bioprocess Biosyst Eng 31:3–9

Delaney DE (2002) Choice of crop species and development of transgenic product lines. In: Hood EE, Howard JA (eds) Plants as factories for protein production. Kluwer, Dordrecht, pp 139–158

Doran PM (2000) Foreign protein production in plant tissue cultures. Curr Opin Biotechnol 11:199–204

Drake PMW, Barbi T, Sexton A, McGowan E, Stadlmann J, Navarre C, Paul MJ, Ma JKC (2009) Development of rhizosecretion as a production system for recombinant proteins from hydroponic cultivated tobacco. FASEB J 23:3581–3589

Elliott S, Lorenzini T, Asher S et al (2003) Enhancement of therapeutic protein in vivo activities through glyco-engineering. Nat Biotechnol 21:414–421

Faye L, Gomord V (2010) Success stories in molecular farming—a brief overview. Plant Biotechnol J 8:525–528

Faye L, Boulaflous A, Benchabane M, Gomord V, Michaud D (2005) Protein modifications in the plant secretory pathway: current status and practical implications in molecular pharming. Vaccine 23:1770–1778

Fischer R, Emans NJ, Twyman RM, Schillberg S (2004) Molecular farming in plants: technology platforms. In: Goodman RB (ed) Encyclopedia of plant and crop science. Marcel Dekker, New York, pp 753–756

Floss DM, Sack M, Arcalis E, Stadlmann J, Quendler H, Rademacher T et al (2009) Influence of elastin-like peptide fusions on the quantity and quality of a tobacco-derived human immunodeficiency virus-neutralizing antibody. Plant Biotechnol J 7:899–913

Franconi R, Demurtas OC, Massa S (2010) Plant-derived vaccines and other therapeutics produced in contained systems. Expert Rev Vaccin 9:877–892

Franklin SE, Mayfield SP (2005) Recent developments in the production of human therapeutic proteins in eukaryotic algae. Expert Opin Biol Ther 5:225–235

Frigerio L, de Virgilio M, Prada A, Faoro F, Vitale A (1998) Sorting of phaseolin to the vacuole is saturable and requires a short C-terminal peptide. Plant Cell 10:1031–1042

Ganapathi TR, Higgs NS, Balint-Kurti PJ, Arntzen CJ, May GD, Van Eck JM (2001) *Agrobacterium* mediated transformation of embryogenic cell suspensions of the banana cultivar *Rasthali* (AAB). Plant Cell Rep 20:157–162

Gasdaska JR, Spencer D, Dickey L (2003) Advantages of therapeutic protein production in the aquatic plant *Lemna*. BioProcess J 2:49–56

Gaume A, Komarnytsky S, Borisjuk N, Raskin I (2003) Rhizosecretion of recombinant proteins from plant hairy roots. Plant Cell Rep 21:1188–1193

Ghosh A, Shekhawat UKS, Ganapathi TR, Bapat VA (2012) Analysis of banana fruit-specific promoters using transient expression in embryogenic cells of banana cultivar *Robusta* (AAA Group). J Plant Biochem Biot 21:189–197

Giritch A, Marillonnet S, Engler C, van Eldik G, Botterman J, Klimyuk V et al (2006) Rapid high yield expression of full-size IgG antibodies in plants coinfected with noncompeting viral vectors. Proc Natl Acad Sci U S A 103:14701–14706

Gleba Y, Klimyuk V, Marillonnet S (2005) Magnifection—a new platform for expressing recombinant vaccines in plants. Vaccine 23:2042–2048

Gomord V, Sourrouille C, Fitchette AC et al (2004) Production and glycosylation of plant-made pharmaceuticals: the antibodies as a challenge. Plant Biotechnol J 2:83–100

Guillon S, Trémouillaux-Guiller J, Kumar Pati P, Rideau M, Gantet P (2006) Hairy root research: recent scenario and exciting prospects. Curr Opin Plant Biol 9:341–346

Hamamoto H, Sugiyama Y, Nakagawa N et al (1993) A new tobacco mosaic virus vector and its use for the systemic production of angiotensin – 1 converting enzyme inhibitor in transgenic tobacco and tomato. Biotechnology 11:930–932

He Z, Jiang XL, Qi Y, Di QL (2008) Assessment of the utility of the tomato fruit-specific E8 promoter for driving vaccine antigen expression. Genetica 133:207–214

Higo K, Saito Y, Higo H (1993) Expression of a chemically synthesized gene for human epidermal growth factor under the control of cauliflower mosaic virus *35S* promoter in transgenic tobacco. Biosci Biotechnol Biochem 57:1477–1481

Hohe A, Decker E, Gorr G, Schween G, Reski R (2002) Tight control of growth and cell differentiation in photoautotrophically growing moss (*Physcomitrella patens*) bioreactor cultures. Plant Cell Rep 20:1135–1140

Hood EE, Witcher DR, Maddock S, Meyer T, Baszczynski C, Bailey M et al (1997) Commercial production of avidin from transgenic maize: characterization of transformant, production, processing, extraction and purification. Mol Breed 3:291–306

Horn ME, Woodard SL, Howard JA (2004) Plant molecular farming: systems and products. Plant Cell Rep 22:711–720

Huang Z, Santi L, LePore K et al (2006) Rapid, high-level production of hepatitis B core antigen in plant leaf and its immunogenicity in mice. Vaccine 24:2506–2513

Jost W, Link S, Horstmann V et al (2005) Isolation and characterization of three moss-derived *betatubulin* promoters suitable for recombinant expression. Curr Genet 47:111–120

Kaiser J (2008) Is the drought over for pharming? Science 320:473–475

Kang JS, Frank J, Kang CH, Kajiura H, Vikram M, Ueda A (2008) Salt tolerance of *Arabidopsis* thaliana requires maturation of N-glycosylated proteins in the Golgi apparatus. Proc Natl Acad Sci U S A 105:5933–5938

Kapila J, De Rycke R, van Montagu M et al (1997) An *Agrobacterium* mediated transient expression system for intact leaves. Plant Sci 122:101–108

Key S, Ma JKC, Drake PMW (2008) Genetically modified plants and human health. J R Soc Med 101:290–298

Kim TG, Baek M, Lee EK, Kwon T-H, Yang M-S (2008) Expression of human growth hormone in transgenic rice cell suspension culture. Plant Cell Rep 27:885–891

Koide Y, Hirano H, Matsuoka K, Nakamura K (1997) The N-terminal propeptide of the precursor to sporamin acts as a vacuole-targeting signal even at the C terminus of the mature part in tobacco cells. Plant Physiol 114:863–870

Koprivova A, Stemmer C, Altmann F, Hoffmann A, Kopriva S, Gorr G et al (2004) Targeted knockouts of *Physcomitrella* lacking plant-specific immunogenic N-glycans. Plant Biotechnol J 2:517–523

Krapp S, Mimura Y, Jefferis R, Huber R, Sondermann P (2003) Structural analysis of human IgGFc glycoforms reveals a correlation between glycosylation and structural integrity. J Mol Biol 325:979–989

Kumar SGB, Ganapathi TR, Bapat VA (2004) Edible vaccines: current status and future prospects. Physiol Mol Biol Plants 10:37–47

Kumar GB, Ganapathi TR, Revathi CJ, Srinivas L, Bapat VA (2005) Expression of hepatitis B surface antigen in transgenic banana plants. Planta 222:484–493

Kumar S, Hahn FM, Baidoo E, Kahlon TS, Wood DF, McMahan CM, Cornish K, Keasling JD, Daniell H, Whalen MC (2012) Remodeling the isoprenoid pathway in tobacco by expressing the cytoplasmic mevalonate pathway in chloroplasts. Metab Eng 14:19–28

Lamphear BJ, Streatfield SJ, Jilka JM, Brooks CA, Barker DK, Turner DD et al (2002) Delivery of subunit vaccines in maize seed. J Control Release 85:169–180

LaVallie ER, McCoy JM (1995) Gene fusion expression systems in *Escherichia coli*. Curr Opin Biotechnol 6:501–506

Lienard D, Sourrouille C, Gomord V, Faye L (2007) Pharming and transgenic plants. Biotechnol Ann Rev 13:115–147

Lu J, Sivamani E, Azhakanandam K, Samadder P, Li X, Qu R (2008) Gene expression enhancement mediated by the *5′ UTR* intron of the rice *rubi3* gene varied remarkably among tissues in transgenic rice plants. Mol Genet Genomics 279:563–572

Lutcke HA, Chow KC, Mickel FS, Moss KA, Kern HF, Scheele GA (1987) Selection of AUG initiation codons differs in plants and animals. EMBO J 6:43–48

Ma JK, Drake PM, Christou P (2003) The production of recombinant pharmaceutical proteins in plants. Nat Rev Genet 4:794–805

Magnuson NS, Linzmaier PM, Reeves R et al (1998) Secretion of biologically active human interleukin- 2 and interleukin-4 from genetically modified tobacco cells in suspension culture. Protein Exp Purif 13:45–52

Mainieri D, Rossi M, Archinti M, Bellucci M, De Marchis F, Vavassori S et al (2004) Zeolin. A new recombinant storage protein constructed using maize gamma-zein and bean phaseolin. Plant Physiol 136:3447–3456

Mason HS, Warzecha H, Mor T, Arntzen CJ (2002) Edible plant vaccines: applications for prophylactic and therapeutic molecular medicine. Trends Mol Med 8:324–329

Matsumoto S, Ikura K, Ueda M et al (1995) Characterization of a human glycoprotein (erythropoietin) produced in cultured tobacco cells. Plant Mol Biol 27:1163–1172

May GD, Kipp PB (1997) Gene promoter sequence from banana. Patent No. WO 9738106-A 1, Zeneca LTD (GB)

Mayfield SP, Franklin SE, Lerner RA (2003) Expression and assembly of a fully active antibody in algae. Proc Natl Acad Sci U S A 100:438–442

McCormick AA, Reddy S, Reinl SJ, Cameron TI, Czerwinkski DK, Vojdani F et al (2008) Plant produced idiotype vaccines for the treatment of non-Hodgkin's lymphoma: safety and immunogenicity in a phase I clinical study. Proc Natl Acad Sci U S A 105:10131–10136

Mishra S, Yadav DK, Tuli R (2006) *Ubiquitin* fusion enhances cholera toxin B subunit expression in transgenic plants and the plant-expressed protein binds GM1 receptors more efficiently. J Biotechnol 127:95–108

Muntz K (1998) Deposition of storage proteins. Plant Mol Biol 38:77–99

Murphy DJ (1993) Structure, function and biogenesis of storage lipid bodies and oleosins in plants. Prog Lipid Res 32:247–280

Nagaya S, Kawamura K, Shinmyo A, Kato K (2010) The *HSP* terminator of *Arabidopsis* thaliana increases gene expression in plant cells. Plant Cell Physiol 51:328–332

Nara Y, Kurata H, Seki M, Taira K (2000) Glucocorticoid-induced expression of a foreign gene by the GVG system in transformed tobacco BY-2 cells. Biochem Eng J 6:185–191

Ni M, Cui D, Einstein J, Narasimhulu S, Vergara CE, Gelvin SB (1995) Strength and tissue specificity of chimeric promoters derived from the *octopine* and *mannopine* synthase genes. Plant J 7:661–676

Nochi T, Takagi H, Yuki Y, Yang L, Masumura T, Mejima M et al (2007) Rice-based mucosal vaccine as a global strategy for cold-chain- and needle-free vaccination. Proc Natl Acad Sci U S A 104:10986–10991

Obembe OO, Popoola JO, Leelavathi S, Reddy SV (2011) Advances in plant molecular farming. Biotechnol Adv 29:210–222

Padidam M (2003) Chemically regulated gene expression in plants. Curr Opin Plant Biol 6:169–177

Palacpac NQ, Yoshida S, Sakai H et al (1999) Stable expression of human 1,4-galactosyltransferase in plant cells modifies N-linked glycosylation patterns. Proc Natl Acad Sci U S A 96:4692–4697

Pelham HR (1990) The retention signal for soluble proteins of the endoplasmic reticulum. Trends Biochem Sci 15:483–486

Peumans WJ, Proost P, Swennen RL et al (2002) The abundant class III chitinase homolog in young developing banana fruits behaves as a transient vegetative storage protein and most probably serves as an important supply of amino acids for the synthesis of

ripening-associated proteins. Plant Physiol 130:1063–1072
Porta C, Lomonossoff GP (2002) Viruses as vectors for the expression of foreign sequences in plants. Biotechnol Genet Eng Rev 19:245–291
Ramessar K, Sabalza M, Capell T, Christou P (2008) Maize plants: an ideal production platform for effective and safe molecular pharming. Plant Sci 174:409–419
Rosales-Mendoza S, Soria-Guerra RE, Moreno-Fierros L, Alpuche-Solís ÁG, Martínez-González L, Korban SS (2010) Expression of an immunogenic F1-V fusion protein in lettuce as a plant-based vaccine against plague. Planta 232:409–416
Ruhlman T, Verma D, Samson N, Daniell H (2010) The role of heterologous chloroplast sequence elements in transgene integration and expression. Plant Physiol 152:2088–2104
Rybicki EP (2010) Plant-made vaccines for humans and animals. Plant Biotechnol J 8:620–637
Salmon V, Legrand D, Slomianny MC et al (1998) Production of human lactoferrin in transgenic tobacco plants. Prot Expr Purif 13:127–135
Santi L, Giritch A, Roy CJ et al (2006) Protection conferred by recombinant *Yersinia pestis* antigen produced by a rapid and highly scalable plant expression system. Proc Natl Acad Sci U S A 103:861–866
Schaefer DG (2002) A new moss genetics: targeted mutagenesis in *Physcomitrella patens*. Annu Rev Plant Biol 53:477–501
Schillberg S, Twyman RM, Fischer R (2005) Opportunities for recombinant antigen and antibody expression in transgenic plants—technology assessment. Vaccine 23:1764–1769
Schillberg S, Raven N, Fischer R, Twyman RM, Schiermeyer A (2013) Molecular farming of pharmaceutical proteins using plant suspension cell and tissue cultures. Curr Pharm Design 19:5531–5542
Schunmann PHD, Surin B, Waterhouse PM (2003) A suite of novel promoters and terminators for plant biotechnology. II. The pPLEX series for use in monocots. Funct Plant Biol 30:453–460
Shaaltiel Y, Bartfeld D, Hashmueli S, Baum G, Brill-Almon E, Galili G et al (2007) Production of glucocerebrosidase with terminal mannose glycans for enzyme replacement therapy of Gaucher's disease using a plant cell system. Plant Biotechnol J 5:579–590
Sharma AK, Sharma MK (2009) Plants as bioreactors: recent developments and emerging opportunities. Biotechnol Adv 27:811–832
Sharma MK, Singh NK, Jani D, Sisodia R, Thungapathra M, Gautam JK et al (2008) Expression of toxin co-regulated pilus subunit A (TCPA) of *Vibrio cholerae* and its immunogenic epitopes fused to choleratoxin B subunit in transgenic tomato (*Solanum lycopersicum*). Plant Cell Rep 27:307–318
Sijmons PC, Dekker BMM, Schrammeijer B et al (1990) Production of correctly processed human serum albumin in transgenic plants. Biotechnology 8:217–221
Smith ML, Keegan ME, Mason HS, Schuler ML (2002) Factors important in the extraction, stability and *in vitro* assembly of the hepatitis B surface antigen derived from recombinant plant systems. Biotechnol Prog 18:538–550
Smith ML, Richter L, Arntzen CJ, Schuler ML, Mason HS (2003) Structural characterization of plant derived hepatitis B surface antigen employed in oral immunization studies. Vaccine 21:4011–4021
Sourrouille C, Marquet-Blouin E, D'Aoust MA, Kiefer-Meyer M-C, Seveno M, Pagny-Salehabadi S et al (2008) Down-regulated expression of plant-specific glycoepitopes in alfalfa. Plant Biotechnol J 6:702–721
Sparrow PA, Irwin JA, Dale PJ, Twyman RM, Ma JK (2007) Pharma-Planta: road testing the developing regulatory guidelines for plant-made pharmaceuticals. Transgenic Res 16:147–161
Spiker S, Thompson WF (1996) Nuclear matrix attachment regions and transgene expression in plants. Plant Physiol 110:15–21
Staub JM, Garcia B, Graves J et al (2000) High yield production of a human therapeutic protein in tobacco chloroplasts. Nat Biotechnol 18:333–338
Stoger E, Sack M, Fischer R, Christou P (2002) Plantibodies: applications, advantages and bottlenecks. Curr Opin Biotechnol 13:161–166
Stoger E, Ma JKC, Fischer R, Christou P (2005) Sowing the seeds of success: pharmaceutical proteins from plants. Curr Opin Biotechnol 16:167–173
Streatfield SJ, Howard JA (2003) Plant-based vaccines. Int J Parasitol 33:479–493
Tremblay R, Wang D, Jevnikar AM, Ma S (2010) Tobacco, a highly efficient green bioreactor for production of therapeutic proteins. Biotechnol Adv 28:214–221
Twyman RM, Schillberg S, Fischer R (2005) Transgenic plants in the biopharmaceutical market. Expert Opin Emerg Drugs 10:185–218
Valenzuela P, Medina A, Rutter W, Ammerer G, Hall B (1982) Synthesis and assembly of hepatitis B virus surface antigen particles in yeast. Nature 298:347–350
Vaquero C, Sack M, Chandler J et al (1999) Transient expression of a tumor specific single chain fragment and a chimeric antibody in tobacco leaves. Proc Natl Acad Sci U S A 96:11128–11133
Vézina L-P, Faye L, Lerouge P, D'Aoust M-A, Marquet-Blouin E, Burel C et al (2009) Transient coexpression for fast and high-yield production of antibodies with human-like N-glycans in plants. Plant Biotechnol J 7:442–455
Voinnet O, Rivas S, Mestre P, Baulcombe D (2003) An enhanced transient expression system in plants based on suppression of gene silencing by the p19 protein of tomato bushy stunt virus. Plant J 33:949–956
Walmsley AM, Arntzen CJ (2003) Plant cell factories and mucosal vaccines. Curr Opin Biotechnol 14:145–150
Wenderoth I, von Schaewen A (2000) Isolation and characterization of plant N-*acetyl glucosaminyltransferase I* (*GntI*) cDNA sequences. Functional analyses in the

Arabidopsis cgl mutant and in antisense plants. Plant Physiol 123:1097–1108

Wilde CD, Peeters K, Jacobs A, Peck I, Depicker A (2002) Expression of antibodies and Fab fragments in transgenic potato plants: a case study for bulk production in crop plants. Mol Breed 9:271–282

Woodleif WG, Chaplin JF, Campbell CR et al (1981) Effect of variety and harvest treatments on protein yield of close grown tobacco. Tob Sci 25:83–86

Yang D, Wu L, Hwang YS, Chen L, Huang N (2001) Expression of the *REB* transcriptional activator in rice grains improves the yield of recombinant proteins whose genes are controlled by a *Reb*-responsive promoter. Proc Natl Acad Sci U S A 98:11438–11443

Yano M, Hirai T, Kato K, Hiwasa-Tanase K, Fukuda N, Ezura H (2010) Tomato is a suitable material for producing recombinant miraculin protein in genetically stable manner. Plant Sci 178:469–473

Yao J, Weng Y, Dickey A, Wang KY (2015) Plants as factories for human pharmaceuticals: applications and challenges. Int J Mol Sci 16:28549–28565

Provitamin A Enrichment for Tackling Malnutrition

19

Navneet Kaur, Shivani, Ashutosh Pandey, and Siddharth Tiwari

Abstract

People in many developing countries are affected by malnutrition due to either less availability of nutritious food or unbalanced diet. The major micronutrient deficiencies in the world are vitamin A, iron, zinc and iodine. Vitamin A deficiency is a leading cause of preventable blindness, growth retardation and decreased resistance to infection. Most of the population in developing countries rely on staple crops; thereby, the biofortification of these crops would be considered as a promising cost-effective and sustainable approach. Banana is an important staple fruit crop that contributes significantly towards the food security in many developing countries where vitamin A deficiency is a major problem. Transgenic banana with up to 20 μg/g dry weight β-carotene has been developed, and field trials are in progress in Uganda. Fruit-specific promoters and carotenoid pathway-related genes in banana were characterized for their role in the development of cisgenic biofortified crop. The recent advancement in precise genome engineering can open the novel prospect for the improvement of banana. These biotechnological approaches could be more advantageous to address regulatory and biosafety challenges for commercialization of biofortified banana. This chapter summarizes the prevalence of vitamin A micronutrient deficiency as well as the highlighted progress in biotechnological approaches for their possible role in the development of provitamin A-enriched banana.

Keywords

Banana • Biofortification • β-carotene • Carotenoids • Malnutrition • Provitamin A • Vitamin A deficiency

Author contributed equally with all other contributors.

N. Kaur • Shivani • A. Pandey • S. Tiwari (✉)
National Agri-Food Biotechnology Institute (NABI), Mohali 160071, Punjab, India
e-mail: siddharth@nabi.res.in; siddharthdna20@yahoo.com

S. Mohandas, K.V. Ravishankar (eds.), *Banana: Genomics and Transgenic Approaches for Genetic Improvement*, DOI 10.1007/978-981-10-1585-4_19

19.1 Introduction

Micronutrients constitute a small fraction of our diet but play a major role in the metabolic process. Dietary intake of micronutrient-enriched food is a natural and affordable way for combating its deficiency. However, most of the staple foods consumed in the developing countries are deficient in micronutrients. The World Health Organization (WHO 2009) considered vitamin A deficiency (VAD) as a priority among various health problems. One of the earliest manifestations of VAD is night blindness. Severe condition leads to keratomalacia (ulceration and sloughing of the cornea) and total blindness. In addition, it has been shown to cause growth retardation, decreased resistance to infections and even death (Burri 2011). Various strategies such as pharmaceutical supplementation, food fortification and altered dietary components can be implemented to alleviate micronutrient deficiencies. According to a report published by Indian National Science Academy(INSA 2011) despite massive investment in many of these interventions, very high levels of specific micronutrient deficiencies persist in many areas. The annual budget for "Micronutrients Initiatives India" in 2009–2010 was worth approximately 32 million US dollars; more than 20 million US dollars were spent on vitamin A procurement and interventions alone. Indian government-supported vitamin A supplementation programme, running for more than 40 years, attained less than 25 % coverage and failed to reach the extreme poor rural population (Burri 2011; INSA 2011). Programmes for vitamin A supplementation among pre-school children in India have resulted in only 5–15 % reduction in malnutrition caused by VAD (Awasthi et al. 2013). A report of the Expert Group of the Indian Council of Medical Research (ICMR 2010) mentioned that 40–60 % of the child population show inadequate biochemical status of vitamin A. Poor bioavailability and low intakes of provitamin A (PVA) from the predominantly vegetarian diet are thought to be the main reasons for the widespread prevalence of VAD. Hence, it is desirable to develop a sustainable and biological safe solution to overcome VAD. An advanced strategy called "biofortification" is being used to develop new varieties of food crops with enhanced nutrient content. The biofortification of major staple crops holds a cost-effective approach because it involves only one-time investment either by selective plant breeding or genetic engineering. This technology has the potential to have a very significant impact on the reduction of micronutrient deficiencies, particularly in rural communities where other interventions showed a limited impact. Further, biofortified crops can be accepted easily because it follows the same traditional eating behaviour, food habits and processing.

Banana and plantain are among the most important fruit crops in the world and known as a poor man's staple food in many developing countries. Banana is ranked fourth in the list of global fruit crops and is still least genetically improved relative to other major crops due to its complex and difficult genetic makeup to generate variability by plant breeding (De et al. 2009). The application of conventional breeding for genetic improvement in banana is more difficult due to the ploidy level and nature of parthenocarpic fruit development. Thus, the development of innovative approaches to complement conventional breeding programmes would be a promising strategy to develop biofortified banana. The application of transgenic approach in banana holds high potential for introducing desirable trait/s and can be relatively fast for the genetic improvement of elite cultivars without affecting their good native traits.

19.2 Food Insecurity, Micronutrient Deficiencies and Their Socio-economic Impacts

Millions of people worldwide are unable to take balanced diet. Food insecurity is defined as the lack of access to sufficient calories and nutrition in food (balanced diet) to maintain a healthy life. A healthy person requires an average minimum balanced diet with 2500 cal/day. These calories come from different food sources consisting of carbohydrates, proteins, fats, vitamins, minerals, etc. (Welch and Graham 2004; Graham et al. 2007). If the intake of calories is less than required, the food is said to be undernourished. According to the recent report of the Food and Agriculture Organization(FAO 2015), nearly 795 million people globally are not eating sufficient food to meet their energy requirements, coming under chronic undernourishment (hunger), while 780 million people from those belong to the developing countries. Micronutrients are normally supplied in sufficient amount through balanced diet. If the food being eaten has a nutritional imbalance due to lack of essential nutrients, it is said to be malnourished. Micronutrients such as minerals and vitamins are vital for human growth and development. They are required in very low amounts in the diet, but their deficiency leads to serious negative impacts on human health and increased risk of death by infectious and chronic diseases (Murgial et al. 2013). Deficiencies of iron, vitamin A, zinc, iodine, vitamin C and vitamin D cause the origin of diseases like anaemia, night blindness, defective immune system, goitre, scurvy and rickets, respectively. Iron, vitamin A and zinc deficiencies are among the top ten leading causes of human death worldwide (WHO 2009). The effect of micronutrient malnutrition also includes intrauterine growth restriction, stunting growth, wasting and suboptimum breastfeeding which caused 45 % of child death and also magnifies the effect of various diseases (Black et al. 2013).

The food insecurity at the household level is one of the main causes of micronutrient malnutrition. Poverty, less inclusive economic growth, political instability, ignorance, low agricultural productivity, environmental conditions and traditional agricultural practices are the other major factors behind the malnourishment of large population. The recent data have shown that the world food availability has risen to meet the per day energy requirement, but still millions of people suffer from malnutrition (FAO 2015). Therefore, the malnutrition is not the problem of undernourished or hungry population. It is even persisting in people who are having sufficient amount of food and is due to either the diet being deficient in nutritious components such as proteins, minerals and vitamins or the bioavailability of nutritious components being poor. Micronutrient deficiency is also referred to as "hidden hunger" because the usual symptoms associated with it do not appear physically in the affected person (Nilson and Piza 1998; Kennedy et al. 2003).

The food and nutritional security should be provided which implies physical, economic and social access to a balanced diet along with safe drinking water, environment and good health care. Micronutrient malnutrition has high economic cost. Productivity losses due to poor nutrition are estimated to be more than 10 % of lifetime earnings for individuals and 2–3 % of gross domestic product (GDP) to the nation (Leadership Agenda 2010). The cost of treating malnutrition is 27-times higher than the investment required for its prevention. According to the Copenhagen Consensus (2008), tackling micronutrient malnutrition by micronutrient supplementation and food fortification is considered as highly beneficial which will generate a high return in socio-economic benefits as compared to the costs.

19.3 Addressing the Problem of Malnutrition Through Biofortification

One of the best ways to tackle malnutrition is by intake of nutrient-rich diet which includes animal and plant food sources. However, numerous factors limit the potential of presently known foods, including high cost, nutrient activity, bioavailability, acceptability of taste, status and prestige, familiarity, seasonality, size of the commonly eaten portion and unacceptability for religious reasons to eat animal foods. The maximum population in the developing countries relies on staple food which are deficient in essential nutrients leading to malnutrition (Zhu et al. 2007). WHO defined fortification as the practice of deliberately increasing the content of essential micronutrients in a food, so as to improve the nutritional quality and provide a public health benefit with minimal risk to health. The biological methods include plant breeding and genetic engineering, which are the more effective and cheaper alternatives to conventional ways for combating micronutrient deficiencies (Bashir et al. 2013). Enrichment of staple food (rice, maize, wheat) and fruit (banana, tomato, citrus) crops by biofortification is the sustainable agricultural approach. Micronutrient-enriched biofortified crops can easily be adapted to relevant environments. Three major crop biofortification strategies worldwide include mineral fertilization, conventional breeding and transgenic approach.

19.3.1 Mineral Fertilization

In this approach plant fertilizers are applied as inorganic minerals to increase the micronutrient content in the crop. Usually trace elements whose accumulation is limited in the edible parts of plant due to the low availability in soil are supplemented with normal fertilization. In some instances, this approach has been quick and successful to increase the availability of zinc and selenium (Lyons et al. 2004). However, it is difficult to apply so many minerals and vitamins as absorption of these compounds depend upon several factors such as the crop species and cultivars, soil composition, mineral mobility, properties of compound and accumulation sites in the plant (Zhu et al. 2007). Application of excess amount of mineral may limit the growth of plant as well as affect the microbial population in the soil. Mineral fertilizers that are required to be applied regularly are costly and also harmful to the environment (Zhu et al. 2007).

19.3.2 Plant Breeding

Plant breeding is a method to improve crops by cross breeding between two different crop species with contrasting traits. It exploits genetic variation among the crop species to introduce new allele from one crop with better trait to another. Improving food crops by breeding is mainly focussed on improving yields and physico-chemical properties rather than nutrient profiles. Conventional breeding exploits inherent properties of the crop and has significant impact on communities and less regulatory restrictions (Welch and Graham 2002; Zhu et al. 2007). The Harvest Plus Biofortification Challenge Programme (www.harvestplus.org) introduced by the Consultative Group on International Agricultural Research (CGIAR) aims at the development of PVA (β-carotene)-rich sweet potato, maize, and cassava;zinc- and iron-rich rice, wheat, maize, pearl millet, and beans. In India, Department of Biotechnology (DBT) network projects on biofortification of rice, wheat and maize are currently being implemented by ICAR institutions, state agriculture universities and the National Institute of Nutrition (NIN). The biofortification of crops by plant breeding is limited either due to the non-existence of the trait within the gene pool or its inability to be transferred by conventional breeding. Moreover, it would take many generations to introgress these traits into local elite breeding varieties (Welch and Graham 2005).

19.3.3 Genetic Engineering

The transgenic approach offers a useful alternative where supplementation and conventional breeding fail to achieve significant improvement in nutritional levels in crop plants. It is one of the most effective and promising tools for increasing productivity with high nutritional content. Genetic engineering has also overcome the limitations of plant breeding and can be used to employ gene of interest even across the taxa. This technology has been already successful in developing many insect-resistant, and herbicide-resistant crop varieties, and is available in the market (James 2014; www.isaaa.org). An excellent example of biofortification by genetic engineering is the Golden Rice (Ye et al. 2000; http://www.goldenrice.org) as the conventional breeding cannot be employed due to unavailability of paddy cultivar with β-carotene in its endosperm (the edible part of the grain). Through genetic engineering strategy, carotenoid enhancement in rice seeds has been successfully employed (Ye et al. 2000). The genetic engineering technique is precise and involves isolation of individual genes from crop species responsible for high production of certain nutrients or important physicochemical property. Staple crops can be biofortified by this strategy which can provide a sustainable, inexpensive, cost-effective and long-term source of micronutrients to the poor (Perez-Massot et al. 2013; Bouis and Welch 2010). Moreover, improved crops provide a feasible means of reaching malnourished rural populations. This approach can also be implemented in the case of genetically sterile staple crops such as banana and cassava where conventional breeding is difficult or less responsive (Zhao and Shewry 2011).

19.4 Vitamin A

Carotenoids, a natural precursor of vitamin A, are synthesized by vascular plants, some blue-green algae, fungi and photosynthetic bacteria but not in the animal kingdom, including humans (Ruiz-Sola and Rodríguez-Concepción 2012). Vitamin A is a group of fat-soluble unsaturated compound with a common retinyl group. This group includes different forms such as retinol, retinoic acid and retinyl esters (Fig. 19.1). One molecule of β-carotene is converted into two molecules of retinal via β-carotene 15,15′-monooxygenase (CMO1). This enzyme cleaves β-carotene centrally to form two molecules of retinal. The other enzyme is β-carotene 9,10-dioxygenase (CDO2) which cleaves β-carotene, α-carotene and β-cryptoxanthin eccentrically to form apocarotenals, and the longer one is oxidized to one molecule of retinal (de Pee and West 1996; Biesalski et al. 2007; Lietz et al. 2010). Retinol is required for the production of rhodopsin in the eye for vision. Retinol is the most usable form of the vitamin A and possesses a β-ionone ring and an unsaturated isoprenoid side chain with an alcohol, aldehyde, a carboxylic acid group or an ester group. The side chain is composed of four isoprenoid units with a series of conjugated double bonds either in cis- or trans-configuration (Gropper et al. 2009). Retinol can be converted into other forms of vitamin A, i.e. retinyl esters, for functional and storage purposes. The two sequential oxidation steps convert retinol to retinaldehyde and further to retinoic acid. Retinoic acid is an important hormonal metabolite for maintenance and differentiation of epithelial and immune cells for growth and development (Fig. 19.1).

19.4.1 Vitamin A Deficiency (VAD)

Vitamin A is an essential micronutrient and required in fewer amounts in the body for several physiological functions. It is required for healthy eyesight (Palczewski 2010; von Lintig 2012), immune function (Cassani et al. 2012; Brown and Noelle 2015), genetic regulation (Blomhoff and Blomhoff 2006; Wolgemuth and Chung 2007; Mark et al. 2009), growth and development (Mark et al. 2009; Kedishvili 2013; Cunningham and Duester 2015). VAD is one of the serious concerns for poor societies and developing countries. The WHO (2009) estimates show 150 million people suffering from VAD worldwide and death of 650,000–700,000 children and pregnant women due to its deficiency (Black et al. 2008; Burri 2011; Awasthi et al. 2013; Bailey et al. 2015). It is prevalent among South-East Asia and Sub-Saharan Africa, where children are at higher risk to death comparative to adults because of poor functional immune system (West 2002; WHO 2009). The ocular manifestation governed by VAD is referred to as xerophthalmia. It includes night blindness, conjuctival xerosis, bitot's spot, corneal xerosis and keratomalacia. In severe condition, night blindness and bitot's spot are the earliest observable ocular manifestations which lead to keratomalacia (ulceration and sloughing of the cornea) and total blindness (xeropthalmia). In case of night blindness, there is deterioration of rod cells (the light-sensitive cells) and the person is not able to see in low light. It is a specific indicator of low serum retinol levels. Serum retinol is vitamin A in the blood stream and is measured as "retinol equivalents". The early symptoms of VAD appear when serum retinol concentration falls below 1.0 μmol/L and get severe when retinol concentration falls

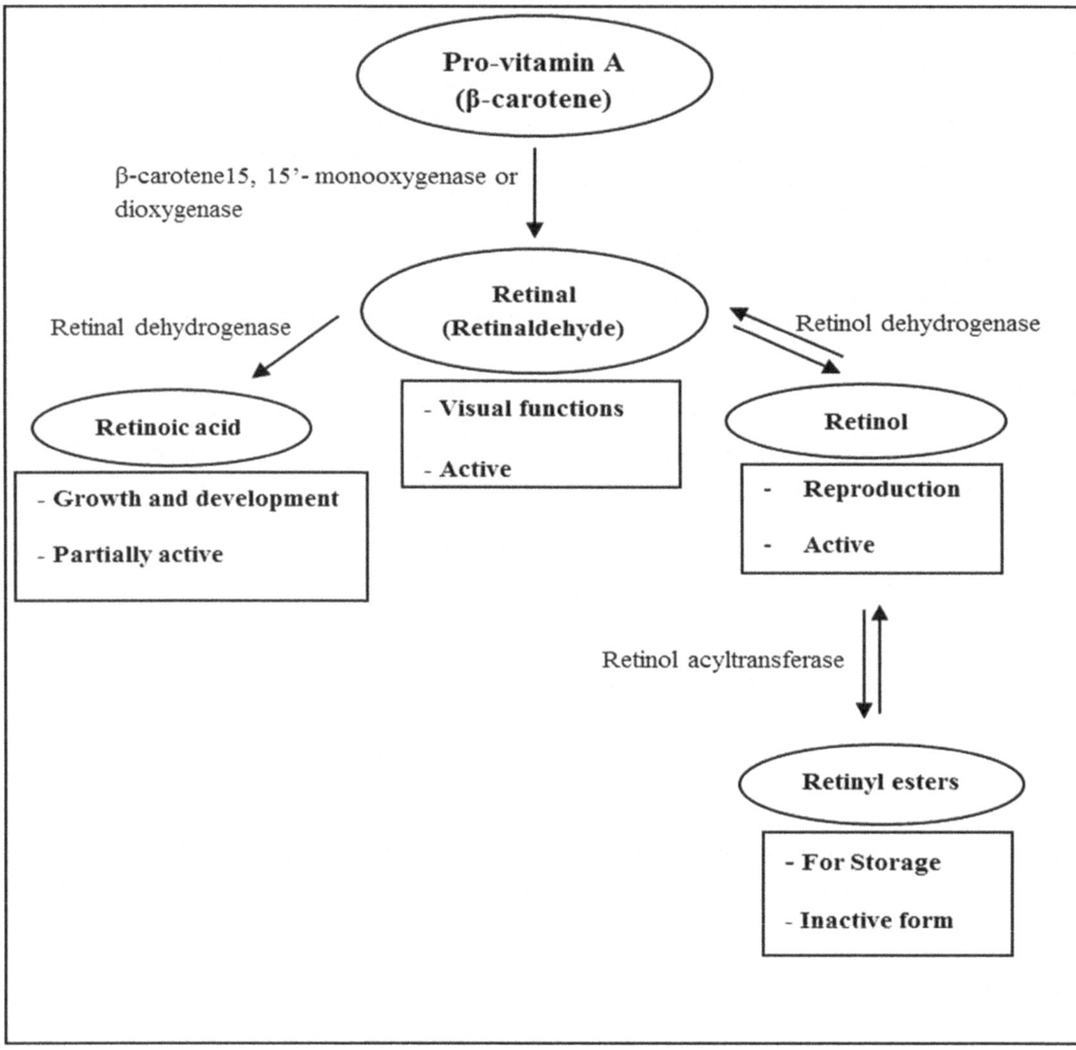

Fig. 19.1 Steps in the synthesis of different forms of vitamin A from provitamin A (PVA)

below 0.35 μmol/L (WHO 2014). Retinal, a form of vitamin A, produces rhodopsin in the eye. It is a photosensitive visual pigment of rods by combining with opsin. VAD leads to the decline in rhodopsin level and hence impaired rod function. Night blindness is prevalent among pregnant women. Bitot's spot is another problem in which opaque whitish spots are deposited on the sclera conjunctiva. Dry and dull spots of keratinized epithelial cells with an appearance of foam are observable symptoms of this condition. Vitamin A also plays multiple apparent roles in iron mobilization, transportation and haematopoiesis. VAD can also lead to anaemia in children and women (West et al. 2007). Vitamin A supplementation can reduce the severity of diseases such as influenza and diarrhoea (Black et al. 2008).

19.4.2 Sources of Vitamin A

Vitamin A is differentiated into two dietary source categories. These are known as retinyl esters (preformed vitamin A) and PVA (vitamin A precursors). Retinyl esters are obtained from the meat and dairy products, whereas PVA is obtained from plant sources in the form of carotenoids. Carotenoids include mainly α-carotene, β-carotene, γ-carotene, ε-carotene, δ- carotene, phytoene, ζ-carotene, lycopene, lutein and xanthophylls (zeaxanthin, antheraxanthin, violaxanthin and neoxanthin). However, only certain forms of carotenoids such as α-carotene, β-carotene and β-cryptoxanthin known as PVA can be converted to vitamin A in the body (Britton et al. 1995; de Pee and West 1996). The β-carotene is considered as nutritionally more useful as it gets converted into retinol very effectively. The α-carotene and β-cryptoxanthin are converted only the half efficiency of the β-carotene. Fruits, vegetables and seeds are the primary dietary sources of carotenoids, but mostly consumed crops are insufficient to meet minimum nutritional requirements; consequently VAD remains prevalent in developing countries (Fitzpatrick et al. 2012). Coloured foods such as carrot, papaya, pumpkin, mango, sweet potato and dark green leafy vegetables like spinach and lettuce are known to be rich sources of β-carotene. One of the richest known sources is the palm oil which has the highest concentration and activity of PVA (Cottrell 1991).

19.4.3 Bioavailability and Recommended Dietary Allowance

Bioavailability can be defined as the fraction of ingested nutrient which is available for utilization in normal physiological functions and for storage (Srinivasan 2001). All the nutrients present in the diet are not bioavailable to the consumer because the plant sources also contain antinutrient compounds which interfere with the absorption or utilization of the nutrients (Welch and Graham 1999). The vitamin A conversion from PVA depends upon regulation in the body as well as on the amount administered. Bioavailability of vitamin A is the percentage of ingested PVA getting converted into a metabolically active form, i.e. retinol, which is available for storage as well as utilization (Van Lieshout et al. 2003). Determining bioavailability of a particular micronutrient from plant food sources to humans is quite complex as multiple factors interact during the process. It depends upon the micronutrient itself, individual, diet as well as the environment (Graham et al. 2001; Degerud et al. 2015; Schweiggert and Carle 2015). The conversion of β-carotene into vitamin A and absorption in the body are variable depending upon the food source, matrix and culinary practices (de Pee et al. 1998; Veda et al. 2006). It was reported that β-carotene can be better absorbed from orange-coloured fruits and vegetables as compared to green leafy vegetables (de Pee et al. 1998; O'Connell et al. 2007). The β-carotene absorption rate is increased when fed with oil as compared to without oil (Failla et al. 2009; Bengtsson et al. 2009). Pullakhandam and Failla (2007) demonstrated high efficiency of carotenoid micellarization and bioaccessibility in drumstick leaves in the presence of peanut oil. Hedrén et al. (2002) demonstrated high in vitro accessibility of β-carotene in the green leaves of amaranth, cow pea, sweet potato, pumpkin and cassava in the presence of oil (39–94 %), compared to the absence of oil (8–29 %). The recommended dietary allowance (RDA) represents the average level of intake of essential nutrients necessary for the maintenance of good health in a particular age and gender. RDA for vitamin A depends upon the bioavailability of PVA and is measured in retinol activity equivalents (RAE). ICMR expert group report (http://icmr.nic.in/final/rda-2010.pdf) considered conversion efficiency of 1:8 and reported RDA of vitamin A for various physiological groups. The RDA of vitamin A and β-carotene for Indian men and women is 600 and 4800 μg/day, respectively (http://icmr.nic.in/final/RDA-2010.pdf).

19.5 Provitamin A Biosynthetic Pathway in Higher Plants

Carotenoids are lipid-soluble tetraterpenoids, synthesized by all photosynthetic organisms, including plants and few non-photosynthetic fungi and bacteria (Ruiz-Sola and Rodríguez-Concepción 2012). The biosynthesis of carotenoids takes place with the synthesis of 5-carbon precursors, isopentenyl pyrophosphate (IPP) and its double-bond isomer dimethylallyl pyrophosphate (DMAPP) (Fig. 19.2). The cytosolic mevalonate (MVA) and plastidial 2-C-methyl-D-erythritol 4-phosphate (MEP) are the two independent pathways which synthesize these precursor molecules in plants (Grassi et al. 2013). Several reports suggested that major flux for carotenoids are derived from a plastid-localized MEP pathway (Flores-Perez et al. 2008; Rodriguez-Concepcion 2010). The MEP pathway utilizes glyceraldehyde 3-phosphate and pyruvate as initial substrate for the biosynthesis of IPP and DMAPP. The first committed step of carotenoid biosynthesis is the condensation of two molecules of geranylgeranyl diphosphate (GGPP) for the production of phytoene by the enzyme phytoene synthase (PSY) (Sandmann et al. 2006; Cazzonelli and Pogson 2010). Multiple PSY catalyse the rate-limiting step in the carotenoid biosynthesis. It is involved in the regulation of carotenoid metabolic flux for downstream enzymatic steps for the biosynthesis of lycopene and α and β-carotene (Sandmann et al. 2006; Cazzonelli and Pogson 2010; Fu et al. 2014; Liu et al. 2014). Phytoene is further desaturated and isomerized into lycopene by the enzyme phytoene desaturase (PDS) and ζ-carotene desaturase (ZDS). The cyclization of the lycopene is the first branch point in the pathway which results in the production of carotenes either with one β-ring and one ε-ring (α-carotene) or with two β-rings (β-carotene) by lycopene β-cyclase (LCYB) and lycopene ε-cyclase (LCYE) enzymes. The hydroxylation of the carotenoids results in the formation of xanthophylls. The hydroxylation of α-carotene results in the formation of lutein. β-carotene is hydroxylated into zeaxanthin which is further converted into abscisic acid (ABA). The complete pathway for the biosynthesis of PVA carotenoids in higher plants is summarized in Fig. 19.2.

19.6 Provitamin A Enhancement by Genetic Engineering in Plants

Many studies reported on the enhanced content of PVA, particularly of β-carotene in constitutive as well as tissue-specific manners is stable in food and fruit crops. Usually stable expression of the candidate gene/s does not interfere with the subsequent propagation of plants either by vegetative or sexual reproduction methods in selected transgenic events. Another transgene/s can also be introduced in previously stable transformed plants to enhance multiple micronutrients. The judicious choice of regulatory elements facilitates tissue-specific expression of desirable gene/s in leaves, fruits, tubers and seeds. A variety of tissues have been targeted for the expression of the carotenoids synthesis gene/s in various plant species. Enhanced expression of β-carotene in food and fruit crops are summarized in Tables 19.1 and 19.2, respectively. Some of the reports published on the enhancement of β-carotene in these crops are described as below.

19.6.1 Food Crops

Developing countries require a robust approach to increase PVA content in crop plants to address the VAD. Food containing PVA carotenoids (β-carotene, β-cryptoxanthin and α-carotene) are the primary source of vitamin A. In all these carotenoids, β-carotene is widely distributed in plants and most efficiently converted into vitamin A. In mid-2005, the Bill and Melinda Gates Foundation funded four projects in biofortification as part of the Grand Challenges in Global Health programme. These four projects involved the biofortification of rice (Golden Rice project), cassava (BioCassava Plus), sorghum (ABS) and bananas (Banana21). The first huge success of genetic engineering for β-carotene biosynthesis

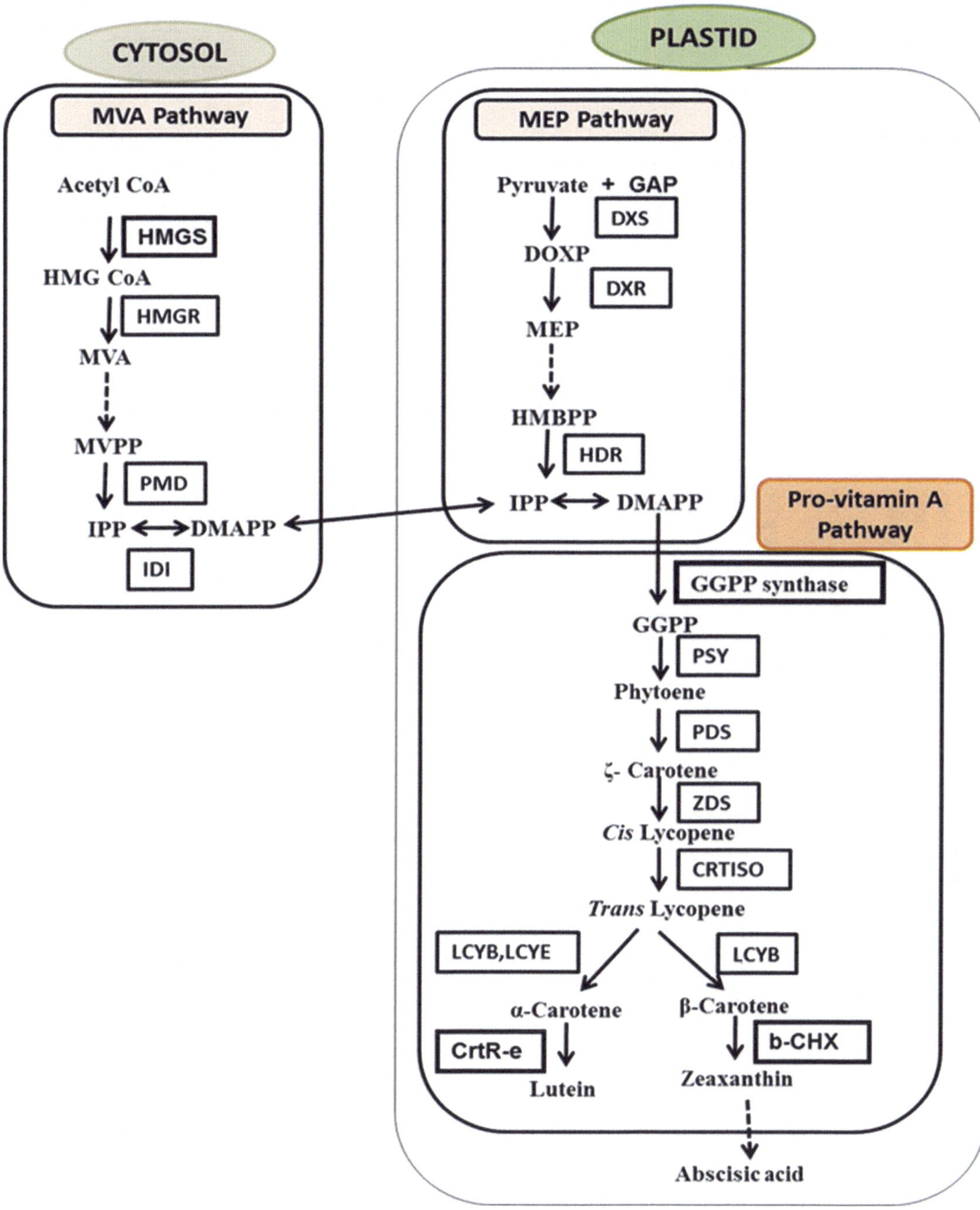

Fig. 19.2 Schematic representation of provitamin A (PVA) biosynthetic pathway in higher plants. Enzymatic reactions are represented by *arrows*; *dashed lines* represent multiple enzymatic steps. Enzymes are represented in the *box*. Enzymes: *HMGS* 3-hydroxy-3-methylglutaryl-CoA synthase, *HMGR* 3-hydroxy-3-methylglutaryl-CoA reductase, *DXS* 1-deoxy-D-xylulose-5-phosphate synthase, *DXR* 1-deoxy-D-xylulose-5-phosphate reductase, *HDR* 1-hydroxy-2-methyl-2-(E)-butenyl-4-diphosphate reductase, *GGPP synthase* geranylgeranyl diphosphate synthase, *PSY* phytoene synthase, *PDS* phytoene desaturase, *ZDS* ζ-carotene desaturase, *LCYB* lycopene β-cyclase, *LCYE* lycopene ε-cyclase, *CRTISO* carotenoid isomerase, *b-CHX* β-ring hydroxylase, *e-CHX* ε-ring hydroxylase, *IDI* IPP isomerase, *PMD* MVPP decarboxylase. Compounds: *HMG-CoA* 3-hydroxy-3-methylglutaryl-CoA, *MVA* mevalonic acid, *MVPP* 5-diphosphomevalonate, *GAP* glyceraldehyde-3-phosphate, *DOXP* 1-deoxy-D-xylulose-5-phosphate, *MEP* 2-C-methyl-D-erythritol 4-phosphate, *HMBPP* 1-hydroxy-2-methyl-2-(E)-butenyl-4-diphosphate, *IPP* isopentenyl diphosphate, *DMAPP* dimethylallyl diphosphate, *GGPP* geranylgeranyl diphosphate

Table 19.1 Provitamin A-rich biofortified food crops developed by genetic engineering

Plant	Targeted gene/s	Gene source	Promoter	Concentration of β-carotene	Targeted tissue	References
Rice	*Phytoene synthase (psy)*	*Zea mays*	*Glutelin (Gt1)*	31 μg/gDW	Endosperm	Paine et al. (2005)
	Phytoene desaturase (crtI)	*Erwinia uredovora*				
	Phytoene synthase (crtB)	*Narcissus pseudonarcissus*	*Glutelin (Gtl)* and *CaMV35S*	1.6 μg/g DW[a]	Endosperm	Ye et al. (2000)
	Phytoene desaturase (crtI)	*Erwinia uredovora*				
Wheat	*Phytoene synthase (psy)*	*Zea mays*	*1Dx5* and *CaMV35S*	4.96 μg/g DW[a]	Endosperm	Cong et al. (2009)
	Phytoene desaturase (crtI	*Erwinia uredovora*				
Canola	*Phytoene synthase (crtB)*	*Erwinia uredovora*	*Napin*	401 μg/g FW	Seed	Shewmaker et al. (1999)
	Phytoene synthase (crtB)	*Erwinia uredovora*	*Napin*	857 μg/g FW		Ravanello et al. (2003)
	Phytoene desaturase (crtI)	*Erwinia uredovora*				
	Lycopene β cyclase (crtY)	*Erwinia uredovora*				
	Lycopene β-cyclase (lcyB)	*Brassica napus*				
	Lycopene ε-cyclase (lcyE)	*Arabidopsis thaliana*	*Napin*	90.76 μg/g FW	Seed	Yu et al. (2008)
	Isopentenyl pyrophosphate isomerase (idi)	*Paracoccus* spp.	*Napin* and *CaMV35S*	214.2 μg/g FW	Seed	Fujisawa et al. (2009)
	β-carotene ketolase (crtW)	*Brevundimonas* spp.				
	β-carotene hydroxylase (crtZ)	*Pantoea ananatis*				
	Phytoene desaturase (crtI)					
	Lycopene β cyclase (crtY)					
	Lycopene ε-cyclase (crtE)					
	Phytoene synthase (crtB)					
	microRNA miR156b	*Arabidopsis thaliana*	*CaMV35S* and *N apin*	0.40 μg/g DW	Seed	Wei et al. (2010)
Potato	*Phytoene synthase (crtB)*	*Pantoea ananatis*	*Patatin*	10.3 μg/g DW	Tuber	Ducreux et al. (2005)
	Lycopene ε cyclase (crtE)	*Pantoea ananatis*	*Patatin*	43.56 ng/g DW	Tuber	Diretto et al. (2006)
	Phytoene synthase (crtB)	*Pantoea ananatis*	*Patatin* and *CaMV35S*	47 μg/g DW	Tuber	Diretto et al. (2007)
	Phytoene desaturase (crtI)					
	Lycopene β cyclase (crtY)					
	β-carotene hydroxylase gene (b-CHX)	*Solanum tuberosum*	*Granule bound starch synthesis (GBSS)* and *CaMV35S*	331 μg/100 g FW	Tuber	Van Eck et al. (2007)
	Orange cauliflower (Or)	*Brassica oleracea*	*GBSS*	5.01 μg/g DW	Tuber	Lopez et al. (2008)

Corn	*Phytoene synthase* (*psy*)	*Zea mays*	Wheat *glutenin*	57.35 μg/g DW	Endosperm	Zhu et al. (2008)
	β-carotene hydroxylase (*b-CHX*)	*Gentiana lutea*	Rice *glutelin-1*			
	Lycopene ε-cyclase (*lcyE*)	*Gentiana lutea*	Rice *prolamin*			
	β-carotene ketolase (*crtW*)	*Paracoccus*	Maize *γ-zein*			
	Phytoene desaturase (*crtI*)	*Pantoea ananatis*	Barley *hordein*			
	Phytoene synthase (*crtB*)	*Pantoea ananatis*	*γ-zein*	9.8 μg/g DW	Endosperm	Aluru et al. (2008)
	Phytoene desaturase (*crtI*)					
	ζ-carotene desaturase (*zds*)					
	Phytoene synthase (*psy*)	*Zea mays*	*Glutenin* (*Gtl*) and barley d-*hordein*	59.32 μg/g DW		Naqvi et al. (2009)
	Phytoene desaturase (*crtI*)	*Pantoea ananatis*				
Carrot	*Ketolase gene*	*Haematococcus pluvialis*	*CaMV35S* and *ubiquitin*	39 μg/g FW	Root	Jayaraj et al. (2008)
	Phytoene synthase (*psy*)	*Arabidopsis thaliana*	*CaMV35S*	1800 μg/g DW[a]	Seed calli	Maass et al. (2009)

Genes derived from plant system represent as: *Phytoene synthase* (*psy*); *Phytoene desaturase* (*pds*); *Lycopene β-cyclase* (*lcyB*)

Genes derived from bacterial system represent as: *Phytoene synthase* (*crtB*); *Phytoene desaturase* (*crtI*); *Lycopene β-cyclase* (*crtY*)

DW: dry weight, FW: fresh weight

[a]Total carotenoid concentration

Table 19.2 Provitamin A-rich biofortified fruit crops developed by genetic engineering

Plant	Targeted gene/s	Gene source	Promoter	Concentration of β-carotene	References
Tomato	*Phytoene desaturase* (*crtI*)	*Erwinia uredovora*	*CaMV35S*	52 μg/g FW	Romer et al. (2000)
	Lycopene β cyclase (*lcyB*)	*Solanum lycopersicum*	*Phytoene desaturase*	57 μg/g FW	Rosati et al. (2000)
	Lycopene β cyclase (*lcyB*)	*Arabidopsis thaliana*	*Phytoene desaturase*	63 μg/g FW	Dharmapuri et al. (2002)
	β-carotene hydroxylase (*b-CHX*)	*Capsicum annuum*			
	Phytoene synthase (*crtB*)	*Erwinia uredovora*	*Polygalacturonase*	82.5 μg/g FW	Fraser et al. (2002)
	Lycopene β cyclase (*lcyB*)	*Solanum lycopersicum*	*CaMV35S*	205 μg/g FW	D'Ambrosio et al. (2004)
	3-hydroxymethylglutaryl CoA (*hmgr*)	*Arabidopsis thaliana*	*CaMV35S* and *fibrillin*	42 μg/g FW	Enfissi et al. (2005)
	1-deoxy- D – xylulose-5-phosphate synthase (*dxs*)	*Escherichia coli*			
	Cryptochrome (*cry1*)	*Solanum lycopersicum*	*CaMV35S*	10.1 μg/g FW	Giliberto et al. (2005)
	Fibrillin	*Capsicum annuum*	*Fibrillin*	150 μg/g FW	Simkin et al. (2007)
	Lycopene β cyclase (*crtY*)	*Narcissus pseudonarcissus*	*rRNA operon*	58 μg/g FW	Apel and Bock (2009)
	SlNCED1	*Solanum lycopersicum*	*Fruit-specific E8*	40 μg/g FW	Sun et al. (2012)
Citrus	*b-carotene hydroxylase* (*b-CHX*)	*Citrus sinensis*	*CaMV35S*	114.0 ng/g of FW	Pons et al. (2014)

Genes derived from plant system represent as: *lycopene β-cyclase* (*lcyB*)
Genes derived from bacterial system represent as: *phytoene synthase* (*crtB*); *phytoene desaturase* (*crtI*); *lycopene β cyclase* (*crtY*)
DW: dry weight, FW: fresh weight

was reported as Golden Rice. In rice the pathway does not exist in the seed but introduced through metabolic engineering by using two genes *phytoene synthase* (*psy*) originating from daffodil (*Narcissus pseudonarcissus*) and bacterial *phytoene desaturase* (*crtI*) originating from *Erwinia uredovora* placed under control of the endosperm-specific *glutelin* (*Gt1*) and the constitutive *Cauliflower mosaic virus* (*CaMV35S*) promoter, respectively (Ye et al. 2000). The technology was escalated to generate Golden Rice 2 where *psy* from maize (*Zea mays*) in combination with bacterial *crtI* under *glutenin* (*Gt1*) promoter produced a preferential accumulation of β-carotene (31 μg/g dry weight). This increase in total carotenoids was up to 23-fold compared to the original Golden Rice (Paine et al. 2005).

Other food crops like wheat, canola, potato, maize, cassava and sorghum have also been genetically engineered to increase the content of PVA. Wheat was transformed with *psy* (maize) and *crtI* (*Erwinia uredovora*) under the control of endosperm-specific *1Dx5*and constitutive *CaMV35S* promoters, respectively. The highest increase of total carotenoids was 4.96 μg/g seed dry weight (DW) and was noted in transgenic wheat line. The best line showed 10.8-fold increases in total carotenoids in wheat endosperm (Cong et al. 2009).

Canola (*Brassica napus*) was overexpressed with bacterial *phytoene synthase* (*crtB*) gene from *Erwinia uredovora* under the seed-specific *napin* promoter and reported enhanced β-carotene up to 949 μg/g fresh weight (FW) in transgenic lines. Total carotenoid concentration also boosted up to 50-fold in comparison with non-transgenic lines (Shewmaker et al. 1999). In another study, *crtB* was expressed with additional bacterial genes, i.e. *geranylgeranyl diphosphate synthase* (*crtE*), *phytoene desaturase* (*crtI*) and *lycopene β-cyclase* (*crtY*) under *napin* promoter. These combinations of different genes along with *crtB* resulted in an overall increase of 857 μg/g FW in canola (Ravanello et al. 2003). In canola seeds, the silencing of *lcyE* gene using RNA interference (RNAi) construct resulted in increased β-carotene up to 91 μg/g FW (Yu et al. 2008). It was also noted that total carotenoid level increased by 42-fold relative to non-transformed control line (Yu et al. 2008). Further attempt to increase carotenoid was done with simultaneous expression of the seven genes, i.e. *β-carotene ketolase* (*crtW*) from *Brevundimonas* spp., *isopentenyl pyrophosphate isomerise* (*idi*) from *Paracoccus* spp., *geranylgeranyl pyrophosphate synthase* (*crtE*) from *Pantoea ananatis*, *β-carotene hydroxylase* (*crtZ*), *crtB*, *crtI* and *crtY* in canola. All genes were expressed under the control of *napin* and *CaMV35S* promoters. This resulted into enhanced β-carotene level, i.e. 214 μg/g FW, and increased total carotenoids by 30-fold compared to the wild type (Fujisawa et al. 2009).

Potato was transformed with *crtB* from *Erwinia uredovora* under the tuber-specific *patatin* promoter. The β-carotene increased up to 10.30 μg/g DW, and total carotenoid improved up to 6.3-fold in potato tubers (Ducreux et al. 2005). The silencing of *lcyE* gene under the control of *patatin* promoter raised β-carotene content to 43.56 ng/g DW, and total carotenoid increased to 2.5-fold in the tubers (Diretto et al. 2006). The combination of *crtB*, *crtI* and *crtY* genes expressed under the *patatin* and *CaMV35S* promoters in potato improved β-carotene to 47.4 μg/g DW and total carotenoids by 20-fold (Diretto et al. 2006). In other study, *β-carotene hydroxylase* (*b-CHX*) gene was used to silence under the control of potato tuber-specific *granule-bound starch synthase* (*GBSS*) and *CaMV35S* promoters by using RNAi constructs. It enhanced β-carotene up to 331 μg/100 g FW in potato tuber (Van Eck et al. 2007). The orange cauliflower (*Or*) gene from *Brassica oleracea* expressed under the *GBSS* promoter produced 5.01 μg/g DW of β-carotene and a 5.7-fold increase in the total carotenoid in potato tubers (Lopez et al. 2008).

Maize was genetically engineered by overexpressing three bacterial genes, i.e. *crtB*, *crtI* and *ζ-carotene desaturase* (*zds*) under the control of endosperm-specific *γ-zein* promoter. It resulted in β-carotene accumulation up to 9.80 μg/g DW, and total carotenoid increased to 34-fold in the endosperm (Aluru et al. 2008). Five different sources derived carotenogenic genes, i.e. *psy1* from *Zea mays*, *crtI* from *Pantoea ananatis*, *lcyB* and *b-CHX* from *Gentiana lutea*, and *crtW* from *Paracoccus*, expressed under different promoters, i.e. wheat *glutenin*, barley *hordein*, rice *prolamin*, rice *glutelin-1*, and maize *γ-zein*, respectively. Maximum β-carotene accumulation was 57.35 μg/g DW, and 142-fold increase in total carotenoid content was reported in the endosperm (Zhu et al. 2008). In another study, enhanced β-carotene was obtained by expressing *psy1* from *Zea mays* and *crtI* gene from *Pantoea ananatis* (formerly *Erwinia uredovora*) under the control of wheat *glutenin* and barley d-*hordein* promoters, respectively. The β-carotene content raised up to 59.32 μg/g DW, and total carotenoid amplified to 112-fold in the best transgenic line (Naqvi et al. 2009).

Carrot was also genetically engineered by expressing *ketolase* gene under the *CaMV35S* and *ubiquitin* promoters which resulted in 39 μg/g FW β-carotene accumulation in roots (Jayaraj et al. 2008).

19.6.2 Fruit Crops

Fruit crops such as tomato, citrus and banana have been genetically engineered for PVA enrichment. Most studies were performed in tomato. *crtI* gene from *Erwinia uredovora* was introduced under the control of the *CaMV35S* promoter. It increased β-carotene up to 520 μg/g DW in tomato (Romer et al. 2000). *lcyB* gene from *Arabidopsis* expressed under the fruit-specific *phytoene desaturase* (*PDS*) promoter from tomato boosted β-carotene level up to 57 μg/g FW (Rosati et al. 2000). *lcyB* and *b-CHX* from pepper expressed under the control of the fruit-specific *PDS* promoter in tomato enhanced β-carotene up to 63 μg/g FW and total carotenoids to 1.7-fold (Dharmapuri et al. 2002). Overexpression of *crtB* from *Erwinia uredovora* with fruit-specific *polygalacturonase* promoter improved β-carotene up to 825 μg/g DW and total carotenoid to 4-fold relative to control in tomato (Fraser et al. 2002). Similar studies were reported by expressing *lcy B* and *cryptochrome* (*cry1*) genes with *CaMV35S* promoter which produced β-carotene up to 205 μg/g FW (D'Ambrosio et al. 2004) and 101 μg/g DW (Giliberto et al. 2005), respectively. Other two genes *3-hydroxymethylglutaryl CoA* (*hmgr*) from *Arabidopsis* under *CaMV35S* and *1-deoxy*-d-*xylulose-5-phosphate synthase* (*dxs*) from *Escherichia coli* with *CaMV35S* and *fibrillin* promoters were also expressed in tomato fruit. Overexpression of *dxs* improved β-carotene up to 42 μg/g FW with *fibrillin* promoter, but total carotenoids increased by only 1.6-fold in tomato fruits (Enfissi et al. 2005). Pepper *fibrillin* gene was introduced in tomato fruit under the control of its own *fibrillin* promoter, which enhanced β-carotene up to 150 μg/g FW and total carotenoid to 2-fold in comparison to non-transgenic line (Simkin et al. 2007). *crtY* from *Narcissus pseudonarcissus* expressed with rRNA operon promoter in the tomato plastid genome increased β-carotene up to 95 μg/g FW in tomato fruits (Apel and Bock 2009). *SlNCED1* gene was silenced using RNAi with fruit-specific *E8* promoter, which improved β-carotene up to 40 μg/g FW and total carotenoid by 2-fold (Sun et al. 2012).

In citrus fruit also β-carotene level increased by metabolic engineering. In the transgenic fruit, *b-CHX* gene from *Citrus sinensis* was placed under *CaMV35S* promoter, and best-performing transgenic lines added β-carotene up to 114 ng/g FW (Pons et al. 2014).

19.7 Biofortification of Banana for Provitamin A Enrichment

Millions of people in the developing countries are suffering from VAD where banana is consumed as a staple crop by the large population. Banana is the tropical and subtropical fruit crop and cultivated in more than 130 countries (De et al. 2009). Banana can be an ideal crop for micronutrient biofortification for developing countries. India is the largest producer with an annual production of nearly 27.57 million tonnes and contributes 25.84 % to the global banana production (FAO STAT 2013). The fruit is available throughout the year at reasonable prices in many developing countries and is the most suitable and choicest fruit in the diets of all age groups. The postharvest life of a banana fruit is fairly good to allow its easy transportation and distribution at the national level. The pulp of banana fruit remains unexposed to light, hence allowing minimal degradation of carotenoids in the edible pulp. This is particularly advantageous as compared to increasing carotenoids like PVA in seeds where carotenoids degrade during storage. The narrow genetic resources for enhanced PVA limit the scope for improving the existing cultivars of banana. Most of the cultivated varieties are triploid in nature, hence sterile, and provide a natural barrier for cross pollination. Therefore, the concerns regarding the effect on existing diversity of banana and transgene outflow are minimal. The genetically modified bananas can be grown alongside non-GM bananas in the same field and will not aggravate heritable mixing in the nature. The cultivated varieties of banana have low PVA, and the biofortification of banana can lead to increased PVA content in the diet. Application of biotechnology for the β-carotene biofortification in widely cultivated varieties would be an ideal approach. In this instance, the Bill and Melinda Gates Foundation funded Banana21 project which is led by Queensland University of Technology (QUT), Australia, with the National Agricultural Research Organization (NARO), Uganda, as the primary developing country partner. The aim of the Banana21 project is to alleviate VAD in Uganda through the development, distribution and consumption of PVA-enriched biofortified bananas. Bananas with up to 20 μg/g of PVA have been developed, and field trials have commenced (Namanya 2011). The Banana21 targets levels of a 4-fold increase in β-carotene for East African Highland bananas in Uganda where there are very high levels of banana consumption, approaching 1 kg per person per day. This is not the situation in many developing countries. Therefore, target levels would need to be much higher. In 2012, the Biotechnology Industry Research Assistance Council (BIRAC), Government of India, has initiated a network project with five collaborating Indian partner institutes and QUT group. The objective of the study is to utilize the experience and achievements of QUT for the improvement, development, validation and transfer of specific traits in selected Indian banana varieties. The success would be remarkable when the technology can be translated into a product in the form of PVA-rich biofortified banana and also achieved RDA for the large population. The successful genetic engineering for PVA enrichment in banana requires identification of one or more rate-limiting genes and selection of promoter/s that would express the genes at a high level in target tissue. The general steps for the development of PVA-enriched biofortified banana are summarized in Fig. 19.3. The effort needs to be devoted for multidisciplinary direction to cope with the challenges of food security and PVA micronutrient malnutrition. We are here summarizing advanced biotechnological approaches that include transgenic, cisgenic and precise genome engineering and their application in the development of genetically improved banana. Some of these advancements may open the wide horizon to implement further for the nutritional improvement of banana.

19.7.1 Application of Transgenic in Banana

Transgenic approach is the introduction of heterologous DNA, i.e. one or several new genes or regulatory element into the plant by different methods. It can be used to alleviate malnutrition by manipulating the relevant metabolic pathway. Different transgenic approaches have been used for the genetic transformation of banana such as electroporation of protoplasts (Sagi et al. 1994), co-cultivating wounded meristem (Tripathi et al. 2008), particle bombardment (Becker et al. 2000) and *Agrobacterium*-mediated transformation (Ganapathi et al. 2001; Khanna et al. 2004; Ghosh et al. 2009). *Agrobacterium*-mediated transformation is the preferred choice because it governs high transformation efficiency and low copy number insertion of transgene into the genome. Banana transformation had been established with constitutive *CaMV35S* promoter (Becker et al. 2000; Khanna et al. 2004; Ghosh

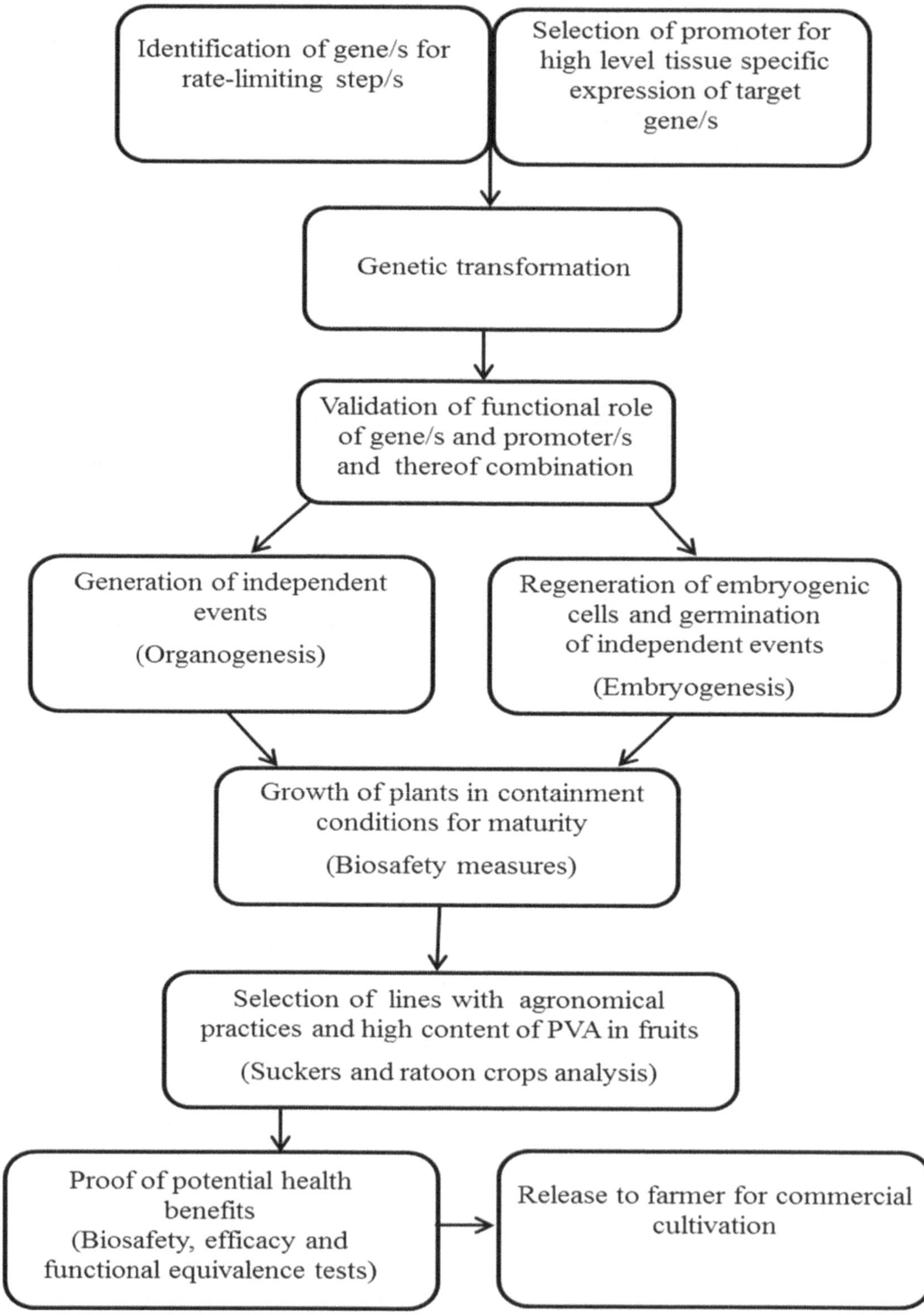

Fig. 19.3 Steps in the production of provitamin A (PVA) biofortified banana plants

et al. 2009) and maize *polyubiqutin* (*Ubi*) promoter (Becker and Dale 2004; Khanna et al. 2007). Banana *actin* promoter (*ACT1*), a sequence of *banana streak badnavirus* (*BSV*) and *banana bunchy top virus* (*BBTV*), has also been analysed for reporter gene expression in banana (Becker et al. 2000; Schenk et al. 2001). Development of transgenic banana for trait-specific manner could be a useful approach. However, some of the unforeseen consequences such as multiple regulatory constraints, ethical issues and the threat of long-term effect of transgenic food on the human body may affect public acceptance and commercial applications. Therefore, some of the recent emerging technologies such as cisgenic and precise genome engineering could be the possible solutions for these limitations of transgenic approach.

19.7.2 Application of Cisgenic in Banana

The concept of cisgenesis was introduced with the expectation that cisgenic crops will be more acceptable to the public and will overcome the concern of unnaturalness. In this approach, genetic material transferred to plant should originate from the plant itself or from closely related species capable of sexual hybridization. Cisgene should be a complete copy of the endogenous gene including promoter, introns and the terminator in the normal sense orientation. In banana, cisgenic approach could be used to improve PVA content by using a range of banana-derived promoters and genes. Using a reporter gene, the function of banana-derived promoters like the *expansin1* (*MaExp1*), *expansin4* (*MaExp4*), *extensin* (*MaExt*), *aminocyclopropane-1-carboxylate oxidase ACO* (*MaACO*), *metallothionein* (*MaMT2a*) and *phytoene synthase* (*APsy2a*) from banana variety Asupina was transiently analysed in an intact banana fruit (Namanya 2011). Homologs of *psy* gene including *psy1*, *psy2* and *psy3* were isolated from high β-carotene containing banana variety Asupina and was characterized. The gene *Psy2a* derived from Asupina (*APsy2a*) was transferred into cooking banana for the improvement of PVA content of the fruit (Namanya 2011). Expansin-related four genes (*MaExpA2*, *MaExpA3*, *MaExpA4*, *MaExpA5*) from fruit of Cavendish and Robusta varieties of banana was reported to express differentially during fruit development and ripening stages (Trivedi and Nath 2004; Asha et al. 2007). Three homologs of metallothioneins (*MT2A*, *MT2B and MT3*) were also expressed in banana fruit (Liu et al. 2002). Cisgenic approach will overcome limitations mostly related to ethical and regulatory constraints. Cisgenesis can also reduce the high cost and time consumed with regulatory process and approval of transgenic plants.

19.7.3 Application of Precise Genome Engineering in Banana

In precise genome engineering, editing of genome is done using site-specific integration and deletion or mutation of interested gene. Editing is carried out at specific site by creating double-strand breaks (DSBs) with the help of site-specific recombinases or nucleases such as zinc-finger nucleases (ZFNs) or transcription activator-like effector nucleases (TALENs). DSBs are then repaired by homologous recombination or by error-prone non-homologous end joining (NHEJ) (Liu et al. 2013). Consequently precise genetic engineering approach allows nutritional traits to be targeted to specific organs (e.g. seeds and fruits) and combination of multiple traits in the same plants without complex breeding programmes. Moreover, this approach unrestricted the access to genetic diversity (Zhu et al. 2007; Naqvi et al. 2009). Genome engineering can be used to enhance PVA content by providing flux in the β-carotene biosynthesis pathway by increasing carotenoid precursor availability, expressing enzymes involved in the early part of the pathway between GGPP and lycopene. Increase in metabolic flux in the β-branch by inhibiting *lycE* activity favours increase in carotenoid accumulation creating carotenoids sink to remove feedback inhibitions

(Farre et al. 2010; Bai et al. 2011). Overexpression of *dxs* in tomato is an example of the first strategy for raising total carotenoids by increasing flux through the pathway (Enfissi et al. 2005). Another useful approach is RNA interference (RNAi), in which short double-stranded RNA molecules provide defence pathway by destroying homologous mRNAs which leads to post-transcriptional gene silencing (Lawrence and Pikaard 2003; Mansoor et al. 2006). Silencing of *β-carotene hydroxylase* or (*b-CHX*) gene was the first report of RNAi approach in citrus fruit. Silencing of *b-CHX* gene involved in the conversion of β-carotene into xanthophylls significantly enhanced β-carotene content up to 36-fold in the pulp of sweet orange (Pons et al. 2014). More recently, ZFNs and TALENs are designed to facilitate genome editing at specific genomic location. ZFNs are fusions of engineered zinc-finger arrays (consist of 3–6 C_2H_2 fingers) to a non-specific DNA-cleavage domain of the *FokI* endonuclease. ZFNs are 6–40 bp in length. Latest report of ZFNs use was in soybean genome where the gene involved in RNA silencing was mutated and resulted in efficient and heritable mutagenesis in the subsequent generation (Curtin et al. 2011). TALENs have also rapidly been used in plant biotechnology. These are fusions of truncated TALEs (containing an amino terminus, a custom-designed DNA-binding domain and a carboxyl terminus with removed activation domain) to a non-specific DNA-cleavage domain of the *FokI* endonuclease. TALENs have been used to knockout four genes in rice and eight genes in *Brachypodium* (Shan et al. 2013). The latest tools in the genome editing are clustered regularly interspaced short palindromic repeats (CRISPRs). CRISPRs are loci containing multiple short direct repeats which are incorporated with short segments of foreign DNA (called spacers) in some bacteria and archaea. When spacers cut and bind to short CRISPR RNAs, CRISPR-associated (Cas) proteins attack sequence of infecting viruses. Co-expression of Cas9 has been used to target the *pds3* gene in *Arabidopsis thaliana* protoplasts or leaves using *Agrobacterium* infection (Li et al. 2013). Among all above discussed genome editing tools, CRISPR-Cas might be the most efficient tool for plant biotechnology. Hence, the use of modern biotechnological approaches could significantly contribute in the nutritional improvement in banana.

19.8 Conclusions and Future Prospects

The world's population at over seven billion (http://esa.un.org) and majority of the poor need affordable and sustainable technological solutions to health. Protection from micronutrient deficiencies is currently one of the most difficult areas to address through supplementation and altered dietary components. Recent progress in plant biotechnology which includes transgenic, cisgenic and precise genome engineering approaches has opened novel opportunities for the safe and cost-effective production of micronutrient-enriched biofortified plants. Several examples cited in this chapter have mentioned that the production of biofortified plants for PVA has a great potential. The development of PVA-enriched biofortified banana would significantly increase its utility of producing cost-effective sustainable solution. However, there are several technical and biological issues which need to be addressed. Some of these are as follows:

(a) The precise knowledge about the structural and regulatory genes involved in biosynthetic pathway of β-carotene is prerequisite for the employment of efficient genetic engineering which is not reported in banana. Further, the selection of efficient promoters that target the expression of pathway enzyme/s in specific plant tissues is required. The selection of fruit tissue-specific promoters, signal sequences, plant-preferred codon usage, stabilization, post-translational modification, etc. can enhance the activity of target enzymes which can lead to increasing accumulation of target metabolite in a tissue-specific manner.

(b) Enhancing the stability of β-carotene in biofortified fruit of banana can give distinct advantage of such technology. Most of the cultivated bananas are triploid in nature and vegetatively propagated, providing simpler natural containment. However, physical containment can provide additional tight regulation to avoid passage of transgene into unwarranted living organism or consumer food stocks.
(c) Standardization of biofortified banana diet to achieve RDA is a major problem as plant to plant and generation to generation content of β-carotene may vary. The expression also depends on where and when the plants are grown. Therefore, a delivery scheme needs to be developed to ensure the delivery of required diet level. Collaborative research between plant and medical scientists may help in resolving these and related issues.
(d) The biofortified transgenic banana should have the same or improve agronomical characteristics as compared to the normal plants.
(e) Issues related to the ethical, social, biosafety and environmental impact which directly/indirectly effect the deployment of genetically modified crops are a concern. However, these will be resolved in the course of time, through enhancing the awareness of merits and the development of regulatory and standard operating procedures.

The need to establish environmental safety, efficacy and functional equivalence of the biofortified crops should guide future development and outcome of the research. However, the deployment of PVA-enriched biofortified banana has the potential to be successful to complement the routine diet to achieve RDA. It could also become a reality when biofortified bananas get incorporated in the midday meal programme in countries like India to achieve nutritional security to school-going children.

Acknowledgements Authors express their gratitude to National Agri-Food Biotechnology Institute (NABI), Department of Biotechnology (Government of India) and Biotechnology Industry Research Assistance Council (BIRAC) (A Government of India Enterprise) for supporting the research work on provitamin A biofortification in Indian banana. NK and S are thankful to Department of Biotechnology, Panjab University, Chandigarh for Ph.D. support and facility.

References

Aluru M, Xu Y, Guo R, Wang Z, Li S (2008) Generation of transgenic maize with enhanced provitamin A content. J Exp Bot 59:3551–3562. doi:10.1093/jxb/ern212

Apel W, Bock R (2009) Enhancement of carotenoid biosynthesis in transplastomic tomatoes by induced lycopene-to-provitamin A conversion. Plant Physiol 151:59–66. doi:10.1104/pp.109.140533

Asha, Sane VA, Sane AP, Nath P (2007) Multiple forms of α-expansin genes are expressed during banana fruit ripening and development. Postharvest Biol Technol 45:184–192. doi:10.1016/j.postharvbio.2007.03.003

Awasthi S, Peto R, Read S, Clark S, Pande V, Bundy D (2013) Vitamin A supplementation every 6 months with retinol in 1 million pre-school children in north India: DEVTA, a cluster-randomised trial. Lancet 381:1469–1477

Bai C, Twyman RM, Farre G, Sanahuja G, Christou P, Capell T, Zhu C (2011) A golden era—pro-vitamin A enhancement in diverse crops. In Vitro Cell Dev Biol Plant 47:205–221

Bailey RL, West KP Jr, Black RE (2015) The epidemiology of global micronutrient deficiencies. Ann Nutr Metab 66:22–33. doi:10.1159/000371618

Bashir K, Takahashi R, Nakanishi H, Nishizawa NK (2013) The road to micronutrient biofortification of rice: progress and prospects. Front Plant Sci 4:15. doi:10.3389/fpls.2013.00015

Becker DK, Dale JL (2004) Transformation of banana using microprojectile bombardment. In: Curtis IS (ed) Transgenic crops of the worldessential protocol. Kluwer Academic Publishers, Dordrecht, pp 131–143

Becker DK, Dugdale B, Smith MK, Harding RM, Dale JL (2000) Genetic transformation of Cavendish banana (*Musa* spp. AAA group) cv 'Grand Nain' via microprojectile bombardment. Plant Cell Rep 19:229–234. doi:10.1007/s002990050004

Bengtsson A, Larsson Alminger M, Svanberg U (2009) In vitro bioaccessibility of β-carotene from heat-processed orange-fleshed sweet potato. J Agric Food Chem 57:9693–9698

Biesalski HK, Chichili GR, Frank J, Von Lintig J, Nohr D (2007) Conversion of β-carotene to retinal pigment. Vitam Horm 75:117–130

Black RE, Allen LH, Bhutta ZA et al (2008) Maternal and child undernutrition: global and regional exposures and health consequences. Lancet 371:243–260. doi:10.1016/S0140-6736(07)61690-0

Black RE, Singhal A, Uaur R (2013) International nutrition: achieving millennium goals and beyond.78th

Nestlé Nutrition Institute Workshop, Muscat, Mar 2013
Blomhoff R, Blomhoff HK (2006) Overview of retinoid metabolism and function. J Neurobiol 66:606–630. doi:10.1002/neu.20242
Bouis EH, Welch MR (2010) Biofortification -a sustainable agricultural strategy for reducing micronutrient malnutrition in the global south. Crop Sci 50:S20–S32. doi:10.2135/cropsci2009.09.0531
Britton JR, Pavord ID, Richards KA, Knox AJ, Wisniewski AF, Lewis SA, Weiss ST (1995) Dietary antioxidant vitamin intake and lung function in the general population. Am J Respir Crit Care Med 151:1383–1387
Brown CC, Noelle RJ (2015) Seeing through the dark: new insights into the immune regulatory functions of vitamin A. Eur J Immunol 45:1287–1295. doi:10.1002/eji.201344398
Burri BJ (2011) Evaluating sweet potato as an intervention food to prevent vitamin A deficiency. Compr Rev Food Sci F 10:118–130
Cassani B, Villablanca EJ, De Calisto J, Wang S, Mora JR (2012) Vitamin A and immune regulation: role of retinoic acid in gut-associated dendritic cell education, immune protection and tolerance. Mol Aspects Med 33:63–76. doi:10.1016/j.mam.2011.11.001
Cazzonelli CI, Pogson BJ (2010) Source to sink: regulation of carotenoid biosynthesis in plants. Trends Plant Sci 15:266–274
Cong L, Wang C, Chen L, Liu H, Yang G, He G (2009) Expression of phytoene synthase1 and carotene desaturase crtI genes result in an increase in the total carotenoids content in transgenic elite wheat (*Triticum aestivum* L.). J Agric Food Chem 57:8652–8660. doi:10.1021/jf9012218
Copenhagen Consensus (2008) http://www.copenhagenconsensus.com/copenhagen-consensus-ii/outcomes
Cottrell RC (1991) Introduction: nutritional aspects of palm oil. Am J Clin Nutr 53:989S–1009S
Cunningham TJ, Duester G (2015) Mechanisms of retinoic acid signalling and its roles in organ and limb development. Nat Rev Mol Cell Biol 16:110–123. doi:10.1038/nrm3932
Curtin SJ, Zhang F, Sander JD, Haun WJ, Starker C, Baltes NJ, Stupar RM (2011) Targeted mutagenesis of duplicated genes in soybean with zinc-finger nucleases. Plant Physiol 156:466–473
D'Ambrosio C, Giorio G, Marino I, Merendino A, Petrozza A, Salfi L, Stigliani AL, Cellini F (2004) Virtually complete conversion of lycopene into β-carotene in fruits of tomato plants transformed with the tomato lycopene β-cyclase (tlcy-b) cDNA. Plant Sci 166:207–214. doi:10.1016/j.plantsci.2003.09.015
de Pee S, West CE (1996) Dietary carotenoids and their role in combating vitamin A deficiency: a review of the literature. Eur J Clin Nutr 50:S38–S53
de Pee S, West CE, Permaesih D, Martuti S, Hautvast JG (1998) Orange fruit is more effective than are dark-green, leafy vegetables in increasing serum concentrations of retinol and beta-carotene in schoolchildren in Indonesia. Am J Clin Nutr 68:1058–1067
De LE, Vrydaghs L, De Maret P, Xavier P, Denham T (2009) Why bananas matter: an introduction to the history of banana domestication. Ethnobot Res Appl 7:165–177
Degerud EM, Manger MS, Strand TA, Dierkes J (2015) Bioavailability of iron, vitamin A, zinc, and folic acid when added to condiments and seasonings. Ann N Y Acad Sci. doi:10.1111/nyas.12947
Dharmapuri S, Rosati C, Pallara P, Pallaraa P, Aquilani R, Bouvierb F, Camarab B, Giulianoa G (2002) Metabolic engineering of xanthophyll content in tomato fruits. FEBS Lett 519:30–34. doi:10.1016/S0014-5793(02)02699-6
Diretto G, Tavazza R, Welsch R, Pizzichini D, Mourgues F, Papacchioli V, Beyer P, Giuliano G (2006) Metabolic engineering of potato tuber carotenoids through tuber-specific silencing of lycopene ε-cyclase. BMC Plant Biol 6:13. doi:10.1038/76523
Diretto G, Al-Babili S, Tavazza R, Papacchioli V, Beyer P, Giuliano G (2007) Metabolic engineering of potato carotenoid content through tuber-specific overexpression of a bacterial mini pathway. PLoS One 2:e350
Ducreux LJM, Morris WL, Hedley PE, Shepherd T, Davies HV, Millam S, Taylor M (2005) Metabolic engineering of high carotenoid potato tubers containing enhanced levels of β-carotene and lutein. J Exp Bot 56:81–89. doi:10.1093/jxb/eri016
Enfissi EMA, Fraser PD, Lois LM, Boronat A, Schuch W, Bramley PM (2005) Metabolic engineering of the mevalonate and non-mevalonate isopentenyl diphosphate-forming pathways for the production of health-promoting isoprenoids in tomato. Plant Biotechnol J 3:17–27. doi:10.1111/j.1467-7652.2004.00091.x
Failla ML, Thakkar SK, Kim JY (2009) In vitro bioaccessibility of β-carotene in orange fleshed sweet potato (*Ipomoea batatas*, Lam.). J Agric Food Chem 57:10922–10927
FAO STAT (2013) Food and Agriculture Organization of the United Nations statistics division. http://faostat3.fao.org
FAO (2015) The state of food insecurity in the world 2015. Meeting the 2015 international hunger targets: taking stock of uneven progress. FAO, Rome. http://www.fao.org/3/a-i4646e.pdf
Farre G, Sanahuja G, Naqvi S, Bai C, Capell T, Zhu C, Christou P (2010) Travel advice on the road to carotenoids in plants. Plant Sci 179:28–48. doi:10.1016/j.plantsci.2010.03.009
Fitzpatrick TB, Basset GJ, Borel P, Carrari F, Della Penna D, Fraser PD, Hellmann H, Osorio S, Rothan C, Valpuesta V, Caris-Veyrat C, Fernie AR (2012) Vitamin deficiencies in humans: can plant science help? Plant Cell 24:395–414
Flores-Pérez Ú, Sauret-Güeto S, Gas E, Jarvis P, Rodríguez-Concepción M (2008) A mutant impaired in the production of plastome-encoded proteins uncovers a mechanism for the homeostasis of isoprenoid biosynthetic enzymes in Arabidopsis plastids. Plant Cell 20:1303–1315

Fraser PD, Romer S, Shipton CA et al (2002) Evaluation of transgenic tomato plants expressing an additional phytoene synthase in a fruit-specific manner. Proc Natl Acad Sci U S A 99:1092–1097. doi:10.1073/pnas.241374598

Fu X, Feng C, Wang C, Yin X, Lu P, Grierson D, Chen K (2014) Involvement of multiple phytoene synthase genes in tissue-and cultivar-specific accumulation of carotenoids in loquat. J Exp Bot 65:4679–4689. doi:10.1093/jxb/eru257

Fujisawa M, Takita E, Harada H, Sakurai N, Suzuki H, Ohyama K, Shibata D, Misawa N (2009) Pathway engineering of *Brassica napus* seeds using multiple key enzyme genes involved in ketocarotenoid formation. J Exp Bot 60:1319–1332. doi:10.1093/jxb/erp006

Ganapathi TR, Srinivas L, Suprasanna P, Bapat VA (2001) Regeneration of plants from alginate-encapsulated somatic embryos of banana CV. Rasthali (*Musa* spp. AAB group). In Vitro Cell Dev Biol-Plant 37:178–181. doi:10.1079IVP2000117

Ghosh A, Ganapathi TR, Nath P, Bapat VA (2009) Establishment of embryogenic cell suspension cultures and Agrobacterium-mediated transformation in an important Cavendish banana cv. Robusta (AAA). Plant Cell Tissue Organ Cult 97:131–139. doi:10.1007/s11240-009-9507-0

Giliberto L, Perrotta G, Pallara P, Weller JL, Fraser PD, Bramley PM, Fiore A, Tavazza M, Giuliano G (2005) Manipulation of the blue light photoreceptor cryptochrome 2 in tomato affects vegetative development, flowering time, and fruit antioxidant content. Plant Physiol 137:199–208. doi:10.1104/pp.104.051987

Graham RD, Welch RM, Bouis HE (2001) Addressing micronutrient malnutrition through enhancing the nutritional quality of staple foods: principles, perspectives and knowledge gaps. Adv Agron 70:77–142. doi:10.1016/S0065-2113(01)70004-1

Graham RD, Welch RM, Saunders DA et al (2007) Nutritious subsistence food systems. Adv Agron 92:1–74. doi:10.1016/S0065-2113(04)92001-9

Grassi S, Piro G, Lee JM, Zheng Y, Fei Z, Dalessandro G, Lenucci MS (2013) Comparative genomics reveals candidate carotenoid pathway regulators of ripening watermelon fruit. BMC Genomics 14:781

Gropper SS, Smith JL, Groff JL (2009) Advanced nutrition and human metabolism, 5th edn. Wadsworth Cengage Learning, Belmont

Hedren E, Mulokozi G, Svanberg U (2002) In vitro accessibility of carotenes from green leafy vegetables cooked with sunflower oil and red palm oil. Int J Food Sci Nutr 53:445–453

ICMR (2010) Nutrient requirements and recommended dietary allowances for Indians. A report of the expert group of the Indian Council of Medical Research. http://icmr.nic.in/final/rda-2010.pdf

INSA (2011) Micro-nutrient security for India–priorities for research and action. www.insaindia.org

James C (2014) Global status of commercialized biotech/GM crops. ISAAA Brief (Vol. 49), Ithaca, New York

Jayaraj J, Devlin R, Punja (2008) Metabollic engineering of novel ketocarotenoid production in carrot plants. Transgenic Res 17:489–501. doi:10.1007/s11248-007-9120-0

Kedishvili NY (2013) Enzymology of retinoic acid biosynthesis and degradation. J Lipid Res 54:1744–1760. doi:10.1194/jlr.R037028

Kennedy G, Nantel G, Shetty P (2003) The scourge of "hidden hunger": global dimensions of micronutrient deficiencies. Food Nutr Agric 32:8–16

Khanna H, Becker D, Kleidon J, Dale J (2004) Centrifugation Assisted Agrobacterium tumefaciens-mediated Transformation (CAAT) of embryogenic cell suspensions of banana (*Musa spp*. Cavendish AAA and Lady finger AAB). Mol Breed 14:239–252. doi:10.1023/B:MOLB.0000047771.34186.e8

Khanna HK, Paul J-Y, Harding RM, Dickman MB, Dale JL (2007) Inhibition of Agrobacterium-induced cell death by antiapoptotic gene expression leads to very high transformation efficiency of banana. Mol Plant Microbe Interact 20:1048–1054. doi:10.1094/MPMI-20-9-1048

Lawrence RJ, Pikaard CS (2003) Transgene-induced RNA interference: a strategy for overcoming gene redundancy in polyploids to generate loss-of-function mutations. Plant J 36:114–121. doi:10.1046/j.1365-313X.2003.01857.x

Leadership agenda (2010) Coalition for sustainable nutrition security for India. http://www.nutritioncoalition.in/publication.asp

Li JF, Norville JE, Aach J, McCormack M, Zhang D, Bush J, Sheen J (2013) Multiplex and homologous recombination mediated genome editing in Arabidopsis and *Nicotiana benthamiana* using guide RNA and Cas9. Nat Biotechnol 31:688–691

Lietz G, Lange J, Rimbach G (2010) Molecular and dietary regulation of β, β-carotene 15, 15′-monooxygenase 1 (BCMO1). Arch Biochem Biophys 502:8–16

Liu P, Goh C, Loh C, Pua E (2002) Differential expression and characterization of three metallothionein-like genes in Cavendish banana (*Musa acuminata*). Physiol Plant 114:241–250

Liu W, Yuan JS, Jr Stewart CN (2013) Advanced genetic tools for plant biotechnology. Nat Rev Genet 14:781–793

Liu JX, Chiou CY, Shen CH, Chen PJ, Liu YC, Jian CD, Yeh KW (2014) RNA interference-based gene silencing of phytoene synthase impairs growth, carotenoids, and plastid phenotype in oncidium hybrid orchid. SpringerPlus 3:478

Lopez AB, Van Eck J, Conlin BJ, Paolillo DJ, O'Neill J, Li L (2008) Effect of the cauliflower or transgene on carotenoid accumulation and chromoplast formation in transgenic potato tubers. J Exp Bot 59:213–223. doi:10.1093/jxb/erm299

Lyons GH, Stangoulis JCR, Graham RD (2004) Exploiting micronutrient interaction to optimize biofortification programs: the case for inclusion of selenium and iodine in the HarvestPlus program. Nutr Rev 62:247–252. doi:10.1111/j.1469-8137.2008.02738.x

Maass D, Arango J, Wüst F, Beyer P, Welsch R (2009) Carotenoid crystal formation in Arabidopsis and carrot roots caused by increased phytoene synthase protein levels. PLoS One 4, e6373. doi:10.1371/journal.pone.0006373

Mansoor S, Amin I, Hussain M, Hussain M, Zafar Y, Briddon RW (2006) Engineering novel traits in plants through RNA interference. Trends Plant Sci 11:559–565. doi:10.1016/j.tplants.2006.09.010

Mark M, Ghyselinck NB, Chambon P (2009) Function of retinoic acid receptors during embryonic development. Nucl Recept Signal 7:e002. doi:10.1621/nrs.07002

Murgial I, Gara LD, Grusak MA (2013) Biofortification: how can we exploit plant science and biotechnology to reduce micronutrient deficiencies. Front Plant Sci 4:429. doi:10.3389/fpls.2013.00429

Namanya P (2011) Towards the biofortification of banana fruit for enhanced micronutrient content. Ph.D. thesis. Queensland University of Technology. http://eprints.qut.edu.au/45785/1/Priver_Namanya_Thesis.pdf

Naqvi S, Zhu C, Farre G et al (2009) Transgenic multivitamin corn through biofortification of endosperm with three vitamins representing three distinct metabolic pathways. Proc Natl Acad Sci U S A 106:7762–7767. doi:10.1073/pnas.0901412106

Nilson A, Piza J (1998) Food fortification: a tool for fighting hidden hunger. Food Nutr Bull 19:49–60

O'Connell OF, Ryan L, O'Brien NM (2007) Xanthophyll carotenoids are more bioaccessible from fruits than dark green vegetables. Nutr Res 27:258–264

Paine J, Shipton C, Chaggar S, Howells R, Kennedy M, Vernon G, Wright S, Hinchliffe E, Adams J, Silverstone A, Drake R (2005) Improving the nutritional value of Golden Rice through increased provitamin A content. Nat Biotechnol 23:482–487. doi:10.1038/nbt1082

Palczewski K (2010) Retinoids for treatment of retinal diseases. Trends Pharmacol Sci 31:284–295. doi:10.1016/j.tips.2010.03.001

Pérez-Massot E, Banakar R, Gómez-Galera S et al (2013) The contribution of transgenic plants to better health through improved nutrition: opportunities and constraints. Genes Nutr 8:29–41. doi:10.1007/s12263-012-0315-5

Pons E, Alquézar B, Rodríguez A, Martorell P, Genoves S, Ramon D, Rodrigo MJ, Zacar L, Pena L (2014) Metabolic engineering of β-carotene in orange fruit increases its in vivo antioxidant properties. Plant Biotechnol J 12:17–27. doi:10.1111/pbi.12112

Pullakhandam R, Failla ML (2007) Micellarization and intestinal cell uptake of beta-carotene and lutein from drumstick (*Moringa oleifera*) leaves. J Med Food 10:252–257

Ravanello MP, Ke D, Alvarez J, Huang B, Shewmaker CK (2003) Coordinate expression of multiple bacterial carotenoid genes in canola leading to altered carotenoid production. Metab Eng 5:255–263

Rodriguez-Concepcion M (2010) Supply of precursors for carotenoid biosynthesis in plants. Arch Biochem Biophys 504:118–122

Romer S, Fraser PD, Kiano JW, Shipton CA, Misawa N, Schuch W, Bramley PM (2000) Elevation of the provitamin A content of transgenic tomato plants. Nat Biotechnol 18:666–669

Rosati C, Aquilani R, Dharmapuri S et al (2000) Metabolic engineering of beta-carotene and lycopene content in tomato fruit. Plant J 24:413–419. doi:10.1046/j.1365-313X.2000.00880.x

Ruiz-Sola MÁ, Rodríguez-Concepción M (2012) Carotenoid biosynthesis in Arabidopsis: a colorful pathway. Arabidopsis Book 10:e0158

Sagi L, Remy S, Panis B, Swennen R, Volckaert G (1994) Transient gene expression in electroporated banana *Musa* spp, cv. 'Bluggoe', ABB group protoplasts isolated from regenerable embryogenic cell suspensions. Plant Cell Rep 13:262–266

Sandmann G, Romer S, Fraser PD (2006) Understanding carotenoid metabolism as a necessity for genetic engineering of crop plants. Metab Eng 8:291–302

Schenk PM, Remans T, Sági L, Adrian L, Elliott R, Dietzgen RG, Swennen R, Ebert PR, Grof CPL, Manners JM (2001) Promoters for pregenomic RNA of banana streak badnavirus are active for transgene expression in monocot and dicot plants. Plant Mol Biol 47:399–412. doi:10.1023/A:1011680008868

Schweiggert RM, Carle R (2015) Carotenoid deposition in plant and animal foods and its impact on bioavailability.Crit Rev Food Sci Nutr. doi:10.1080/10408398.2015.1012756

Shan Q, Wang Y, Chen K, Liang Z, Li J, Zhang Y, Gao C (2013) Rapid and efficient gene modification in rice and Brachypodium using TALENs. Mol Plant 6:1365–1368

Shewmaker C, Sheehy J, Daley M et al (1999) Seed-specific overexpression of phytoene synthase: increase in carotenoids and other metabolic effects. Plant J20:401–412X.doi:10.1046/j.1365-313X.1999.00611.x

Simkin AJ, Gaffé J, Alcaraz JP et al (2007) Fibrillin influence on plastid ultrastructure and pigment content in tomato fruit. Phytochemistry 68:1545–1556. doi:10.1016/j.phytochem.2007.03.014

Srinivasan VS (2001) Bioavailability of nutrients: a practical approach to in vitro demonstration of the availability of nutrients in multivitamin-mineral combination products. J Nutr 131:1349S–1350S

Sun L, Yuan B, Zhang M, Wang L, Cui M, Wang Q, Leng P (2012) Fruit-specific RNAi-mediated suppression of SlNCED1 increases both lycopene and β-carotene contents in tomato fruit. J Exp Bot 63:3097–3108. doi:10.1093/jxb/ers026

Tripathi L, Tripathi JN, Tushemereirwe WK (2008) Rapid and efficient production of transgenic East African

Highland Banana (Musa spp.) using intercalary meristematic tissues. Afr J Biotechnol 7:1438–1445. doi:10.4314/ajb.v7i10.58691
Trivedi PK, Nath P (2004) MaExp1, an ethylene-induced expansin from ripening banana fruit. Plant Sci 167:1351–1358. doi:10.1016/j.plantsci.2004.07.005
Van Eck J, Conlin B, Garvin DF (2007) Enhancing beta-carotene content in potato by RNAi-mediated silencing of the beta-carotene hydroxylase gene. Am J Potato Res 84:331–342. doi:10.1007/BF02986245
Van Lieshout M, West CE, Van Breemen RB (2003) Isotopic tracer techniques for studying the bioavailability and bioefficacy of dietary carotenoids, particularly beta-carotene, in humans: a review. Am J Clin Nutr 77:12
Veda S, Kamath A, Platel K, Begum K, Srinivasan K (2006) Determination of bioaccessibility of β-carotene in vegetables by in vitro methods. Mol Nutr Food Res 50:1047–1052. doi:10.1002/mnfr.200600076
von Lintig J (2012) Metabolism of carotenoids and retinoids related to vision. J Biol Chem 287:1627–1634. doi:10.1074/jbc.R111.303990
Wei S, Yu B, Gruber MY, Khachatourians GG, Hegedus DD, Hannoufa A (2010) Enhanced seed carotenoid levels and branching in transgenic Brassica napus expressing the Arabidopsis miR156b gene. J Agric Food Chem 58:9572–9578. doi:10.1021/jf102635f
Welch RM, Graham RD (1999) A new paradigm for world agriculture: meeting human needs: productive, sustainable, nutritious. Field Crops Res 60:1–10
Welch RM, Graham RD (2002) Breeding crops for enhanced micronutrient content. Plant Soil 245:205–214
Welch RM, Graham RD (2004) Breeding for micronutrients in staple food crops from a human nutrition perspective. J Exp Bot 55:353–364. doi:10.1093/jxb/erh064
Welch RM, Graham RD (2005) Agriculture: the real nexus for enhancing bioavailable micronutrients in food crops. J Trace Elem Med Biol 18(4):299–307
West KP (2002) Extent of vitamin A deficiency among preschool children and women of reproductive age. J Nutr 132:2857S–2866S
West KP Jr, Gernand A, Sommer A (2007) Vitamin A in nutritional anemia. In: Kraemer K, Zimmermann MB (eds) Nutritional anemia. Sight and Life Press, Basel
WHO (2009) Global health risks: mortality and burden of disease attributable to selected major risks. http://www.who.int/healthinfo/global_burden_disease/GlobalHealthRisks_report_full.pdf
WHO (2014) Xerophthalmia and night blindness for the assessment of clinical vitamin A deficiency in individuals and populations. WHO/NMH/NHD/EPG/14.4
Wolgemuth DJ, Chung SSW (2007) Retinoid signaling during spermatogenesis as revealed by genetic and metabolic manipulations of retinoic acid receptor alpha. Soc Reprod Fertil Suppl 63:11–23. http://www.ncbi.nlm.nih.gov/pmc/articles/PMC3796155/
Ye X, Al-Babili S, Kloti A, Zhang J, Lucca P, Beyer P, Potrykus I (2000) Engineering the pro-vitamin A (β-carotene) biosynthetic pathway into (carotenoid-free) rice endosperm. Science 287:303–305. doi:10.1126/science.287.5451.303
Yu B, Lydiate DJ, Young LW et al (2008) Enhancing the carotenoid content of *Brassica napus* seeds by downregulating lycopene epsilon cyclase. Transgenic Res 17:573–585. doi:10.1007/s11248-007-9131-x
Zhao F-J, Shewry PR (2011) Recent developments in modifying crops and agronomic practice to improve human health. Food Policy 36:S94–S101
Zhu C, Naqvi S, Gomez-Galera S, Pelacho AM, Capell T, Christou P (2007) Transgenic strategies for the nutritional enhancement of plants. Trends Plant Sci 12:1360–1385. doi:10.1016/j.tplants.2007.09.007
Zhu C, Naqvi S, Breitenbach J, Sandmannb G, Christoua P, Capell T (2008) Combinatorial genetic transformation generates a library of metabolic phenotypes for the carotenoid pathway in maize. Proc Natl Acad Sci U S A 105:18232–18237. doi:10.1073/pnas.0809737105

Biofortification for Alleviating Iron Deficiency Anemia

20

Prashanti Patel, Karuna Yadav, and T.R. Ganapathi

Abstract

Iron deficiency anemia (IDA) is a global problem, with women and children of lower strata of society bearing the brunt of its effects. Several measures to rectify this exist, of which biofortification of food crops is increasingly gaining importance as it provides a safe and palatable way to supply adequate iron. Work on this aspect involves the physiology of iron uptake, translocation, storage, and redistribution in plants. While several food crops such as rice, wheat, and maize are already popular model plants, banana, a widely consumed fruit as well as infant food, is considered as a potential fruit crop to become "micronutrient enriched." Fortification of banana is more advantageous over other plants owing to its ploidy, parthenocarpic fruit development, its reach to the masses at large, and availability throughout the year. Bioavailability studies on the absorption of iron from banana cultivars give an indication on the time frame needed to achieve the desired levels of iron content in banana fruit. The recent advances made in understanding the mechanisms of iron uptake in humans, plants, and homeostasis along with biofortification of crop plants including banana for increasing the iron content as described in this article appear to alleviate IDA in the future.

Keywords

Iron deficiency anemia (IDA) • Biofortification • Banana • Transgenic • Iron acquisition

P. Patel • K. Yadav • T.R. Ganapathi (✉)
Plant Cell Culture Technology Section, Nuclear Agriculture and Biotechnology Division, Bhabha Atomic Research Centre,
Trombay, Mumbai 400 085, India
e-mail: trgana@barc.gov.in

S. Mohandas, K.V. Ravishankar (eds.), *Banana: Genomics and Transgenic Approaches for Genetic Improvement*, DOI 10.1007/978-981-10-1585-4_20

20.1 Introduction

20.1.1 Micronutrient Deficiencies: An Overview

Micronutrient deficiencies are the most prevalent nutritional deficiency disorders worldwide. As of 2015, around 795 million people (FAO 2015) around the world are suffering from chronic undernourishment according to FAO food security statistics. This stems from a lack of purchasing power (WHO 2007), unavailability of food, and unequal food distribution, the health effects of which are compounded by unsanitary conditions and inadequate care. The most vulnerable are women and children, particularly girls due to socioeconomic status and societal prejudices. In this scenario, while food security for every person remains to be achieved, a more ominous challenge is ensuring the quality of food. One of the parameters to measure the wholesomeness of food is the contribution by micronutrients. Micronutrients are essential components of the diet that are required in only trace amounts in an organism's physiology. However, the roles they play are far from minor, with several, e.g., manganese, copper, iron, and zinc, being important cofactors of diverse enzymes and involved in structural development (Fraga 2005).

Considering the very small quantities in which they are required, it is ironic that most of the manifested nutrient deficiencies in the world are that of micronutrients like iodine, vitamin A, and iron (WHO 2007; Stoltzfus 2003; Zimmermann and Hurrell 2007). In fact, one of the food security indicators as per FAO guidelines is the prevalence of anemia, iodine, and vitamin A deficiency in children, indicating their indispensability to human health. Vitamin A deficiency is a leading cause for xerophthalmia progressing to blindness and, in the advanced stages, ulceration and keratomalacia leading to loss of the eye (Sommer 2008; McLaren 1999). This deficiency is also associated with an increased severity and death risk in illnesses, collectively termed VADD (vitamin A deficiency disorders). Similarly, iodine deficiency presents a spectrum of iodine deficiency disorders with stillbirths, cretinism, and goitrous thyroid due to metabolic disorders (Hetzel 1983). The recent advances made in alleviating iron deficiency, the mechanisms of iron uptake and homeostasis, and the biofortification of plants, including banana for increasing iron content, are described in the review. The focus in the following section will be upon iron deficiency, the most common deficiency in the world (Black 2003).

Micronutrient impoverishment, aptly called "hidden hunger," is relevant in the current scenario for two major reasons:

1. *Crop yield*: At present, the world's population is burgeoning with limited land space and resources. As a result, there is overcultivation of available arable land with concomitant abuse of fertilizers, thus causing leaching and loss of micronutrients. This leads to a compromise on quality and quantity of food produced.
2. *Human health*: A heavy dependence of humans on crop-based diets implies "what affects the farmer affects the consumer too" (DellaPenna 1999). The impact on human health can be measured in the number of DALYs (disability-adjusted life years), which is the summation of the number of years of productive life lost due to mortality as well as those who lived with disability. In a 2010 comprehensive study to determine the DALYs for 291 diseases and injuries carried out over 187 countries (Murray et al. 2013), IDA (iron deficiency anemia) was found to be predominant. A model designed to estimate the economics and human costs of illness due to micronutrient deficiencies in Filipino children put the figure as high as lakh DALYs and attendant medical costs in the millions (Wieser et al. 2013). These were more likely to be worse for the population in the lower socioeconomic strata.

20.1.2 Importance of Iron in Plants

Typical concentration of iron in plant tissue is <200 ug/g dry weight (Grusak and DellaPenna

Table 20.1 Nutritional iron requirements for humans based on dietary vitamin C, animal protein, and cereal content

Food article contribution in percentages	Children (1–3 years) mg/day	Children (4–6 years) mg/day	Women (19–50 years) mg/day	Women during pregnancy (second trimester) mg/day	Women during breastfeeding (0–3 months of lactation) mg/day	Men (19–50 years) mg/day
15 %	3.9	4.2	19.6	>50.0	10.0	9.1
10 %	5.8	6.3	29.4	>50.0	15.0	13.7
5 %	11.6	12.6	58.8	>50.0	30.0	27.4

Adapted from WHO (2003)
%: recommended daily intake for iron depends on the relative contribution by different food articles
15 %: C diet rich in animal protein + vitamin
10 %: diet rich in cereals + vitamin C, low in animal protein
5 %: diet poor in animal protein + vitamin

1999). Fe is essential for plants because of its widespread role in enzyme structure and activity, e.g., heme- and Fe-dependent enzymes (such as those involved in sulfate assimilation, ethylene synthesis, and respiration), Fe-S clusters in mitochondria, and chloroplasts.

20.1.3 Banana: Food for the Masses

As a fruit crop, banana holds sway over much of Africa, Asia, the Pacific, and Central and South America where it is either a staple or supplements the diet. Along with plantains it is eaten raw and can be cooked, boiled, baked, or even pulverized into powder to be consumed with other foods (Chandler 1995). The rest of the plant itself has potential use in pharmaceutical, packaging, and other industrial applications. These find root in some beneficial characteristics such as the high sugar content, the durable fibrous nature of the pseudostem, and the presence of antimicrobial lectins for human consumption (Mohapatra et al. 2010; Swanson et al. 2010), besides many others. As described later in Sect. 20.6, many crops have been considered for biofortification such as rice, tobacco, wheat, and maize. However, banana, which is largely indispensable to a chunk of the world's population and enjoys the status of being an infant food, has been so far unexplored in terms of its potential for biofortification (Kumar et al. 2011). It has widespread popularity because of the following characteristics:

1. High acceptance and consumption due to palatability, aroma, and instant energy source
2. Wider reach to the masses, specifically the poorer sections of society, owing to its availability at a cheaper rate
3. Is a cheap and good source of macronutrients especially potassium (Hardisson et al. 2001)
4. No loss of micronutrients upon cooking and also dessert banana being consumed raw

Attempts have been made to characterize different banana cultivars in terms of their nutrient quantities (Hardisson et al. 2001; Wall 2006). However, the average iron content of banana fruit (0.3 mg/100 g) (García et al. 2015) is too low to meet daily needs as detailed in Table 20.1. In this scenario, enriching the fruit with iron seems a promising approach to combating IDA. An understanding, therefore, of the iron homeostatic mechanisms in humans and plants is essential to work out effective strategies for this purpose, and these are dealt with in the following sections.

20.2 Iron Malnutrition: Prevalence and Its Effect

Iron malnourishment is considerably widespread, especially in women, menstruating girls, infants, and children in developing countries. It is usually more severe in vegetarians than in those who eat meat and seafood because the absorption of heme iron from animal sources is better than the nonheme iron of plants (Shayeghi et al. 2005). Iron bioavailability from mixed diets has been reported to be 14–18 % compared to 5–12 % from vegetarian diets (Hurrell and Egli 2010). However, complexation by phytate, tannins, and polyphenols in vegetarian diets renders bioavailability of minerals like zinc and nonheme iron very low. Some proteins like egg white and casein and conglycinin of soybean also inhibit iron absorption (Hurrell and Egli 2010). People with impaired acid secretion (achlorhydria) may suffer loss of iron as gastric acid solubilizes and stabilizes ferrous complexes (Jacobs and Miles 1969; Zimmermann and Hurrell 2007). Gastrointestinal hookworm infestations also exacerbate the problem by causing blood loss (Crompton and Nesheim 2002). Hence, in order to address these problems effectively, the understanding of iron uptake and homeostasis is essential.

20.2.1 Iron Uptake and Homeostasis in the Human Body

Iron uptake and its absorption need to be fine-tuned to prevent blood and intracellular iron toxicity as well as allocate the precious metal in a balanced manner to every pathway requiring it (Donovan et al. 2005). Both heme and nonheme iron are taken up by duodenal enterocytes through microvilli. Electron microscopic studies reveal that iron can travel only through the enterocyte and out through the basolateral membrane into the lumen (Wyllie and Kaufman 1982). A putative intestinal heme iron transporter heme carrier protein 1 (HCP1) identified by Shayeghi et al. (2005) transports heme into the enterocyte and is induced by hypoxia. Its subcellular localization is influenced by iron status. Once inside, the HCP-heme iron complex is endocytosed and the heme is destroyed by heme oxygenase, liberating iron (Raffin et al. 1974). This iron may join the free iron pool as nonheme iron or be sequestered by ferritin in the enterocyte (Fig. 20.1). Nonheme iron absorption through the gut requires reduction of Fe^{+3} to Fe^{+2}, facilitated by a duodenal plasma membrane two-heme-containing ferrireductase belonging to the cytochrome B_{561} family (DCYTB) (McKie et al. 2001); however, *Dcytb*$^{-/-}$ mice do not show iron status abnormalities, indicating redundancy of function (Andrews and Schmidt 2007). The membrane divalent metal transporter DMT1 then transports Fe^{+2} into the cell, driven by the inward proton gradient (Gunshin et al. 1997). Intracellular iron not directly transported into the bloodstream is bound by ferritin and lost when the cell sloughs off. Efflux in the blood through the basolateral membrane is mediated by the transporter ferroportin 1 (FPN1/IREG1) (Donovan et al. 2005) and the iron oxidase hephaestin (Vulpe et al. 1999). In the blood, ferrous iron is oxidized and bound by circulating transferrin (Fig. 20.1).

20.2.1.1 Regulation

Hepcidin, a peptide hormone synthesized by the liver, negatively regulates both iron absorption and release from macrophages and other cell types. It binds to FPN1, causing the latter to get internalized and degraded. Thus iron entry into the blood decreases but is also lost due to sloughing off of cells (Nemeth et al. 2004). Iron deficiency decreases hepcidin release from the liver and thus allows maximal iron absorption (Nicolas et al. 2002).

Gut iron uptake is also controlled by the IRP-IRE system. The IREs are iron-responsive elements found in the 5′ and 3′ UTRs of the mRNAs of these genes. IRP1 and IRP2 are iron-regulated proteins that bind to IREs in iron-starved cells (Rouault 2006; Muckenthaler et al. 2008). IRP1 is the enterocyte Fe sensor. Under iron replete conditions, a 4Fe-4S cluster is formed in the cytosol and loaded into the IRP1 protein, which decreases the binding ability of IRP1. Under iron deficiency, the cluster cannot form and IRP1 binds to the 5′ UTR preventing the translation. On the other hand, IRP2 is regulated by proteasomal degradation through the activity of the IRP2-targeting E3 ubiquitin ligase subunit, F-box and leucine-rich repeat protein (FBXL5) (Guo et al. 1995 and reviewed by Rouault 2006 and Lane et al. 2015). IRP1 targets hypoxia-inducible factor 2α (HIF2α). In normoxic and high iron conditions, prolyl hydroxylases (PHDs) degrade this protein, while the converse conditions

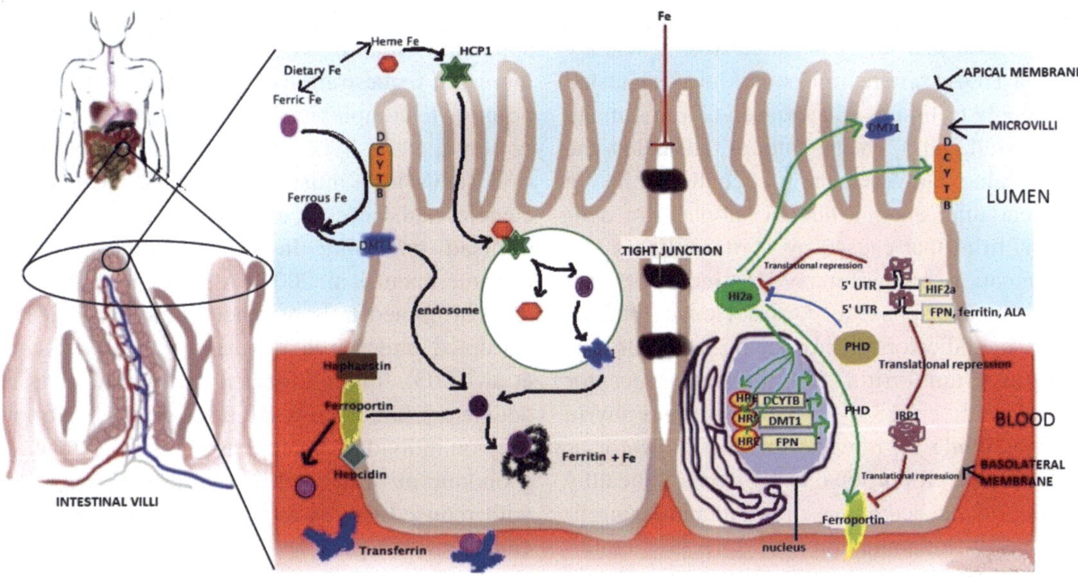

Fig. 20.1 Absorption and regulation of iron uptake from duodenal intestine in humans. *Green arrows* (→) represent activation, *while red curvy arrows* with crossbar (⊣) represent inhibition by translation repression. The *purple curvy arrow* with crossbar (⊣) represents proteasomal degradation by PHDs. ( ) represents the RNA stem loop sequence in the 5′ UTR, which is bound by IRP1. The crossbarred *arrow* from Fe between the cells shows that Fe cannot travel paracellularly owing to tight junctions (Figure adapted from Zimmermann and Hurrell 2007; Andrews and Schmidt 2007; Lane et al. 2015)

increase transcript levels. The HIF2α-HIF1β complex enters the nucleus and regulates cognate downstream genes containing hypoxia-regulated elements (HREs), such as inducing the gene-coding erythropoietin (EPO) as well as the duodenal CYTB-DMT1-FPN1 system (reviewed by Mole 2010; Lane et al. 2015). Under low iron, HIF2α induces transcription of the genes coding for DCYTB, DMT1, and FPN1. IRP1, due to reasons mentioned above, binds to the 5′ UTR IRE in the mRNA of HIF2α, so translational decrease of HIF2α follows. This leads to a decrease in EPO production in renal fibroblasts so that it does not prematurely use up what little store of iron is present. However, there is also decreased activity of the PHDs. Thus overall, HIF2α activity increases and so maximal iron is taken up by the DCYTB-DMT1-FPN1 system in the gut. This IRP-HIF2α axis thus evokes heightened iron uptake in the gut but checks EPO production in the kidney, thus ensuring replenishment of iron stores till usable levels. Contrastingly, under high iron conditions, the IRP1 block is removed and HIF2α levels increase; thus EPO production is stimulated. However because of the simultaneously active proteolysis of HIF2α by the PHDs, the enterocyte levels of HIF2α decrease, thus decreasing iron uptake into the gut (reviewed by Zhang et al. 2014; Lane et al. 2015) (Fig. 20.1).

20.2.2 Iron Deficiency: Indicators and Repercussions

Iron deficiency has quite debilitating effects on the growth of infants, mental development, cognitive functions, and performance at work, attentiveness, and general fitness. The severity and frequency of respiratory diseases are also high in anemic children. In fact, several published reports based on studies of brain development in rats (Beard 2003) have stressed the importance of iron nourishment at a "critical" phase in infant development, failing which there is irreversible delay in achieving brain growth milestones and consequent abnormalities especially causing defects in dopamine secretion. An exacerbated form of iron deficiency is called iron deficiency anemia (IDA) where iron stores are exhausted and supply to tissues is decreased (Zimmermann and Hurrell 2007). The hematocrit falls below the cutoff value. However, the real challenge lies in differentiating anemia due to simple iron deficiency, from that caused by chronic illness and inflammation, because many IDA indicators also change in acute phase responses. Hence a combination of indicators is now being used. Briefly, they are serum ferritin (SF), transferrin receptor (TfR), and erythrocyte zinc protoporphyrin (ZnPP). While SF remains the best indicator as its levels are proportional to Fe stores in healthy people, it is an acute phase protein and is unreliable under inflammation and infection conditions. In contrast serum TfR is not that affected by inflammation, but may be by malaria, age, and race (Zimmermann and Hurrell 2007). ZnPP is also used as an indicator, as during iron deficiency, the enzyme ferrochelatase substitutes zinc instead of iron, in the protoporphyrin IX molecule.

The absorption of other nutrients is also affected by iron deficiency. Efforts to reinstate iodine status in goitrous children with iron deficiency are not successful probably because of reduced activity of the heme-requiring thyroid peroxidase (Hess et al. 2002; Hess 2010; Zimmermann et al. 2000). Dual fortification of salt with iodine and ferric pyrophosphate led to better iron and iodine status as measured by increased Hb and urinary iodine levels. Thyroid volume also increased indicating improved function (Zimmermann et al. 2002, 2004). Iron and vitamin A seem to be inextricably interrelated as each influences the other's uptake and metabolism. Supplementing iron with vitamin A bettered hematological parameters such as hemoglobin (Hb), serum Fe, and percent serum transferrin saturation in young anemic children (Mejía and Chew 1988), as compared to those who received Fe or vitamin A alone. Correspondingly, low retinol concentrations in the serum can increase if iron is given to iron-deficient children (Muñoz et al. 2000), though another study by Wieringa et al. in 2003 surprisingly found that iron supplementation induced apparent vitamin A deficiency because of increased sequestration of the vitamin in the liver, pointing to complexities in the cycling of vitamin A between the liver and circulation. The body iron status also impacts secondary metal deficiencies and excesses, e.g., there is increased risk of lead poisoning in iron-deficient children (Zimmermann et al. 2006). Also, the delicate balance of homeostasis of other essential metals is disrupted under iron deficiency anemia (Keen et al. 2003). Taken together, the burden of iron deficiency has wide-reaching repercussions not only on individual health but also on the general economic prosperity of a country that struggles with reduced fitness as a cause for loss of precious human capital.

20.3 Remedial Measures

With the realization that focusing on only protein energy malnutrition (PEM) in developing countries as done in the 1980s did not address the root of malnourishment, there was a paradigm shift toward boosting the micronutrient content of food (Allen 2003). However, Solomons and Ruz in 1998 suggested that creating a fixed universal set of guidelines for micronutrient intake would not be fruitful, given that people of diverse geographic and ecological areas would differ in body type, race, genetic makeup, diet, and environmental stress exposure. Even so, it was well known that simple fortification of table salt with iodine could prevent goiter, whereas adding vitamin A to margarine as 25 mg vitamin A along with 3.5 mg β-carotene/kg was shown to be effective in the Philippines (Dary and Mora 2002). Thus implementation programs for different micronutrients began in earnest.

Micronutrient intake is generally adequate in the economically sound parts of the world, on account of their being able to reinforce their diet with varied animal-sourced food items such as eggs, milk, poultry, meat, and fish as well as diverse fruits and vegetables. This, however, is not the case for the poor, who rely on a monotonous diet with heavy inclusion of foods such as wheat, rice, cassava corn, tubers, and roots, with very little variation (DellaPenna 1999). Such foods, termed "staples," are often lacking in many essential micronutrients and also low in fat, which further hinders absorption of whatever meager fat-soluble vitamins are present. Based on a study, Nigerian children subsisting on a staple diet of cassava were found to be at high risk for insufficient intake of zinc, vitamin A, and iron that could predispose them to the respective deficiencies (Gegios et al. 2010).

Some plants become staple to an area on the basis of their availability and adaptability, ease of cultivation, and domestication of a particular variety which may have been introduced through population migration. This together with ingrained traditional practices of preparing food and natural reluctance to accepting new food items due to cost factors and caution necessitates innovative interventions. Hence improvement of diet through different means (plant and animal sources) is imperative to ameliorate the associated repercussions of the corresponding deficiency in the population. There are various strategies to mitigate iron deficiency, as outlined below.

20.3.1 Supplementation

Iron is provided externally as ferrous gluconate or ferrous sulfate capsules. Though highly bioavailable, their faithful intake is difficult to meticulously implement because of distribution problems and lack of compliance among the users (Galloway and McGuire 1994). The standard dosage followed is 60 mg of elemental iron (as 300 mg tablet of ferrous sulfate) for an adult, given three to four times a day (Zimmermann and Hurrell 2007). There are inherent disadvantages like epigastric pain, nausea, and constipation, when taken on an empty stomach, which also, however, allows optimal absorption of iron. Reports argue for or against this method, with some hailing it as having a wider outreach than fortification (Baltussen et al. 2004) based on model estimates, having a probable prophylactic effect when given to initially iron replete pregnant women in the USA (Cogswell et al. 2003) because it led to significantly higher birth weight and lesser incidence of low birth weight, as well as improving cognition, verbal learning, and memory among adolescent girls with non-anemic iron deficiency (Bruner et al. 1996). Vitamin A-supplemented iron capsules maximally improved hemoglobin levels in anemic pregnant women of Indonesia, rather than either Fe or vitamin A alone (Suharno et al. 1993). However, in the same breath, Baltussen et al. in 2004 also perceive this method to be less cost-effective than fortification, and there are dangers of overdosing, because Fe is not actively excreted from the body and can thus cause hemochromatosis (Iannotti et al. 2006). Sazawal et al. in 2006 dosed children in Zanzibar, a highly endemic area for malaria, with Fe and folic acid supplements. They revealed an alarming propensity for severe illness and death, suggesting that interventional programs in malarial areas be carried out with caution as they may accelerate the risk of disease.Sazawal et al. (2006), Curie et al. (2009), Koike et al. (2004) are cited in text but not given in the reference list. Please provide details in the list.These references have been inserted in the list

20.3.2 Fortification

By far the safest, most practical way of maintaining iron stores in the body, this strategy involves adding edible forms of iron to regular foods, like milk and cereals. Various forms of iron are available, such as salt forms (ferrous sulfates, gluconates, lactates, and citrates), which are water soluble and readily absorbed but are colored and oxidize the food matrix. In contrast, they are water insoluble and hence much less bioavailable forms (elemental iron, ferric pyro- and orthophosphates) which however are inert with respect to reactivity and color (Hoppe et al. 2006). An intermediate trade-off is the dilute acid-soluble forms (ferrous fumarate, succinate, saccharate, and glycerophosphate) which are low in reactivity but still retain appreciable bioavailability (Hurrell 2002). A novel compound now in use is sodium ferrous ethylenediaminetetraacetic acid (Na-FeEDTA), an effective measure when added to sugar, curry powder, fish sauce, soy sauce, and maize sauce

(Zimmermann and Hurrell 2007). It does not oxidize fats and does not precipitate peptides in fish sauce, but allows two- to threefolds more iron absorption from high-phytate diets than ferrous sulfates. Other forms of use that are gaining popularity are encapsulated ferrous sulfates, micronized dispersible ferric pyrophosphates, and bisglycinates that are used to fortify liquids (reviewed by Zimmermann and Hurrell 2007). A generally accepted strategy is to fortify food with forms of iron that don't react unfavorably with food. However these forms have less bioavailability. An advantage of this method is that it does not put consumers at risk of infection.

There are a number of reports on the fortification of food with the abovementioned iron forms (Zimmermann et al. 2005; Van Thuy et al. 2003). A comprehensive review of 54 interventional studies (longitudinal and randomized controlled trials) that had used various iron forms – food matrix combinations – was conducted by Das et al. in 2013, to measure the impacts of iron fortification on women and children. They concluded that the fortification strategies used significantly improved hematologic parameters in both women and children. Zimmermann et al. in 2003 administered iodine and micronized iron-fortified salt to Moroccan children and reported increased hemoglobin, urinary iodine, and thyroid volumes and decrease in the prevalence of IDA. A multiple micronutrient-fortified milk intervention study by Sazawal et al. in 2010 also yielded encouraging results.

As mentioned earlier, vegetarian populations are more prone to iron deficiencies, and hence today improvement of available plant resources is one of the most researched areas and also the most challenging. It is imperative to improve the nutritional status of the plant, as it is directly linked and impacts the micronutrient status of the consuming population at large, especially with increasing numbers of people turning to vegetarianism.

20.3.3 Biofortification

This involves the enriching of a staple food crop with a nutrient, which is limiting in it, either through conventional breeding or transgenic approaches (Haas et al. 2005), and has been reviewed by Graham et al. (2001), White and Broadley (2009), and Welch and Graham (2004), with a focus on improving agronomic practices and understanding nutrient fluxes in the plant, to target micronutrient accumulation in the edible parts. A feeding trial of 192 religious sisters in the Philippines, with high iron rice, led to better iron stores and proved the efficacy of this approach.

Breeding strategies are thought to improve the iron status by 400–500 % than the existing varieties (Haas et al. 2005; Pfeiffer and McClafferty 2007). Under the HarvestPlus project, crops (Cantrell and Reeves 2002) like rice, wheat, and maize (Lung'aho et al. 2011), cassava (Ariza-Nieto et al. 2006; Leyva-Guerrero et al. 2012), and beans (Welch et al. 2000; Ariza-Nieto et al. 2007; Blair et al. 2009) have been developed with higher content of micronutrients like iron, zinc, and vitamin A (Nestel et al. 2006; Pfeiffer and McClafferty 2007). Common cultivars have 12 mg of iron and 25 mg of zinc per kilogram; double of these values are seen in some cultivated traditional varieties. The grains of IR68144-3B-2-2-3, an aromatic elite variety, have ~21 mg/kg in the unpolished rice. This variety is also tolerant to rice tungro virus and to mineral-deficient soils with excellent grain qualities (Gregorio et al. 2000).

Conventional breeding has its own pros and cons and so does transgenic technology. This is because not in all cases will one particular methodology work. It depends on the complexity of the genome as well as the pathway to be manipulated. A review by Sperotto et al. in 2012 shows how rice, which possesses both strategies, does not show root proton extrusion or root ferric reductase activity under Fe deficiency, but does have IRT1 activity. The FRO activity in rice is exclusively in shoots for uptake of xylem Fe^{+3}.

The basic tenet of conventional breeding is the absolute necessity of intraspecific, interspecific, and intergeneric variations in the wild. This approach involves making crosses between the plant to be improved and the elite wild donor enriched with the particular micronutrient. Thus, it is imperative to have as exhaustive as possible a collection of the germplasm of all the available wild relatives of a particular variety. Several studies have found that micronutrient content traits have quantitative trait loci (QTLs) associated with them. For example, Stangoulis et al., in 2007, identified QTLs for grain phytate, Zn, and Fe content, located on different chromosomes, in the rice IR64 X Azucena double haploid line. The QTL for phytate was situated on chromosome 5 different from the chromosome harboring the QTLs for micronutrient traits. This hints at the potential to improve the line with regard to decreasing the phytate levels without affecting the micronutrient content, at the same time improving bioavailability. However, while successful with *Arabidopsis* and some plant species (Shi et al. 2008; Sankaran et al. 2009; Ozkan et al. 2007; Distelfeld et al. 2007; Peleg et al. 2009; Ghandilyan et al. 2009), it requires extensive knowledge of markers, a high density map of QTLs, and the issue of reproductive compatibility with the wild relatives. It is also time and labor intensive with an inherent penalty of linkage drag transferring undesirable genes as well. On the other hand, more often than not, it doesn't require an extensive knowledge of the pathways involved to be modified and thus may be the best option in some cases. Furthermore, as no foreign genes are being introduced, the variety can directly be field tested and released without any regulatory hurdles.

In transgenic approaches, gene manipulation is easier owing to well-elucidated available pathways for the trait and the influx of annotated genomic sequences of model and economically important crop plants. The gene for the trait in question is either overexpressed or the inhibitory genes affecting the same are knocked down or silenced. It is a relatively fast process and is useful for obtaining the functional significance of each probable player in a pathway (Thorneycroft et al. 2001). However all plants are not amenable to transformation and thus this technology remains a yet unexplored research avenue for those crops. Since foreign genes and selectable markers are introduced, safety and regulatory concern haunts quick commercial release of such crops, magnified by the support from anti-GMO lobbies (Potrykus 2015).

Ultimately, an integrated approach is required for the best understanding of the problems like development of abiotic and biotic stress tolerance, plants with heightened levels of micronutrients like iron, etc. This way, the strengths of both the approaches can be effectively harnessed (Cattivelli et al. 2008) for mankind.

20.3.3.1 Banana: Why Transgenics?

While rice has a higher content of iron when compared to banana, the latter is the poster child of biofortification now because of its wide use as infant food and the fact that there is a critical developmental period as mentioned above during which iron is to be mandatorily provided to children (Vasconcelos et al. 2003).

However there is a caveat to biofortifying banana, which is the seedless nature in most of the edible varieties. This does not allow conventional breeding approaches to be used. Efforts are underway to biofortify banana through genetic engineering to enrich it for several traits like resistance to abiotic and biotic challenges and alleviation of IDA. The following reasons make banana particularly suitable for transgenic work:

1. The cultivated edible forms of banana fruit develop parthenocarpically, and hence the spread of a transgene through the pollen is barely minimum. This will ensure that there is no cross-pollination and thus mixing of transgenic and wild species in nature, thus preserving the biodiversity of banana cultivars (Waltz 2014).
2. The present marketed bananas are monocultures of the Cavendish variety as the earlier variety Gros Michel was devastated due to *Fusarium* outbreak.
3. Current methods are available for tissue culture propagation and regeneration through embryogenic cell suspensions to develop disease-resistant banana, as it is susceptible to a range of pathogens like *Fusarium*, *Xanthomonas*, BBTV, banana streak virus, etc.
4. Since banana is such an important cash and food crop, energy rich but micronutrient poor, efforts have been made to elevate the micronutrient levels in the fruit. Towards this aim,

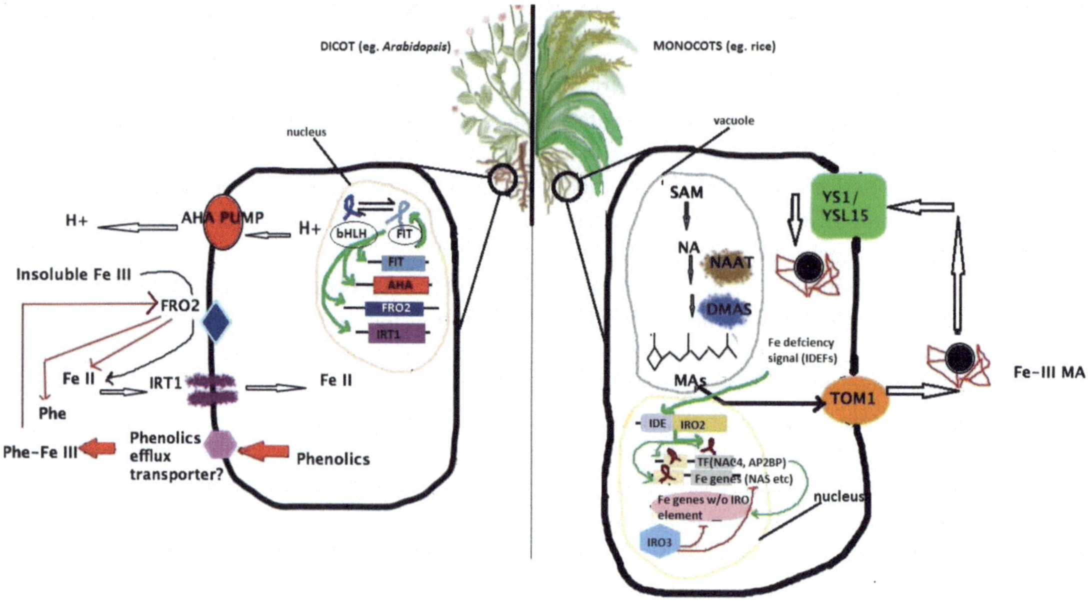

Fig. 20.2 Iron uptake strategies in dicot and monocot roots. *Green arrows* (→) represent activation, while *red* with crossbar (⊣) represent inhibition. ( ) is the IRO2 TF which binds to its cognate-conserved sequence (5′-CACGTGG-3) in the promoters of Fe-responsive genes, e.g., the MA synthesis and secretion pathway. However, though IRO2 is shown to regulate 59 genes, not all have the IRO2-binding element and thus may be regulated indirectly by IRO2 as shown through the TFs NAC4 and AP2-binding protein. Abbreviations: *AHA pump* H^+ extrusion pump, *FRO2* ferric chelate reductase, *IRT1* iron-regulated transporter, *FIT* Fer-like iron deficiency induced, *bHLH* basic helix loop helix, *SAM* S-adenosylmethionine, *NA* nicotianamine, *NAAT* nicotianamine aminotransferase, *DMAS* deoxymugeneic acid synthase, *MA* mugeneic acid, *TOM* transporter of MA, *YS1/YSL15* yellow stripe (maize)/yellow stripe-like proteins, *IDE* iron deficiency-responsive elements (Adapted from Ivanov et al. 2012; Ogo et al. 2007; Kobayashi and Nishizawa 2012)

the effect, if any, of the banana fruit matrix upon the absorption of iron was studied by Garcia et al. in 2015 using extrinsically labeled stable isotopes of iron. The group estimated iron absorption from raw and cooked bananas in rural Mexican women. They found that there was no significant difference between absorptions from raw and cooked banana, proving noninterference of the matrix in the absorption process.

20.4 Iron Acquisition by Plants

Iron acquisition is well evolved in the plant kingdom on account of the sessile nature of plants which forces them to exploit the immediate soil environment as efficiently as possible. However, uptake, storage, and transport of this metal have to be meticulously controlled because of its redox activity, which can create havoc through the Fenton reaction. This has been taken care of by both or either of two strategies described below. The authors attempt to provide a road map and not an exhaustive detailing of the various participants in the iron physiology of the plant, based on elegant studies as cited in the text. Figure 20.2 diagrammatically explains the two well-studied strategies.

20.4.1 Strategy 1

A characteristic of non-graminaceous dicots like tomato and *Arabidopsis* mainly features the extrusion of protons by an H^+-ATPase in the root plasma membrane. The soil and apoplast are acidified and ferric ions are solubilized. FRO2, a membrane ferric chelate reductase reduces Fe^{+3} ions to Fe^{+2}, which then enter into the root symplast through the transporter IRT1. All three proteins show increased activity when the plant is challenged with iron deficiency.

Several different H^{+}-ATPases belonging to the *Arabidopsis H⁺-ATPase* (*AHA*) family have been implicated in the response to Fe deficiency, for instance, the *Arabidopsis* AHA 7 and cucumber CsHA1, which are both upregulated in Fe scarcity (Santi et al. 2005, Santi and Schmidt 2008). First reported in *Arabidopsis* (Robinson et al. 1999) and then pea (Waters et al. 2002), FRO2 belongs to a family of seven ferric chelate reductase genes that differently localize throughout the plant body for tissue- and organ-specific Fe uptake. The protein is an 8-transmembrane domain (8-TD)-containing enzyme that uses two intramembrane heme groups and has a cytosolic binding site for nucleotide cofactors that transfer electrons (Robinson et al. 1999). FRO2, FRO3, and FRO5 are found in roots, FRO3 in the vascular cylinder and shoot, and FRO6, 7, and 8 in the shoot (Mukherjee et al. 2006). The *Arabidopsis frd1* is severely chlorotic when Fe is limiting. This phenotype is however rescued by native FRO2 and plants overexpressing FRO2 are tolerant to low iron. IRT1 is an iron-regulated transporter 1, belonging to the ZIP (ZRT, *I*RT-like proteins) family of transporters, which is induced under iron-limited conditions (Vert et al. 2002). This transporter is essential for iron uptake in *Arabidopsis* as evidenced by knockout mutants showing chlorosis with lethality as well as developmental defects in chloroplasts and vascular bundles (Henriques et al. 2002). The uptake of other divalent cations is also compromised hinting at the capability of IRT1 to also transport ions like Zn and Cd (Korshunova et al. 1999). In addition to the above inducible proteins, plant root morphology is altered, and root hair architecture significantly changed to permit better uptake by cells.

Transcriptome and proteome interrogations with bioinformatic tools have revealed the functional roles of proteins that are now widely recognized as being the master regulators for strategy 1 in non-graminaceous plants. Dependence on two transcription factor families is to date studied. In brief these are as follows.

20.4.1.1 The FIT and Associated Transcription Factors

The low-Fe-inducible FIT (FER-LIKE IRON DEFICIENCY-INDUCED TRANSCRIPTION FACTOR) was first identified in *Arabidopsis*, as a functional homologue of the tomato FER protein (Ling et al. 2002; Yuan et al. 2005), expressed in root epidermal cells. It is essential to the induction and maintenance of the iron deficiency response in plants, because *fit1* mutants (Colangelo and Guerinot 2004) are chlorotic lethal and can only be rescued by externally supplying iron. Thus it is evident that FIT controls some aspect of the iron uptake pathway, as substantiated by the finding that FIT regulates both FRO2 and IRT1. FIT upregulates FRO2 transcript levels, while it activates both transcription and translation of IRT1. Under Fe scarcity, FIT transcription itself is increased and so is FIT protein turnover, presumably to enable the plant to quickly adjust to change in Fe status (from deficiency to replete conditions) (Fig. 20.2).

However FIT overexpression alone could not increase FRO2 and IRT1 levels. Subsequently, the bHLH TFs, namely, *bHLH38, 39, 100, and 101* (Yuan et al. 2008; Wang et al. 2013), were also discovered by yeast two-hybrid screening and bimolecular fluorescence complementation. The action of FIT in conjunction with transcription factors has been deduced from overexpression studies of FIT and bHLH38/39 together (Yuan et al. 2008). The mentioned genes together help in higher uptake of Fe in plants, simultaneously upregulating downstream genes FRO2 and IRT1.

Heavy metal contamination is a major problem in iron-deficient soil due to the somewhat nonspecific nature of transporters. Plants resist heavy metal dissemination through the shoot by sequestering the corresponding metals in vacuoles. This process is mediated by the proteins HMA3 (heavy metal ATPase 3), IRT2 (Fe II transporter), MTP3 (metal tolerance protein 3), and IREG2/FPN2 (ferroportin). In this regard,

FIT may probably protect against heavy metal toxicity. Combined overexpression of FIT and bHLH38/39 leads to elevated levels of cadmium in root vacuoles thereby preventing its transport into the shoots (Wu et al. 2012). In addition to this, NAS levels were also increased thus allowing better iron translocation through the shoots. Owing to these phenomena, the plant system overcomes the competitive uptake of nonessential toxic metals under iron deficiency.

20.4.1.2 The POPEYE Network

Dinneny et al. in 2008 profiled the transcriptomic responses of each cell layer of the *Arabidopsis* root to Fe deficiency in a temporal manner. Proteins with DNA-binding domains suggesting their roles as transcription factors were selected to identify novel regulators of the Fe deficiency response. Screening through the corresponding mutant lines having abnormal Fe deficiency responses revealed a protein *PYE*, which was highly upregulated in Fe deficiency (Long et al. 2010).

Upon Fe deficiency, PYE mRNA is expressed mainly in the pericycle, but the protein is found in the nuclei of all root cells, indicating mobility throughout the root.

The pericycle is essential in the response to Fe starvation. Surrounding the vasculature, it is the transit layer for iron to enter the xylem, from the root. The pericycle FRD3 citrate transporter mediates efflux of citrate into the xylem to enable chelation of ferric iron (Rogers and Guerinot 2002; Green and Rogers 2004). Comparative transcriptional analysis during Fe deficiency revealed that several genes encoding transcription factors were co-expressed with PYE, in the pericycle. This points to the importance of the pericycle as a probable "hub" in controlling the Fe deficiency response. *pye* mutants could not display wild-type responses to Fe deficiency. Root growth and root hair growth were both inhibited; root cortical and epidermal cells were swollen under iron deficiency. Mutants had chlorotic leaves and could not efficiently utilize Fe from calcareous soils. In addition, Fe reductase activity and rhizosphere acidification decreased, indicating attenuated Fe deficiency responses. Also, under iron deficiency, plants tend to take up more Zn, Cd, Co, and Mn; however this was not as pronounced in *pye-1* mutants as in wild type. Perl's staining and ICP-OES revealed that the roots of *pye-1* mutants accumulated more iron than *wt* in both Fe-sufficient and Fe-deficient conditions. The same was seen for the shoot, however, only during Fe deficiency, indicating that *Pye* is required for regular iron homeostasis, but is even more indispensable for the deficiency response. Some of these phenotypic changes were reversed when the PYE gene was introduced into the *pye* mutants under the control of the native promoter. Thus loss in PYE activity definitely alters the response to Fe deficiency (Long et al. 2010).

Transcriptional profiling of pye-1 mutants against wild-type controls identified the targets of PYE in plants. Relevant ones are outlined below:

MRH2 and MRH6: MRH2 is a kinesin-related protein required for maintaining straight root tip growth and a single growing tip, as mutants have wavy roots and a wide base (Jones et al. 2006). MRH6 is similar to the *Escherichia coli* universal stress protein A. It is required for root hair initiation and ensuring that only a single root hair develops on each hair-bearing cell. *mrh6* mutants have multiple root hairs from a single cell (Jones et al. 2006). Both were repressed in Pye mutants under sufficiency and deficiency.

Ferritins and FSD1: Both are involved in Fe homeostasis as ferritins are Fe storage proteins and FSD1 is required to quench free radicals produced under oxidative stress. Repressed in both Pye and *wt* under Fe deficiency, they are relatively more highly expressed in *pye*. This may be due to the fact that *pye* plants are under higher oxidative stress stemming from increased iron accumulation in roots.

IRT1: It was upregulated both in *wt* and *pye* plants under iron deficiency, but to a lesser extent in the mutant, indicating decreased Fe uptake from the soil.

Metal ion transporters and facilitators: The NRAMP4 (Lanquar et al. 2005), OPT3 (Zhai et al. 2014), and FRD3 proteins all translocate Fe through different plant tissues as described later in the chapter. Similarly, the ZIF1, NAS4, and FRO3 genes are also involved in metal transport. ZIF1 (ZINC-INDUCED FACILITATOR 1) is a zinc transporter which mediates transport of zinc-chelate complexes into the vacuole, and zif1 mutants are sensitive to Zn and Cd (Haydon and Cobbett 2007). NAS4 encodes nicotianamine synthase, which synthesizes NA required for metal transport between cells (Klatte et al. 2009). FRO3 expresses in the vascular cylinder and may function in mitochondrial iron uptake (Jeong and Connolly 2009). These were induced in Fe deficiency in wt and *pye* mutants but even higher in the latter, suggesting that both inter- and intracellular movement of Fe is increased in *pye* mutants because PYE actually acts to repress them in the wild state.

The basic helix-loop-helix transcription factors, bHLH104, bHLH115, and ILR3 (IAA-leucine resistant 3), are PYE homologues (Long et al. 2010). Knockout mutants of ILR3 and bHLH104 cannot tolerate Fe deficiency and develop chlorosis and shorter roots, in addition to having very low FRO2, FIT, and IRT1 levels. Overexpressing lines on the other hand tolerate Fe deficiency (Zhang et al. 2015), and bHLH104 overexpression causes Fe to accumulate in the phloem. ILR3 modulates auxin conjugate hydrolysis in an iron-dependent manner. It controls Fe availability for auxin homeostasis, probably by transcriptionally controlling Fe transporters (Rampey et al. 2006).

Through Y2H screening it was found that bHLH104 interacts with ILR3 and bHLH115 (Zhang et al. 2015). Both the former also bind to the promoters of POPEYE and bHLH38, 39, 100, and 101, activating them. On the other hand, the iron-binding hemerythrin (HHE) containing E3 ubiquitin ligase BRUTUS (BTS) is a negative regulator as it directly interacts with and targets the PYELs for proteasomal degradation (Selote et al. 2015). BTS is itself induced in iron deficiency. Thus a model was proposed whereby in Fe deficiency BTS is induced and degrades PYELs. Thus even though Fe acquisition is increased by the action of PYE and PYELs, it is fine-tuned by BTS so that just sufficient iron is taken up without toxicity being induced. When Fe is sufficient, Fe binds to the HHE domain of BTS and destabilizes it thus reducing its activity. In addition the PYE and PYELs are also down-regulated. In all, this provides for quick adaptive response under differential changing iron conditions (Selote et al. 2015; Kobayashi et al. 2013).

20.4.2 Strategy 2 for Iron Uptake

This strategy operates in the grass orders Restionales, Eriocaulales, or Commelinales. Here, phytosiderophores (PSs) are extruded by roots to sequester iron in its insoluble form as Fe^{+3} (Römheld and Marschner 1986). Chemically, the PS molecules are derived from the methionine cycle which generates nicotianamine (NA) and is further metabolized to give mugeneic and avenic acids (Ma and Nomoto 1996). This ferric PS complex is then internalized via transporters available in the root membrane.

The major participants of this strategy are described below:

Chelators These are the downstream products of the methionine cycle (Ma et al. 1995). The typical ones are NA, deoxymugeneic acid (DMA), and mugeneic acid (MA).

Enzymes Three molecules of S-adenosyl methionine (SAM) are trimerized by nicotianamine synthase (NAS), to form the nonproteinic amino acid nicotianamine (NA) (Higuchi et al. 1999). Nicotianamine aminotransferase (NAAT) brings about oxidative deamination producing a keto intermediate (Takahashi et al. 1999). This is further reduced to deoxymugeneic acid (DMA), by DMA synthase (DMAS) (Bashir et al. 2006). Further, in barley but not in rice, the dioxygenase iron deficiency-specific clones 2 and 3 (IDS2 and

IDS3) (Nakanishi et al. 2000) were found to catalyze the last step of hydroxylation of DMA at 2′ and 3′ positions, to give MA and epihydroxy MA, respectively (Kobayashi et al. 2001).

Other enzymes shown to be regulated by Fe deficiency are glutathione reductase which may act to scavenge ROS under iron deficiency (Bashir et al. 2007) and an as-yet uncharacterized rice-specific mitochondrial protein whose knockout mutant was highly sensitive to Fe deficiency (Ishimaru et al. 2009).

Transporters The maize yellow stripe 1 (YS1) family members are H^+-coupled symporters for Fe-PS complexes. The original founder, maize yellow stripe 1 (YS1) transporter (Curie et al. 2001; Schaaf et al. 2004), takes up Fe III-phytosiderophore (Fe III-PS) complexes from the soil in strategy II plants. Maize plants with loss of function in the YS1 gene show "yellow-striped" interveinal chlorosis (von Wirén et al. 1994). Subsequent members were named YSLs for yellow stripe-like based on their homology to YS1 – eight members in *Arabidopsis* (Curie et al. 2001; Waters et al. 2006), three members in *Thlaspi caerulescens* (Gendre et al. 2007), and 18 members in rice (DiDonato et al. 2004; Inoue et al. 2009; Ishimaru et al. 2010) and barley (Araki et al. 2011; Zheng et al. 2011). These YSLs internalize Fe complexed with NA, MA, DMA, or citrate and at different spatial locations. Rice YSL15 transports Fe-DMA from rhizosphere into roots (Inoue et al. 2009). YSL18 distributes Fe-DMA into reproductive organs (Aoyama et al. 2009). Rice YSL2 nonspecifically transports both Fe-NA and Mn-NA (Ishimaru et al. 2010).

Nozoye et al., in 2011, identified a major missing link in the picture of strategy II, i.e., the PS efflux transporters, namely, TOM from barley and rice, as well as the nicotianamine efflux transporter ENA (*effluxer of nicotianamine*). Heterologous expression in eggs of *Xenopus laevis* confirmed that TOM is the exporter of only MA and not NA, while ENA secretes only NA and not mugeneic acids.

Transcription Factors Under iron deficiency, the strategy 2 response is mediated through positive regulator TFs like IRO2 and IDEF1/2 and negative regulators like IRO3 in rice. IRO2 is differentially expressed under iron deficiency, in both root and shoots (Ogo et al. 2006). It recognizes the consensus sequence CACGTGG in the sequences of downstream genes; however, not all genes regulated by IRO2 have this sequence. Thus IRO2 may act by interacting with other TFs as well (Ogo et al. 2007). Overexpression of IRO2 leads to a higher tolerance to low levels of iron better performance than the corresponding wild-type plants on alkaline soil (Ogo et al. 2011). RNAi knockdown of IRO2 greatly decreases the ability of rice plants to withstand iron deficiency in soil, below that of the control. As expected, PS synthesis genes including NAS are also induced in IRO2 overexpressing lines. In addition, the expression patterns of IRO2 and downstream genes like NAS are synchronous, as evidenced by promoter-GUS staining of seeds and vascular tissue. Notably, levels of IRT1 were unchanged in IRO2 overexpressing lines, showing that this TF is specific to the strategy 2 response (Ogo et al. 2007).

The IDEF1/2 system (IDE-binding factor) also positively regulates the iron deficiency response of strategy 2 plants. These were originally identified as the factors that regulate the barley root-expressed IDS2 gene, which is a dioxygenase in the PS synthesis pathway, by binding to conserved sequences in the promoter. These sequences are called iron deficiency-responsive elements (IDE) (Kobayashi et al. 2007). OsIDEF1 overexpression renders rice plants more tolerant to Fe deficiency though as such IDEF1 is constitutive and not regulated by iron levels (Kobayashi et al. 2007). The IDE-like elements are found in the promoters of several effector genes like OsNAS1/2/3, OsYSL2/15, rice, and *Arabidopsis* IRT1 and *Arabidopsis* FRO2. As evidenced from time course studies, IDEF1 appears to regulate the early response to

Fe deficiency as the abovementioned genes are upregulated in synchrony with IDEF1; however this transcription factor ceases to affect them in the later part of deficiency response (Kobayashi et al. 2009). IDEF1 contains proline- and asparagine-rich regions which were shown in vitro, to bind zinc and iron (Kobayashi et al. 2012). In contrast IDEF2 appears on the scene at a later stage of Fe deficiency, but its effects are more or less constant on downstream genes (Kobayashi et al. 2010).

IRO3 acts as a dampener to the Fe deficiency-responsive effectors like OsNAS1/2/3, OsIRT1, and OsYSL2/15 by negatively regulating them under iron starvation (Zheng et al. 2010). Under these conditions, it is dramatically induced in both roots and shoots. Overexpression of this bHLH TF makes the plant hypersensitive to Fe deficiency but with less accumulation of iron in the shoot compared to wild plants. The downstream effectors as mentioned above are also suppressed.

Both these strategies also intersect with extraneous signaling molecules like nitric oxide (NO), ethylene, jasmonate, and the phytohormones auxin (Schmidt et al. 2000) and cytokinin. They regulate the Fe status of the plant either by positive or by negative modulation. To the best of our knowledge, there have been no reports of Fe sensors in plants though corresponding ones exist in bacteria and mammals. However, these signaling entities integrate aspects of plant development and stress response with Fe acquisition strategies.

Auxin is responsible for modulating root system architecture in response to micronutrient fluxes (Giehl et al. 2012). It seems to be involved in both strategy 1 and strategy 2, though the mechanistics are better elucidated in strategy 1 plants. Synthesis of auxin is increased under Fe deficiency and also increases activity of FIT and FRO of strategy 1 plants, thus accumulating more iron. The *yucca* mutant of *Arabidopsis* is deregulated to overproduce auxin and shows higher levels of FIT and FRO2 expression (Chen et al. 2010). It is known that nutrient availability affects root system architecture (López-Bucio et al. 2003), and thus experiments were carried out to determine how iron affects this.

Localized iron application to the root strongly induced lateral root formation (Giehl et al. 2012). Symplastic iron is required for this induction, as was seen in the case of the *irt1* mutant, which was unable to produce lateral roots under low iron conditions, but could be rescued by the application of iron to the shoot, because of phloem transport of iron to roots. This is dependent on the distribution of auxin by AUX1 transporter, in the direction of roots. Support for this came in the form of AUX1 promoter induction under Fe application and the fact that mutants defective for auxin transport are unable to show lateral root elongation under localized Fe treatment. A model was proposed to explain the cross talk between auxin and iron, whereby symplastic iron accumulation in the root (through the action of IRT1) creates a gradient of iron toward increasingly Fe-deficient root tissues. This local symplastic iron induces AUX1 expression and auxin is further redirected to these tissues, where it itself induces AUX1 and elongation of lateral roots. Thus here, symplastic iron may stimulate the cascade of signals that ultimately lead to lateral rooting from the deficient tissue (Giehl et al. 2012; Hindt and Guerinot 2012).

In rice, a strategy 2 plant, auxin plays a similar role in that it is required for root architectural adaptation in response to iron deficiency. However, no mechanism as yet has been uncovered. An auxin-responsive transcription factor, OsARF12, has been identified that enables the expression of the auxin-responsive promoter construct DR5::GFP containing an auxin response element. The auxin-insensitive *osarf12* rice mutant shows altered root morphology and lowered iron content together with decreased levels of IRT1 and increased levels of *mitochondrial iron-regulated* (MIR) gene (Qi et al. 2012).

Ethylene as a signaling molecule in the plant response to stress has also been implicated in the response to nutrient stress (Lynch and Brown 1997). Root ethylene is produced by strategy 1 plants in response to iron deficiency. The application to iron-sufficient plants such as tomato, cucumber, and pea induced typical iron starvation responses like increased FRO2 activity and swollen root tips (Romera et al. 1999). Similar to auxin, ethylene also induces the expression of FIT/FER which in turn upregulates the FRO- and IRT1-mediated strategy I response genes. Lingam et al. in 2011 identified the factors ETHYLENE INSENSITIVE 3(EIN3) and ETHYLENE INSENSITIVE3-LIKE 1 (EIL1) as direct interacting partners (Brumbarova et al. 2015) of the master regulator FIT, which were absolutely essential for full-level FIT expression under Fe deficiency conditions. The *ein3 eil1* mutant transcriptome was more sensitive to iron deficiency as compared to wild type, though FRO2 and IRT1 were still induced. Also, while there was no significant difference in FIT levels between mutants and wild type under Fe sufficiency, FIT levels were nearly halved in the mutants in Fe-starved conditions. This demonstrated that EIN3/EIL1 is positive regulators of the Fe deficiency response. The application of aminoethoxyvinylglycine (an ethylene synthesis inhibitor), in planta, reduces FIT levels of *ein3 eil1* mutants, while adding a protease inhibitor MG132 restored FIT levels. This led to a model in which interaction with EIN3/EIL1 stabilized molecule the FIT protein by stalling its turnover by proteasomal degradation and thus allowing it to induce its downstream effectors FRO2 and IRT1 (Lingam et al. 2011).

The effects of ethylene have not been demonstrated in any strategy 2 plant other than rice so far, utilizing both strategies 1 and 2. Here also, ethylene production was induced under iron deficiency, and the exogenously applied synthetic ethylene analogue aminocyclopropane (ACC) could impart tolerance to Fe deficiency (Wu et al. 2010). The transcripts of OsNAS1/2, OsIRO2, OsYSL15, and OsIRT1 were elevated suggesting positive effect of ethylene on members of both the strategies. IDEF1 also is slightly induced upon the application of ACC, while the promoters of OsIRO2, EILs, and EREBPs (ethylene response element-binding proteins) contain ethylene cis-acting elements, suggesting that they probably be influenced by ethylene.

Nitric oxide is also a positive regulator of the iron deficiency response, in that it controls lateral root formation and may stabilize FIT (Correa-Aragunde et al. 2004). Exogenous NO application induces FER, FRO2, and IRT1, while NO inhibitors decrease the same. However, supplying proteasome inhibitors to these plants reverses the effect of the NO inhibitors, thus indicating a probable proteasome-mediated action of NO (Méndez-Bravo et al. 2010). Reports suggest that auxin acts upstream of NO (Chen et al. 2010), while NO and ethylene produce a convergent mode of action as they upregulate similar sets of genes under Fe deficiency (García et al. 2010).

Further, negative regulators for the deficiency response, cytokinin and jasmonate, have also been investigated. Cytokinins have been reported to downregulate genes at the root level, i.e., IRT1, FRO2, and FIT, irrespective of the Fe status of the plant (Séguéla et al. 2008). However, they don't require FIT for this suppression but the cytokinin receptors CRE1 and AHK3 are essential. Also, it has been proposed that cytokinins mediate their roles via root growth suppression because cytokinin treatment inhibits primary root elongation (García et al. 2011).

Jasmonate is an oxylipin derivative and a plant hormone involved in systemic signaling pathways in response to abiotic and biotic stress (e.g., UV, wounding, and pathogen attack) (Wasternack 2007; Bari and Jones 2009). Through studies on the jasmonate-insensitive mutant *jar1-1*and the application of the lipoxygenase inhibitor ibuprofen, it has been implicated as a negative regulator in the Fe deficiency response. Like cytokinins, upon the application of exogenous methyl jasmonate, it downregulates FIT, FRO, and IRT1 but does not seem to act via FIT. However it may act to fine-tune the deficiency response rather than systemically downregulating it, because it is not required for shoot signaling of Fe sufficiency to the roots (Maurer et al. 2011).

20.4.3 Which Strategy Used When?

From the viewpoint of the plant, strategy I is not as advantageous as strategy II because of the inhibitory effect of high pH on FRO2 reductase and auto-oxidation of ferrous ions (Römheld 1987). Strategy II is less affected by pH vagaries and can grow efficiently even in calcareous soils. Grasses are hence more resistant to iron deficiency than strategy I plants (Römheld 1987). It is also highly specific for iron uptake from phytosiderophores and does not utilize microbial siderophores (Römheld and Marschner 1986). Unlike other plants, which are constrained in the kind of strategy they use, experimental results suggest that rice can use both the approaches (Ishimaru et al. 2006). Although, rice is a graminaceous plant, the shallow roots are not very efficient in secreting phytosiderophores (Mori et al. 1991) unlike barley, and hence young rice plants are highly vulnerable to iron deficiency. However, under the anoxic conditions in the paddy field in which iron exists predominantly as ferrous forms, strategy I predominates. It's not clear if this provides an evolutionary advantage to the rice system, but it probably helps it to take both the forms of iron from rhizosphere and then change their oxidation states within the cell when needed. This might hint at the possibility of soil conditions dictating the kind of strategy plants use to survive under Fe-deficient conditions.

Qualitative and quantitative differences in the strategy of Fe uptake exist within and between plant species. Even among plants following the same strategy, there are intensity variations in the response to iron deficiency (Römheld 1987). Regarding banana, the authors have found through homology searches that banana possesses homologues of both strategy I and II players. However, which strategy is activated may in all likelihood depend on soil conditions (Kokot and Phuong 1999). Unlike *Arabidopsis*, banana is as yet not a thoroughly investigated plant because of the attendant difficulties as mentioned earlier. The genome of *Musa acuminata* was (D'Hont et al. 2012) released only as recently as 2012 and the draft genome of *Musa balbisiana* in 2013 (Davey et al. 2013). Also, there are very few crystal structures of *Musa* proteins deposited in the Protein Data Bank, and there is no crystal structure for any of the genes involved in iron homeostasis in banana. Therefore, the authors have made an attempt here to predict the phylogeny of the major transcriptional regulators of the iron acquisition strategies using MEGA6.0 software (Tamura et al. 2013). Upon homology search, it is seen that *AtPOPEYE* and *AtFIT* have homologues in banana with significant query coverage. Phylogenetic relationship of these two sequences with other plant species has been depicted below (Figs. 20.3 and 20.4).

20.5 Translocation of Iron from Root to Shoot and Whole Plant

Iron is a double-edged sword in terms of its function and toxicity; this calls for tight regulation of the iron transport system. Iron requirements for plants are generally in the range 10^{-4}–10^{-8}M, but only 10^{-17} M is available at pH 7 (Mori 1999). While strategy 1 and strategy 2 mechanisms are quite distinct in roots of their respective plants, the aerial parts commonly use homologues of cross-strategies to achieve the intracellular homeostatic levels of iron. This hints at the merging boundaries of the two strategies with the advent of voluminous data coming from genomic, transcriptomic, and proteomic studies.

Iron is first taken from soil and travels through apoplastic and symplastic routes (Eddings and Brown 1967) through the root cellular layers. From here, it makes its way into the aerial parts via the xylem to the phloem and finally into the sink tissues – fruit and seeds (Kim and Guerinot 2007). In addition, Fe may also be retro-transported from senescing tissue to young and juvenile tissues coordinated by members of the NAC TF family (Ricachenevsky et al. 2013). In brief, Fe is transported through the plant as follows.

It enters the root epidermis through the concerted action of IRT1 and FRO2 (strategy 1) or as PS complexes taken up by TOM transporters (strategy 2). In the cell it exists as Fe^{+2}complexed with unknown chelators. The movement between root cell layers to the stele is via the symplast, as the endodermis prevents apoplastic leakage because of the hydrophobic Casparian strip (Kobayashi and Nishizawa 2012). At this point, there are two pathways: one for the efflux of chelators and the other for the efflux of the Fe

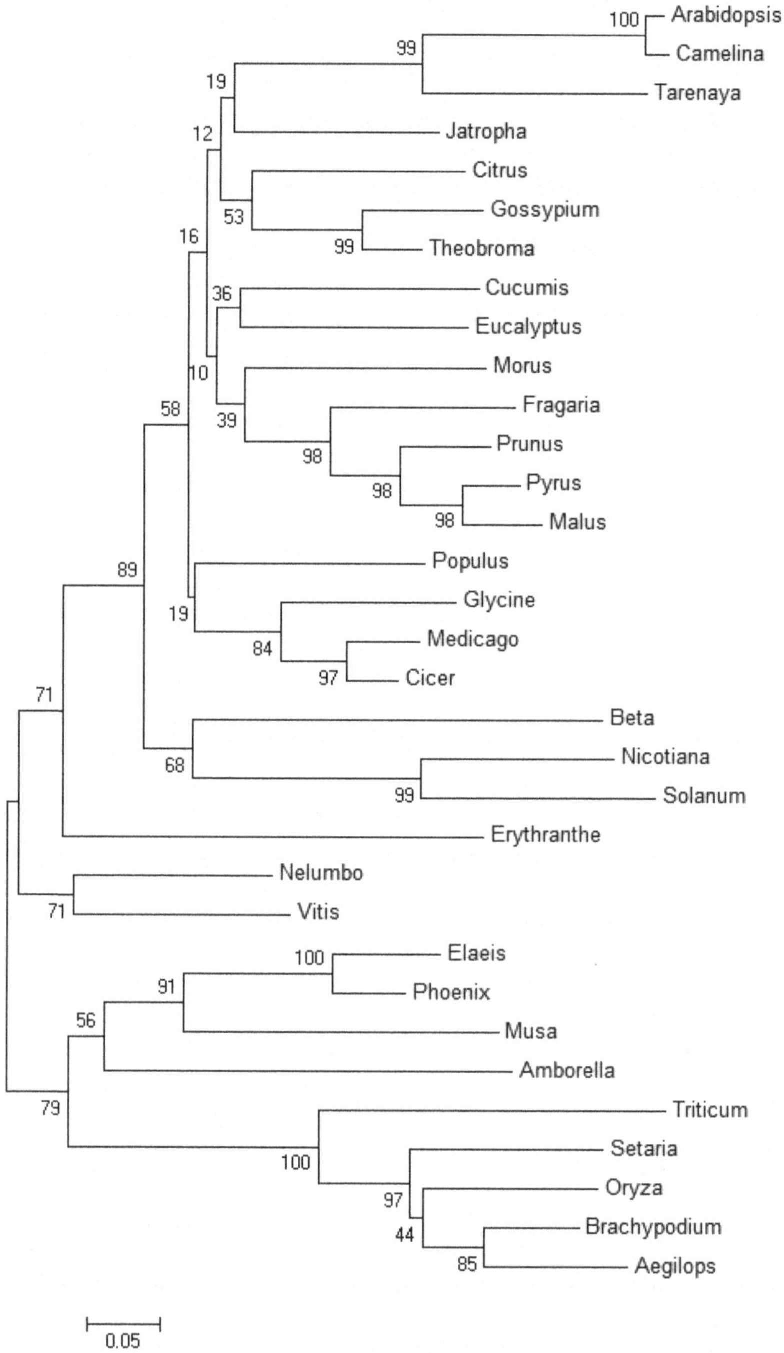

Fig. 20.3 Phylogenetic relationship of *Musa bHLH47* (putative POPEYE) with other closely related POPEYE sequences with Bootstrap test (10,000 replicates) was generated using Clustal Omega, and the evolutionary analyses were conducted in MEGA6 using neighbor-joining method (Tamura et al. 2013)

ion. The multidrug and toxin efflux transporter FRD3 transports citrate into the xylem (Rogers and Guerinot 2002; Durrett et al. 2007) as evidenced by chlorotic frd-3 plants which accumulated less shoot ferric but could be rescued by supplying exogenous citrate. As for the transport of Fe into the xylem, the most likely exporter for this is the *Arabidopsis* ferroportin 1/iron-regulated 1 (FPN1/IREG1) protein, homologous to mammalian FPN1 and expressed in the stele (Morrissey et al. 2009).

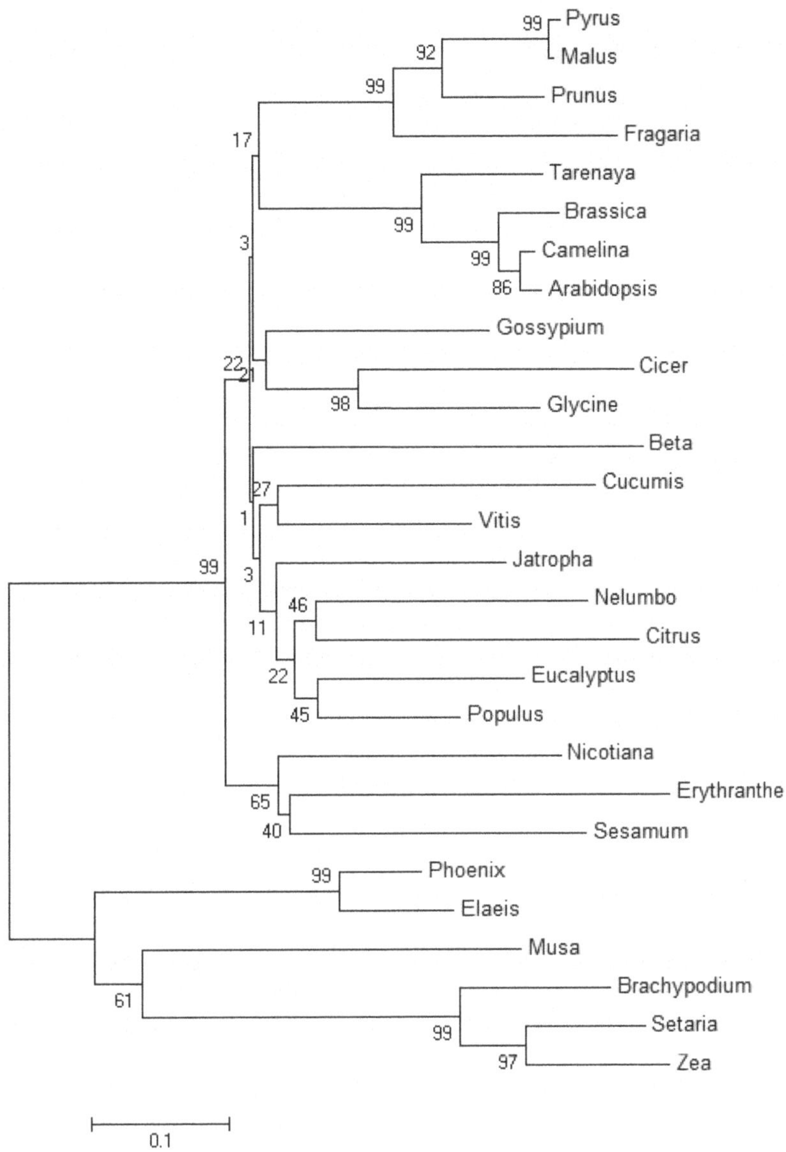

Fig. 20.4 Phylogenetic relationship of *MusaFER-LIKE IRON DEFICIENCY-INDUCED TRANSCRIPTION FACTOR* (putative FIT) with other closely related FIT sequences with Bootstrap test (10,000 replicates) was generated using Clustal Omega, and the evolutionary analyses were conducted in MEGA6 using neighbor-joining method (Tamura et al. 2013)

The transpiration stream carries the chelated Fe^{+3} into the source tissue (leaf). At the acidic pH (5.5) of the xylem, Fe^{+3} is better chelated by citrate (Rellán-Álvarez et al. 2010a, b). Intercellularly, iron is off-loaded from the complex and remains in oxidized form. Such precipitated apoplastic Fe is also precious and is salvaged by complexation with phenolics such as protocatechuic acid and caffeic acid, secreted into the xylem by the PHENOLICS EFFLUX ZERO (PEZ) transporter (Bashir et al. 2011a). This transporter was originally discovered in rice through a mutant deficient in xylem phenolics and is expressed on the plasma membrane in root tips and leaf mesophyll and vasculature. A number of reducing agents like ascorbate, light, as well as membrane FRO homologues (FRO6) (Brüggemann et al. 1993; Feng et al. 2006; Jeong and Connolly 2009) reduce Fe^{+3} to Fe^{+2}, and in this form it is transported into the mesophyll cells by a member of the IRT family. First discovered in an *Arabidopsis halleri* Zn hyperaccumulator, *At*IRT3 seems to be a potential candidate because it is highly expressed in the xylem and mesophyll cells (Lin et al. 2009) and mediates both Fe and Zn uptake into *Arabidopsis*. Once inside, Fe moves passively between cells through plasmodesmata as Fe^{+2} complexed with the non-proteinaceous amino acid nicotianamine (Takahashi et al. 2003). Any Fe^{+2}-NA complexes formed in the apoplast are taken up by cells through the members of the YSL transporter family described earlier. Once thought to be specific to strategy II plants only, they are now also known to translocate Fe across tissues like vasculature, phloem, seed (Ishimaru et al. 2010), and reproductive tissues such as the pollen tube and phloem of lamina joints (Aoyama et al. 2009) in both dicots and non-graminaceous monocots (reviewed by Curie et al. 2009). In fact *Arabidopsis ysl1ysl3* double mutants are defective in reproduction as these transporters are required for fertility and normal seed loading (Chu et al. 2010). These also mediate xylem-to-phloem translocation of this iron complex, e.g., OsYSL2 (Koike et al. 2004) and OsYSL15 in phloem transport of iron to seeds, flowers, and scutellar epithelial cells (Inoue et al. 2009). The phloem pH is alkaline (pH 7.5) which favors the formation of Fe-NA complexes.

For recirculation of iron through the phloem cells, the membrane oligopeptide transporter OPT3 has been implicated in the movement of iron. Related to YSLs, OPT3 is essential for the mobilization of Fe from leaves via the phloem and redistribution of Fe, Zn, and Cd in the plant (Zhai et al. 2014; Mendoza-Cózatl et al. 2014). However the identity of the chelate it transports is unknown. *opt3* mutants are embryo lethal, and *opt3* knockdown plants exhibit constitutive Fe deficiency response, hyperaccumulating Fe in leaves. In all this movement, pH plays a pivotal role in determining the stability and direction in which the complexes are mobilized.

Fe from the phloem, i.e., from the sieve tube, is remobilized and distributed through both the apoplast and the symplast. Another protein which is upregulated in Fe deficiency is the OPT7 transporter of rice, expressed in the shoot, root tips, and vascular tissue of roots. As yet its function has yet not been identified, for neither does it rescue the yeast *fet3fet4* mutant nor does it transport Fe III-DMA or Fe II-NA in frog oocytes. However in *opt7* knockouts, the Fe deficiency response is permanently heightened, though there are high iron and ferritin levels in shoot tissues, thus reflecting apparent iron starvation because of impairment of physiological availability of iron (Bashir et al. 2015). Conversely overexpressing OPT7 plants show comparable levels of iron as control, without any observable phenotypic changes.

Iron Translocation to Sink Seed iron comes from xylem and phloem sources (Grillet et al. 2013). Xylem sources are the transpiration stream which carries ferric citrate, while the phloem mainly gains its iron from remobilization of stores in aging leaves. Waters and Grusak in 2008 estimated 60–70 % of seed iron to be of xylem origin, while 30–40 % is phloem imported. Members of the NAC transcription factor family are upregulated in both abiotic and biotic stress (reviewed by Ricachenevsky et al. 2013). They are also responsible for reverse loading of Fe from vegetative tissues into grain as demonstrated for wheat (Uauy et al. 2006; Waters et al. 2009), where the NAM-B1 locus was identified as a QTL associated with increased grain protein, iron, and zinc. In order to ensure the safe accumulation of iron in sinks and other tissues, the role of the vacuole cannot be understated (Lanquar et al. 2005). It acts as a storehouse of available Fe which is used by developing seedlings until Fe uptake systems are well established. In this regard the vacuolar iron transporter (VIT1) shuttles iron into the vacuole from the cytosol (Kim et al. 2006). The *vit1* mutant has disrupted Fe localization around the provascular tissue, and seedlings develop poorly, though overall iron content is not altered. Overexpression of VIT1 in cassava under the patatin promoter increased the iron content of cassava storage roots. But it also led to interveinal chlorosis of young leaves and lowered iron content in them, while there was a high concentration of iron in the stem tissue. This suggested that iron homeostasis was altered between the root and shoot (Narayanan et al. 2015). Conversely, the natural resistance-associated macrophage proteins, *At*NRAMP3 and

*At*NRAMP4, transport Fe out of the vacuole. NRAMPs were first identified in humans as the NRAMP2/DMT1 transporter in the gut. *nramp3nramp4* double mutants show unhealthy seedlings on Fe-deficient soil though seed iron levels are the same as wild-type seeds. These plants are not able to mobilize Fe from vacuolar globoids under Fe deficiency but are completely rescued upon supplying them iron, confirming the indispensability of these transporters in allowing utilization of vacuolar resources. The chloroplast and mitochondria are also hubs of Fe activity due to its role in the photosynthetic and respiratory pathways, respectively, and hence also the most vulnerable to oxidative damage. A brief summary of the iron uptake and homeostasis pathways is presented below.

Chloroplasts Iron is required in the photosystems PSI and PSII (12 atoms of Fe per monomer), the cytochrome b_6f, and the ferredoxins. It is also a component of the FeSOD enzyme which detoxifies superoxide radicals. The enzyme which synthesizes chlorophyll from Mg-protoporphyrin IX has a diiron catalytic center, and Fe deficiency may thus affect chlorophyll synthesis, causing chlorosis. In all, 80–90 % of cellular iron is chloroplastic (reviewed by Nouet et al. 2011).

When plants encounter below-optimal iron and high light, the electron transfer chain in both chloroplasts and mitochondria is unable to mediate safe passage of electrons through the chain. This leads to uncontrolled electron reactivity which creates free radicals from oxygen, in turn creating toxic tetrapyrrole intermediates as well as damaged proteins, lipids, and carbohydrates. Accordingly, Fe deficiency leads to a decrease in all photosynthetic proteins (Calvin cycle enzymes, plastocyanin, etc.) in *Chlamydomonas reinhardtii* (Hsieh et al. 2013), while in plants, the relative amounts of electron transfer complexes were reduced and carbon-linked reaction proteins increased as determined from the chloroplast proteome of sugar beet exposed to Fe deficiency (Andaluz et al. 2006). In response to Fe deficiency, barley, a tolerant crop, modifies its chloroplastic PSII system, by altering the levels of the monomer of Lhcb1 protein. This light-harvesting complex protein is essential for thermal dissipation of excitation energy (non-photosynthetic quenching) (Saito et al. 2010), and though the plants are chlorotic, they can survive long-term Fe deficiency.

The PERMEASE IN CHLOROPLASTS (PIC), originally called Tic21, a member of the protein translocon family (Teng et al. 2006), functions as a permease in the chloroplastic membrane. *Pic* knockout plants are chlorotic and have the same amount of leaf iron as wild type, but it is sequestered by increased levels of ferritin and is thus unutilized. This leads to poor chloroplast development and no thylakoids, as well as decrease in root IRT1. The evidence implicates an iron allocative role for PIC beyond that of protein translocation (Duy et al. 2007) and also that the chloroplast can act as a signaling medium to communicate the iron status of the shoot to the root. Similar to the other family members, FRO7 is a ferric chelate reductase in the chloroplast membrane which reduces Fe^{+3} and allows its utilization by the organelle. Plants knocked out for *fro7* are chlorotic and die on alkaline soils, but can be rescued by watering with high amounts of soluble Fe. In addition the chloroplasts have reduced iron and ferric chelate reductase activity (Jeong et al. 2008). This demonstrated that the organelle is capable of using both forms of Fe and a substantial proportion is from ferric iron. To avoid toxicity due to free iron, the chloroplastic ferritins (*At*FER1, *At*FER2, *At*FER3, and *At*FER4) sequester iron. Only FER2 is found in seeds and is not regulated by Fe, while FER1 and FER3 are induced by iron excess (Petit et al. 2001). Ferritins do not constitute the main iron source for seedling germination and photosynthesis machinery. However, the triple mutant *fer1fer3fer4* has high free-radical levels and shows altered flower development, with iron misaccumulation in reproductive organs and deregulated metal homeostasis between the stem and flower, besides elevated antioxidant mechanisms (Ravet et al. 2009). Taken together, ferritins, therefore, are oxidative protectants and may constitute part of the stem-flower iron signaling.

Mitochondria Iron is integral to the respiratory complexes I–IV as Fe-S clusters and heme, alternative oxidase, biotin synthase, and ferredoxins (reviewed by Nouet et al. 2011). Besides causing structural alterations in mitochondria (Vigani et al. 2009), iron deficiency decreases energy production and respiratory rate because the complexes are downregulated and heightens glycolytic and pentose phosphate pathways (Vigani and Zocchi 2009). However, NADPH oxidation still continues unimpaired because of the induction of alternative dehydrogenases (Vigani and Zocchi 2010). This is probably to ensure availability of ATP and reducing equivalents for proton extrusion and FRO activity in the Fe deficiency response (reviewed by Nouet et al. 2011).

The mitochondrial carrier family (MCF) member called mitochondrial iron transporter (MIT) in rice mediates iron uptake in the mitochondrion. Its orthologs are found in yeast, mammals, and zebra fish. MITs are required for proper seedling growth because knockout mutants are embryo lethal, while knockdown mutants have poor growth, fertility, and yield together with disturbed Fe regulation and less iron in the mitochondrion than wild type (Bashir et al. 2011a, b). In these plants, the VIT1 transporter is upregulated, suggesting that iron is being diverted to vacuoles for safety. Also, loss of MIT1 causes decrease in Fe-S clusters and the aconitase enzyme activity. As in the chloroplast, the localization of *At*FRO3 and *At*FRO8 in the mitochondrion suggests their involvement in the utilization of ferric iron by this organelle (reviewed by Jeong and Connolly 2009; Jain and Connolly 2013).

20.6 Iron Biofortification in Different Crops

In recent times genome sequencing of various plants has unveiled many genes in iron homeostasis. The genes related to iron metabolism have been overexpressed in both model and commercially important plants to understand their distribution, spatiotemporal expression, and functions. The table below provides a crisp gist of the major candidates used in these studies (Table 20.2).

The pleiotropic effects of Fe deficiency have been well characterized by transcriptome studies in *Arabidopsis* root and leaf. The data suggests maximum changes in the stele of the root and other changes which correlate well at both transcript and protein levels as briefed before in the case of POPEYE. Enzymes involved in tetrapyrrole synthesis and chlorophyll conversion from *b* to *a* were downregulated in Fe-deficient leaves, while xanthophyll biosynthesis was upregulated, possibly conferring protection from oxidative stress due to toxic tetrapyrrole intermediates. Interestingly, the composition of ribosomes was differentially altered in both sets of transcriptomes, in an organ-specific manner. In leaves, proteins belonging to the large subunit were downregulated, while in roots, it was true for those of the small subunit. Also, topological changes in the root are observed such as increased root length (Müller and Schmidt 2004), swelling of root tips, and extensive branching (Schmidt 1999) along with the increase in riboflavin amounts. Proteomic responses to iron deficiency have also been investigated in plants. In wild-type tomato challenged with Fe deficiency (Li et al. 2008), the proteome revealed upregulated starch degradation, elevated TCA cycle, and enhanced ascorbate cycling. This was accompanied by decrease in sucrose and glucan metabolism. The authors propose that the increased demand for iron under deficiency conditions may lead to starch breakdown in the face of lowered photosynthesis, to continue to provide energy via the TCA cycle and ascorbate for antioxidant protection. The TCA cycle also provides abundant organic acids (citrate, malate), which are employed by the plant to chelate iron both within the xylem and as extrusions from the root. Methionine cycle enzymes were also upregulated; pointing to an increased requirement for nicotianamine which complexed with Fe-II transports it through the phloem into source and sink tissues. Heat shock chaperonins were also

Table 20.2 Genes involved in iron homeostasis

Gene	Plant	Function	References
IRT1/2	*Arabidopsis (functional complementation in yeast)*	Fe^{+2} uptake in roots	Vert et al. (2002) and Korshunova et al. (1999)
			Vert et al. (2001, 2009)
IRT3	*Arabidopsis (functional complementation in yeast)*	Fe^{+2}/Zn^{+2} transporter for uptake in root and shoot	Lin et al. (2009)
FRO2	*Arabidopsis*	Ferric chelate reductase for iron uptake in roots	Robinson et al. (1999) and Connolly et al. (2003)
AHA PUMP	*Arabidopsis*, cucumber	Proton extrusion from roots to acidify rhizosphere and solubilize Fe	Santi et al. (2005) and Santi and Schmidt (2008)
PEZ	Rice	Efflux of protocatechuic acid from strategy I plant roots to chelate iron	Bashir et al. (2011a)
POPEYE	*Arabidopsis*	Negative regulation of iron homeostasis genes	Long et al. (2010)
FIT/FER	Tomato, *Arabidopsis*	Positive regulator of the Fe deficiency response	Ling et al. (2002) and Yuan et al. (2005)
NAS	Rice, barley	Synthesis of NA a metal chelator for phloem transport of iron and other metals	Ling et al. (1999), Higuchi et al. (1999), and Inoue et al. (2003)
NAAT	Rice, barley	Oxidative deamination of NA to produce mugeneic acid (MA)	Takahashi et al. (1999) and Inoue et al. (2008)
DMAS	Rice, barley, wheat, maize	Reduction of MA to give DMAs	Bashir et al. (2006)
IDS2, IDS3	Barley	Dioxygenases in phytosiderophore synthesis act upon DMA and epihydroxy DMA	Nakanishi et al. (2000)
TOM1	Rice, barley	Effluxer of PS in graminaceous plants	Nozoye et al. (2011)
YS1/YSL	Maize (YS1), rice (YSL)	Uptake of Fe^{+3}-PS chelates from soil	von Wirén et al. (1994) and Curie et al. (2001)
IRO 2	Rice	Positive regulator of iron homeostasis genes	Ogo et al. (2006, 2007, 2011)
IDEF1, IDEF2	Rice	Positive regulator of iron homeostasis genes and late embryogenesis-abundant proteins, in early and late phase of response, respectively	Kobayashi et al. (2007, 2009, 2012) and Ogo et al. (2008)
IRO3	Rice	Negative regulator	Zheng et al. (2010)
Ferritin	*Arabidopsis*, tobacco, lettuce, rice, wheat, maize, banana, pineapple	Fe storage protein, protection from oxidative stress	Briat et al. (1995), Goto et al. (2000), Vasconcelos et al. (2003), Drakakaki et al. (2000, 2005), Kumar et al. (2011), and Mhatre et al. (2011)

Adapted from Kobayashi and Nishizawa (2012)

upregulated, presumably to aid in proper folding of proteins under nutrient stress. On the other hand, when similar analysis of the *fer* mutant roots was carried out, it yielded ten extra divergent proteins of most transcription factor and signaling roles as well as redox enzymes, thus confirming the important role of FER in the Fe deficiency response (Li et al. 2008).

Similarly, Brumbarova et al. in 2008 included FER overexpressing lines in their comparison of tomato plants exposed to Fe paucity and observed changes in stress proteins, signaling, TCA cycle, glycolysis, and amino acid metabolism. It is a known fact that Fe deficiency alters root architecture and physiology, a finding confirmed in the root metabolome-proteome study by Rellan Alvarez et al. in 2010a, b. Sugar beet roots subjected to iron deficiency in hydroponics were found to substantiate many of the changes seen in previous proteomic studies, with the inclusion of two novel characters: the DMRL (dimethyl-8-ribitylumazine) synthase and RFOs (raffinose family of oligosaccharides). The DMRL synthase is a key enzyme in riboflavin biosynthesis and required for the synthesis of the nucleotide redox equivalents FMN and FAD. FAD is a cofactor for the FRO2 reductase in the roots of strategy I plants like sugar beet (Robinson et al. 1999), and flavins accumulate in roots of Fe-deficient plants (Susin et al. 1993, 1994) hinting at a role in the management of root Fe in such plants. Additionally, the concentrations of raffinose and galactinol were increased in these roots. These function in the transport and storage of carbon via the phloem and antioxidant function, acting as compatible solutes for osmoprotection that is needed to withstand water fluxes because of ion disturbances. Thus they may represent a link in the signaling of Fe status of roots through the phloem to the aerial tissues, simultaneously serving protective roles in the root.

The results of such studies provide inroads into the complex world of interconnected pathways in the regulation of iron homeostasis, by identifying which genes change their expression profile in response to fluctuations in the metal supply. Traditional approaches to increasing iron content through transgenic means involve consideration of three aspects: sequestration, transport, and accumulation of iron in vacuoles of sinks like seed and fruits. In the former category, plant ferritins are among the popular candidates for iron biofortification in different economically important plants. Each molecule has a capacity to chelate up to 4500 atoms in the core which is formed from the assembly of 24 subunits (Briat et al. 1995). Thus, under iron deficiency plants rely upon this store of iron. Under unstressed conditions, ferritin levels are low but are induced in the presence of high iron, abscisic acid, and other abiotic and biotic stressors, revealing another essential role in the protection against oxidative stress (Proudhon et al. 1989; Deák et al. 1999). Heterologous expression studies have been done on *Arabidopsis*, lettuce, rice, cassava, tobacco, wheat, and banana, using ferritin of soybean origin. Overexpression of soybean ferritin (*SoyFer*) increased Fe in transgenic leaves of tobacco and lettuce by 30 % and 70 %, respectively. This also favored enhanced growth of the transgenic plants during early development stages (Goto et al. 2000). Transgenic tobacco transformed with *SoyFer* under 35S constitutive promoter not only exhibited increased iron levels in leaves but also created a pseudo-iron-deficient condition because of heightened sequestration of iron, corroborated by increased ferric reductase activity. An advantage conferred on transgenic plants was the resistance to methyl viologen treatment as well as to pathogens (Van Wuytswinkel et al. 1999; Deák et al. 1999). In the same line of thought, Djennane et al. in 2011, overexpressed pea ferritin in pear, in order to test whether the fruit would become resistant to fire blight, caused by the pathogen *Erwinia amylovora*. However, they were not able to report any significant resistance to the disease in transgenic plants as well as any changes in iron homeostasis genes. The ferritin protein also accumulated in young seedlings but not in dry seeds. This was consistent with the observations made by Drakakaki et al. in 2000 that under the constitutive maize ubiquitin promoter, Fe levels were high in vegetative tissues but not in seeds. Overexpressing alfalfa ferritin tobacco lines under CaMV 35S and Rubisco small subunit promoters (Deák et al. 1999) had increased ferritin levels in the leaves and stem. The study by Mhatre et al. in 2011 reported 3.6-fold and twofold increases in Fe and Zn, respectively, in soybean ferritin overexpressing lines of pineapple. Similarly, the first study carried out in banana, by Kumar et al. in 2011, reports a sixfold increase in Fe and a fourfold increase in Zn levels in soybean ferritin overexpressing lines.

Overexpression of soybean ferritin in maize plants has effects on the intrinsic gene and protein levels as well as the concentrations of different minerals. After transformation of SoyFer, along with the expected increase in *SoyFer*, only NAAT1 levels were further increased, while in leaf samples, ferredoxin mRNA-binding levels were increased 6.6-fold after the introduction of soybean ferritin, though other endogenous mRNA levels were not changed. However, the most changes were evident in seed endosperm in which the 19 kDa and the 22 kDa zeins were reduced by eightfold each, but 27, 16, 15, and 18 kDa genes were not affected (Kanobe et al. 2013). Also, ferredoxin, NAS3, and both maize and soybean ferritin were upregulated in the seeds of transgenic plants. As for the metals, the Ca, Mg, and Fe levels in grain endosperm were found to be elevated, whereas the Cu, Mn, and Zn levels were not significantly different. The authors of the study propose that the increased level of NAAT in the root may reflect a pseudo-iron-deficient condition created due to sequestration of iron in the endospermic ferritin core. Explaining the effect on ferredoxin, they hypothesize that this may not be directly induced by ferritin overexpression but rather due to iron itself. This is because ferredoxin is a 4Fe-4S cluster-containing protein, which may be regulated by Fe just as ferritin is also affected by iron status. As for the effect on zeins, this may have happened because of the usage of the super gamma zein endosperm-specific promoter. This study exemplifies the need to comprehensively assess the changes that ensue from the introduction of a transgene and the explicit effects on endogenous transcriptome and proteome. This could potentially aid in the consumer acceptance and release of transgenic crops.

The scenario is different when heterologous expression under endosperm-specific promoters is used. Goto et al. in 1999 reported threefold greater increase of Fe in seeds of transgenic rice. This "ferritin rice," as they opine, would be sufficient to account for 30–50 % of the daily adult requirement of iron. Seconding this was a study by Qu et al. in 2005 where promoters of seed storage proteins, namely, glutelin and globulin, were tested for their ability to drive seed-specific and high-level expression of *SoyFer.* In their study of the accumulation of iron, each promoter alone behaved as expected accumulating higher iron than control in subaleurone layer and endosperm respectively, but the expected additive effect was not seen with both promoters together due to possible co-suppression. Elemental imaging studies such as those using X-ray synchrotron fluorescence have proved valuable in depicting the layout of metals in the rice grain (Takahashi et al. 2009). Their study revealed iron accumulation in the aleurone layer, integument, and scutellum of the grain. Such tissues are lost during milling, necessitating the need for targeted approaches to increasing iron content of the endosperm proper, so that the beneficial effect of fortification is not negated.

Another avenue explored to increase plant iron levels is the usage of a metal chelator to facilitate iron transport within the plant. Nicotianamine (NA) is believed to be an iron-chelating agent thus maintaining iron solubility within the plant. Synthesized from three molecules of S-adenosylmethionine (SAM) (Shojima et al. 1989), it is also the precursor for the phytosiderophore biosynthesis pathway (Higuchi et al. 1999; Lee et al. 2009). It has been shown to bind both ferrous and ferric ions and may probably serve a protective role because the Fe-NA complexes are not Fenton reactive. The tomato *chloronerva* mutant (Ling et al. 1999) had severe interveinal chlorosis and a hyperactive strategy I iron deficiency response, mapped to a mutation in the gene for nicotianamine synthase (NAS). The activity of NAS differs among crops (Higuchi et al. 2001), as seen in barley (Herbik et al. 1999) and rice. This gene has been cloned, purified, and assayed in both plants, but it functions differently in each, the two crops being differently tolerant to Fe deficiency (Von Wirén et al. 1999). Barley, the tolerant one, secretes more MAs than rice and thus has higher root levels of NA. Unlike in strategy II plants, NAS is not upregulated in strategy I plants upon iron deficiency, but is all the same, required for both types of responses to function normally. Transferring the NAAT gene from barley into tobacco (a strategy I plant which does not produce MAs) consumed endogenous NA, making the transgenic tobacco develop interveinal chlorosis and sterile abnormal flowers, pointing to the importance of this polyamine in intercellular metal transport (Masuda et al. 2012).

Rice OsNAS1, OsNAS2, and OsNAS3 are three genes with functional and expression pattern differences (Inoue et al. 2003). NAS1 and 2 are mainly expressed in root stele in Fe replete conditions, with widening expression throughout the root and shoot when challenged with deficiency (Wirth et al. 2009). On the other hand, NAS3 is expressed in shoots under sufficiency and changes to a root-specific pattern under deficiency (Takahashi et al. 2003). This probably reflects the variable requirements of the shoot and root in deficient and sufficient conditions. These genes were independently constitutively overexpressed (Johnson et al. 2011) in the plant, and the expression of OsNAS2 altered the distribution of Fe in the grain as compared to WT.

20.6.1 Combinatorial Strategy Approach for Increasing Iron Content

From overexpression of single genes, the focus today is to use combinations of genes to obtain, store, and channelize the iron in the plant system. Goto et al. in 1999 suggest that in order to obtain a higher level of iron, co-expressing ferric reductase and ferritin will aid in effective uptake of iron from root level and storage, respectively. This therefore makes this strategy more effective than expression of ferritin alone.

Upon treating ferric salts with dietary extracts of beef, egg albumin, and organic acids like ascorbate, it was found that meat extracts and ascorbate imparted higher bioavailability to iron by increasing the proportion of ferrous form of iron (Kapsokefalou and Miller 1991). Another study indicated an inverse correlation between iron absorption and phytic acid content in cereals studied by Cook et al. 1997. Previous studies on phytase of *Aspergillus niger* have revealed its ability to increase bioavailability of Fe, when added to wheat rolls, through its degradative activity on the antinutrient phytic acid (Sandberg et al. 1996). Lucca et al. in 2001a, b integrated these concepts with the conventional overexpression of ferritin alone and obtained a higher iron level in rice grains.

The combined and targeted expression of *Pvferritin*, *AtNAS1*, and *Afphytase* in rice endosperm was studied, and a sixfold increase in iron content in the endosperm of transgenic rice plants was obtained. Drakakaki et al. in 2005 also overexpressed phytase, alone or in combination with ferritin in maize, and obtained significant levels of iron in seeds as well as better bioavailability as measured by in vitro digestibility studies. Pleiotropic effects due to integrated genes were found to be nil as there was no observable difference in manganese and other cation levels of control and transgenic plants (Wirth et al. 2009).

Three iron homeostasis genes working at three different levels were introduced by Masuda et al. in 2012 who found it to be a more effective strategy to increase iron in the endosperm by targeted overexpression. The genes they used were ferritin under endosperm-specific promoter, increased translocation was achieved by overexpression of nicotianamine, and the iron flux was diverted into the endosperm by overexpression of nicotianamine transporter OsYSL2 put under endosperm-specific promoter and sucrose transporter promoter. Drawing upon the ideology of Lucca et al. (2001a, b), Wirth et al. (2009) overexpressed *AtNAS*, *Pvferritin*, and *Afphytase* in rice. The latter two genes were under the control of endosperm-specific globulin promoter, while NAS was constitutively expressed under the CaMV 35S promoter throughout the plant. They obtained up to sixfold increase in iron content of the transgenic endosperms as well as increased zinc, but no effects on other divalent metal ions.

Banana is increasingly finding importance as an experimental crop for transgenic studies. Sagi et al., in 1995, first reported the transformation of banana with the beta-glucuronidase gene by particle bombardment (Becker et al. 2000) of embryogenic suspension cells (ECS). Subsequently, various parts of the plant such as suckers, shoot tips, and ECS are generated from male flower buds in conjunction with other methods such as *Agrobacterium*-mediated transformation (Ganapathi et al. 2001; Khanna et al. 2004; Sreeramanan et al. 2006; Subramanyam et al. 2011) have been tested with success. Transgenic banana resistant to various biotic and abiotic stress conditions (Shekhawat et al. 2011; Shekhawat and Ganapathi 2013; Wairegi et al. 2010; Ndungo et al. 2006; Studholme et al. 2010) has been developed. Also, an attempt to express hepatitis B surface antigen in banana for the purpose of edible vaccine production is evident from the study by Kumar et al. in 2005. As far as nutrition is concerned, Kumar et al. in 2011 first reported the iron biofortification of banana using soybean ferritin (SF) transformed into the Rasthali (AAB) cultivar of banana. They cloned the gene into the modified expression binary vector pBI121 under two promoters: a constitutive ubiquitin promoter of *Arabidopsis* (pSF) and an inducible ethylene-forming enzyme promoter (of the gene aminocyclopropane-1-carboxylic acid oxidase) (pEFE-SF). Both constructs were introduced into banana embryogenic cell suspensions via *Agrobacterium*-mediated transformation, and embryos were placed on successive rounds of selection on geneticin. Successfully regenerated plantlets were hardened in the greenhouse and leaves tested for Fe and Zn levels by atomic absorption spectroscopy. A dramatic increase in iron of up to 667 ug/g in the pSF plants and up to 264 ug/g in pEFE-SF plants was observed. Zinc levels were also correspondingly increased. These amounted to a sixfold increase of Fe and a 4.6-fold increase in Zn, respectively, in plants constitutively expressing SF. The increase, though to a smaller extent, was also observed in pEFE-SF expressing lines and possibly could be due to a difference in promoter strength. Although fruit levels of Fe and Zn remain to be analyzed, this study also brings to light the potential of inducible promoters such as EFE, to restrict expression to fruit and thus prevent unwanted development and growth defects in the whole plant due to the transgene. Elsewhere, micronutrient biofortification of banana is underway with field trials of vitamin A- and iron-biofortified banana showing promising results (Mlalazi et al. 2012; Waltz 2014).

20.7 Conclusion

Though various strategies have been worked out to increase micronutrient contents especially iron and zinc in food crops, the results obtained so far may not be enough to alleviate micronutrient deficiency. Among the different experimental crops tried for this purpose, banana seems to be an attractive choice as mentioned earlier. As described above in a preliminary report by Kumar et al. (2011) who described the development of transgenic banana plants overexpressing soybean ferritin gene and reported a 6.32-fold increase of iron in the leaves of transgenic banana plants. If such an increase can be obtained in the fruits, the authors suggest that consumption of 30 g banana fruit per day can give required RDA of iron. Though it looks attractive, achieving this magnitude of accumulation in banana fruits is a challenge. This would require overexpression of iron storage genes along with strategic transporters under fruit-specific promoters like that of expansin and reduction of antinutrient levels. Also, transgenic banana in addition to regulatory norms will have to pass the human consumption nutritional test trials. However, Chong-Pérez et al. in 2013 succeeded in generating 41.7 % of marker-free transgenic banana plants providing hope that this technology in the time to come could be used in a widespread manner. Other technologies gaining importance for genome editing of transgenic plants include the usage of zinc finger nucleases (Urnov et al. 2010), transcription activator-like effector nucleases (TALEN) (Gaj et al. 2013), and the recently introduced highly specific CAS-CRISPER (Cong et al. 2013) technology. As mentioned before, bioavailability studies were carried out recently (García et al. 2015) with non-transgenic banana fruits, but it is essential to establish the same with transgenic banana fruits as well. Also extensive studies on different networks of genes and/or transcription factors controlling iron chelation, absorption, and transport within the plant system need to be researched further to achieve the set target of alleviating iron deficiency anemia. Toward this goal of biofortifying bananas, the Department of Biotechnology, Government of India, through the Biotechnology Industry Research Assistance Council funded a technology transfer project from Queensland University of Technology, Australia, to India in a network project mode with five Indian laboratories including ours. The results of this project will be available in a couple of years from now. It is hoped that these banana plants will prove to be important contributors to alleviating iron deficiency anemia and possibly other micronutrient deficiencies as well.

Acknowledgment The authors thank Biotechnology Industry Research Assistance Council (a Government of India Enterprise of DBT) for the financial assistance in the form of banana biofortification project.

References

Allen LH (2003) Interventions for micronutrient deficiency control in developing countries: past, present and future. J Nutr 133(11):3875S–3878S

Andaluz S, López-Millán AF, De las Rivas J, Aro EM, Abadía J, Abadía A (2006) Proteomic profiles of thylakoid membranes and changes in response to iron deficiency. Photosynth Res 89(2–3):141–155

Andrews NC, Schmidt PJ (2007) Iron homeostasis. Annu Rev Physiol 69:69–85

Aoyama T, Kobayashi T, Takahashi M et al (2009) OsYSL18 is a rice iron (III)–deoxymugineic acid transporter specifically expressed in reproductive organs and phloem of lamina joints. Plant Mol Biol 70(6):681–692

Araki R, Murata J, Murata Y (2011) A novel barley yellow stripe 1-like transporter (HvYSL2) localized to the root endodermis transports metal–phytosiderophore complexes. Plant Cell Physiol 52(11):1931–1940

Ariza-Nieto M, Sanchez MT, Heller LI, Hu Y, Welch RM, Glahn RP (2006) Cassava (*Manihot esculenta*) has high potential for iron biofortification. FASEB J 20(4):A624

Ariza-Nieto M, Blair MW, Welch RM, Glahn RP (2007) Screening of iron bioavailability patterns in eight bean (Phaseolus vulgaris L.) genotypes using the Caco-2 cell in vitro model. J Agric Food Chem 55(19):7950–7956

Baltussen R, Knai C, Sharan M (2004) Iron fortification and iron supplementation are cost-effective interventions to reduce iron deficiency in four subregions of the world. J Nutr 134(10):2678–2684

Bari R, Jones JD (2009) Role of plant hormones in plant defence responses. Plant Mol Biol 69(4):473–488

Bashir K, Inoue H, Nagasaka S, Takahashi M, Nakanishi H, Mori S, Nishizawa NK (2006) Cloning and characterization of deoxymugineic acid synthase genes from graminaceous plants. J Biol Chem 281(43):32395–32402

Bashir K, Nagasaka S, Itai RN et al (2007) Expression and enzyme activity of glutathione reductase is upregulated by Fe-deficiency in graminaceous plants. Plant Mol Biol 65(3):277–284

Bashir K, Ishimaru Y, Shimo H, Kakei Y, Senoura T, Takahashi R, Satob Y, Satob Y, Uozumib N, Nakanishis H, Nishizawa NK (2011a) Rice phenolics efflux transporter 2 (PEZ2) plays an important role in solubilizing apoplasmic iron. Soil Sci Plant Nutr 57(6):803–812

Bashir K, Ishimaru Y, Shimo H et al (2011b) The rice mitochondrial iron transporter is essential for plant growth. Nat Commun 2:322. doi:10.1038/ncomms1326

Bashir K, Ishimaru Y, Itai RN, Senoura T, Takahashi M, An G, Oikawa T, Ueda M, Sato A, Uozumi N, Nakanishi H, Nishizawa NK (2015) Iron deficiency regulated OsOPT7 is essential for iron homeostasis in rice. Plant Mol Biol 88(1–2):165–176

Beard J (2003) Iron deficiency alters brain development and functioning. J Nutr 133(5):1468S–1472S

Becker DK, Dugdale B, Smith MK, Harding RM, Dale JL (2000) Genetic transformation of Cavendish banana (Musa spp. AAA group) cv 'Grand Nain' via microprojectile bombardment. Plant Cell Rep 19(3):229–234

Black MM (2003) Micronutrient deficiencies and cognitive functioning. J Nutr 133(11):3927S–3931S

Blair MW, Astudillo C, Grusak MA, Graham R, Beebe SE (2009) Inheritance of seed iron and zinc concentrations in common bean (*Phaseolus vulgaris* L.). Mol Breed 23(2):197–207

Briat JF, Labouré AM, Laulhére JP et al (1995) Molecular and cellular biology of plant ferritins. In: Iron nutrition in soils and plants. Springer, Dordrecht, pp 265–276

Brüggemann W, Maas-Kantel K, Moog PR (1993) Iron uptake by leaf mesophyll cells: the role of the plasma membrane-bound ferric-chelate reductase. Planta 190(2):151–155

Brumbarova T, Matros A, Mock HP, Bauer P (2008) A proteomic study showing differential regulation of stress, redox regulation and peroxidase proteins by iron supply and the transcription factor FER. Plant J 54(2):321–334

Brumbarova T, Bauer P, Ivanov R (2015) Molecular mechanisms governing *Arabidopsis* iron uptake. Trends Plant Sci 20(2):124–133

Bruner AB, Joffe A, Duggan AK, Casella JF, Brandt J (1996) Randomised study of cognitive effects of iron supplementation in non-anaemic iron-deficient adolescent girls. Lancet 348(9033):992–996

Cantrell RP, Reeves TG (2002) The cereal of the world's poor takes center stage. Science 296(5565):53. doi:10.1126/science.1070721

Cattivelli L, Rizza F, Badeck FW et al (2008) Drought tolerance improvement in crop plants: an integrated view from breeding to genomics. Field Crop Res 105(1):1–14

Chandler S (1995) The nutritional value of bananas. In: Gowen S (ed) Bananas and plantains. Chapman & Hall, London, pp 468–480. SE1 8HN. ISBN: 0-412-36870-6

Chen WW, Yang JL, Qin C, Jin CW, Mo JH, Ye T, Zheng SJ (2010) Nitric oxide acts downstream of auxin to trigger root ferric-chelate reductase activity in response to iron deficiency in *Arabidopsis*. Plant Physiol 154(2):810–819

Chong-Pérez B, Reyes M, Rojas L, Ocaña B, Ramos A, Kosky RG, Angenon G (2013) Excision of a selectable marker gene in transgenic banana using a Cre/lox system controlled by an embryo specific promoter. Plant Mol Biol 83(1–2):143–152

Chu HH, Chiecko J, Punshon T, Lanzirotti A, Lahner B, Salt DE, Walker EL (2010) Successful reproduction requires the function of *Arabidopsis* yellow Stripe-Like1 and yellow Stripe-Like3 metal-nicotianamine transporters in both vegetative and reproductive structures. Plant Physiol 154(1):197–210

Cogswell ME, Parvanta I, Ickes L, Yip R, Brittenham GM (2003) Iron supplementation during pregnancy, anemia, and birth weight: a randomized controlled trial. Am J Clin Nutr 78(4):773–781

Colangelo EP, Guerinot ML (2004) The essential basic helix-loop-helix protein FIT1 is required for the iron deficiency response. Plant Cell 16:3400–3412

Cong L, Ran FA, Cox D et al (2013) Multiplex genome engineering using CRISPR/Cas systems. Science 339(6121):819–823

Connolly EL, Campbell NH, Grotz N, Prichard CL, Guerinot ML (2003) Overexpression of the FRO2 ferric chelate reductase confers tolerance to growth on low iron and uncovers posttranscriptional control. Plant Physiol 133(3):1102–1110

Cook JD, Reddy MB, Burri J, Juillerat MA, Hurrell RF (1997) The influence of different cereal grains on iron absorption from infant cereal foods. Am J Clin Nutr 65(4):964–969

Correa-Aragunde N, Graziano M, Lamattina L (2004) Nitric oxide plays a central role in determining lateral root development in tomato. Planta 218(6):900–905

Crompton DWT, Nesheim MC (2002) Nutritional impact of intestinal helminthiasis during the human life cycle. Annu Rev Nutr 22(1):35–59

Curie C, Panaviene Z, Loulergue C, Dellaporta SL, Briat JF, Walker EL (2001) Maize yellow stripe1 encodes a membrane protein directly involved in Fe (III) uptake. Nature 409(6818):346–349

Curie C, Cassin G, Couch D, Divol F, Higuchi K, Le Jean M et al (2009) Metal movement within the plant: contribution of nicotianamine and yellow stripe 1-like transporters. Ann Bot 103(1):1–11

D'Hont A, Denoeud F, Aury JM, Baurens et al (2012) The banana (*Musa acuminata*) genome and the evolution of monocotyledonous plants. Nature 488(7410):213–217

Dary O, Mora JO (2002) Food fortification to reduce vitamin A deficiency: international vitamin a consultative group recommendations. J Nutr 132(9):2927S–2933S

Das JK, Salam RA, Kumar R, Bhutta ZA (2013) Micronutrient fortification of food and its impact on woman and child health: a systematic review. Syst Rev 2(1):67. doi:10.1186/2046-4053-2-67

Davey MW, Gudimella R, Harikrishna JA, Sin LW, Khalid N, Keulemans J (2013) A draft *Musa balbisiana* genome sequence for molecular genetics in polyploid, inter-and intra-specific Musa hybrids. BMC Genomics 14(1):683. doi:10.1186/1471-2164-14-683

Deák M, Horváth GV, Davletova S, Török K, Sass L, Vass I, Barna B, Király Z, Dudits D (1999) Plants ectopically expressing the ironbinding protein, ferritin, are tolerant to oxidative damage and pathogens. Nat Biotechnol 17(2):192–196

DellaPenna D (1999) Nutritional genomics: manipulating plant micronutrients to improve human health. Science 285(5426):375–379

DiDonato RJ, Roberts LA, Sanderson T, Eisley RB, Walker EL (2004) *Arabidopsis* yellow Stripe-Like2 (YSL2): a metal-regulated gene encoding a plasma membrane transporter of nicotianamine–metal complexes. Plant J 39(3):403–414

Dinneny JR, Long TA, Wang JY, Jung JW, Mace D, Pointer S, Barron C, Brady SM, Schiefelbein J, Benfey PN (2008) Cell identity mediates the response of *Arabidopsis* roots to abiotic stress. Science 320(5878):942–945

Distelfeld A, Cakmak I, Peleg Z et al (2007) Multiple QTL-effects of wheat Gpc-B1 locus on grain protein and micronutrient concentrations. Physiol Plant 129(3):635–643

Djennane S, Cesbron C, Sourice S, Cournol R, Dupuis F, Eychenne M, Loridon K, Chevreau E (2011) Iron homeostasis and fire blight susceptibility in transgenic

pear plants overexpressing a pea ferritin gene. Plant Sci 180(5):694–701
Donovan A, Lima CA, Pinkus JL, Pinkus GS, Zon LI, Robine S, Andrews NC (2005) The iron exporter ferroportin/Slc40a1 is essential for iron homeostasis. Cell Metab 1(3):191–200
Drakakaki G, Christou P, Stöger E (2000) Constitutive expression of soybean ferritin cDNA intransgenic wheat and rice results in increased iron levels in vegetative tissues but not in seeds. Transgenic Res 9(6):445–452
Drakakaki G, Marcel S, Glahn RP, Lund EK, Pariagh S, Fischer R et al (2005) Endosperm-specific co-expression of recombinant soybean ferritin and *Aspergillus* phytase in maize results in significant increases in the levels of bioavailable iron. Plant Mol Biol 59(6):869–880
Durrett TP, Gassmann W, Rogers EE (2007) The FRD3-mediated efflux of citrate into the root vasculature is necessary for efficient iron translocation. Plant Physiol 144(1):197–205
Duy D, Wanner G, Meda AR, von Wirén N, Soll J, Philippar K (2007) PIC1, an ancient permease in *Arabidopsis* chloroplasts, mediates iron transport. Plant Cell 19(3):986–1006
Eddings JL, Brown AL (1967) Absorption and translocation of foliar-applied iron. Plant Physiol 42(1):15–19
Feng H, An F, Zhang S, Ji Z, Ling HQ, Zuo J (2006) Light-regulated, tissue-specific, and cell differentiation-specific expression of the *Arabidopsis* Fe (III)-chelate reductase gene AtFRO6. Plant Physiol 140(4):1345–1354
Food and Agriculture Organization of the United Nations (FAO) (2015) The state of food insecurity in the world. http://www.fao.org/hunger/glossary/en/
Fraga CG (2005) Relevance, essentiality and toxicity of trace elements in human health. Mol Aspects Med 26(4):235–244
Gaj T, Gersbach CA, Barbas CF (2013) ZFN, TALEN, and CRISPR/Cas-based methods for genome engineering. Trends Biotechnol 31(7):397–405
Galloway R, McGuire J (1994) Determinants of compliance with iron supplementation: supplies, side effects, or psychology? Soc Sci Med 39(3):381–390
Ganapathi TR, Higgs NS, Balint-Kurti PJ, Arntzen CJ, May GD, Van Eck JM (2001) Agrobacterium-mediated transformation of embryogenic cell suspensions of the banana cultivar Rasthali (AAB). Plant Cell Rep 20(2):157–162
García MJ, Lucena C, Romera FJ, Alcántara E, Pérez-Vicente R (2010) Ethylene and nitric oxide involvement in the up-regulation of key genes related to iron acquisition and homeostasis in *Arabidopsis*. J Exp Bot 61(14):3885–3899. doi:10.1093/jxb/erq203
García MJ, Suárez V, Romera FJ, Alcántara E, Pérez-Vicente R (2011) A new model involving ethylene, nitric oxide and Fe to explain the regulation of Fe-acquisition genes in strategy I plants. Plant Physiol Biochem 49(5):537–544
García OP, Martínez M, Romano D et al (2015) Iron absorption in raw and cooked bananas: a field study using stable isotopes in women. Food Nutr Res 59:25976. doi:10.3402/fnr.v59.25976
Gegios A, Amthor R, Maziya-Dixon B et al (2010) Children consuming cassava as a staple food are at risk for inadequate zinc, iron, and vitamin A intake. Plant Foods Hum Nutr 65(1):64–70
Gendre D, Czernic P, Conéjéro G et al (2007) TcYSL3, a member of the YSL gene family from the hyper-accumulator Thlaspi caerulescens, encodes a nicotianamine-Ni/Fe transporter. Plant J 49(1):1–15
Ghandilyan A, Ilk N, Hanhart C et al (2009) A strong effect of growth medium and organ type on the identification of QTLs for phytate and mineral concentrations in three *Arabidopsis thaliana* RIL populations. J Exp Bot 60(5):1409–1425
Giehl RF, Lima JE, von Wirén N (2012) Localized iron supply triggers lateral root elongation in *Arabidopsis* by altering the AUX1-mediated auxin distribution. Plant Cell 24(1):33–49
Goto F, Yoshihara T, Shigemoto N, Toki S, Takaiwa F (1999) Iron fortification of rice seed by the soybean ferritin gene. Nat Biotechnol 17(3):282–286
Goto F, Yoshihara T, Saiki H, Takaiwa F, Shigemoto N (2000) Iron accumulation in transgenic plants expressing the soybean ferritin gene. Acta Hortic 521:101–110
Graham RD, Welch RM, Bouis HE (2001) Addressing micronutrient malnutrition through enhancing the nutritional quality of staple foods: principles, perspectives and knowledge gaps. Adv Agron 70:77–142
Green LS, Rogers EE (2004) FRD3 controls iron localization in *Arabidopsis*. Plant Physiol 136(1):2523–2531
Gregorio GB, Senadhira D, Htut H, Graham RD (2000) Breeding for trace mineral density in rice. Food Nutr Bull 21(4):382–386
Grillet L, Mari S, Schmidt W (2013) Iron in seeds–loading pathways and subcellular localization. Front Plant Sci 4:535. doi:10.3389/fpls.2013.00535
Grusak MA, DellaPenna D (1999) Improving the nutrient composition of plants to enhance human nutrition and health. Annu Rev Plant Biol 50(1):133–161
Gunshin H, Mackenzie B, Berger UV et al (1997) Cloning and characterization of a mammalian proton-coupled metal-ion transporter. Nature 388(6641):482–488
Guo B, Phillips JD, Yu Y, Leibold EA (1995) Iron regulates the intracellular degradation of iron regulatory protein 2 by the proteasome. J Biol Chem 270(37):21645–21651
Haas JD, Beard JL, Murray-Kolb LE, del Mundo AM, Felix A, Gregorio GB (2005) Iron-biofortified rice improves the iron stores of nonanemic Filipino women. J Nutr 135(12):2823–2830
Hardisson A, Rubio C, Baez A, Martin M, Alvarez R, Diaz E (2001) Mineral composition of the banana (Musa acuminata) from the island of Tenerife. Food Chem 73(2):153–161

Haydon MJ, Cobbett CS (2007) A novel major facilitator superfamily protein at the tonoplast influences zinc tolerance and accumulation in *Arabidopsis*. Plant Physiol 143(4):1705–1719

Henriques R, Jásik J, Klein M, Martinoia E, Feller U, Schell J, Pais MS, Koncz C (2002) Knock-out of *Arabidopsis* metal transporter gene IRT1 results in iron deficiency accompanied by cell differentiation defects. Plant Mol Biol 50(4–5):587–597

Herbik A, Koch G, Mock HP, Dushkov D, Czihal A, Thielmann J, Stephan UW (1999) Isolation, characterization and cDNA cloning of nicotianamine synthase from barley. Eur J Biochem 265(1):231–239

Hess SY (2010) The impact of common micronutrient deficiencies on iodine and thyroid metabolism: the evidence from human studies. Best Pract Res Clin Endocrinol Metab 24(1):117–132

Hess SY, Zimmermann MB, Arnold M, Langhans W, Hurrell RF (2002) Iron deficiency anemia reduces thyroid peroxidase activity in rats. J Nutr 132(7):1951–1955

Hetzel B (1983) Iodine deficiency disorders (IDD) and their eradication. Lancet 322(8359):1126–1129

Higuchi K, Suzuki K, Nakanishi H, Yamaguchi H, Nishizawa NK, Mori S (1999) Cloning of nicotianamine synthase genes, novel genes involved in the biosynthesis of phytosiderophores. Plant Physiol 119(2):471–480

Higuchi K, Watanabe S, Takahashi M, Kawasaki S, Nakanishi H, Nishizawa NK, Mori S (2001) Nicotianamine synthase gene expression differs in barley and rice under Fe-deficient conditions. Plant J 25(2):159–167

Hindt MN, Guerinot ML (2012) Getting a sense for signals: regulation of the plant iron deficiency response. Biochim Biophys Acta (BBA) Mol Cell Res 1823(9):1521–1530

Hoppe M, Hulthén L, Hallberg L (2006) The relative bioavailability in humans of elemental iron powders for use in food fortification. Eur J Nutr 45(1):37–44

Hsieh SI, Castruita M, Malasarn D et al (2013) The proteome of copper, iron, zinc, and manganese micronutrient deficiency in chlamydomonas reinhardtii. Mol Cell Proteomics 12(1):65–86

Hurrell R (2002) How to ensure adequate iron absorption from iron-fortified food. Nutr Rev 60(7):S7–S15. doi: http://dx.doi.org/10.1301/002966402320285137

Hurrell R, Egli I (2010) Iron bioavailability and dietary reference values. Am J Clin Nutr 91(5):1461S–1467S

Iannotti LL, Tielsch JM, Black MM, Black RE (2006) Iron supplementation in early childhood: health benefits and risks. Am J Clin Nutr 84(6):1261–1276

Inoue H, Higuchi K, Takahashi M, Nakanishi H, Mori S, Nishizawa NK (2003) Three rice nicotianamine synthase genes, OsNAS1, OsNAS2, and OsNAS3 are expressed in cells involved in long-distance transport of iron and differentially regulated by iron. Plant J 36(3):366–381

Inoue H, Takahashi M, Kobayashi T, Suzuki M, Nakanishi H, Mori S, Nishizawa NK (2008) Identification and localisation of the rice nicotianamine aminotransferase gene OsNAAT1 expression suggests the site of phytosiderophore synthesis in rice. Plant Mol Biol 66(1–2):193–203

Inoue H, Kobayashi T, Nozoye T et al (2009) Rice OsYSL15 is an iron-regulated iron (III)-deoxymugineic acid transporter expressed in the roots and is essential for iron uptake in early growth of the seedlings. J Biol Chem 284(6):3470–3479

Ishimaru Y, Suzuki M, Tsukamoto T et al (2006) Rice plants take up iron as an Fe3+-phytosiderophore and as Fe2+. Plant J 45(3):335–346

Ishimaru Y, Bashir K, Fujimoto M et al (2009) Rice-specific mitochondrial iron-regulated gene (MIR) plays an important role in iron homeostasis. Mol Plant 2(5):1059–1066

Ishimaru Y, Masuda H, Bashir K et al (2010) Rice metal-nicotianamine transporter, OsYSL2, is required for the long-distance transport of iron and manganese. Plant J 62(3):379–390

Ivanov R, Brumbarova T, Bauer P (2012) Fitting into the harsh reality: regulation of iron-deficiency responses in dicotyledonous plants. Mol Plant 5(1):27–42

Jacobs A, Miles PM (1969) Role of gastric secretion in iron absorption. Gut 10(3):226–229

Jain A, Connolly EL (2013) Mitochondrial iron transport and homeostasis in plants. Front Plant Sci 4:348. doi:10.3389/fpls.2013.00348

Jeong J, Connolly EL (2009) Iron uptake mechanisms in plants: functions of the FRO family of ferric reductases. Plant Sci 176(6):709–714

Jeong J, Cohu C, Kerkeb L, Pilon M, Connolly EL, Guerinot ML (2008) Chloroplast Fe (III) chelate reductase activity is essential for seedling viability under iron limiting conditions. Proc Natl Acad Sci 105(30):10619–10624

Johnson AA, Kyriacou B, Callahan DL, Carruthers L, Stangoulis J, Lombi E, Tester M (2011) Constitutive overexpression of the OsNAS gene family reveals single-gene strategies for effective iron-and zinc-biofortification of rice endosperm. PLoS One 6(9):e24476. doi:10.1371/journal.pone.0024476

Jones MA, Raymond MJ, Smirnoff N (2006) Analysis of the root-hair morphogenesis transcriptome reveals the molecular identity of six genes with roles in root-hair development in *Arabidopsis*. Plant J 45(1):83–100

Kanobe MN, Rodermel SR, Bailey T, Scott MP (2013) Changes in endogenous gene transcript and protein levels in maize plants expressing the soybean ferritin transgene. Front Plant Sci 4:196. doi:10.3389/fpls.2013.00196

Kapsokefalou M, Miller DD (1991) Effects of meat and selected food components on the valence of nonheme iron during in vitro digestion. J Food Sci 56(2):352–355

Keen CL, Clegg MS, Hanna LA et al (2003) The plausibility of micronutrient deficiencies being a significant

contributing factor to the occurrence of pregnancy complications. J Nutr 133(5):1597S–1605S
Khanna H, Becker D, Kleidon J, Dale J (2004) Centrifugation assisted *Agrobacterium tumefaciens*-mediated transformation (CAAT) of embryogenic cell suspensions of banana (Musa spp. Cavendish AAA and lady finger AAB). Mol Breed 14(3):239–252
Kim SA, Guerinot ML (2007) Mining iron: iron uptake and transport in plants. FEBS Lett 581(12):2273–2280
Kim SA, Punshon T, Lanzirotti A et al (2006) Localization of iron in *Arabidopsis* seed requires the vacuolar membrane transporter VIT1. Science 314(5803):1295–1298
Klatte M, Schuler M, Wirtz M, Fink-Straube C, Hell R, Bauer P (2009) The analysis of *Arabidopsis* nicotianamine synthase mutants reveals functions for nicotianamine in seed iron loading and iron deficiency responses. Plant Physiol 150(1):257–271
Kobayashi T, Nishizawa NK (2012) Iron uptake, translocation, and regulation in higher plants. Annu Rev Plant Biol 63:131–152
Kobayashi T, Nakanishi H, Takahashi M, Kawasaki S, Nishizawa NK, Mori S (2001) In vivo evidence that Ids3 from *Hordeum vulgare* encodes a dioxygenase that converts 2-deoxymugineic acid to mugineic acid in transgenic rice. Planta 212:864–871
Kobayashi T, Ogo Y, Itai RN, Nakanishi H, Takahashi M, Mori S, Nishizawa NK (2007) The transcription factor IDEF1 regulates the response to and tolerance of iron deficiency in plants. Proc Natl Acad Sci 104(48):19150–19155
Kobayashi T, Itai RN, Ogo Y, Kakei Y, Nakanishi H, Takahashi M, Nishizawa NK (2009) The rice transcription factor IDEF1 is essential for the early response to iron deficiency, and induces vegetative expression of late embryogenesis abundant genes. Plant J 60(6):948–961
Kobayashi T, Ogo Y, Aung MS et al (2010) The spatial expression and regulation of transcription factors IDEF1 and IDEF2. Ann Bot 105(7):1109–1117
Kobayashi T, Itai RN, Aung MS, Senoura T, Nakanishi H, Nishizawa NK (2012) The rice transcription factor IDEF1 directly binds to iron and other divalent metals for sensing cellular iron status. Plant J 69(1):81–91
Kobayashi T, Nagasaka S, Senoura T, Itai RN, Nakanishi H, Nishizawa NK (2013) Iron-binding haemerythrin RING ubiquitin ligases regulate plant iron responses and accumulation. Nat Commun 4:2792. doi:10.1038/ncomms3792
Koike S, Inoue H, Mizuno D, Takahashi M, Nakanishi H, Mori S et al (2004) OsYSL2 is a rice metal-nicotianamine transporter that is regulated by iron and expressed in the phloem. Plant J 39(3):415–424
KoiKokot S, Phuong TD (1999) Elemental content of Vietnamese rice part 2.† Multivariate data analysis. Analyst 124(4):561–569
Korshunova YO, Eide D, Clark WG, Guerinot ML, Pakrasi HB (1999) The IRT1 protein from *Arabidopsis thaliana* is a metal transporter with a broad substrate range. Plant Mol Biol 40(1):37–44
Kumar GBS, Ganapathi TR, Revathi CJ, Srinivas L, Bapat VA (2005) Expression of hepatitis B surface antigen in transgenic banana plants. Planta 222(3):484–493
Kumar GBS, Srinivas L, Ganapathi TR (2011) Iron fortification of banana by the expression of soybean ferritin. Biol Trace Elem Res 142(2):232–241
Lane DJ, Bae DH, Merlot AM, Sahni S, Richardson DR (2015) Duodenal cytochrome b (DCYTB) in iron metabolism: an update on function and regulation. Nutrients 7(4):2274–2296
Lanquar V, Lelièvre F, Bolte S et al (2005) Mobilization of vacuolar iron by AtNRAMP3 and AtNRAMP4 is essential for seed germination on low iron. EMBO J 24(23):4041–4051
Lee S, Jeon US, Lee SJ, Kim YK, Persson DP, Husted S et al (2009) Iron fortification of rice seeds through activation of the nicotianamine synthase gene. Proc Natl Acad Sci 106(51):22014–22019
Leyva-Guerrero E, Narayanan NN, Ihemere U, Sayre RT (2012) Iron and protein biofortification of cassava: lessons learned. Curr Opin Biotechnol 23(2):257–264
Li J, Wu XD, Hao ST, Wang XJ, Ling HQ (2008) Proteomic response to iron deficiency in tomato root. Proteomics 8(11):2299–2311
Lin YF, Liang HM, Yang SY, Boch A, Clemens S, Chen CC, Wu J-F, Huang J-L, Yeh KC (2009) *Arabidopsis* IRT3 is a zinc-regulated and plasma membrane localized zinc/iron transporter. New Phytol 182(2):392–404
Ling HQ, Koch G, Bäumlein H, Ganal MW (1999) Map-based cloning of chloronerva, a gene involved in iron uptake of higher plants encoding nicotianamine synthase. Proc Natl Acad Sci 96(12):7098–7103
Ling HQ, Bauer P, Bereczky Z, Keller B, Ganal M (2002) The tomato fer gene encoding a bHLH protein controls iron-uptake responses in roots. Proc Natl Acad Sci U S A 99:13938–13943
Lingam S, Mohrbacher J, Brumbarova T, Potuschak T, Fink-Straube C, Blondet E, Genschik P, Bauer P (2011) Interaction between the bHLH transcription factor FIT and ETHYLENE INSENSITIVE3/ ETHYLENE INSENSITIVE3-LIKE1 reveals molecular linkage between the regulation of iron acquisition and ethylene signaling in *Arabidopsis*. Plant Cell 23(5):1815–1829
Long TA, Tsukagoshi H, Busch W, Lahner B, Salt DE, Benfey PN (2010) The bHLH transcription factor POPEYE regulates response to iron deficiency in *Arabidopsis* roots. Plant Cell 22:2219–2236
López-Bucio J, Cruz-Ramírez A, Herrera-Estrella L (2003) The role of nutrient availability in regulating root architecture. Curr Opin Plant Biol 6(3):280–287
Lucca P, Hurrell R, Potrykus I (2001a) Genetic engineering approaches to improve the bioavailability and the level of iron in rice grains. Theor Appl Genet 102(2–3):392–397

Lucca P, Ye X, Potrykus I (2001b) Effective selection and regeneration of transgenic rice plants with mannose as selective agent. Mol Breed 7(1):43–49

Lung'aho MG, Mwaniki AM, Szalma SJ et al (2011) Genetic and physiological analysis of iron biofortification in maize kernels. PLoS One 6(6):e20429. doi:10.1371/journal.pone.0020429

Lynch J, Brown KM (1997) Ethylene and plant responses to nutritional stress. Physiol Plant 100(3):613–619

Ma JF, Nomoto K (1996) Effective regulation of iron acquisition in graminaceous plants. The role of mugineic acids as phytosiderophores. Physiol Plant 97(3):609–617

Ma JF, Shinada T, Matsuda C, Nomoto K (1995) Biosynthesis of phytosiderophores, mugineic acids, associated with methionine cycling. J Biol Chem 270(28):16549–16554

Masuda H, Ishimaru Y, Aung MS, Kobayashi T, Kakei Y, Takahashi M, Higuchi K, Nakanishi H, Nishizawa NK (2012) Iron biofortification in rice by the introduction of multiple genes involved in iron nutrition. Sci Rep 2:543. doi:10.1038/srep00543

Maurer F, Müller S, Bauer P (2011) Suppression of Fe deficiency gene expression by jasmonate. Plant Physiol Biochem 49(5):530–536

McKie AT, Barrow D, Latunde-Dada GO et al (2001) An iron-regulated ferric reductase associated with the absorption of dietary iron. Science 291(5509):1755–1759

McLaren DS (1999) Vitamin A deficiency disorders. J Indian Med Assoc 97(8):320–323

Mejía LA, Chew F (1988) Hematological effect of supplementing anemic children with vitamin A alone and in combination with iron. Am J Clin Nutr 48(3):595–600

Méndez-Bravo A, Raya-González J, Herrera-Estrella L, López-Bucio J (2010) Nitric oxide is involved in alkamide-induced lateral root development in *Arabidopsis*. Plant Cell Physiol 51(10):1612–1626

Mendoza-Cózatl DG, Xie Q, Akmakjian GZ et al (2014) OPT3 is a component of the iron-signaling network between leaves and roots and misregulation of OPT3 leads to an over-accumulation of cadmium in seeds. Mol Plant 7(9):1455–1469

Mhatre M, Srinivas L, Ganapathi TR (2011) Enhanced iron and zinc accumulation in genetically engineered pineapple plants using soybean ferritin gene. Biol Trace Elem Res 144(1–3):1219–1228

Mlalazi B, Welsch R, Namanya P et al (2012) Isolation and functional characterisation of banana phytoene synthase genes as potential cisgenes. Planta 236(5):1585–1598

Mohapatra D, Mishra S, Sutar N (2010) Banana and its by-product utilization: an overview. J Sci Ind Res 69(5):323–329

Mole DR (2010) Iron homeostasis and its interaction with prolyl hydroxylases. Antioxid Redox Signal 12(4):445–458

Mori S (1999) Iron acquisition by plants. Curr Opin Plant Biol 2(3):250–253

Mori S, Nishizawa N, Hayashi H, Chino M, Yoshimura E, Ishihara J (1991) Why are young rice plants highly susceptible to iron deficiency? In: Chen Y, Hadar Y (eds) Iron nutrition and interactions in plants. Kluwer Academic Publishers, Netherlands, pp 175–188

Morrissey J, Baxter IR, Lee J, Li L, Lahner B, Grotz N, Kaplan J, Salt DE, Guerinot ML (2009) The ferroportin metal efflux proteins function in iron and cobalt homeostasis in *Arabidopsis*. Plant Cell 21(10):3326–3338

Muckenthaler MU, Galy B, Hentze MW (2008) Systemic iron homeostasis and the iron-responsive element/iron-regulatory protein (IRE/IRP) regulatory network. Annu Rev Nutr 28:197–213

Mukherjee I, Campbell NH, Ash JS, Connolly EL (2006) Expression profiling of the *Arabidopsis* ferric chelate reductase (FRO) gene family reveals differential regulation by iron and copper. Planta 223(6):1178–1190

Müller M, Schmidt W (2004) Environmentally induced plasticity of root hair development in *Arabidopsis*. Plant Physiol 134(1):409–419

Muñoz EC, Rosado JL, López P, Furr HC, Allen LH (2000) Iron and zinc supplementation improves indicators of vitamin A status of Mexican preschoolers. Am J Clin Nutr 71(3):789–794

Murray CJL, Vos T, Lozano R et al (2013) Disability-adjusted life years (DALYs) for 291 diseases and injuries in 21 regions, 1990–2010: a systematic analysis for the global burden of disease study 2010. Lancet 380(9859):2197–2223

Nakanishi H, Yamaguchi H, Sasakuma T, Nishizawa NK, Mori S (2000) Two dioxygenase genes, Ids3 and Ids2, from hordeum vulgare are involved in the biosynthesis of mugineic acid family phytosiderophores. Plant Mol Biol 44:199–207

Narayanan N, Beyene G, Chauhan RD, Gaitán-Solis E, Grusak MA, Taylor N, Anderson P (2015) Overexpression of *Arabidopsis* VIT1 increases accumulation of iron in cassava roots and stems. Plant Sci 240:170–181

Ndungo V, Eden-Green S, Blomme G, Crozier J, Smith JJ (2006) Presence of banana xanthomonas wilt (Xanthomonas campestris pv. musacearum) in the Democratic Republic of Congo (DRC). Plant Pathol 55(2):294

Nemeth E, Tuttle MS, Powelson J et al (2004) Hepcidin regulates cellular iron efflux by binding to ferroportin and inducing its internalization. Science 306(5704):2090–2093

Nestel P, Bouis HE, Meenakshi JV, Pfeiffer W (2006) Biofortification of staple food crops. J Nutr 136(4):1064–1067

Nicolas G, Chauvet C, Viatte L et al (2002) The gene encoding the iron regulatory peptide hepcidin is regulated by anemia, hypoxia, and inflammation. J Clin Invest 110(7):1037–1044

Nouet C, Motte P, Hanikenne M (2011) Chloroplastic and mitochondrial metal homeostasis. Trends Plant Sci 16(7):395–404

Nozoye T, Nagasaka S, Kobayashi T et al (2011) Phytosiderophore efflux transporters are crucial for iron acquisition in graminaceous plants. J Biol Chem 286(7):5446–5454

Ogo Y, Itai RN, Nakanishi H et al (2006) Isolation and characterization of IRO2, a novel iron-regulated bHLH transcription factor in graminaceous plants. J Exp Bot 57(11):2867–2878

Ogo Y, Itai RN, Nakanishi H, Kobayashi T, Takahashi M, Mori S, Nishizawa NK (2007) The rice bHLH protein OsIRO2 is an essential regulator of the genes involved in Fe uptake under Fe-deficient conditions. Plant J 51(3):366–377

Ogo Y, Kobayashi T, Itai RN et al (2008) A novel NAC transcription factor, IDEF2, that recognizes the iron deficiency-responsive element 2 regulates the genes involved in iron homeostasis in plants. J Biol Chem 283(19):13407–13417

Ogo Y, Itai RN, Kobayashi T, Aung MS, Nakanishi H, Nishizawa NK (2011) OsIRO2 is responsible for iron utilization in rice and improves growth and yield in calcareous soil. Plant Mol Biol 75(6):593–605

Ozkan H, Brandolini A, Torun A, Altintas S, Eker S, Kilian B, Cakmak I (2007) Natural variation and identification of microelements content in seeds of einkorn wheat (*Triticum monococcum*). In: Buck HT, Nisi JE, Salomon N (eds) Wheat production in stressed environments, vol 12. Springer, Dordrecht, pp 455–462

Peleg Z, Cakmak I, Ozturk L et al (2009) Quantitative trait loci conferring grain mineral nutrient concentrations in durum wheat× wild emmer wheat RIL population. Theor Appl Genet 119(2):353–369

Petit JM, van Wuytswinkel O, Briat JF, Lobréaux S (2001) Characterization of an iron-dependent regulatory sequence involved in the transcriptional control of *AtFer1*and *ZmFer1* plant ferritin genes by iron. J Biol Chem 276(8):5584–5590

Pfeiffer WH, McClafferty B (2007) HarvestPlus: breeding crops for better nutrition. Crop Sci 47(Supplement_3):S88–S105. doi:10.2135/cropsci2007.09.0020IPBS

Potrykus I (2015) From the concept of totipotency to biofortified cereals. Plant Biol 66(1):1–22

Proudhon D, Briat JF, Lescure AM (1989) Iron induction of ferritin synthesis in soybean cell suspensions. Plant Physiol 90(2):586–590

Qi Y, Wang S, Shen C et al (2012) OsARF12, a transcription activator on auxin response gene, regulates root elongation and affects iron accumulation in rice (Oryza sativa). New Phytol 193:109–120. doi:10.1111/j.1469-8137.2011.03910.x

Qu LQ, Yoshihara T, Ooyama A, Goto F, Takaiwa F (2005) Iron accumulation does not parallel the high expression level of ferritin in transgenic rice seeds. Planta 222(2):225–233

Raffin SB, Woo CH, Roost KT, Price DC, Schmid R (1974) Intestinal absorption of hemoglobin iron-heme cleavage by mucosal heme oxygenase. J Clin Invest 54:1344–1352

Rampey RA, Woodward AW, Hobbs BN, Tierney MP, Lahner B, Salt DE, Bartel B (2006) An *Arabidopsis* basic helix-loop-helix leucine zipper protein modulates metal homeostasis and auxin conjugate responsiveness. Genetics 174:1841–1857

Ravet K, Touraine B, Boucherez J, Briat JF, Gaymard F, Cellier F (2009) Ferritins control interaction between iron homeostasis and oxidative stress in *Arabidopsis*. Plant J 57(3):400–412

Rellán-Álvarez R, Andaluz S, Rodríguez-Celma J, Wohlgemuth G, Zocchi G, Álvarez-Fernández A et al (2010a) Changes in the proteomic and metabolic profiles of beta vulgaris root tips in response to iron deficiency and resupply. BMC Plant Biol 10(1):120. doi:10.1186/1471-2229-10-120

Rellán-Álvarez R, Giner-Martínez-Sierra J, Orduna J, Orera I, Rodríguez-Castrillón JÁ, García-Alonso JI, Abadía J, Álvarez-Fernández A (2010b) Identification of a tri-iron (III), tri-citrate complex in the xylem sap of iron-deficient tomato resupplied with iron: new insights into plant iron long-distance transport. Plant Cell Physiol 51(1):91–102

Ricachenevsky FK, Menguer PK, Sperotto RA (2013) kNACking on heaven's door: how important are NAC transcription factors for leaf senescence and Fe/Zn remobilization to seeds? Front Plant Sci 4:226. doi:10.3389/fpls.2013.00226

Robinson NJ, Procter CM, Connolly EL, Guerinot ML (1999) A ferric-chelate reductase for iron uptake from soils. Nature 397(6721):694–697

Rogers EE, Guerinot ML (2002) FRD3, a member of the multidrug and toxin efflux family, controls iron deficiency responses in *Arabidopsis*. Plant Cell 14(8):1787–1799

Romera FJ, Alcantara E, De La Guardia MD (1999) Ethylene production by Fe-deficient roots and its involvement in the regulation of Fe-deficiency stress responses by strategy I plants. Ann Bot 83(1):51–55

Römheld V (1987) Different strategies for iron acquisition in higher plants. Physiol Plant 70(2):231–234

Römheld V, Marschner H (1986) Evidence for a specific uptake system for iron phytosiderophores in roots of grasses. Plant Physiol 80(1):175–180

Rouault TA (2006) The role of iron regulatory proteins in mammalian iron homeostasis and disease. Nat Chem Biol 2(8):406–414

Sagi L, Panis B, Remy S, Schoofs H, De Smet K, Swennen R, Cammue BPA (1995) Genetic transformation of banana and plantain (*Musa* spp.) via particle bombardment. Biotechnology 13(5):481–485

Saito A, Iino T, Sonoike K, Miwa E, Higuchi K (2010) Remodeling of the major light-harvesting antenna protein of PSII protects the young leaves of barley (Hordeum vulgare L.) from photoinhibition under prolonged iron deficiency. Plant Cell Physiol 51(12):2013–2030

Sandberg AS, Hulthen LR, Turk M (1996) Dietary *Aspergillus niger* phytase increases iron absorption in humans. J Nutr 126(2):476–480

Sankaran RP, Huguet T, Grusak MA (2009) Identification of QTL affecting seed mineral concentrations and content in the model legume Medicago truncatula. Theor Appl Genet 119(2):241–253

Santi S, Schmidt W (2008) Laser microdissection-assisted analysis of the functional fate of iron deficiency-induced root hairs in cucumber. J Exp Bot 59(3):697–704

Santi S, Cesco S, Varanini Z, Pinton R (2005) Two plasma membrane H+-ATPase genes are differentially expressed in iron-deficient cucumber plants. Plant Physiol Biochem 43(3):287–292

Sazawal S, Black RE, Ramsan M, Chwaya HM, Stoltzfus RJ, Dutta A et al (2006) Effects of routine prophylactic supplementation with iron and folic acid on admission to hospital and mortality in preschool children in a high malaria transmission setting: community-based, randomised, placebocontrolled trial. Lancet 367(9505):133–143

Sazawal S, Dhingra U, Dhingra P et al (2010) Micronutrient fortified milk improves iron status, anemia and growth among children 1–4 years: a double masked, randomized, controlled trial. PLoS One 5(8):e12167. doi:10.1371/journal.pone.0012167

Schaaf G, Ludewig U, Erenoglu BE, Mori S, Kitahara T, von Wirén N (2004) ZmYS1 functions as a proton-coupled symporter for phytosiderophore-and nicotianamine-chelated metals. J Biol Chem 279(10):9091–9096

Schmidt W (1999) Mechanisms and regulation of reduction-based iron uptake in plants. New Phytol 141(1):1–26

Schmidt W, Tittel J, Schikora A (2000) Role of hormones in the induction of iron deficiency responses in *Arabidopsis* roots. Plant Physiol 122(4):1109–1118

Séguéla M, Briat JF, Vert G, Curie C (2008) Cytokinins negatively regulate the root iron uptake machinery in *Arabidopsis* through a growth-dependent pathway. Plant J 55(2):289–300

Selote D, Samira R, Matthiadis A, Gillikin JW, Long TA (2015) Iron-binding E3 ligase mediates iron response in plants by targeting basic helix-loop-helix transcription factors. Plant Physiol 167(1):273–286

Shayeghi M, Latunde-Dada GO, Oakhill JS et al (2005) Identification of an intestinal heme transporter. Cell 122(5):789–801

Shekhawat UK, Ganapathi TR (2013) *MusaWRKY71* overexpression in banana plants leads to altered abiotic and biotic stress responses. PLoS One 8(10):e75506. doi:10.1371/journal.pone.0075506

Shekhawat UKS, Ganapathi TR, Srinivas L (2011) Cloning and characterization of a novel stress-responsive WRKY transcription factor gene (*MusaWRKY71*) from Musa spp. cv. Karibale Monthan (ABB group) using transformed banana cells. Mol Biol Rep 38(6):4023–4035

Shi R, Li H, Tong Y, Jing R, Zhang F, Zou C (2008) Identification of quantitative trait locus of zinc and phosphorus density in wheat (Triticum aestivum L.) grain. Plant and Soil 306(1–2):95–104

Shojima S, Nishizawa NK, Mori S (1989) Establishment of a cell-free system for the biosynthesis of nicotianamine. Plant Cell Physiol 30(5):673–677

Solomons NW, Ruz M (1998) Trace element requirements in humans: an update. J Trace Elem Exp Med 11(2–3):177–195

Sommer A (2008) Vitamin A deficiency and clinical disease: an historical overview. J Nutr 138(10):1835–1839

Sperotto RA, Ricachenevsky FK, de Abreu WV, Fett JP (2012) Iron biofortification in rice: it's a long way to the top. Plant Sci 190:24–39

Sreeramanan S, Mahmood M, Abdullah MP, Meon S, Xavier R (2006) Transient expression of gusA and gfp gene in Agrobacterium-mediated banana transformation using single tiny meristematic bud. Asian J Plant Sci 5(3):468–480

Stangoulis JC, Huynh BL, Welch RM, Choi EY, Graham RD (2007) Quantitative trait loci for phytate in rice grain and their relationship with grain micronutrient content. Euphytica 154(3):289–294

Stoltzfus RJ (2003) Iron deficiency: global prevalence and consequences. Food Nutr Bull 24(Supplement 2):99–103

Studholme DJ, Kemen E, MacLean D et al (2010) Genome-wide sequencing data reveals virulence factors implicated in banana Xanthomonas wilt. FEMS Microbiol Lett 310(2):182–192

Subramanyam K, Subramanyam K, Sailaja KV, Srinivasulu M, Lakshmidevi K (2011) Highly efficient Agrobacterium-mediated transformation of banana cv. Rasthali (AAB) via sonication and vacuum infiltration. Plant Cell Rep 30(3):425–436

Suharno D, Karyadi D, West CE, Hautvast JGA (1993) Supplementation with vitamin A and iron for nutritional anaemia in pregnant women in west Java, Indonesia. Lancet 342(8883):1325–1328

Susin S, Abian J, Sanchez-Baeza F, Peleato ML, Abadia A, Gelpi E, Abadia J (1993) Riboflavin 3′-and 5′-sulfate, two novel flavins accumulating in the roots of iron-deficient sugar beet (Beta vulgaris). J Biol Chem 268(28):20958–20965

Susín S, Abián J, Peleato ML, Sánchez-Baeza F, Abadía A, Gelpi E, Abadía J (1994) Flavin excretion from roots of iron-deficient sugar beet (Beta vulgaris L.). Planta 193(4):514–519

Swanson MD, Winter HC, Goldstein IJ, Markovitz DM (2010) A lectin isolated from bananas is a potent inhibitor of HIV replication. J Biol Chem 285(12):8646–8655

Takahashi M, Yamaguchi H, Nakanishi H, Shioiri T, Nishizawa NK, Mori S (1999) Cloning two genes for nicotianamine aminotransferase, a critical enzyme in iron acquisition (Strategy II) in graminaceous plants. Plant Physiol 121(3):947–956

Takahashi M, Terada Y, Nakai I, Nakanishi H, Yoshimura E, Mori S, Nishizawa NK (2003) Role of nicotianamine in the intracellular delivery of metals and plant reproductive development. Plant Cell 15(6):1263–1280

Takahashi M, Nozoye T, Kitajima N et al (2009) In vivo analysis of metal distribution and expression of metal transporters in rice seed during germination process by microarray and X-ray fluorescence imaging of Fe, Zn, Mn, and Cu. Plant and Soil 325(1–2):39–51

Tamura K, Stecher G, Peterson D, Filipski A, Kumar S (2013) MEGA6: molecular evolutionary genetics analysis version 6.0. Mol Biol Evol 30:2725–2729

Teng YS, Su YS, Chen LJ, Lee YJ, Hwang I, Li HM (2006) Tic21 is an essential translocon component for protein translocation across the chloroplast inner envelope membrane. Plant Cell 18(9):2247–2257

Thorneycroft D, Sherson SM, Smith SM (2001) Using gene knockouts to investigate plant metabolism. J Exp Bot 52(361):1593–1601

Uauy C, Distelfeld A, Fahima T, Blechl A, Dubcovsky J (2006) A NAC gene regulating senescence improves grain protein, zinc, and iron content in wheat. Science 314(5803):1298–1301

Urnov FD, Rebar EJ, Holmes MC, Zhang HS, Gregory PD (2010) Genome editing with engineered zinc finger nucleases. Nat Rev Genet 11(9):636–646

Van Thuy P, Berger J, Davidsson L et al (2003) Regular consumption of NaFeEDTA-fortified fish sauce improves iron status and reduces the prevalence of anemia in anemic Vietnamese women. Am J Clin Nutr 78(2):284–290

Van Wuytswinkel O, Vansuyt G, Grignon N, Fourcroy P, Briat JF (1999) Iron homeostasis alteration in transgenic tobacco overexpressing ferritin. Plant J 17(1):93–97

Vasconcelos M, Datta K, Oliva N et al (2003) Enhanced iron and zinc accumulation in transgenic rice with the ferritin gene. Plant Sci 164(3):371–378

Vert G, Briat JF, Curie C (2001) *Arabidopsis* IRT2 gene encodes a root-periphery iron transporter. Plant J 26(2):181–189

Vert G, Grotz N, Dédaldéchamp F, Gaymard F, Guerinot ML, Briat JF, Curie C (2002) IRT1, an *Arabidopsis* transporter essential for iron uptake from the soil and for plant growth. Plant Cell 14(6):1223–1233

Vert G, Barberon M, Zelazny E, Séguéla M, Briat JF, Curie C (2009) *Arabidopsis* IRT2 cooperates with the high-affinity iron uptake system to maintain iron homeostasis in root epidermal cells. Planta 229(6):1171–1179

Vigani G, Zocchi G (2009) The fate and the role of mitochondria in Fe-deficient roots of strategy I plants. Plant Signal Behav 4(5):375–379

Vigani G, Zocchi G (2010) Effect of Fe deficiency on mitochondrial alternative NAD (P) H dehydrogenases in cucumber roots. J Plant Physiol 167(8):666–669

Vigani G, Maffi D, Zocchi G (2009) Iron availability affects the function of mitochondria in cucumber roots. New Phytol 182(1):127–136

von Wirén N, Mori S, Marschner H, Romheld V (1994) Iron inefficiency in maize mutant ys1 (Zea mays L. cv Yellow-Stripe) is caused by a defect in uptake of iron phytosiderophores. Plant Physiol 106(1):71–77

Von Wirén N, Klair S, Bansal S, Briat JF, Khodr H, Shioiri T, Leigh RA, Hider RC (1999) Nicotianamine chelates both Fe III and Fe II. Implications for metal transport in plants. Plant Physiol 119:1107–1114

Vulpe CD, Kuo YM, Murphy TL et al (1999) Hephaestin, a ceruloplasmin homologue implicated in intestinal iron transport, is defective in the Sla mouse. Nat Genet 21(2):195–199

Wairegi LW, van Asten PJ, Tenywa MM, Bekunda MA (2010) Abiotic constraints override biotic constraints in east African highland banana systems. Field Crop Res 117(1):146–153

Wall MM (2006) Ascorbic acid, vitamin A, and mineral composition of banana (Musa sp.) and papaya (Carica papaya) cultivars grown in Hawaii. J Food Compos Anal 19(5):434–445

Waltz E (2014) Vitamin A super banana in human trials. Nat Biotechnol 32(9):857

Wang N, Cuia Y, Liua Y, Fana H, Dua J, Huangd Z, Yuana Y, Wua H, Linga HQ (2013) Requirement and functional redundancy of Ib subgroup bHLH proteins for iron deficiency responses and uptake in *Arabidopsis thaliana*. Mol Plant 6(2):503–513

Wasternack C (2007) Jasmonates: an update on biosynthesis, signal transduction and action in plant stress response, growth and development. Ann Bot 100(4):681–697

Waters BM, Blevins DG, Eide DJ (2002) Characterization of FRO1, a pea ferric-chelate reductase involved in root iron acquisition. Plant Physiol 129(1):85–94

Waters BM, Chu HH, DiDonato RJ et al (2006) Mutations in *Arabidopsis* yellow stripe-like1 and yellow stripe-like3 reveal their roles in metal ion homeostasis and loading of metal ions in seeds. Plant Physiol 141(4):1446–1458

Waters BM, Uauy C, Dubcovsky J, Grusak MA (2009) Wheat (*Triticum aestivum*) NAM proteins regulate the translocation of iron, zinc, and nitrogen compounds from vegetative tissues to grain. J Exp Bot 60(15):4263–4274

Welch RM, Graham RD (2004) Breeding for micronutrients in staple food crops from a human nutrition perspective. J Exp Bot 55(396):353–364

Welch RM, House WA, Beebe S, Cheng Z (2000) Genetic selection for enhanced bioavailable levels of iron in bean (Phaseolus vulgaris L.) seeds. J Agric Food Chem 48(8):3576–3580

White PJ, Broadley MR (2009) Biofortification of crops with seven mineral elements often lacking in human diets–iron, zinc, copper, calcium, magnesium, selenium and iodine. New Phytol 182(1):49–84

WHO (2003) Diet, nutrition and the prevention of chronic diseases. World Health Organ Tech Rep Ser 916:i–viii. http://www.who.int/dietphysicalactivity/publications/trs916/en/

Wieringa FT, Dijkhuizen MA, West CE, Thurnham DI, Van der Meer JW (2003) Redistribution of vitamin A after iron supplementation in Indonesian infants. Am J Clin Nutr 77(3):651–657

Wieser S, Plessow R, Eichler K, Malek O, Capanzana MV, Agdeppa I, Bruegger U (2013) Burden of micronutrient deficiencies by socio-economic strata in children aged 6 months to 5 years in the Philippines. BMC Public Health 13(1):1167

Wirth J, Poletti S, Aeschlimann B, Yakandawala N, Drosse B, Osorio S, Tohge T, Fernie AR, Günther D, Gruissem W, Sautter C (2009) Rice endosperm iron biofortification by targeted and synergistic action of nicotianamine synthase and ferritin. Plant Biotechnol J 7(7):631–644

World Health Organization (2007) Joint statement by the World Health organization, the World Food Programme and the United Nations Children's Fund: preventing and controlling micronutrient deficiencies in populations affected by an emergency. WHO, Geneva. http://www.unicef.org/nutrition/index_iodine.html

Wu J, Wang C, Zheng L, Wang L, Chen Y, Whelan J, Shou H (2010) Ethylene is involved in the regulation of iron homeostasis by regulating the expression of iron-acquisition-related genes in *Oryza sativa*. J Exp Bot 62(2):667–674

Wu H, Chen C, Du J, Liu H, Cui Y, Zhang Y, Ling HQ et al (2012) Co-overexpression FIT with AtbHLH38 or AtbHLH39 in *Arabidopsis*-enhanced cadmium tolerance via increased cadmium sequestration in roots and improved iron homeostasis of shoots. Plant Physiol 158(2):790–800

Wyllie JC, Kaufman N (1982) An electron microscopic study of heme uptake by rat duodenum. Labo Investig; J Tech Methods Pathol 47(5):471–476

Yuan YX, Zhang J, Wang DW, Ling HQ (2005) AtbHLH29 of *Arabidopsis thaliana* is a functional ortholog of tomato FER involved in controlling iron acquisition in strategy I plants. Cell Res 15:613–621

Yuan Y, Wu H, Wang N, Li J, Zhao W, Du J, Wang D, Ling HQ (2008) FIT interacts with AtbHLH38 and AtbHLH39 in regulating iron uptake gene expression for iron homeostasis in *Arabidopsis*. Cell Res 18:385–397

Zhai Z, Gayomba SR, Jung HI et al (2014) OPT3 is a phloem-specific iron transporter that is essential for systemic iron signaling and redistribution of iron and cadmium in *Arabidopsis*. Plant Cell Online 26(5):2249–2264

Zhang DL, Ghosh MC, Rouault TA (2014) The physiological functions of iron regulatory proteins in iron homeostasis-an update. Front Pharmacol 13(5):124. doi:10.3389/fphar.2014.00124

Zhang J, Liu B, Li M, Feng D, Jin H, Wang P, Liu J, Xiong F, Wang J, Wang HB (2015) The bHLH Transcription Factor bHLH104 Interacts with IAA-LEUCINE RESISTANT3 and Modulates Iron Homeostasis in *Arabidopsis*. Plant Cell 27(3):787–805

Zheng L, Ying Y, Wang L, Wang F, Whelan J, Shou H (2010) Identification of a novel iron regulated basic helix-loop-helix protein involved in Fe homeostasis in *Oryza sativa*. BMC Plant Biol 10:166. http://www.biomedcentral.com/1471 2229/10/166

Zheng L, Fujii M, Yamaji N et al (2011) Isolation and characterization of a barley yellow stripe-like gene, HvYSL5. Plant Cell Physiol 52(5):765–774

Zimmermann MB, Hurrell RF (2007) Nutritional iron deficiency. Lancet 370(9586):511–520

Zimmermann M, Adou P, Torresani T, Zeder C, Hurrell R (2000) Persistence of goiter despite oral iodine supplementation in goitrous children with iron deficiency anemia in Cote d'Ivoire. Am J Clin Nutr 71(1):88–93

Zimmermann MB, Zeder C, Chaouki N, Torresani T, Saad A, Hurrell RF (2002) Addition of microencapsulated iron to iodized salt improves the efficacy of iodine in goitrous, iron-deficient children: a randomized, double-blind, controlled trial. Eur J Endocrinol 147(6):747–753

Zimmermann MB, Zeder C, Chaouki N, Saad A, Torresani T, Hurrell RF (2003) Dual fortification of salt with iodine and microencapsulated iron: a randomized, double-blind, controlled trial in Moroccan schoolchildren. Am J Clin Nutr 77(2):425–432

Zimmermann MB, Wegmueller R, Zeder C et al (2004) Dual fortification of salt with iodine and micronized ferric pyrophosphate: a randomized, double-blind, controlled trial. Am J Clin Nutr 80(4):952–959

Zimmermann MB, Winichagoon P, Gowachirapant S et al (2005) Comparison of the efficacy of wheat-based snacks fortified with ferrous sulfate, electrolytic iron, or hydrogen-reduced elemental iron: randomized, double-blind, controlled trial in Thai women. Am J Clin Nutr 82(6):1276–1282

Zimmermann MB, Muthayya S, Moretti D, Kurpad A, Hurrell RF (2006) Iron fortification reduces blood lead levels in children in Bangalore, India. Pediatrics 117(6):2014–2021

Index

A
Abiotic stress
  agroecosystems, 182
  definition, 78, 79
  dynamic molecular phenotyping, 81
  feature selection, 79, 80
  gene and cell function, 78
  germplasm, 182
  inter- and intraspecific hybridization, 182
  molecular banana research, 78
  molecular variation, genotypes, 80
  Musa species, 182
  next-generation sequencing, 82
  sterility, 81
  tissue culture, 186–187
  tolerance, 78
Abscisic acid (ABA), 190, 284
Acetosyringone, 133
Aequorea victoria, 160
AF119096.1, 168
Agrobacterium, 128, 291
Agrobacterium tumefaciens, 132
Agrobacterium-mediated gene transfer
  Agrobacterium, 133
  biotic and abiotic stresses, 133
  cocultivation media, 133
  post-transformation, 133
  practical applications, 132
Agroinfiltration method, 263
Agronomically superior diploids, 27
Allene oxide synthase (AOS), 215
Allium cepa antimicrobial protein1 (Ace-AMP1), 221
Allopolyploid clones, 52
Allozymes, 38
Alpha-amylase inhibitors, 254
Amalgam, 54
Amplified Fragment Length Polymorphism (AFLP), 42, 53, 54
  marker technology, 45
Anthracnose, 66
Antibiosis mechanisms, 250
Antimicrobial peptides, 232–233
AOS. *See* Allene oxide synthase (AOS)
Aquaporins, 187
Arabidopsis, 187, 290
Arabidopsis BAG7 protein (AtBAG7), 220
Arabidopsis thaliana, 160
Ascorbic acid, 96
AS-PCRs, 53, 54
Australimusa., 14
Available reference genome (A-genome), 55

B
Bacterial diseases, 65
Banana, 94, 212, 223, 303, 309–310
  breeding, 182–183
  chromosomes, 6
  cooling treatments and controlled atmosphere storage, 94
  cultivars, 94
  medicinal properties, 108
  medicinal uses, 109–110
  metabolite profiling, 108–110
  and plantains (*Musa* spp.), 3, 94
  ripening (*see* Ripening)
  seedless edible bananas, 6
  tissue culture, 183–186
  volatile aroma, 117
Banana bract mosaic virus (BBrMV), 68, 239
Banana breeding
  annual world production, 14
  balbisiana, 15
  bananas and plantains, 14
  breeding targets, 18–19
  cavendish cultivars, 16
  challenges in Musa, 19–20
  CIRAD, 25
  cultivars and landraces, 16
  diploid breeding, 20–22
  FHIA programme, 22
  fruit development, 16
  global production, 14
  IIHR, 27–28
  IITA, 23–24
  KAU, 26–27
  molecular marker methods, 28
  Mysore subgroup, 17

S. Mohandas, K.V. Ravishankar (eds.), *Banana: Genomics and Transgenic Approaches for Genetic Improvement*, DOI 10.1007/978-981-10-1585-4

Banana breeding (*cont.*)
NRCB, 27
Pisang Awak, 17
plantain subgroup, 16
polyploid crop, 19
Pome subgroup, 17
Silk subgroup, 17
taxonomic method, 15
taxonomy, 14
tetraploid hybrids, 22
Banana bunchy top virus (BBTV), 18, 63, 68, 69, 238–239, 293
Banana corm weevil, 18
Banana crop improvement
AFLP, 42
biochemical markers, 38
DArT, 43
ecotilling, 43
family musaceae, 35
gene, 36
genetic marker, 38
genome, 36
introduce/transfer, 37
marker-assisted breeding, 46
molecular marker, 39
morphological marker, 38
RAPD marker, 40
RFLP, 39
SSR marker, 42
VNTR marker, 39
Banana diseases
bacterial diseases, 65
banana streak disease, 68–69
cultivation, 64
fungal diseases, 65–67
infection, 64
production, 64
symptoms, 64
viral diseases, 68
Banana genome
characterization, 52
cytogenetics, 52
flow cytometry, 52
Gros Michel variety, 52
landraces, 52
MAS, 52
molecular marker-based linkage map, 52–54
Musa chromosome numbers, 52
Mycosphaerella fijiensis, 52
NGS, 54–56
production, 51
sterile and vegetatively propagated, 52
tropical race 4, 52
Banana leaf, 69
Banana streak badnavirus (*BSV*), 293
Banana streak disease, 68–69
Banana streak virus (BSV), 18, 69, 239
Banana weevil
AA genome progenitor, 250
banana and plantain production, 249
genes, 253–255
nAChR-binding peptide, 251
R. similis, 253
rhizomes, 249
toxic compounds, 250
Banana *Xanthomonas* wilt (BXW) disease, 65, 198
disease transmission, 199–200
management, 200–202
symptoms, 199
Banana-infecting *Badnavirus* (BIB), 68
Bcl-2-associated athanogenes (BAG), 220
Benzimidazoles, 231
Biofortification
banana, 291
biological methods, 279
β-carotene, 291
cisgenic approac, 293
genetic engineering, 280–281
micronutrient-enriched biofortified crops, 280
mineral fertilization, 280
nutrient-rich diet, 279
plant breeding, 280
PVA-enriched biofortified banana, 291
staple food and fruit, 280
transgenic approach, 291
Biofortification, iron
endosperm-specific promoters, 325
functional and expression pattern, 326
genome sequencing, 322
hydroponics, 324
iron homeostasis genes, 326
pleiotropic effects, 322
soybean ferritin in maize plants, 325
sugar beet roots, 324
transgenic studies, 327
Bioinformatics
cis-acting elements, 163
one-hybrid and EMSA, 163
PlantCARE/PLACE, 163
Biolistic-mediated transformation, 150–151
BioProspector, 163
Bio-Rad PDS-1000/He™, 131
Biotechnology Industry Research Assistance Council (BIRAC), 291
Biotic stress, 64–71
banana diseases (*see* Banana diseases)
banana virus diseases, 63
diseases, 62
environmental conditions, 62
fungal and bacterial diseases, 62
genomics, 63–64
host and the pathogen, 62
insect outbreaks, 63
insect pests (*see* Pests)
mechanism and tolerance, 63–64
natural environment, 61
pathogen/herbivore attack, 62
pests, 62
plant disease triangle, 62
plant-pathogen interaction, 63

range of diseases, 62
scab moths, 63
Black leaf streak disease (BLSD), 18, 228
Black Sigatoka (BS), 65
BLSD. *See* Black leaf streak disease (BLSD)
Blumeria graminis, 233
Brachypodium, 294
Brassica napus, 289
Bremia lactucae, 233
Brevundimonas spp., 289
BSV Goldfinger (BSGfV), 69
BSV-expressed locus (BEL), 69
Bugtok disease, 198

C
Caffeoyl-CoA O-methyltransferase (CCOMT), 214
Canola, 289
β-carotene hydroxylase or (*b-CHX*), 294
Carotenoid, 110, 113–114
Cauliflower mosaic virus (CaMV), 158, 289
Cavendish resistance, 52
CEBiP. *See* Chitin elicitor-binding protein (CEBiP)
Cell strain, 147
Centrifugation assisted somatic embryogenesis organogenesis
acetosyringone, 134
micropropagation technique, 134
Centrifugation-Assisted *Agrobacterium tumefaciens*-Mediated Transformation (CAAT), 133–134
Chitin elicitor-binding protein (CEBiP), 216
Chitinase, 95, 233
Chitinase enzymes, 254
Chitinase promoter, 100
Chlamydomonas reinhardtii, 266
Chloroplast transformation, 262–263
Chlorothalonil, 231
Citrus sinensis, 290
Clustered regularly interspaced short palindromic repeats (CRISPRs), 294
Comparative transcriptomics, 90
Consultative Group on International Agricultural Research (CGIAR), 280
Cowpea mosaic virus, 263
CRISPRs. *See* Clustered regularly interspaced short palindromic repeats (CRISPRs)
Critical nutrient levels (CNLs), 70
Cry5B protein, 251
Cry6A, 254
Cucumber mosaic virus (CMV), 18, 240
Cultivation-independent and cultivation-dependent methods, 67
Cytogenetics, 52

D
DAD. *See* Defender against death (DAD)
Dahlia merckii, 232
DArTs, 53, 54
Defender against death (DAD), 220
Dehydrin (DHN), 187
Delayed fruit ripening, 98–102
genetic transformation of banana
black patches, 98
comparative analysis of fruit firmness, 99, 100
ECS, 98
ethylene-treated transgenic fruit, 100
ethylene-untreated transgenic fruits, 100
expansin and MADSrin genes, 98
expansin transgenic fruits, 100–102
firmness, 100
MADSrin transgenic fruits, 98, 99
non-transgenic fruits, 98
penetrometer, 99
promoters, 98
promoters, 100–103
Department of Biotechnology (DBT), 280
Diploid Breeding, 20–22
Dithiocarbamates, 231
Diversity arrays technology (DArT), 43
DMAPP. *See* Double-bond isomer dimethylallyl pyrophosphate (DMAPP)
Domestication, 4–7
Double-bond isomer dimethylallyl pyrophosphate (DMAPP), 284
Double-strand breaks (DSBs), 293
Double-stranded RNA (dsRNA), 252
DSBs. *See* Double-strand breaks (DSBs)

E
East African highland banana (EAHB), 65, 70
Ecotilling, 43–44
Effective number of codons (ENC), 55
Electroporation
banana cell transformation, 130
somatic embryo, 129
transformation, 129
transient GUS expression, 130
Embryogenic cell suspension (ECS), 96, 144–145, 147, 184, 241, 242
Engineering resistance, viruses
banana viruses, 238
BBrMV, 239
BBTV, 238
BSV, 239
CMV, 240
transgenic approaches, 240–242
ERFs. *See* Ethylene-responsive transcription factors (ERFs)
Erwinia uredovora, 289, 290
Ethylene, 94, 95
Ethylene-responsive transcription factors (ERFs), 215
Ethylene-treated transgenic fruit, 100
Ethylene-untreated transgenic fruits, 100
Eumusa, 15
Eumusa cultivars, 4
Explants, 142, 143

F

F1-based linkage map, 53
Ferredoxin-like amphipathic protein (PFLP), 203
Fertility, 17, 20, 22
Floral hand positions, 144
Flow cytometry, 52
Food and agriculture organization, 279
Fruit beetle, 69
Fruit ripening, 96–103
  ABA, 95
  ACC synthase gene, 96
  biochemical and physiological attributes, 94
  biotechnological strategies, 103
  categorization of *rin*, 95
  cell wall metabolism, 96
  cell wall modifying and structural proteins, 94
  cell wall softening, 95
  characterization, 94
  chitinase, 95
  ethylene synthesis, 94, 95
  fruit size, 95
  gene-silencing agents, 103
  IAA, 95
  IFR genes, 96
  methyl jasmonate (MJ), 95
  methyl salicylate (MS), 95
  MS treatment, 95
  MuMADS1, 95, 96
  rapid strides in genomics, 103
  revenue generation and maintenance, 103
  rin mutation, 95
  somatic embryogenesis and plant regeneration (*see* Somatic embryogenesis)
  tomato ripening mutants, 94
  transgenic plants, 103
Fungal diseases
  anthracnose, 66
  banana-associated microorganisms, 67
  characterisation of tolerance, 67
  fungal pathogens, 67
  Fusarium wilt (Panama disease), 66
  GPI protein family, 66
  Mycosphaerella fijiensis, 65
  NGS, 66
  pre-and postharvest production losses, 65
  RFO, 67
  Sigatoka disease, 65
  SSH, 66
  SSR, 66
  transcriptome studies, 67
  virulence-associated genes, 67
Fusarium, 18, 120, 220–223
  banana, 223
  banana and plantains, 212
  cell wall fortification, 214
  disease resistance
    native cell death genes, 220
    resistance gene (R Gene), 220
  Foc, 213
  jasmonic acid/ethylene (JA/ET), 215
  national and international scenario, 212–213
  pathogen and disease cycle, 213
  pathogenesis-associated genes and defence, 216–217
  pathogenicity mechanism, Foc, 217–218
  reactive oxygen species, 215
  salicylic acid, 215
  signal transduction, 215–216
  transgenic approach
    antiapoptosis-related proteins, 222
    antimicrobial peptide gene, 221–222
    PR protein genes, 220–221
    RNAi-mediated host-induced silencing, 222–223
Fusarium oxysporum, 233
Fusarium oxysporum f.sp. *cubense*, 67, 212
Fusarium wilt (Panama disease), 66

G

Galanthus nivalis agglutinin (GNA), 254
Gel-based technique, 87
Gel-free technique, 87, 88
GenBank, 166
Gene action in *Musa*, 24
Gene expression, 159
Gene transfer
  biological vector, 128
  biotechnology, 128
  diploid bananas, 128
  electroporation, 129
  foreign DNA, 128
  plant transformation, 128
  recombinant DNA, 128
  somaclonal variations, 129
  totipotent cells, 129
Generation Challenge Programme (GCP), 56
Genetic engineering, 218, 222
  abiotic stress, 188
  identification and cloning, 187
  intronic sequences, 189
  malondialdehyde levels, 189
  mannitol-induced drought stress, 189
  *Musami*RNA156, 190
  MusaPIP2;6, 189
  plant-associated proteins, 190
  PR-10 proteins, 190
  SK-3-type dehydrin, 187
Genetic transformation of banana, 145–151
  delayed fruit ripening, 98–100
Genetically modified bananas, 167–175
Genome composition determination, 45
Genome editing, 192–193
Genomics, 61–64
  biotic stress (*see* Biotic stress)
  comparative, 83
  gene expression, 84
  high throughput profiling, 84
  microarrays, 84, 85
  RNA-sequencing, 85
  in silico analyses, 82
  transcriptomics, 83

Genotype, 44, 143–144, 147
Genotyping by sequencing (GBS), 55
Geranylgeranyl diphosphate (GGPP), 284
Germplasm collection, 44
GGPP. *See* Geranylgeranyl diphosphate (GGPP)
Global Musa Genomics Consortium (GMGC), 56
Glucanase, 233
Glucanase promoter with ethephon, 101
β-glucuronidase, 160
Glycosylation, 268
Glycosylphosphatidylinositol (GPI) protein family, 66
Green fluorescent protein (GFP), 189
Gros Michel variety, 52
Gross domestic product (GDP), 279
Gut iron uptake, 304

H
Hairy root-based recombinant production, 265
Harvest Plus Biofortification Challenge Programme, 280
Heat shock proteins (Hsp), 133
Helium pressure-driven gene gun, 131
Hepcidin, 304
Heterologous protein expression, 158
Histidine kinase (HK), 217
History and domestication, 4–7
Hypersensitive response-assisting protein (HRAP), 203

I
IFR promoter, 102
Illumina HiSeq 2000 II technology, 55
Imperial College of Tropical Agriculture (ICTA), 20
Indian National Science Academy, 278
Institute of Horticultural Research (IIHR), 27
Internal Transcribed Spacers (ITS), 41
International *Musa* Testing Programme (IMTP), 29
Inter-retrotransposon amplified polymorphism (IRAP), 41–42
Intersimple sequence repeats (ISSR), 40–41
IPP. *See* Isopentenyl pyrophosphate (IPP)
Iron acquisition, 311–315, 317, 320–322
  by plants
    auxin, 315
    chelators, 313
    chloroplasts, 321
    enzymes, 313, 314
    ethylene, 315
    FIT, 311–312
    function and toxicity, 317
    mitochondria, 322
    from phloem, 320
    TFs, 314
    transporters, 314
  enzyme structure and activity, 303
  genes, homeostasis, 323
  iron uptake strategies, 310
  iron uptake, duodenal intestine, 305
  nutritional requirements for humans, 303
  phylogenetic relationship, 318, 319
Iron deficiency anemia (IDA)
  bioavailability from mixed diets, 304
  biofortification, 308, 309
  fortification, 307, 308
  indicators and repercussions, 305, 306
  iron uptake and its absorption, 304
  micronutrient intake, 306
  supplementation, 307
Isoflavone reductase (IFR) genes, 96
Isopentenyl pyrophosphate (IPP), 284
Isozymes, 38

J
JF320825.1, 167

K
Kanamycin pressure, 160
Kerala Agricultural University (KAU), 26–27

L
Linkage groups (LGs), 52, 53
Linkage map, 52, 53, 56, 58
Lipoxygenase (LOX), 215
Low reproductive fertility, 23
LOX. *See* Lipoxygenase (LOX)
Lycopene β-cyclase (LCYB), 284, 289
Lycopene ε-cyclase (LCYE), 284

M
M. eumusae, 228, 229
M. fijiensis, 229, 233
M. musicola, 228, 229
MADS-box transcription factor gene (MuMADS1), 95, 96
Magnifection technology, 263
Marker-assisted breeding, 191–192
Marker-assisted selection (MAS), 52
Matrix attachment regions (MAR), 267
McIntosh apples ABA treatment, 95
Medicago truncatula, 160
Metabolites
  banana, 111, 112
  chilling injury, 117–118
  nematode infection, 120, 121
  in pulp, 118
Metabolomics
  banana, 108
  environmental challenges, 108
  high-throughput technologies, 108
Methyl jasmonate (MJ), 95
Methyl salicylate (MS), 95
Methylation-sensitive amplification polymorphism (MSAP), 28
Micronutrient deficiencies, 302

Microprojectile bombardment
biolistic method, 131
biolistic transformation, 132
cell survival, 131
DNA into banana cell, 130
frequency of transformation, 131
macrocarrier flight distance, 132
microcarriers, 131
osmoticum, 132
Molecular farming
agroinfiltration method, 263
chloroplast transformation, 262–263
glycosylation, 268
industrial and medicinal purposes, 261
magnifection technology, 263
nuclear transformation, 262
plant cell suspension, 263
plant production systems, 266–268
RNA viruses, 263
tobacco, 264
Molecular marker methods, 28
Molecular marker-based linkage map, 52–54
Morpholines, 231
Multiplex reverse transcription-PCR (mRT-PCR) technique, 68
Musa, 5
MBMEDA, 163, 167
motif sequences, 163, 168–173
Musa acuminata, 7, 15, 53, 186, 187, 192, 212, 216
Musa Asr gene, 187
Musa balbisiana, 52, 187, 212
Musa breeding, 20
Musa Genome Resource Centre (MGRC), 56
Musa Germplasm Information System, 18
Musa spp. domestication, 5
Musaceae, 8
Mutagenesis, 186
Mutation breeding, 30
Mycosphaerella, 228
Mycosphaerella fijiensis, 52, 65, 212
Mycosphaerella musicola, 212, 229
Mycosphaerellaceae, 228
Mysore Subgroup, 17

N
Narcissus pseudonarcissus, 289, 290
National Agricultural Research Organization (NARO), 291
National Institute of Nutrition (NIN), 280
National Research Centre for Banana (NRCB), 27, 28
Natural resistance genes, 252
Nematicides, 249
Nematode and weevil resistance
biosafety, 255–256
chitinase enzymes, 254
genetic transformation, 255
RNAi-based approaches, 255
Nematode resistance genes, 250–253
Nematodes, 70–71, 248
Next-generation sequencers (NGS)
A-genome, 55, 56
agronomically, 54
amalgam, 54
banana databases and bioinformatics tools, 56
B-genome, 56
DH pahang, 57
DNA/RNA strands, 54
effective number of codons, 55
GBS, 55
GC3, 55
gene-based SSR markers, 56
genic markers, 56
GMGC, 56
Musa acuminata – DH Pahang, 54, 55
Musa genomic nuclear DNA, 54
Pahang linkage groups, 56
paired-end sequencing, 55
PKW genome, 55
RNA-Seq, 55
Sanger sequencing, 54
SGS, 54
SNPs, 55
SSRs, 55
stress adaptation, 55
transcriptome analysis, 54
WGRS, 54
WGS, 54, 55
Next-generation sequencing (NGS), 192
fungal diseases, 66
NHEJ. *See* Non-homologous end joining (NHEJ)
Nicotinic acetylcholine receptors (nAChRs), 251
Non-homologous end joining (NHEJ), 293
Nuclear ribosomal genes (rDNA), 41

O
Omics integration, 88–90
Organogenesis, 136, 137, 183
Oryza sativa, 160

P
Panama disease, 66
Parthenocarpy, 15, 16, 19, 26, 29, 53
Pathogen/herbivore attack, 62
Pathogenesis-related class 10 (PR-10) genes, 189
PCD. *See* Plant cell-death (PCD)
PDS-1000/He™ system, 131
Penetrometer, 99
Pentalonia nigronervosa, 68
Peroxidase isoenzyme, 38
Pests
banana leaf, 69
corm and root system, 69
fruit beetle, 69
nematodes, 70–71
weevils, 70
Phenolics, 112, 117, 120
Photinus pyralis, 160

Physcomitrella, 266
Physcomitrella patens, 266
Phytoene desaturase (*crtI*), 289
Phytoene desaturase (PDS), 284
Phytoene synthase (PSY), 284
Pisang Awak, 17
Pisang Klutuk Wulung (PKW) (B-genome), 55
Plant cell-death (PCD), 220
Plant growth regulators, 144
Plant lectins, 254
Plant lectins and insecticidal proteins, 254–255
Plant production systems
  recombinant protein, 266
  regulatory promoters, 267
  stability of foreign protein, 267
Plant regeneration, 142
Plant-derived pharmaceuticals, 270
Plant-parasitic nematodes, 70
Plant-pathogen interaction, 63, 64
Pluronic F68, 133
Polygalacturonase, 290
Polymorphism information content (PIC), 68
Pome subgroup, 17
Popeye network, 312, 313
Pro-embryogenic mass, 143
Programmed cell death (PCD), 133
Promoter tagging
  banana, 164–165
  DNA sequences, 162
  genetically modified bananas, 174
  heterologous regulatory sequences, 158
  mRNA differential expression, 158
  in silico analysis, 158
  TATA box and TSS, 162
Promoters
  cell wall hydrolases, 100
  chitinase, 100, 103
  climacteric nature, 102
  expansin, 102, 103
  expression, 102
  glucanase, ethephon, 101
  hydrolases, 102
  IFR, 102, 103
  reporter gene *gusA*, 100
  short shelf life, 102
Proteinase inhibitors (PI), 250, 253–254
Protein-coding gene model, 36
Proteomics
  high-throughput proteomic profiling, 87
  protein activity assessment, 86
Protoplast culture, 185
Provitamin A (PVA)
  banana and plantain, 278
  biofortification, 278
  biofortified crops, 278
  carotenoids, 283
  DSBs, 293
  food crops, 284–290
  food insecurity, 278–279
  fruit crops, 290
  micronutrients, 278–279
  parthenocarpic fruit, 278
  RNAi, 294
  technical and biological issues, 294, 295
Pro-vitamin A carotenoid (pVAC), 37
Pseudocercospora fijiensis, 229
PSY. *See* Phytoene synthase (PSY)

Q

Queensland University of Technology (QUT), 291

R

Radopholus similis, 70, 120, 121, 252
Ralstonia solanacearum, 198
Random Amplified Polymorphic DNA (RAPD), 40, 66
RDA. *See* Recommended dietary allowance (RDA)
Recombinant protein, 264, 269, 270
Recommended dietary allowance (RDA), 283
Reporter genes, 160, 161
Resistance gene (R Gene), 220
Resistance gene analogue (RGA), 68
Response regulator (RR), 217
Restriction Fragment Length Polymorphism (RFLP), 39, 53, 54
Rhodochlamys, 14
Rin mutation, 95
Ripening, 94–96
  fruit (*see* Fruit ripening)
  lipophilic constituents, 115–116
  metabolites, 119
  volatile constituents, 118–119
RNA interference, 240
RNA interference (RNAi), 68, 71, 252, 294
ROS-scavenging genes, 67

S

Sanger sequencing, 54
Second division restitution (SDR), 23
Second-generation sequencers (SGS), 54
Seedless edible bananas, 6
Sigatoka, 232–233
  abiotic and biotic factors, 228
  banana and plantains, 228
  BLSD, 228
  disease symptoms, 229
  life cycle, 229–231
  management, 231–232
  Mycosphaerella, 228
  transgenic approach
    antimicrobial peptides, 232–233
    RNAi-mediated host-induced gene silencing, 233
Sigatoka disease, 65, 119
Silk Subgroup, 17
Simple sequence repeat (SSR), 42–43, 66, 68
Somaclonal variation, 185

Somatic embryogenesis, 135, 184–185
  abiotic and biotic factors, 142
  cellular totipotency, 142
  plantlet germination, 145
Somatic embryogenesis and plant regeneration
  ascorbic acid, 96
  characterization, 96, 98
  cultures, 98
  ECS, 96
  genetic transformation, 96
  genetic Transformation, 98–100
  morphological development, 98
  PCV, 98
  promoters, 100–103
  stages, 96, 97
Sterility, 15, 20, 23, 25, 53
Stilbene synthase (StSy), 232
Stress adaptation, 55
Stress-associated proteins (SAP), 190
Strobilurins, 231
Suppression subtractive hybridisation (SSH), 66

T

TALENs. *See* Transcription activator-like effector nucleases (TALENs)
Tamil Nadu Agricultural University (TNAU), 25–26
Targeted induced local lesions in genomes (TILLING), 192
TATA-less promoters, 159
T-DNA tagging, 160
T-DNA transfer, 133
Tetraploid plantain hybrid, 24
Tobacco mosaic virus, 263
Tomato ripening mutants, 94
Total dietary fiber (TDF), 119
Transcription activator-like effector nucleases (TALENs), 293
Transcription start site (TSS), 158
Transcriptome analysis, 54
Transcriptome studies, 67
Transcriptomic approach, 56
Transgenic
  bacterial diseases, 198
  banana, 136, 137, 204–205
  BBTV, 240, 241
  causal agent, 199
  ECS, 241
  environmental factors, 198
  ethylene-treated/untreated transgenic fruit, 100
  expansin transgenic fruits, 101, 102
  field evaluations, 242
  firmness, 100
  MADS gene, 98
  MADSrin, 99, 100
  MADSrin and *MaExpansin*, 98
  plant regeneration, 103
  resistance to bacterial diseases, 202–204
Triazoles, 231
Trichoderma harzianum, 232
Tricyclazole, 119
Triploid hybrids, 29
TSSP plant promoter prediction program, 163

V

VAD. *See* Vitamin A deficiency (VAD)
Variable number of tandem repeats (VNTR), 39
Viral diseases, 68
Virulence-associated genes, 67
Virus resistance, 240–243
Virus-induced gene silencing (VIGS), 71
Vitamin A
  bioavailability, 283
  α-carotene, 283
  β-carotene, 283
  carotenoids, 281
  β-cryptoxanthin, 283
  RDA, 283
  retinol, retinoic acid and retinyl esters, 281
  retinyl esters, 282
  VAD, 278, 281–282
Vitamin A deficiency (VAD), 278, 281–282

W

Weevils, 70
Western blot analysis, 88
Whole genome re-sequencing (WGRS), 54
Whole genome sequencing (WGS), 54

X

Xanthomonadaceae, 199
Xanthomonas campestris pv. musacearum, 198, 202
Xanthomonas campestris pv. *musacearum* (*Xcm*), 65

Z

Zea mays, 289
Zinc finger nucleases (ZFNs), 192
Zingiberales genes, 54

The manufacturer's authorised representative in the EU is Springer Nature Customer Service Centre GmbH, Europaplatz 3, 69115 Heidelberg, Germany. If you have any concerns regarding our products, please contact ProductSafety@springernature.com

Printed and bound by CPI Group (UK) Ltd, Croydon, CR0 4YY
22/09/2026
02220948-0020